Infinite Crossed Products

This is Volume 135 in
PURE AND APPLIED MATHEMATICS

H. Bass, A. Borel, J. Moser, and S.-T. Yau, editors
Paul A. Smith and Samuel Eilenberg, founding editors

A list of titles in this series appears at the end of this volume.

Infinite Crossed Products

Donald S. Passman

Mathematics Department
University of Wisconsin – Madison
Madison, Wisconsin

ACADEMIC PRESS, INC.
Harcourt Brace Jovanovich, Publishers

Boston   San Diego   New York
Berkeley   London   Sydney
Tokyo   Toronto

ACADEMIC PRESS, INC.
1250 Sixth Avenue, San Diego, CA 92101

United Kingdom Edition published by
ACADEMIC PRESS INC. (LONDON) LTD.
24–28 Oval Road, London NW1 7DX

Library of Congress Cataloging-in-Publication Data

Passman, Donald S., Date—
    Infinite crossed products / Donald Passman.
        p.   cm. — (Pure and applied mathematics : v. 135)
    Bibliography: p.
    Includes index.
    ISBN 0-12-546390-1
    1. Von Neumann algebras—Crossed products.   2. Group rings.
3. Galois theory.   I. Title.   II. Series:
Pure and applied mathematics : 135.
QA3.P8 vol. 135
[QA326]
510 s—dc 19
[512'.55]   88-7597
                                                 CIP

Printed in the United States of America
89 90 91 92   9 8 7 6 5 4 3 2 1

This book is dedicated to

my mother, Fanny Passman,

and to the memory of

my father, Julius Passman.

Contents

Preface

Crossed products are another meeting place for group theory and ring theory. Historically, they first occurred in the study of finite dimensional division algebras and central simple algebras. More recently, they have become closely related to the study of infinite group algebras, group-graded rings and the Galois theory of noncommutative rings. This book is mainly concerned with these newer developments.

During the past few years, there have been a number of major achievements in this field. These include:

1. Cohen-Montgomery duality, a machine to translate crossed product results into the context of group-graded rings;

2. Understanding and computing the symmetric Martindale ring of quotients of prime and semiprime rings;

3. Classifying prime and semiprime crossed products and, more generally, the prime ideals in certain important special cases;

4. The Galois theory of prime and semiprime rings, along with skew group ring applications to the subject;

5. Determining the Grothendieck group of a Noetherian crossed product to settle the zero divisor and Goldie rank conjectures.

Indeed, these topics form the core of the book and the reason for its existence.

Chapter 1 is introductory in nature. It contains many of the basic definitions and proves duality and various versions of Maschke's theorem. Chapter 2 uses Delta methods, a coset counting technique, to classify the prime and semiprime crossed products. Chapter 3 discusses the left and symmetric Martindale ring of quotients and X-inner automorphisms of rings. Numerous examples are computed. Chapters 4 and 5 study prime ideals in crossed products $R*G$ with either G finite or with G polycyclic-by-finite and R right Noetherian. Chapters 6 and 7 are concerned with group actions on rings. Topics include the existence of fixed points, integrality, prime ideals and the Galois theory of prime rings. Finally, Chapters 8 and 9 consider the Grothendieck groups of Noetherian crossed products. In particular this material includes the induction theorem, the zero divisor and Goldie rank conjectures, the Zalesskii-Neroslavskii example and some specific computations.

The book is written in a reasonably self-contained manner. Nevertheless, some facts (usually concerning infinite groups, group algebras or homological algebra) have to be quoted. At such points, the necessary prerequisites are at least precisely stated. The book contains over 200 exercises and a challenge to the reader to overcome certain notational inconsistencies. For example, modules are usually right, but not always. Functions are sometimes written on the left, sometimes on the right and sometimes as exponents and, of course, this effects the way they multiply. In any case, $\supset$ always stands for strict inclusion, while $\supseteq$ allows for equality.

In closing, I would like to thank the following colleagues and friends for their many helpful comments and suggestions: Gerald Cliff, Martin Lorenz, Susan Montgomery, Jim Osterburg and Declan Quinn. Of course, thanks also to the National Science Foundation for its support of my research. Finally, I would like to express my love and appreciation to my family Marjorie, Barbara and Jonathan for their encouragement and support of this project.

Donald S. Passman

Madison, Wisconsin
June 1988

1   Crossed Products and Group-Graded Rings

1. Crossed Products

Let K be a field and let H be a multiplicative group. Then the *group algebra* $K[H]$ is an associative K-algebra with basis $\{\, x \mid x \in H \,\}$ and with multiplication defined distributively using the group multiplication in H. If N is a subgroup of H, then certainly $K[N] \subseteq K[H]$. Furthermore, if $N \triangleleft H$, then it is natural to expect that $K[H]$ is somehow constructed from the subgroup ring $K[N]$ and the quotient group H/N.

To this end, let $R = K[N]$ and $G = H/N$. For each $x \in G$ let $\bar{x} \in H$ be a fixed inverse image. Then the disjoint union $H = \bigcup_x \bar{x}N$ implies that

$$K[H] = \oplus \sum_x \bar{x}K[N] = \oplus \sum_x \bar{x}R$$

so $\bar{G} = \{\, \bar{x} \mid x \in G \,\}$ is an R-basis for $K[H]$.

Since $N \triangleleft H$, $\bar{x}^{-1}N\bar{x} = N$ so $\bar{x}^{-1}K[N]\bar{x} = K[N]$ and $\bar{x}$ induces a conjugation automorphism $\sigma(x)$ on R. Thus we have a map, although not necessarily a group homomorphism, $\sigma\colon G \to \operatorname{Aut}(R)$.

Furthermore, if $x \in G$ and $r \in R$ then

$$r\bar{x} = \bar{x}(\bar{x}^{-1}r\bar{x}) = \bar{x}r^{\bar{x}} = \bar{x}r^{\sigma(x)}.$$

Note that σ is trivial if N is central in H.

Next for $x, y \in G$ we have $\bar{x}N \cdot \bar{y}N = \overline{xy}N$ so $\bar{x}\bar{y} = \overline{xy}\tau(x, y)$ for some $\tau(x, y) \in N \subseteq U(R)$, the group of units of R. Thus we have a map $\tau \colon G \times G \to U(R)$ which is called the twisting. Note that τ is trivial if we can choose a consistent set of coset representatives, that is if $H = N \rtimes G$.

What we have shown is that $K[H] = R*G = K[N]*(H/N)$ is a crossed product of H/N over the ring $K[N]$.

Definition. Let R be a ring with 1 and let G be a group. Then a *crossed product* $R*G$ of G over R is an associative ring which contains R and has as an R-basis the set $\bar{G}$, a copy of G. Thus each element of $R*G$ is uniquely a finite sum $\sum_{x \in G} \bar{x}r_x$ with $r_x \in R$. Addition is as expected and multiplication is determined by the two rules below. Specifically for $x, y \in G$ we have

(twisting) $\bar{x}\bar{y} = \overline{xy}\tau(x, y)$

where $\tau \colon G \times G \to U = U(R)$, the group of units of R. Furthermore for $x \in G$ and $r \in R$ we have

(action) $r\bar{x} = \bar{x}r^{\sigma(x)}$

where $\sigma \colon G \to \operatorname{Aut}(R)$.

Note that, by definition, a crossed product is merely an associative ring which happens to have a particular structure relative to R and G. We will usually assume that $R*G$ is given. However for the rare occasions when we wish to construct such rings, the following is crucial.

Lemma 1.1. *The associativity of $R*G$ is equivalent to the assertions that for all $x, y, z \in G$*

i. $\tau(xy, z)\tau(x, y)^{\sigma(z)} = \tau(x, yz)\tau(y, z)$

ii. $\sigma(y)\sigma(z) = \sigma(yz)\eta(y, z)$ where $\eta(y, z)$ denotes the automorphism of R induced by the unit $\tau(y, z)$.

Proof. The associativity of $R*G$ is clearly equivalent to the equality

$$[(\bar{x}r)(\bar{y}s)](\bar{z}t) = (\bar{x}r)[(\bar{y}s)(\bar{z}t)]$$

for all $r, s, t \in R$ and $x, y, z \in G$. Simple computation shows that the left-hand expression equals

$$\overline{xyz}\tau(xy, z)\tau(x, y)^{\sigma(z)}r^{\sigma(y)\sigma(z)}s^{\sigma(z)}t$$

while the right-hand expression becomes

$$\overline{xyz}\tau(x, yz)r^{\sigma(yz)}\tau(y, z)s^{\sigma(z)}t$$
$$= \overline{xyz}\tau(x, yz)\tau(y, z)r^{\sigma(yz)\eta(y,z)}s^{\sigma(z)}t.$$

The result follows by first setting $r = s = t = 1$. ∎

Equation (i) above asserts that τ is a 2-*cocycle* for the action of G on U.

Unlike group rings, crossed products do not have a natural basis. Indeed if $d: G \to U$ assigns to each element $x \in G$ a unit d_x, then $\tilde{G} = \{\tilde{x} = \bar{x}d_x \mid x \in G\}$ yields an alternate R-basis for $R*G$ which still exhibits the basic crossed product structure. We call this a *diagonal change of basis*.

Now it is easy to see (Exercise 2) that the identity element of $R*G$ is of the form $1 = \bar{1}u$ for some $u \in U$. Thus, via a diagonal change of basis, we can and will assume that $\bar{1} = 1$. The embedding of R into $R*G$ is then given by $r \mapsto \bar{1}r$. On the other hand, G is in general not contained in $R*G$. Rather, each $\bar{x}$ is a unit of the ring and $\mathcal{G} = \{\bar{x}u \mid x \in G, u \in U\}$, the *group of trivial units* of $R*G$, satisfies $\mathcal{G}/U \cong G$. Note that $\mathcal{G}$ acts on both $R*G$ and R by conjugation and that for $x \in G$, $r \in R$ we have $\bar{x}^{-1}r\bar{x} = r^{\sigma(x)}$.

Certain special cases of crossed products have their own names. If there is no action or twisting, that is if $\sigma(x) = 1$ and $\tau(x, y) = 1$ for all $x, y \in G$, then $R*G = R[G]$ is an ordinary *group ring*. If the action is trivial, then $R*G = R^t[G]$ is a *twisted group ring*. Finally if the twisting is trivial, then $R*G = RG$ is a *skew group ring*. We frequently construct the latter rings using the following immediate consequence of Lemma 1.1.

Lemma 1.2. *The associativity of RG is equivalent to the map $\sigma\colon G \to \mathrm{Aut}(R)$ being a group homomorphism.*

Note that since the twisting is trivial in RG we have $\bar{x}\bar{y} = \overline{xy}$. Thus we can drop the overbars here and assume that $RG \supseteq G$.

Historically, crossed products arose in the study of division rings. Let K be a field and let D be a division algebra finite-dimensional over its center K. If F is a maximal subfield of D, then $\dim_K D = (\dim_K F)^2$. Suppose that F/K is normal, although this is not always true. If $x \in \mathrm{Gal}(F/K) = G$, then the Skolem-Noether theorem implies that there exists $\bar{x} \in D \setminus 0$ with $\bar{x}^{-1}f\bar{x} = f^x$ for all $f \in F$. Furthermore, $\bar{x}\bar{y}$ and $\overline{xy}$ agree in their action on F so $\bar{y}^{-1}\bar{x}^{-1}\overline{xy} \in \mathbf{C}_D(F) = F$. Once we show (Exercise 5) that the elements $\bar{x}$ are linearly independent over F, we can then conclude by computing dimensions that $D = \oplus \sum_{x\in G} \bar{x}F = F*G$.

More generally, suppose A is a central simple algebra over K so that A is a simple ring, finite dimensional over its center K. Then $A = \mathrm{M}_n(D)$ for some n and division ring D with $\mathbf{Z}(D) = K$. Two such algebras A and B are equivalent if they have the same D. The equivalence classes then form a group, the *Brauer group*, under tensor product $\otimes_K$. Now given A, one can show that there exists $B \sim A$ with $B = F*G$. But $F*G$ is determined by the twisting function $\tau\colon G \times G \to F$, a 2-cocycle. Thus in this way we obtain the homological characterization of the Brauer group as the 2nd-*cohomology group*. See [**73**, Chapter 4] for more details.

The thrust of this book is in other directions; namely we are concerned here with the ring theoretic structure of crossed products. For example we will consider when such rings are prime or semiprime and we will discuss the nature of their prime ideals and modules. Furthermore we will describe crossed product applications to problems in group algebras and to the Galois theory of rings.

We remark that the notation $R*G$ for a crossed product is certainly ambiguous since it does not convey the full σ, τ-structure. Nevertheless it is simpler and hence preferable to something like (R, G, σ, τ). Moreover it rarely if ever causes confusion. For example if $\alpha = \sum_{x \in G} \bar{x} r_x \in R*G$, then the *support* of α is given by

$$\text{Supp } \alpha = \{\, x \in G \mid r_x \neq 0 \,\}.$$

It follows that if H is a subgroup of G, then

$$\{\, \alpha \in R*G \mid \text{Supp } \alpha \subseteq H \,\} = R*H$$

is the naturally embedded sub-crossed product. Furthermore the argument at the beginning of this section yields

Lemma 1.3. *Let $R*G$ be given and let $N \triangleleft G$. Then*

$$R*G = (R*N)*(G/N)$$

where the latter is some crossed product of the group G/N over the ring $R*N$.

Recall that conjugation by the elements of $\mathcal{G}$ yields a homomorphism $\mathcal{G} \to \text{Aut}(R)$. Therefore $\mathcal{G}$ permutes the ideals of R and, since U obviously fixes all such ideals, we obtain a well-defined permutation action of $G \cong \mathcal{G}/U$ on the set of these ideals. If I is a G-stable ideal of R we write $I*G = (R*G)I$.

Lemma 1.4. *Let $R*G$ be given.*
i. If $J \triangleleft R*G$, then $J \cap R$ is a G-stable ideal of R and $J \supseteq (J \cap R)*G$.
ii. If I is a G-stable ideal of R, then $I*G \triangleleft R*G$ with $(I*G) \cap R = I$. Moreover $(R*G)/(I*G) \cong (R/I)*G$ where the latter is a suitable crossed product of G over R/I.

Proof. (i) This is clear since both R and J are stable under the action of $\mathcal{G}$.

(ii) Since I is $\mathcal{G}$-stable, it is clear that $I\mathcal{G} = \mathcal{G}I$ and hence that $I*G = (R*G)I$ is an ideal of $R*G$. Furthermore since

$$I*G = \left\{ \sum_{x \in G} \bar{x}r_x \,\middle|\, r_x \in I \text{ for all } x \in G \right\}$$

we have $(I*G) \cap R = I$. Finally let $\phi\colon R*G \to (R*G)/(I*G)$ be the natural homomorphism. Then $\phi(R*G)$ is an associative ring containing $\phi(R) \cong (R/I)$ and $\phi(\bar{G})$. Since the action and twisting equations in $R*G$ map to similar equations in $\phi(R*G)$ and since $\phi(\bar{G})$ is clearly a free basis for $\phi(R*G)$ over $\phi(R)$, the result follows. ∎

Suppose $N \triangleleft G$. Then $R*G = (R*N)*(G/N)$ so G/N and hence also G permute the ideals of $R*N$. If I is a G-stable ideal of $R*N$, we deduce from the above that $(R*G)I \triangleleft R*G$ and that $(R*G)/(R*G)I = [(R*N)/I]*(G/N)$.

The closure properties of crossed products exhibited in the previous two lemmas are crucial. The first allows us to study $R*G$ by lifting information from various $R*N$ with $N \triangleleft G$; the second allows us to study ideals J of $R*G$ under the simplifying assumption that $J \cap R = 0$.

Now let H be a subgroup of G. Then there is a natural projection $\pi_H\colon R*G \to R*H$ given by

$$\pi_H \left(\sum_{x \in G} \bar{x}r_x \right) = \sum_{x \in H} \bar{x}r_x.$$

Note that both $R*G$ and $R*H$ are $R*H$-bimodules under left and right multiplication and then clearly

Lemma 1.5. *The map* $\pi_H\colon R*G \to R*H$ *is an* $R*H$-*bimodule homomorphism.*

Two classes of groups play key roles in the study of crossed products. First, of course, there are the finite groups; next there are the polycyclic-by-finite ones. Recall that G is *polycyclic-by-finite* if G has a subnormal series

$$1 = G_0 \triangleleft G_1 \triangleleft \ldots \triangleleft G_n = G$$

with each G_{i+1}/G_i either infinite cyclic or finite. The following is a simple extension of the Hilbert Basis Theorem.

Proposition 1.6. *If R is right Noetherian and G is polycyclic-by-finite, then $R*G$ is right Noetherian.*

Proof. By induction and Lemma 1.3, it suffices to assume that G is either infinite cyclic or finite.

If $G = \langle x \rangle$ is infinite cyclic, then $R*G = \langle R, \bar{x}, \bar{x}^{-1} \rangle$ is generated by R, $\bar{x}$ and $\bar{x}^{-1}$ with $\bar{x}^{-1} R \bar{x} = R$. It follows from [**161**, Theorem 10.2.6(iii)] that $R*G$ is right Noetherian.

If G is finite, then $R*G$ is a finitely generated R-module and hence a Noetherian right R-module. Thus $R*G$ is also Noetherian as a right $R*G$-module. ∎

In rare cases we will consider *semigroup crossed products*. These have the same action and twisting structure as ordinary crossed products, but G is allowed to be a multiplicative semigroup. Thus for example, if G is the infinite cyclic semigroup $G = \{ 1, x, x^2, \dots \}$ we obtain $R*G = R[x; \sigma]$ a *skew polynomial ring*. If G is the free abelian semigroup on n variables, then $R*G$ is just a noncommutative analog of a polynomial ring in n variables. The next result follows from [**161**, Theorem 10.2.6(ii)] by induction on n.

Lemma 1.7. *If R is right Noetherian and G is the free abelian semigroup on n generators, then $R*G$ is right Noetherian.*

Obviously Proposition 1.6 and Lemma 1.7 have left analogs. Returning to groups G, we close this section by observing that any crossed product $R*G$ has an untwisted extension. Specifically, there exists an overring $S \supseteq R$ such that $S*G \supseteq R*G$ and $S*G = SG$ is a skew group ring. The proof of this result requires an extremely large ring extension of R.

If R is given and $\{ u_i \mid i \in I \}$ is a collection of symbols, then we let

$$S = \langle R, u_i, u_i^{-1} \mid i \in I \rangle$$

be the ring freely generated by R and the various u_i and u_i^{-1}. Furthermore we insist that $1 \in R$ is the identity of S. This ring can be constructed by first taking $F = \langle u_i \mid i \in I \rangle$ to be the free group generated by the u_i and then forming the free product of the ring R with the integral group ring of F amalgamating the identity elements of the two rings. We will not use any specific structure theorem for S. We will only need the fact that S exists and satisfies an appropriate universal property. To be precise, suppose T is a ring and $\eta \colon R \to T$ is an embedding. If t_i is a unit of T for each $i \in I$, then η extends to a homomorphism $\eta \colon S \to T$ with $u_i \mapsto t_i$.

The following result generalizes the standard untwisting of 2-cocycles in the commutative case.

Lemma 1.8. *Let $R*G$ be given and set $S = \langle R, u_x, u_x^{-1} \mid x \in G \backslash \{1\} \rangle$. Then there exists a crossed product $S*G$ containing $R*G$ such that $S*G = SG$ is a skew group ring.*

Proof. We recall the basic crossed product definitions. Thus in $R*G$ we have $\bar{x}\bar{y} = \overline{xy}\tau(x,y)$ and $r\bar{x} = \bar{x}r^{\sigma(x)}$ where $\tau(x,y) \in U(R)$ and $\sigma(x) \in \text{Aut}(R)$. Since we assume that $\bar{1} = 1$, we have in addition $\tau(x,1) = \tau(1,x) = 1$ and $\sigma(1) = 1$. Furthermore, by Lemma 1.1, associativity in $R*G$ is equivalent to

$$\tau(xy,z)\tau(x,y)^{\sigma(z)} = \tau(x,yz)\tau(y,z) \tag{$*$}$$

$$\sigma(y)\sigma(z) = \sigma(yz)\eta(y,z) \tag{$**$}$$

for all $x, y, z \in G$, where $\eta(y,z)$ is the automorphism of R induced by the unit $\tau(y,z)$.

Now let S be as given and set $u_1 = 1 \in S$. For each $y \in G$ we extend $\sigma(y) \in \text{Aut}(R)$ to an endomorphism of S by $r \mapsto r^{\sigma(y)}$ for $r \in R$ and $u_x \mapsto \tau(x,y)^{-1}u_{xy}u_y^{-1}$. Note that each $\tau(x,y)^{-1}u_{xy}u_y^{-1}$ is a unit of S and that when $x = 1$ we have $u_1 \mapsto 1$. Thus, by the universal property of S, $\sigma(y)$ is indeed an endomorphism of S.

We first show that these extended endomorphisms also satisfy $(**)$. This is certainly the case when applied to any $r \in R$ so we need only apply these maps to u_x. We have

$$(u_x)^{\sigma(y)\sigma(z)} = \left(\tau(x,y)^{-1}u_{xy}u_y^{-1}\right)^{\sigma(z)}$$

$$= \tau(x,y)^{-\sigma(z)}\tau(xy,z)^{-1}u_{xyz}u_{yz}^{-1}\tau(y,z)$$

and

$$(u_x)^{\sigma(yz)\eta(y,z)} = \left(\tau(x,yz)^{-1}u_{xyz}u_{yz}^{-1}\right)^{\eta(y,z)}$$
$$= \tau(y,z)^{-1}\tau(x,yz)^{-1}u_{xyz}u_{yz}^{-1}\tau(y,z)$$

so, by (*), these two are equal. Furthermore note that $\sigma(1)\colon u_x \mapsto u_x$ so $\sigma(1) = 1$. Hence setting $z = y^{-1}$ in (**) for the extended σ's, we get $\sigma(y)\sigma(y^{-1}) = \eta(y,y^{-1})$. But the latter is an automorphism of S and hence we conclude that $\sigma(y) \in \text{Aut}(S)$.

We can now form $S*G$ using the same twisting and the extended map $\sigma\colon G \to \text{Aut}(S)$. Since (*) is inherited by $S*G$, it follows from Lemma 1.1 that $S*G$ is a crossed product containing $R*G$. Finally set $\tilde{x} = \bar{x}u_x \in S*G$. Then

$$\tilde{x}\tilde{y} = \bar{x}u_x\bar{y}u_y = \bar{x}\bar{y}(u_x)^{\sigma(y)}u_y$$
$$= \overline{xy}\tau(x,y) \cdot \tau(x,y)^{-1}u_{xy}u_y^{-1} \cdot u_y$$
$$= \overline{xy}u_{xy} = \widetilde{xy}.$$

Since this is a diagonal change of basis, we conclude that $S*G = SG$ is a skew group ring. ∎

Note that, in the above, if R is commutative we could take S to be the group ring $S = R[u_x, u_x^{-1} \mid x \in G \setminus \{1\}]$. Furthermore in the usual case with R a field, we could just take S to be the purely transcendental extension of R generated by the u_x.

EXERCISES

1. Show that under a diagonal change of basis, $R*G$ still retains the basic crossed product structure. Determine the new action and twisting functions in terms of the old. Note that τ will be changed by a factor which is called a 2-*coboundary*.

2. Suppose $R*G$ was defined without assuming that it contains R as a subring. In other words, the elements of $R*G$ would then be formal finite sums $\sum_{x \in G} \bar{x}r_x$ with multiplication given by

$$(\bar{x}r)(\bar{y}s) = \overline{xy}\tau(x,y)r^{\sigma(y)}s.$$

Show first that there exists a unit $u \in U$ such that $e = \bar{1}u$ is an idempotent. Then prove that $R*G = e(R*G) = (R*G)e$. Deduce that e is the identity element of $R*G$ and that the map $r \mapsto er$ embeds R into $R*G$.

3. Let S be a ring containing R and a group G of units. Suppose that $g^{-1}Rg = R$ for all $g \in G$. Show that this action of G on R gives rise to a homomorphism $\sigma: G \to \mathrm{Aut}(R)$ and that the embeddings of R and G into S extend to a unique ring homomorphism of RG into S. This yields a universal property of the skew group ring RG.

4. Let $\mathcal{G}$ be the group of trivial units of $R*G$ so that $\mathcal{G}$ acts on R by conjugation. Show that there is a homomorphism from the skew group ring $R\mathcal{G}$ onto $R*G$ obtained by identifying $R \cap \mathcal{G}$ with U. Obtain a universal property of the crossed product $R*G$. There are several possibilities.

5. Let D be the division ring discussed after Lemma 1.2. Use a shortest length argument to show that the elements $\bar{x}$ are F-linearly independent. To this end, suppose $\sum_x \bar{x}f_x = 0$ and that both $y, z \in G$ occur in this expression. Multiply the equation on the left by an appropriate $f \in F$ and on the right by f^y and then subtract to obtain a shorter expression.

6. If $R*G$ is commutative and G is free abelian, show that $R*G \cong R[G]$. If G is any cyclic group, determine when $R*G = RG$ via a diagonal change of basis.

2. Group-Graded Rings and Duality

Let $S = R*G$ be a crossed product and for each $x \in G$ set $S_x = \bar{x}R$. Then $S = \oplus \sum_{x \in G} S_x$ and $S_x S_y = S_{xy}$. Thus S is a strongly G-graded ring.

Definition. Let G be a multiplicative group. An associative ring S is G-graded if $S = \oplus \sum_{x \in G} S_x$ is a direct sum of additive subgroups S_x indexed by the elements $x \in G$ and if $S_x S_y \subseteq S_{xy}$ for all $x, y \in G$. Clearly $R = S_1$ is a subring of S, its base ring, and each S_x is an R-bimodule under left and right multiplication. We say that S is strongly G-graded if $S_x S_y = S_{xy}$ for all $x, y \in G$.

It is easy to verify (Exercise 1) that the identity element $1 \in S$ is contained in R and we will assume this throughout. Furthermore S is strongly graded if and only if $1 \in S_x S_{x^{-1}}$ for all $x \in G$.

If X is any subset of G, we write $R(X) = \sum_{x \in X} S_x$. Thus $R(G) = S$ and if H is a subgroup of G, then $R(H)$ is the naturally contained H-*graded subring*. Furthermore if $N \triangleleft G$, then S is also (G/N)-graded with components $S_{Nx} = R(Nx)$. Hence the base ring here is $R(N)$ and we have $S = R(G) = R(N)(G/N)$, the analog of Lemma 1.3. If $g \in G$, we sometimes write $R(g)$ for $R(\{g\}) = S_g$.

As we mentioned above, $S = R*G$ is strongly G-graded with base ring R. Indeed it is easy to see (Exercise 2) that a G-graded ring S is a crossed product if and only if each component S_x contains a unit. An example of a strongly graded ring which is not a crossed product is given in Exercise 3.

Another well-known example comes from the theory of Lie algebras. Let L be a Lie algebra over K and let $U(L)$ be its *universal enveloping algebra*. If $\lambda \colon L \to K$ is a linear functional, let

$$S_\lambda = \left\{ u \in U(L) \mid [\ell, u] = \lambda(\ell)u \text{ for all } \ell \in L \right\}.$$

These are the *semi-invariants* corresponding to λ. Now set $S = \sum_\lambda S_\lambda$. Then one knows that the sum is direct and that $S_\lambda S_\mu \subseteq S_{\lambda + \mu}$. Thus the *semicenter* S of $U(L)$ is graded by the additive group $\mathrm{Hom}_K(L, K)$.

We remark that G-graded rings without additional assumptions can frequently have no relationship to the group G. For example, let R be a ring, let M be an R-bimodule and let $G = \{1, g, \dots\}$ be any nonidentity group. If S is the ring $S = R \oplus M$ with $M^2 = 0$, then S becomes G-graded by setting $S_1 = R$, $S_g = M$ and all other components zero. Similarly if S is G-graded and T is H-graded, then $S \oplus T$ is naturally a W-graded ring for any group W containing G and H as disjoint subgroups. Therefore mild assumptions must sometimes be imposed on graded rings to enable the group structure to play its appropriate role.

Group-graded rings were introduced in [45] as a formal way to deal with finite group representation problems. Indeed the standard module arguments carried over immediately to that context. In addition, group-graded rings occur naturally in certain Galois theory situations and, of course, they are related to crossed products. For the most part, our interest in group-graded rings will center on their relationship to crossed products. We will not go much further afield.

It is of course tempting to try to extend all crossed product results to group-graded rings. One soon discovers however that the old techniques do not carry over. Fortunately there exists a *duality* machine begun in [**39**] and extended in [**180**] which translates many of the crossed product results directly to this new context. We will use the more concrete construction of the latter paper and we will deal with both finite and infinite groups at the same time. Because of this, the infinite results we list here are slightly less precise than those of [**180**].

Note that if $s \in S$, we let s_x be its *x-component* so that $s_x \in S_x$ and $s = \sum_{x \in G} s_x$.

Definition. Let S be a G-graded ring with base ring $R = S_1$ and use $|G|$ to denote the cardinality of the set G. Let $M_G(S)$ denote the ring of row and column finite $|G| \times |G|$ matrices over S with rows and columns indexed by the elements of G. In particular if $\alpha \in M_G(S)$ and $x, y \in G$ then we use $\alpha(x, y)$ to denote the (x, y)-entry of α.

Let $M_G^b(S)$ denote the set of all matrices in $M_G(S)$ having only finitely many nonzero entries. It is clear that $M_G^b(S)$ is an essential right and left ideal of $M_G(S)$.

Now for each $g \in G$ we let $\bar{g} \in M_G(S)$ be the permutation matrix $\bar{g} = [\delta_{g^{-1}x,y}]$ which has a 1 in the $(x, g^{-1}x)$-positions and zeros elsewhere. In addition for each $s \in S$ we define $\tilde{s} \in M_G(S)$ to be the matrix satisfying $\tilde{s}(x, y) = s_{x^{-1}y}$ for all $x, y \in G$.

Lemma 2.1. *With the above notation we have*

    i. *The map* $^-: G \to \bar{G} = \{\bar{g} \mid g \in G\}$ *is a group isomorphism embedding G into $M_G(S)$.*

    ii. *The map* $^\sim: S \to \tilde{S} = \{\tilde{s} \mid s \in S\}$ *is a ring isomorphism embedding S into $M_G(S)$.*

    iii. *If $g \in G$, $s \in S$ then $\bar{g}\tilde{s} = \tilde{s}\bar{g}$.*

This is just a simple matrix computation. We now come to a crucial

Definition. Let H be any subgroup of G. We define $S\{H\} \subseteq \mathrm{M}_G(S)$ by

$$S\{H\} = \left\{\, \alpha \in \mathrm{M}_G(S) \mid \alpha(x,y) \in R(x^{-1}Hy) \text{ for all } x, y \in G \,\right\}.$$

Furthermore, $\mathrm{D}(H)$ and $\mathrm{D}^{-1}(H)$ are the subsets of $\mathrm{diag}\ \mathrm{M}_G(S)$, the set of diagonal matrices in $\mathrm{M}_G(S)$, given by

$$\mathrm{D}(H) = \left\{\, \alpha \in \mathrm{diag}\ \mathrm{M}_G(S) \mid \alpha(x,x) \in R(Hx) \text{ for all } x \in G \,\right\}$$

and

$$\mathrm{D}^{-1}(H) = \left\{\, \alpha \in \mathrm{diag}\ \mathrm{M}_G(S) \mid \alpha(x,x) \in R(x^{-1}H) \text{ for all } x \in G \,\right\}.$$

In particular $S\{G\} = \mathrm{M}_G(S)$ and $\mathrm{D}(G) = \mathrm{D}^{-1}(G) = \mathrm{diag}\ \mathrm{M}_G(S)$.

The applications to group-graded rings come from having alternate descriptions of $S\{H\}$. First we have

Proposition 2.2. [180] *If H is a subgroup of G, then $S\{H\}$ is a subring of $\mathrm{M}_G(S)$. Furthermore*

$$\mathrm{D}^{-1}(H) \cdot \mathrm{M}_G(R(H)) \cdot \mathrm{D}(H) \subseteq S\{H\}$$

and

$$\mathrm{D}(H) \cdot S\{H\} \cdot \mathrm{D}^{-1}(H) \subseteq \mathrm{M}_G(R(H)).$$

This is immediate from the formula $R(X) \cdot R(Y) \subseteq R(XY)$ for any subsets $X, Y \subseteq G$. Note that the above two equations say that $S\{H\}$ and $\mathrm{M}_G(R(H))$ are *generalized conjugates* via certain well-understood sets of diagonal matrices. In particular this yields a correspondence between the ideals of $S\{H\}$ and of $R(H)$ which we will consider at the end of this section.

For each $x \in G$, let $e_x \in \mathrm{M}_G^\flat(S) \subseteq \mathrm{M}_G(S)$ denote the idempotent with 1 in the (x,x)-position and zeros elsewhere. Since $1 \in R(1)$ we have $e_x \in S\{H\}$ and it follows that $S\{H\}^\flat = \mathrm{M}_G^\flat(S) \cap S\{H\}$ is an essential right and left ideal of $S\{H\}$. When $H = 1$, a more precise description of $S\{H\}$ can be given.

Lemma 2.3. $S\{1\}^\flat = \oplus \sum_{x \in G} e_x \tilde{S}$ is a free right $\tilde{S}$-module with basis $\{ e_x \mid x \in G \}$. Furthermore if $s = \sum_z s_z \in S$ and $x, y \in G$ then

$$e_x \widetilde{s_{x^{-1}y}} = e_x \tilde{s} e_y = \widetilde{s_{x^{-1}y}} e_y.$$

Proof. Note that $\tilde{S} \subseteq S\{1\}$ and since $e_x \in S\{1\}^\flat$ we have $S\{1\}^\flat = \oplus \sum_x e_x S\{1\}$. Let $\alpha \in e_x S\{1\}$, the x-th row of $S\{1\}$. If $t = \sum_y \alpha(x, y) \in S$, then $t_{x^{-1}y} = \alpha(x, y)$ and hence $e_x \tilde{t} = \alpha$. Thus $S\{1\}^\flat = \oplus \sum_x e_x \tilde{S}$.

Finally let $s = \sum_z s_z \in S$ so that $e_x \tilde{s} e_y = e_x s_{x^{-1}y} e_y$. In particular, if $e_x \tilde{s} = 0$ then $s = 0$. Furthermore since $\widetilde{s_{x^{-1}y}}$ has only one nonzero entry in each row and column we conclude that

$$e_x \widetilde{s_{x^{-1}y}} = e_x \tilde{s} e_y = \widetilde{s_{x^{-1}y}} e_y$$

as required. ∎

In case G is finite, we recognize the above structure as coming from the theory of Hopf algebras; it is the smash product of the G-graded ring $\tilde{S} \cong S$ by the dual of the group algebra of G. Specifically if S is a G-graded ring with G finite, then the *smash product* $S\#G^*$ is an associative ring with 1 having S as a subring. Furthermore there exists a decomposition of 1 into orthogonal idempotents p_x, one for each $x \in G$, such that $\{ p_x \mid x \in G \}$ is a free right S-basis for $S\#G^*$ with

$$p_x s_{x^{-1}y} = p_x s p_y = s_{x^{-1}y} p_y$$

for all $x, y \in G$ and $s \in S$. Since the above assertions uniquely determine the arithmetic in $S\#G^*$, we conclude from Lemmas 2.1 and 2.3 that, for G finite, $S\#G^* \cong S\{1\}$ via the map given by $\tilde{\ }: S \to \tilde{S}$ and $p_x \mapsto e_x$. (We remark that the notation $S\#G^*$ is not standard; we include the * to indicate the presence of the Hopf algebra dual.)

The second description of $S\{H\}$ is contained in the following.

Lemma 2.4. *Let S be a G-graded ring.*

   i. If $g, x \in G$ then $\bar{g}^{-1} e_x \bar{g} = e_{g^{-1}x}$ and hence $\bar{G}$ acts as automorphisms on $S\{1\}$ centralizing $\tilde{S}$.

   ii. $S\{G\} \supseteq \oplus \sum_{g \in G} \bar{g} S\{1\} = S\{1\} \bar{G} \supseteq S\{G\}^{\flat}$ where $S\{1\} \bar{G}$ is a skew group ring of $\bar{G}$ over $S\{1\}$.

   iii. If H is a subgroup of G, then $S\{1\} \bar{G} \cap S\{H\} = S\{1\} \bar{H}$ is the naturally embedded sub-skew group ring.

Proof. (i) The formula $\bar{g}^{-1} e_x \bar{g} = e_{g^{-1}x}$ is a simple matrix computation. Since the x-th row of $S\{1\}$ is $e_x S\{1\} = e_x S\{1\}^{\flat} = e_x \tilde{S}$, by Lemma 2.3, it follows from Lemma 2.1(iii) that $\bar{G}$ permutes these rows and centralizes $\tilde{S}$. Thus clearly $\bar{G}$ normalizes $S\{1\}$.

   (ii)(iii) We assign a grade to certain elements of $M_G(S)$ as follows. We let $\alpha \in M_G(S)$ have grade $g \in G$ if and only if for all $x, y \in G$ the entry $\alpha(x, y)$ satisfies $\alpha(x, y) \in S_z$ with $xzy^{-1} = g$. If T_g denotes the set of elements of $M_G(S)$ of grade g, then it follows easily that $M_G(S) \supseteq T = \oplus \sum_{g \in G} T_g \supseteq M_G^{\flat}(S)$ and that $T_g T_h \subseteq T_{gh}$. In particular, T is a G-graded ring.

   Moreover $T_1 = S\{1\}$ and $\sum_{h \in H} T_h = T \cap S\{H\}$ since $xzy^{-1} \in H$ if and only if $z \in x^{-1} H y$. Note also that $\bar{g} \in T$ has grade g. Thus $\bar{g}^{-1} T_g \subseteq T_1$ so $T_g = \bar{g} T_1$ and hence $T = \oplus \sum_{g \in G} T_g = \oplus \sum_{g \in G} \bar{g} T_1$. The result now follows from Lemma 2.1(i) and the above observations. ∎

   It is now a simple matter to obtain the *duality theorem*.

Theorem 2.5. [39] *Let S be a G-graded ring with G finite and let $S \# G^* = \sum_{x \in G} p_x S$ be the smash product of G over S. Then G acts on $S \# G^*$ as automorphisms via $(p_x s)^g = p_{g^{-1}x} s$ for all $x, g \in G$ and $s \in S$. Furthermore, with respect to this action, the skew group ring $(S \# G^*) G$ satisfies $(S \# G^*) G \cong M_G(S)$.*

Proof. Since G is finite, $M_G^{\flat}(S) = M_G(S)$ and therefore a number of the inclusions above become equalities. In view of Lemma 2.4(i) and the isomorphisms $G \cong \bar{G}$ and $S \# G^* \cong S\{1\}$ we see that G does indeed act as automorphisms on $S \# G^*$ in the indicated manner. Furthermore by Lemma 2.4(ii)(iii) we have

$$(S \# G^*) G \cong S\{1\} \bar{G} = S\{G\} = M_G(S)$$

as required. ∎

There is another duality theorem for G finite. If $R*G$ is a crossed product, then it is a G-graded ring with $(R*G)_x = \bar{x}R$ and we can form the smash product $(R*G)\#G^*$. The theorem of [39] and [199] then asserts that $(R*G)\#G^* \cong \mathrm{M}_G(R)$ (see Exercises 7 and 8 of Section 3). In fact if $S = R(G)$ is strongly G-graded, then more generally we have $S\#G^* \cong \mathrm{End}(_R S)$. There are also some analagous results when G is infinite. In this case, if S is G-graded, then it is more difficult to define $S\#G^*$ in the sense of Hopf algebras because of the infinite dimensional dual. Furthermore even when it can be defined (see [25]), it is not in general equal to $S\{1\}$.

We close this section by describing the ideal correspondence determined by the generalized conjugation in Proposition 2.2. If S is a G-graded ring, we say that S is *component regular* if, for each $x \in G$, the component S_x has right and left annihilator in S equal to zero. The following result uses the notation of Lemma 2.4. Furthermore if T is any ring we let $\mathcal{I}(T)$ denote its family of two-sided ideals.

Proposition 2.6. [180] *Let S be a G-graded ring with base ring $R = S_1$ and let H be a subgroup of G. Then there exist inclusion preserving maps*

$$\phi \colon \mathcal{I}(R(H)) \to \mathcal{I}(S\{1\}\bar{H})$$

and

$$\xi \colon \mathcal{I}(S\{1\}\bar{H}) \to \mathcal{I}(R(H))$$

such that

    i. *If $A, B \lhd R(H)$, then $A^\phi B^\phi \subseteq (AB)^\phi$ and $A^{\phi\xi} = A$.*

    ii. *If $I, J \lhd S\{1\}\bar{H}$, then $I^\xi J^\xi \subseteq (IJ)^\xi$. Furthermore if S is component regular or if $H = G$, then $I^\xi = 0$ implies that $I = 0$.*

Proof. For convenience write $D = \mathrm{D}(H), D^{-1} = \mathrm{D}^{-1}(H)$ and let $e_{x,y}$ denote the matrix with 1 in the (x, y)-position and zeros elsewhere. We first observe that $D \cdot S\{H\}^\flat \subseteq \mathrm{M}_G^\flat(R(H)) \cdot D$. To this end, let $\alpha \in S\{H\}^\flat$ have only one nonzero entry, namely $\alpha(x, y) \in R(x^{-1}Hy)$, and let $\delta \in D$. We note that

$$d' = (\delta\alpha)(x, y) = \delta(x, x)\alpha(x, y) \in R(Hx)R(x^{-1}Hy) \subseteq R(Hy).$$

Then $e_{x,y} \in M^\flat_G(R(H))$, $\delta' = d'e_{y,y} \in D$ and

$$\delta\alpha = e_{x,y}\delta' \in M^\flat_G(R(H)) \cdot D$$

as required. Similarly we have $S\{H\}^\flat \cdot D^{-1} \subseteq D^{-1} \cdot M^\flat_G(R(H))$.

Now let $A \lhd R(H)$ and define $A^\phi = D^{-1} \cdot M^\flat_G(A) \cdot D$. Then by Proposition 2.2 and Lemma 2.4,

$$A^\phi \subseteq D^{-1} \cdot M^\flat_G(R(H)) \cdot D \subseteq S\{H\}^\flat \subseteq S\{1\}\bar{H} \subseteq S\{H\}.$$

Furthermore $A^\phi \lhd S\{H\}$ since for example

$$A^\phi \cdot S\{H\} = A^\phi \cdot S\{H\}^\flat = D^{-1} \cdot M^\flat_G(A) \cdot D \cdot S\{H\}^\flat$$
$$\subseteq D^{-1} \cdot M^\flat_G(A) \cdot M^\flat_G(R(H)) \cdot D = A^\phi.$$

Thus $A^\phi \lhd S\{1\}\bar{H}$. In addition $DD^{-1} \subseteq M_G(R(H))$ yields $A^\phi B^\phi \subseteq (AB)^\phi$ and since $e_{1,1}D^{-1} = e_{1,1}R(H) = De_{1,1}$ we have $e_{1,1}A^\phi e_{1,1} = e_{1,1}A$.

In the other direction, let $I \lhd S\{1\}\bar{H} \subseteq S\{H\}$ and define the set $I^\xi = \{\alpha(1,1) \mid \alpha \in I\}$. Since $I \subseteq S\{H\}$ and $e_{1,1}R(H) \subseteq S\{H\}^\flat \subseteq S\{1\}$, it follows that $I^\xi \lhd R(H)$. Furthermore if $A \lhd R(H)$, then $A^{\phi\xi} = A$ since $e_{1,1}A^\phi e_{1,1} = e_{1,1}A$. Finally suppose S is component regular and that $I \neq 0$. Choose $\alpha \in I$ with $\alpha(x,y) \neq 0$ and note that $e_{1,x}R(Hx)$, $e_{y,1}R(y^{-1}H) \subseteq S\{H\}^\flat \subseteq S\{1\}\bar{H}$. Thus

$$e_{1,1}R(Hx)\alpha(x,y)R(y^{-1}H) = e_{1,x}R(Hx) \cdot \alpha \cdot e_{y,1}R(y^{-1}H) \subseteq I$$

and hence $R(Hx)\alpha(x,y)R(y^{-1}H) \subseteq I^\xi$. We conclude that $I^\xi \neq 0$ since $\alpha(x,y) \neq 0$ and since either S is component regular or $H = G$ and $1 \in R(Hx) = R(y^{-1}H)$. ∎

Corollary 2.7. *Let S be a G-graded ring.*

i. If $S\{1\}\bar{H}$ is prime or semiprime, then so is $R(H)$.

ii. Suppose that either S is component regular or $H = G$. If $R(H)$ is prime or semiprime, then so is $S\{1\}\bar{H}$.

This is an immediate consequence of Proposition 2.6. It will be used to transfer results on prime or semiprime crossed products to the context of group-graded rings. Finally we mention a simpler result which holds when $H = G$.

Proposition 2.8. [31] Let S be a G-graded ring. If P is a prime ideal of S, then $P' = M_G(P) \cap S\{1\}\bar{G}$ is a prime of $S\{1\}\bar{G}$. Furthermore, the map $P \mapsto P'$ is one-to-one.

Proof. By Lemma 2.4(ii), $S\{1\}\bar{G} \supseteq M_G^\flat(S)$ so it is clear that the map $P \mapsto P'$ is one-to-one. Moreover, for all $x, y \in G$, $S\{1\}\bar{G}$ contains the matrix unit $e_{x,y}$ which has a 1 in the (x,y)-position and zeros elsewhere.

Now let $A, B \triangleleft S\{1\}\bar{G}$ with $AB \subseteq P'$ and write $e_{1,1}Ae_{1,1} = e_{1,1}A''$ and $e_{1,1}Be_{1,1} = e_{1,1}B''$ where A'' and B'' are ideals of S. Then

$$e_{1,1}A''B'' = e_{1,1}Ae_{1,1} \cdot e_{1,1}Be_{1,1}$$
$$\subseteq e_{1,1}P'e_{1,1} = e_{1,1}P$$

so $A''B'' \subseteq P$. Therefore since P is prime, one of these factors, say A'', is contained in P. It follows that if $\alpha \in A$ and $x, y \in G$, then $e_{1,1}\alpha(x,y) = e_{1,x}\alpha e_{y,1} \in A$ so $\alpha(x,y) \in A'' \subseteq P$. Thus $\alpha \in M_G(P) \cap S\{1\}\bar{G} = P'$ so $A \subseteq P'$ and P' is indeed a prime ideal. ∎

EXERCISES

1. Let S be a G-graded ring with 1. Show that $1 \in S_1$ and that S is strongly G-graded if and only if $1 \in S_x S_{x^{-1}}$ for all $x \in G$.

2. Let S be G-graded. If $u \in S_x$ is a unit of S, prove that $u^{-1} \in S_{x^{-1}}$. Deduce that $S = R*G$ if and only if each S_x contains a unit of S.

3. Let $S = M_3(K)$ and let $G = \{1, x\}$ be a group of order 2. Define

$$S_1 = \begin{pmatrix} K & K & 0 \\ K & K & 0 \\ 0 & 0 & K \end{pmatrix} \quad \text{and} \quad S_x = \begin{pmatrix} 0 & 0 & K \\ 0 & 0 & K \\ K & K & 0 \end{pmatrix}.$$

Show that this makes S a strongly G-graded ring, but that S is not a crossed product. For the latter, compute $\dim_K S_1$ and $\dim_K S_x$.

homomorphism $\rho: R \rightarrow \text{End}_Z(V)$. We can then obtain a different structure by composing ρ with σ^{-1} to yield the representation

$$R \xrightarrow{\sigma^{-1}} R \xrightarrow{\rho} \text{End}_Z(V).$$

Alternately we can let $V^\sigma = \{ v^\sigma \mid v \in V \}$ be an isomorphic copy of V as an additive abelian group and we can define the R-module structure on V^σ via the formula $v^\sigma r^\sigma = (vr)^\sigma$. It is clear that V and V^σ have the same lattice of submodules. Hence V is irreducible, completely reducible, indecomposable or Noetherian if and only if V^σ is. Furthermore if E ess V, then E^σ ess V^σ.

We now consider crossed products.

Lemma 3.3. *Let $S = R*G$ be given and let V be an R-module. Then $V^{|S} = \oplus \sum_{x \in G} V \otimes \bar{x}$ is an R-module direct sum with $V \otimes \bar{x} \cong V^{\sigma(x)}$.*

Proof. Since S is a free left R-module with basis $\bar{G}$ we have $V^{|S} = V \otimes_R S = \oplus \sum_{x \in G} V \otimes \bar{x}$, a direct sum of additive abelian groups. Furthermore, by definition of the module structure, it follows that for $v \in V$, $x \in G$ and $r \in R$ we have

$$(v \otimes \bar{x}) r^{\sigma(x)} = v \otimes (\bar{x} r^{\sigma(x)}) = v \otimes (r\bar{x}) = (vr) \otimes \bar{x}.$$

In other words if we write $v^{\sigma(x)} = v \otimes \bar{x}$, then $v^{\sigma(x)} r^{\sigma(x)} = (vr)^{\sigma(x)}$ and $V \otimes \bar{x}$ is an R-module isomorphic to $V^{\sigma(x)}$. ∎

Note that if $N \triangleleft G$ then $S = R*G = (R*N)*(G/N)$. Thus if V is an $R*N$-module, then there is an analogous formula for $V^{|S}$.

The following is a simple extension of the Hilbert Basis Theorem and we just sketch its proof.

Proposition 3.4. *Let $S = R*G$ be a crossed product where G is a polycyclic-by-finite group. If V is a Noetherian R-module, then $V^{|S}$ is a Noetherian S-module.*

Proof. In view of Lemma 1.3 and the transitivity of induction given in Lemma 3.1(v), it suffices to assume that G is either infinite cyclic or finite.

In the latter case, observe that each $V \otimes \bar{x} \cong V^{\sigma(x)}$ is also a Noetherian R-module. Hence since $V^{|S} = \oplus \sum V \otimes \bar{x}$ is a finite sum, we see that $V^{|S}$ is Noetherian as an R-module and therefore as an S-module.

Now let $G = \langle g \rangle$ be infinite cyclic. Via a diagonal change of basis, we may assume that $\bar{G} = \{ \bar{g}^i \mid i = \ldots, -1, 0, 1, \ldots \}$ so that $W = V^{|S} = \oplus \sum_{i=-\infty}^{\infty} V \otimes \bar{g}^i$. Set $W^+ = \sum_{i=0}^{\infty} V \otimes \bar{g}^i$.

Let U be an S-submodule of W. The goal is to show that U is finitely generated. To this end, we define for each integer $i \geq 0$

$$U_i = \left\{ v \in V \mid v \otimes \bar{g}^i + \text{ lower degree terms } \in U \cap W^+ \right\}.$$

It follows that U_i is an R-submodule of V and that $U_i \subseteq U_{i+1}$ since $U\bar{g} = U$. In particular, since V is Noetherian, the ascending series $U_0 \subseteq U_1 \subseteq \ldots \subseteq V$ must stabilize, say at $i = n$.

Now each of $U_0, U_1, \ldots, U_n$ is a finitely generated R-module and if $v_{i,1}, v_{i,2}, \ldots, v_{i,k(i)}$ generate U_i, we can choose $u_{i,j} \in U \cap W^+$ with

$$u_{i,j} = v_{i,j} \otimes \bar{g}^i + \text{ lower degree terms }.$$

It is easy to show (Exercise 2), by induction on the degree of the element $u \in U \cap W^+$, that $U \cap W^+ \subseteq \sum_{i,j} u_{i,j} S$. Finally if $u \in U$, then $u\bar{g}^m \in U \cap W^+$ for some m so $u \in (U \cap W^+)\bar{g}^{-m}$. We conclude that U is generated as an S-module by the finite set $\{ u_{i,j} \}$. ∎

We remark that if R is right Noetherian, then the above result with $M = R$ implies that $S = R \otimes_R S$ is also Noetherian. This is Proposition 1.6.

Our goal now is to discover an analog of this result in the context of group-graded rings. Suppose S is a G-graded ring with $R = S_1$. If V is an R-module, then $V^{|S}$ has a special structure here; it is a graded S-module.

Definition. Let $S = R(G)$ be a G-graded ring. An S-module M is said to be a *graded module* if $M = \oplus \sum_{x \in G} M_x$ is the direct sum of the additive subgroups M_x, indexed by the elements $x \in G$, with $M_x S_y \subseteq M_{xy}$ for all $x, y \in G$. In particular, each M_x is an R-submodule of M. Furthermore S itself is a graded right S-module.

If $N = \oplus \sum_{x \in G} N_x$ is another graded module, then N is a *graded submodule* of M if $N \subseteq M$ and $N_x = N \cap M_x$ for all $x \in G$. We say that M is *graded simple* if 0 and M are the unique graded submodules of M. Similarly M is *graded Noetherian* if the lattice of graded submodules of M satisfies the ascending chain condition. Finally an S-homomorphism $\theta: M \to N$ is a *graded homomorphism* if $\theta(M_x) \subseteq N_x$ for all $x \in G$. It is clear that the kernel of a graded homomorphism is a graded submodule of M.

Lemma 3.5. *Let $S = R(G)$ be a G-graded ring. If V is an R-module, then $V^{|S}$ is a graded S-module. Conversely if S is strongly G-graded and if M is a graded S-module, then $M = (M_1)^{|S}$. Hence, in the latter case, there is an isomorphism between the lattice of graded S-submodules of M and the lattice of R-submodules of M_1.*

Proof. Since $S = \oplus \sum_{x \in G} S_x$, it follows immediately that $W = V^{|S} = \oplus \sum_{x \in G} (V \otimes S_x)$ and that W is graded with $W_x = V \otimes S_x$.

Now let S be strongly G-graded and let $M = \oplus \sum M_x$ be any graded S-module. Then

$$M_x = M_x S_1 = (M_x S_{x^{-1}}) S_x \subseteq M_1 S_x$$

and thus clearly $M_x = M_1 S_x$. In particular, $M_1 = 0$ implies that $M = 0$. Furthermore we have a graded epimorphism $\theta: (M_1)^{|S} \to M$ given by $m_1 \otimes s \mapsto m_1 s$ for all $m_1 \in M_1$ and $s \in S$. Note that the kernel K of θ is a graded submodule of $(M_1)^{|S}$ and that $K_1 = 0$. Thus $K = 0$ and θ is an isomorphism. ∎

Thus in the above we see that M_1 is a Noetherian R-module if and only if M is graded Noetherian. The graded analog of Proposition 3.4 is therefore: If M is a graded Noetherian S-module and if G is a polycyclic-by-finite group, then M is a Noetherian S-module. This was essentially proved in [**13**] for S strongly graded (see Exercise 3) and in [**148**] for G a finitely generated nilpotent group. The general result requires the duality machine.

Let $S = R(G)$ be a G-graded ring. We use the notation of the previous section so that $M_G(S)$ is the ring of row and column finite

$|G| \times |G|$ matrices over S. Then $M_G(S)$ contains the subring

$$S\{1\} = \left\{\, \alpha \mid \alpha(x,y) \in R(x^{-1}y) = S_{x^{-1}y} \,\right\},$$

the subgroup $\bar{G} = \left\{\, \bar{g} = [\delta_{g^{-1}x,y}] \mid g \in G \,\right\}$ and the skew group ring $S\{1\}\bar{G}$.

Definition. Let M be an S-module. We define $\mathrm{Row}_G(M)$ to be the set of all row finite $1 \times |G|$ row matrices with entries in M. Thus if $v \in \mathrm{Row}_G(M)$ then v has only finitely many nonzero entries and its entry in the x-column is denoted by $v(x)$.

If $M = \oplus \sum_{x \in G} M_x$ is also a graded module, then we can define $\tilde{M} \subseteq \mathrm{Row}_G(M)$ by

$$\tilde{M} = \left\{\, v \in \mathrm{Row}_G(M) \mid v(x) \in M_x \text{ for all } x \in G \,\right\}.$$

Key properties of these sets are as follows.

Lemma 3.6. *Let S be a G-graded ring and let M be an S-module. Assume M is graded in parts (ii) and (iii) below.*

i. $\mathrm{Row}_G(M)$ is a right $S\{1\}\bar{G}$-module and all submodules are of the form $\mathrm{Row}_G(N)$ for N an S-submodule of M.

ii. $\tilde{M}$ is a right $S\{1\}$-module and all submodules are of the form $\tilde{N}$ for N a graded S-submodule of M.

iii. $\mathrm{Row}_G(M)$ is the induced module

$$\mathrm{Row}_G(M) = \tilde{M}^{|S\{1\}\bar{G}}.$$

Proof. We first introduce some notation. For each $x, y \in G$ let $e_{x,y} \in M_G(S)$ denote the matrix with a 1 in the (x,y)-position and zeros elsewhere. Note that $e_{x,x} \in S\{1\}$ for all $x \in G$ and that $S\{1\}\bar{G} \supseteq M_G^b(S)$ by Lemma 2.4(ii).

For each $x \in G$, let $\pi_x: \mathrm{Row}_G(M) \to M$ be the projection map into the x-coordinate so that $\pi_x(v) = v(x)$. Furthermore, for convenience, let f_x denote the $1 \times |G|$ row matrix with a 1 in the x-coordinate and zeros elsewhere. Thus if $v \in \mathrm{Row}_G(M)$ we can write

$$v = \sum_{x \in G} v(x)f_x = \sum_{x \in G} \pi_x(v)f_x.$$

In particular, $\mathrm{Row}_G(M) = \sum_x M f_x$ and $\tilde{M} = \sum_x M_x f_x$.

(i) It is clear that $\mathrm{Row}_G(M)$ is a right $\mathrm{M}_G(S)$-module and hence a right $S\{1\}\bar{G}$-module. Let W be a submodule. Since $e_{x,x}S \subseteq S\{1\}\bar{G}$ it follows that $W = \sum_x \pi_x(W) f_x$ and that each $\pi_x(W)$ is an S-submodule of M. Furthermore since $e_{x,y} \in S\{1\}\bar{G}$ we see that $\pi_x(W) = \pi_y(W)$ for all x, y. Thus $W = \sum_x \pi_1(W) f_x = \mathrm{Row}_G(\pi_1(W))$.

(ii) If $v \in \tilde{M}$ and $\alpha \in S\{1\}$, then $v(x)\alpha(x,y) \in M_x S_{x^{-1}y} \subseteq M_y$ and it follows that $v\alpha \in \tilde{M}$. Thus $\tilde{M}$ is an $S\{1\}$-module. Now let W be a submodule of $\tilde{M}$ and set $N = \sum_x \pi_x(W) \subseteq M$. Since $e_{x,y}S_{x^{-1}y} \subseteq S\{1\}$ we see that $\pi_x(W)S_{x^{-1}y} \subseteq \pi_y(W)$ for all $x, y \in G$ and hence that $N = \oplus \sum_x \pi_x(W)$ is a graded submodule of M with $N_x = \pi_x(W)$. Furthermore since $e_{x,x} \in S\{1\}$ we conclude that

$$W = \sum_x \pi_x(W) f_x = \sum_x N_x f_x = \tilde{N}$$

as required.

(iii) Finally since $f_x \bar{g} = f_{g^{-1}x}$ and $\tilde{M} = \sum_x M_x f_x$ we have

$$\sum_g \tilde{M}\bar{g} = \oplus \sum_{g,x} M_x f_{g^{-1}x} = \oplus \sum_{g,y} M_{gy} f_y$$

$$= \oplus \sum_y M f_y = \mathrm{Row}_G(M).$$

It now follows easily from

$$\tilde{M}^{|S\{1\}\bar{G}} = \oplus \sum_g \tilde{M} \otimes \bar{g} = \oplus \sum_g (\tilde{M} \otimes 1)\bar{g}$$

and the above that $\tilde{M}^{|S\{1\}\bar{G}}$ is isomorphic to $\mathrm{Row}_G(M)$. ∎

We remark that parts (i) and (ii) above yield lattice isomorphisms between the lattice of submodules of various modules. In particular, one module is Noetherian if and only if the other is. It is now a simple matter to prove

Theorem 3.7. [31] *Let S be a G-graded ring with G polycyclic-by-finite and let M be a graded S-module. Then M is Noetherian if and only if it is graded Noetherian.*

Proof. Certainly if M is Noetherian then it is graded Noetherian. Conversely assume that M is graded Noetherian and form $\tilde{M} \subseteq \mathrm{Row}_G(M)$. Then, by Lemma 3.6(ii), $\tilde{M}$ is a Noetherian $S\{1\}$-module and hence Proposition 3.4 and Lemma 3.6(iii) imply that $\tilde{M}^{|S\{1\}\bar{G}} \cong \mathrm{Row}_G(M)$ is a Noetherian $S\{1\}\bar{G}$-module since G is polycyclic-by-finite. We conclude from Lemma 3.6(i) that M is a Noetherian S-module. ∎

As an immediate consequence of the above and Lemma 3.5 we obtain

Corollary 3.8. [13] *Let S be a strongly G-graded ring where G is a polycyclic-by-finite group. If V is a Noetherian S_1-module, then the induced module $V^{|S}$ is also Noetherian.*

Other applications of duality come from the following observation of [**39**].

Lemma 3.9. *Let S be a G-graded ring with G finite. If M is a right $S\#G^*$-module, then M becomes a graded S-module by defining $M_x = Mp_x$ for all $x \in G$. Conversely if M is a graded S-module, then M becomes an $S\#G^*$-module by defining $mp_x = m_x$ for all $m \in M$, $x \in G$.*

Proof. Let M be a right $S\#G^*$-module and define $M_x = Mp_x$ so that $M = \oplus \sum_x M_x$ since $1 = \sum_x p_x$ is an orthogonal decomposition of 1. Furthermore, $p_x S_y = S_y p_{xy}$ implies that $M_x S_y \subseteq M_{xy}$.

Conversely let M be a graded S-module and let $\pi_x : M \to M_x$ be the natural projection. For any $s \in S$ it follows easily that

$$\pi_x s_{x^{-1}y} = \pi_x s \pi_y = s_{x^{-1}y} \pi_y$$

and thus M becomes an $S\#G^*$-module with p_x acting like π_x. ∎

Therefore we have a one-to-one inclusion preserving correspondence between graded S-modules and arbitrary $S\#G^*$-modules.

We return to crossed products $R*G$ and close this section by briefly considering induced modules from sub-crossed products. The following is known as the *Mackey decomposition*.

Lemma 3.10. Let $R*G$ be a crossed product and suppose that H and A are subgroups of G and that V is an $R*H$-module. Let $\mathcal{D}$ be a complete set of (H, A)-double coset representatives in G. If $d \in \mathcal{D}$, then $V \otimes \bar{d}$ is an $R*H^d$-module and we have

$$(V^{|R*G})_{|R*A} \cong \oplus \sum_{d \in \mathcal{D}} \left[(V \otimes \bar{d})_{|R*(H^d \cap A)} \right]^{|R*A}.$$

Proof. If X is any subset of G, then we let

$$R*X = \{\, \alpha \in R*G \mid \text{Supp } \alpha \subseteq X \,\}.$$

Since G is the disjoint union $G = \bigcup_{d \in \mathcal{D}} HdA$, it follows that

$$V^{|R*G} = \oplus \sum_{d \in \mathcal{D}} V \otimes R*(HdA).$$

Note that the above tensor $\otimes$ is over the ring $R*H$ and that $V_d = V \otimes R*(HdA)$ is clearly an $R*A$-submodule of $V^{|R*G}$ since $R*(HdA)$ is an $(R*H, R*A)$-bimodule. It remains to describe each V_d.

To this end, let T be a right transversal for $H^d \cap A$ in A. It follows (Exercise 6) that dT is a right transversal for H in $D = HdA$. Thus $R*D$ is a free left $R*H$-module with basis $\bar{d}\bar{T}$ and we have

$$V_d = \oplus \sum_{t \in T} V \otimes \overline{dt}.$$

Now note that $V \otimes \bar{d}$ is an $R*H^d$-module since $\bar{d}(R*H^d) = (R*H)\bar{d}$. Thus $(V \otimes \bar{d})_{|R*(H^d \cap A)}$ is an $R*(H^d \cap A)$-module and we can induce it to $R*A$. By definition of T this yields

$$\left[(V \otimes \bar{d})_{|R*(H^d \cap A)} \right]^{|R*A} = \oplus \sum_{t \in T} (V \otimes \bar{d}) \otimes' \bar{t}$$

where $\otimes'$ denotes tensor over $R*(H^d \cap A)$. It is now a simple exercise to show that V_d is $R*A$-isomorphic to this induced module via the map $v \otimes \overline{dt} \mapsto (v \otimes \bar{d}) \otimes' \bar{t}$ for all $v \in V$, $t \in T$. ∎

EXERCISES

1. Verify the equivalence of the two definitions for the module structure of the conjugate module V^σ. Furthermore if $\sigma, \tau \in \mathrm{Aut}(R)$ prove that $(V^\sigma)^\tau \cong V^{\sigma\tau}$.

2. In the proof of Proposition 3.4, first verify that each U_i is an R-submodule of V. Next suppose that $u \in U \cap W^+$ with $u = v \otimes \bar{g}^i +$ lower degree terms. If $i \leq n$ show that $u - \sum_j u_{i,j} r_j \in U \cap W^+$ has degree less than i for suitable $r_j \in R$. If $i > n$ show that $u - \sum_j u_{n,j} r_j \bar{g}^{i-n} \in U \cap W^+$ has degree less than i for suitable $r_j \in R$. Deduce that $U \cap W^+ \subseteq \sum_{i,j} u_{i,j} S$.

3. Let S be a strongly G-graded ring with base ring R. Let $x \in G$ and write $1 = \sum_i \alpha_i \beta_i$ with $\alpha_i \in S_x$, $\beta_i \in S_{x^{-1}}$. Show that $S_x = \sum_i \alpha_i R$ and $S_{x^{-1}} = \sum_i R\beta_i$. Furthermore, prove that S_x is a finitely generated projective right and left R-module.

4. Suppose V is a right S-module where S is G-graded with G finite. Then $S \subseteq S\#G^*$ and we consider the induced module $V^{|S\#G^*}$. Show that $V^{|S\#G^*} = \oplus \sum_x V \otimes p_x$ and that $(v \otimes p_x)s_y = vs_y \otimes p_{xy}$ for all $v \in V$, $x, y \in G$ and $s_y \in S_y$. In particular if $S = K[G]$ is a group algebra and $\dim_K V = 1$ conclude that $\left(V^{|S\#G^*}\right)_{|S} \cong S_S$. This example is from [**24**].

5. Let G act as permutations on the set Ω. If K is a field, then the vector space $K\Omega$, with K-basis Ω, becomes in a natural manner a $K[G]$-module; it is called a *permutation module*. Show that $K\Omega$ is the direct sum of the permutation modules corresponding to the orbits of G on Ω. Furthermore suppose G is transitive on Ω, with H the stabilizer of some point. Prove that $K\Omega$ is induced from a 1-dimensional $K[H]$-module.

6. In Lemma 3.10, let $a, b \in A$. Show that $Hda = Hdb$ if and only if $ab^{-1} \in H^d \cap A$ and deduce that dT is a right transversal for H in HdA. Furthermore prove that the map given by $v \otimes \bar{dt} \mapsto (v \otimes \bar{d}) \otimes' \bar{t}$ is an $R*A$-isomorphism. To this end, let $a \in A$, $r \in R$ and define $\alpha \in R*(H^d \cap A)$ and $\beta \in R*H$ by $\bar{t}ar = \alpha\bar{t_1}$ for some $t_1 \in T$ and $\bar{d}\alpha = \beta\bar{d}$. Compute $(v \otimes \bar{dt})\bar{a}r$ and $\left[(v \otimes \bar{d}) \otimes' \bar{t}\right]\bar{a}r$.

7. If S is G-graded, use Lemma 3.9 to show that S is an $(R, S\#G^*)$-bimodule and that $S\#G^* \subseteq \mathrm{End}(_R S)$ if S is component

regular. In addition, if $S = R*G$, prove that $\operatorname{End}(_RS) \cong \operatorname{M}_G(R)$.

8. We continue with the preceding notation and assume further that S is strongly graded. The goal is to show that $S\#G^* = \operatorname{End}(_RS)$. Let $\theta \in \operatorname{End}(_RS)$ and choose $s \in S_x$, $t \in S_{y^{-1}}$ for $x, y \in G$. Observe that

$$R \xrightarrow{s} S_x \xrightarrow{p_x\theta p_y} S_y \xrightarrow{t} R$$

is an R-endomorphism of R. Deduce that $s \cdot p_x\theta p_y \cdot t = p_x r$ for some $r \in R$. Finally use the facts that $\sum_x p_x = 1$ and $1 \in S_{x^{-1}}S_x$ to conclude that $\theta \in S\#G^*$. This is a result of [**199**].

4. Maschke's Theorem

Maschke's theorem is probably the first major theorem proved about group algebras of finite groups. It shows the strong effect of $|G|$-torsion, or the lack of it, on the structure of these algebras. As we see below, it converts directly to a result on crossed products. Furthermore there is an essential version of this theorem which is particularly useful for proving the nonexistence of nilpotent ideals. Finally, much of this translates, via duality, into results on group-graded rings.

Recall that an abelian group V is said to have *n-torsion* for some integer n if there exists $0 \neq v \in V$ with $vn = 0$. We begin with the classical version of *Maschke's theorem*.

Theorem 4.1. [**120**] *Given $R*G$ with G finite. Let $W \subseteq V$ be $R*G$-modules with no $|G|$-torsion and let $V = W \oplus U$ where U is a complementary R-submodule. Then there exists an $R*G$-submodule U' of V with $V \cdot |G| \subseteq W \oplus U'$.*

Proof. We note that G permutes the R-submodules of V. Indeed if U is any R-submodule of V and if $x \in G$, then the formula $\bar{x}r^{\sigma(x)} = r\bar{x}$ implies that $U\bar{x}$ is an R-submodule isomorphic to the conjugate module $U^{\sigma(x)}$.

Now let $V = W \oplus U$ be as given. Then for all $x \in G$, $V = W\bar{x} \oplus U\bar{x} = W \oplus U\bar{x}$ and we let $\pi_x\colon V \to W$ be the R-homomorphism determined by this decomposition. Note that if $v = w + u\bar{x}$, then $v\bar{y} = w\bar{y} + u\bar{x}\bar{y}$ so clearly

$$\pi_{xy}(v\bar{y}) = w\bar{y} = \pi_x(v)\bar{y}.$$

It follows that $\pi = \sum_x \pi_x$ is an $R*G$-homomorphism from V to W since

$$\pi(v\bar{y}) = \sum_x \pi_x(v\bar{y}) = \sum_x \pi_{xy}(v\bar{y})$$

$$= \sum_x \pi_x(v)\bar{y} = \pi(v)\bar{y}$$

for all $v \in V$, $y \in G$. Set $U' = \mathrm{Ker}(\pi)$ so that U' is an $R*G$-submodule of V.

Now if $w \in W$, then $\pi(w) = w \cdot |G|$ so $W \cap U' = 0$ since V has no $|G|$-torsion. Finally if $v \in V$, then $v \cdot |G| - \pi(v) \in \mathrm{Ker}(\pi)$ and hence $v \cdot |G| \in W \oplus U'$ as required. ∎

In particular, if $V \cdot |G| = V$, then $V = W \oplus U'$. As a consequence we see that if $V \cdot |G| = V$ and if $V_{|R}$ is completely reducible, then so is V. This is immediate since V is completely reducible if and only if every submodule is a direct summand. Note that $V \cdot |G| = V$ is automatically satisfied if $|G|^{-1} \in R$. We now have the following result on $\mathrm{J}(R*G)$, the *Jacobson radical* of $R*G$.

Theorem 4.2. [201] *Let $R*G$ be a crossed product with G finite. Then*

$$\mathrm{J}(R*G)^{|G|} \subseteq \mathrm{J}(R)*G \subseteq \mathrm{J}(R*G).$$

Furthermore if $|G|^{-1} \in R$ then $\mathrm{J}(R*G) = \mathrm{J}(R)*G$.

Proof. Let V be an irreducible $R*G$-module. Then V is a cyclic $R*G$-module and hence $V_{|R}$ is finitely generated. It now follows from Nakayama's lemma that $V\mathrm{J}(R) \neq V$. But $V\mathrm{J}(R)$ is easily seen to be an $R*G$-submodule so $V\mathrm{J}(R) = 0$. We conclude that $\mathrm{J}(R) \subseteq \mathrm{J}(R*G)$ and hence that $\mathrm{J}(R)*G \subseteq \mathrm{J}(R*G)$.

In the other direction, let W be an irreducible R-module and form the induced $R*G$-module $V = W^{|R*G}$. By Lemma 3.3, $V_{|R} = \oplus \sum_{x \in G} W \otimes \bar{x}$ is the direct sum of $n = |G|$ irreducible R-modules.

Thus V must have composition length $\leq n$ and $V \cdot J(R*G)^n = 0$. Now let $\alpha = \sum_{g \in G} r_g \bar{g} \in J(R*G)^n$. Then for any $w \in W$ we have $0 = (w \otimes 1)\alpha = \sum wr_g \otimes \bar{g}$ so $wr_g = 0$ and $Wr_g = 0$. Since this is true for all such W we have $r_g \in J(R)$ and therefore $J(R*G)^n \subseteq J(R)*G$.

Finally if $|G|^{-1} \in R$, then since $V_{|R}$ is completely reducible it follows from Theorem 4.1 that V is completely reducible. Hence in this case $V J(R*G) = 0$ for all W and the above argument yields $J(R*G) \subseteq J(R)*G$. ∎

To deal with nilpotent ideals, we require the following *essential version* of Maschke's theorem.

Proposition 4.3. [112] [165] *Given $R*G$ with G finite. Let $W \subseteq V$ be $R*G$-modules with no $|G|$-torsion. Then W ess V if and only if $W_{|R}$ ess $V_{|R}$.*

Proof. If $W_{|R}$ ess $V_{|R}$ then surely W ess V. Conversely assume W ess V.

Suppose first that $V = W \oplus U$ where U is an R-submodule. Then, by Theorem 4.1, there exists an $R*G$-submodule U' of V with $V \cdot |G| \subseteq W \oplus U'$. But W ess V so $U' = 0$ and $U \cdot |G| \subseteq U \cap W = 0$. Thus $U = 0$ and $W = V$.

Now for the general case. Choose $U_R \subseteq V$ maximal satisfying $U_R \cap W = 0$. Then $(W_{|R} \oplus U)$ ess $V_{|R}$ and we set $E = \bigcap_{x \in G}(W \oplus U)\bar{x}$. Note that $(W \oplus U)\bar{x}$ ess $V_{|R}$ (see Exercise 1). Thus since the intersection is finite, it follows that E is an $R*G$-submodule of V with $E_{|R}$ ess $V_{|R}$. Furthermore, $W \subseteq E \subseteq W \oplus U$ so $E = W \oplus (U \cap E)$. The result of the preceding paragraph now implies that $W = E$ so $W_{|R}$ ess $V_{|R}$. ∎

This yields the important

Theorem 4.4. [58] *Let R be a semiprime ring with no $|G|$-torsion. Then $R*G$ is semiprime.*

Proof. Let $N \lhd R*G$ with $N^2 = 0$. If $L = \ell_{R*G}(N)$, then $L \lhd R*G$ and L ess $R*G$ as right ideals. Since $R*G$ has no $|G|$-torsion, Proposition 4.3 now implies that $L_{|R}$ ess $R*G_{|R}$ so $(L \cap R)$ ess R_R. Since

R is semiprime, we conclude that $\mathrm{r}_R(L \cap R) = 0$ and then $N \subseteq$ $\mathrm{r}_{R*G}(L \cap R) = 0$ by the freeness of $R*G$ over R (see Exercise 2). ∎

The proof given above is from [165] but the original techniques of [58] are still needed for certain generalizations which we will consider in Section 18.

In the case of group-graded rings, nilpotent ideals can also occur because of certain degeneracies in the structure. Let S be a G-graded ring. If B is a subring of S, possibly without 1, then we say that B is a *graded subring* if $B = \sum_{x \in G} B_x$ with $B_x = B \cap S_x$. We have the following pigeon-hole argument.

Lemma 4.5. [74] [40] *Let $S = R(G)$ be a G-graded ring with G finite and let B be a graded subring (without 1). Then B is nilpotent if and only if B_1 is nilpotent. Indeed if $B_1^n = 0$, then $B^{n|G|} = 0$.*

Proof. If B is nilpotent, then so is B_1. Conversely suppose $B_1^n = 0$ and set $m = n|G|$. Consider any product

$$C = B_{x_1} B_{x_2} \cdots B_{x_m}$$

with $x_1, x_2, \ldots, x_m \in G$ and define $y_0, y_1, \ldots, y_m \in G$ by $y_i = x_1 x_2 \cdots x_i$ with $y_0 = 1$. By the definition of m, at least one value of y must occur $n+1$ times here. But observe that if $y_i = y_j$ with $j > i$ then $x_{i+1} \cdots x_{j-1} x_j = 1$ so $B_{x_{i+1}} \cdots B_{x_{j-1}} B_{x_j} \subseteq B_1$. It follows that $C \subseteq SB_1^n S = 0$ and hence, since B is graded, we have $B^m = 0$. ∎

This of course applies with B any graded left or right ideal of S.

Definition. Let S be a G-graded ring. We say that S is *graded semiprime* if S has no nonzero nilpotent graded ideal. When this occurs, then certainly S has no nonzero nilpotent graded right or left ideal.

We say that S is *nondegenerate* if for all $x \in G$ and all $0 \neq s \in S_x$ we have $sS_{x^{-1}} \neq 0$ and $S_{x^{-1}}s \neq 0$.

We remark that nondegeneracy is a slightly weaker assumption than component regularity. Indeed $S = R(G)$ is component regular if and only if it is nondegenerate with each component S_x faithful as a left and right R-module (see Exercise 6). A key relationship between the above two definitions is as follows.

Lemma 4.6. *Let* $S = R(G)$ *be a group-graded ring with* G *finite. Then* S *is graded semiprime if and only if* S *is nondegenerate and* R *is semiprime.*

Proof. Suppose S is graded semiprime. If $0 \neq s \in S_x$, then sS is a nonzero graded right ideal and hence is not nilpotent. By Lemma 4.5, $(sS)_1 = sS_{x^{-1}}$ is not zero. Similarly $S_{x^{-1}}s \neq 0$ and S is nondegenerate. Moreover if I is a nilpotent ideal of R, then IS is a right ideal of S with $(IS)_1 = IS_1 = I$. Thus again by Lemma 4.5, IS is nilpotent so $IS = 0$ and $I = 0$.

Conversely let S be nondegenerate and let R be semiprime. If I is a graded nilpotent ideal of S, then I_1 is a nilpotent ideal of R and hence $I_1 = 0$. Finally $I_x S_{x^{-1}} \subseteq I_1 = 0$ so since S is nondegenerate we have $I_x = 0$ for all $x \in G$. $\blacksquare$

The goal now is to obtain a graded version of Theorem 4.4. For this we use duality. Recall that the smash product $S\#G^*$ is given by $S\#G^* = \sum_x p_x S$ with $p_x s_{x^{-1}y} = p_x s p_y = s_{x^{-1}y} p_y$. Furthermore G acts as automorphisms on $S\#G^*$ via $(p_x s)^g = p_{g^{-1}x} s$.

Lemma 4.7. *Let* S *be a* G*-graded ring with* G *finite and let* $I \triangleleft S\#G^*$. *Then* $I \cap S$ *is a graded ideal of* S. *Furthermore* I *is* G*-stable if and only if* $I = (I \cap S)(S\#G^*) = (S\#G^*)(I \cap S)$.

Proof. If $s \in I \cap S$ then

$$s_x = \sum_y p_{yx^{-1}} s p_y \in I \cap S$$

for all $x \in G$. Thus $I \cap S$ is graded.

If $I = (I \cap S)(S\#G^*)$, then I is surely G-stable. Conversely let I be G-stable and let $\alpha = \sum \alpha(x) p_x \in I$ with $\alpha(x) \in S$. Then $\alpha(x) p_x = \alpha p_x \in I$ and, since I is G-stable, we have

$$\alpha(x) = \sum_{g \in G} (\alpha p_x)^g \in I \cap S$$

as required. For the other equality, write $\alpha = \sum p_x \alpha'(x)$. ∎

Theorem 4.8. [39] *Let S be a G-graded ring with G finite. If S is graded semiprime, then $S\#G^*$ is semiprime. If in addition S has no $|G|$-torsion, then S is semiprime.*

Proof. If J is a nilpotent ideal of $S\#G^*$, then so is $I = \sum_{x\in G} J^x$ and I is G-stable. Since S is graded semiprime, it follows from the preceding lemma that $J \subseteq I = 0$. Thus $S\#G^*$ is semiprime. If in addition S has no $|G|$-torsion, then Theorems 2.4 and 4.4 imply that $\mathrm{M}_G(S)$ is semiprime. We conclude that S is semiprime. ∎

In particular, in view of Lemma 4.6, S is semiprime if it is non-degenerate, has no $|G|$-torsion and if S_1 is semiprime. Thus we have an appropriate graded analog of Theorem 4.4.

Now we consider the structure of the Jacobson radical $\mathrm{J}(S)$ with S a G-graded ring. This also follows fairly directly from duality and does not require the next two results. Nevertheless, we include them because of their intrinsic interest.

Lemma 4.9. *Let $S = R(G)$ be a G-graded ring with G finite and let $V \neq 0$ be an S-module. Suppose W is an R-submodule of V with W ess $V_{|R}$. Then W contains a nonzero S-submodule of V.*

Proof. Since $V \neq 0$ we have $W \neq 0$. Note that $w \in W \backslash 0$ implies that $wR \subseteq W$. Now choose X a subset of G of maximal size such that there exists $w \in W \backslash 0$ with $wR(X) \subseteq W$. We claim that $X = G$. If not, let $g \in G \backslash X$. If $wR(g) = 0$ then $wR(X \cup \{g\}) \subseteq W$, a contradiction. Thus $wR(g) \neq 0$.

Now $wR(g)$ is a nonzero R-submodule of V and W ess $V_{|R}$. Thus there exists $w' \in wR(g) \cap W$ with $w' \neq 0$. Observe that

$$w'R(g^{-1}X) \subseteq wR(g)R(g^{-1}X) \subseteq wR(X) \subseteq W$$

and $w' \in W$ so $w'R(g^{-1}X \cup \{1\}) \subseteq W$, again a contradiction. Thus $X = G$ and $wR(G)$ is a nonzero S-submodule contained in W. ∎

Proposition 4.10. [68] *Let $S = R(G)$ be G-graded with G finite and let V be a completely reducible S-module. Then $V_{|R}$ is completely reducible.*

Proof. We can assume that V is irreducible. The preceding lemma then implies that $V_{|R}$ has no proper essential submodules and hence is completely reducible.

We now begin our work on $J(S)$.

Definition. If S is a G-graded ring we let $J_G(S)$ denote its *graded Jacobson radical*, that is $J_G(S)$ is the intersection of the annihilators of all graded simple right S-modules. It is clear that $J_G(S)$ is a graded ideal of S. By standard arguments (see Exercise 7) $J_G(S)$ is the intersection of the maximal graded right ideals of S; it is the largest graded ideal with quasi-regular identity component, and its definition is right-left symmetric.

Lemma 4.11. *Let $S = R(G)$ be G-graded with G finite.*
    i. $J(S) \cap R = J(R)$.
    ii. $J(S\#G^*) \cap S \subseteq J(S)$.

Proof. (i) Let R/M be an irreducible R-module. Then MS is a graded right ideal of S with $MS \cap R = (MS)_1 = M$ so $MS \neq S$. Choose M' a maximal right ideal of S containing MS. Then $M' \cap R = M$ so $R/M \hookrightarrow S/M'$. Since $J(S)$ annihilates S/M', it follows easily by varying the maximal right ideal M that $J(S) \cap R \subseteq J(R)$.

Conversely, let V be an irreducible S-module. Then $V_{|R}$ is completely reducible by the preceding proposition so $VJ(R) = 0$ and $J(R) \subseteq J(S) \cap R$.

(ii) If M is a maximal right ideal of S, then $M(S\#G^*) \neq S\#G^*$ by the freeness of $S\#G^*$ over S. The argument now proceeds as in (i) above. ∎

Part of the above proof could of course be replaced by a quasi-regularity argument. Indeed suppose $S \supseteq R$ are any rings with the same 1 and with the property that any element of R invertible in S

is invertible in R. Then $J(S) \cap R$ is a quasi-regular ideal of R and hence is contained in $J(R)$.

We now come to the graded analog of Theorem 2.2. Here part (iii) is an observation of [**68**].

Theorem 4.12. [39] *Let $S = R(G)$ be a G-graded ring with G finite.*

   i. $J_G(S)$ is the largest graded ideal contained in $J(S)$.

   ii. $J(S)^{|G|} \subseteq J_G(S) \subseteq J(S)$. Furthermore $J_G(S) = J(S)$ if $|G|^{-1} \in S$.

   iii. $J_G(S)^{|G|} \subseteq SJ(R)S \subseteq J_G(S)$.

Proof. (i) Let L be the largest graded ideal of S contained in $J(S)$. If V is a graded simple S-module, then V is a cyclic S-module so $VL \neq V$ by Nakayama's lemma. But VL is graded, so $VL = 0$ and we conclude that $L \subseteq J_G(S)$.

For the converse we use the one-to-one correspondence between the graded S-modules and the $S\#G^*$-modules given by Lemma 3.9. It follows that the graded simple S-modules must correspond to the simple $S\#G^*$-modules and therefore that $J(S\#G^*) \cap S = J_G(S)$. Since $J_G(S)$ is a graded ideal, Lemma 4.11(ii) yields the result.

(ii) Let $\bar{S} = S/J_G(S)$ so that $\bar{S}$ is G-graded and $J(\bar{S})$ contains no nonzero graded ideal. Since $J(\bar{S}\#G^*)$ is a characteristic ideal of $\bar{S}\#G^*$ it follows from Lemmas 4.7 and 4.11(ii) that $J(\bar{S}\#G^*) = 0$. Now $(\bar{S}\#G^*)G \cong M_G(S)$ by Theorem 2.5. Hence, since $J(\bar{S}\#G^*) = 0$, Theorem 2.2 implies that $J(M_G(\bar{S}))^{|G|} = 0$ and that $J(M_G(\bar{S})) = 0$ if $|G|^{-1} \in \bar{S}\#G^*$. Since $J(M_G(\bar{S})) = M_G(J(\bar{S}))$, we conclude from the definition of $\bar{S}$ that $J(S)^{|G|} \subseteq J_G(S)$ and that $J(S) \subseteq J_G(S)$ if $|G|^{-1} \in S$.

(iii) Set $I = SJ(R)S$ so that I is a graded ideal of S. By Lemma 4.11(i) and part (i) above, $I \subseteq J_G(S) \subseteq J(S)$. Moreover

$$J(R) \subseteq I \cap R \subseteq J(S) \cap R = J(R)$$

so $J_G(S)/I$ is a graded ideal of S/I which has identity component 0. By Lemma 4.5 we have $J_G(S)^{|G|} \subseteq I$. This completes the proof. ∎

We close this section with the *graded version* of Maschke's theorem.

Proposition 4.13. **[148] [180]** *Let $S = R(G)$ be a G-graded ring with G finite and let $W \subseteq V$ be S-modules with no $|G|$-torsion. Assume that W ess V. If $0 \neq v \in V$, then for some $y \in G$ we have $vS_y \cap W \neq 0$. If in addition $\ell_V(S_x) = 0$ for all $x \in G$, then $W_{|R}$ ess $V_{|R}$.*

Proof. We use duality and recall that $\mathrm{Row}_G(V)$ is an $S\{1\}\bar{G}$-module. Moreover, by Lemma 3.6, it follows that $\mathrm{Row}_G(W)$ is an essential $S\{1\}\bar{G}$-submodule of $\mathrm{Row}_G(V)$. Thus since $\mathrm{Row}_G(V)$ has no $|\bar{G}|$-torsion, Proposition 4.3 implies that

$$\mathrm{Row}_G(W)_{|S\{1\}} \text{ ess } \mathrm{Row}_G(V)_{|S\{1\}}.$$

Now let $0 \neq v \in V$ and let $\bar{v} \in \mathrm{Row}_G(V)$ have 1-component v and all other entries zero. Then the essentiality implies that $\bar{v}S\{1\} \cap \mathrm{Row}_G(W) \neq 0$; say $0 \neq \bar{v}\alpha \in \mathrm{Row}_G(W)$ with $\alpha \in S\{1\}$. By definition of $\bar{v}$ and α, the x-component of $\bar{v}\alpha$ is given by $(\bar{v}\alpha)(x) = v\alpha(x,1) \in vS_{x^{-1}}$. Thus for some $y \in G$ we have $vS_y \cap W \neq 0$ as required.

Finally suppose $\ell_V(S_x) = 0$ for all $x \in G$ and choose $0 \neq w \in vS_y \cap W$. Then

$$0 \neq wS_{y^{-1}} \in vS_y S_{y^{-1}} \cap WS_{y^{-1}} \subseteq vR \cap W$$

and hence $W_{|R}$ ess $V_{|R}$. ∎

We remark that the condition $\ell_V(S_x) = 0$ is of course a module analog of component regularity.

EXERCISES

1. Let $S = R(G)$ be a strongly graded ring and let V be an S-module. If U is an R-submodule of V, define $U^x = US_x \subseteq V$. Show that this yields an inclusion preserving permutation representation of G on the R-submodules of V. In particular U ess W if and only if U^x ess W^x.

2. Let R be a semiprime ring with ideals A and B. Show that the conditions $AB = 0$, $BA = 0$ and $A \cap B = 0$ are equivalent. Now assume that $A = \ell_R(B)$. Prove that R/A is semiprime and that if R has no n-torsion, then neither does R/A.

For any ring R, let $\mathrm{N}(R)$ denote the sum of its nilpotent ideals and let $\mathrm{P}(R)$ be its *prime radical*. Thus $\mathrm{P}(R)$ is the intersection of all prime ideals of R. Furthermore $\mathrm{P}(R)$ is the last term in the upper N-series defined transfinitely by $N_0 = 0$, $N_\alpha/N_{\alpha-1} = \mathrm{N}(R/N_{\alpha-1})$ if α has a predecessor $\alpha - 1$, and $N_\alpha = \bigcup_{\beta < \alpha} N_\beta$ if α is a limit ordinal.

3. If R has no n-torsion, prove that neither does $R/\mathrm{N}(R)$ or $R/\mathrm{P}(R)$.

4. If $S = R(G)$ is G-graded, show that $\mathrm{N}(S) \cap R \subseteq \mathrm{N}(R)$ and $\mathrm{P}(S) \cap R \subseteq \mathrm{P}(R)$ with equality if G is finite. For the latter use Lemma 4.5.

5. Let $S = R*G$ with G finite and assume that R has no $|G|$-torsion. Show that $\mathrm{P}(S) = \mathrm{P}(R)*G$.

6. Let $S = R(G)$ be a group-graded ring. Prove that S is component regular if and only if it is nondegenerate with each component S_x faithful as a right and as a left R-module. Define left and right degenerate in the obvious manner. Show by example with $|G| = 2$ that these concepts are not equivalent in general. What happens if G is finite and R is semiprime?

7. Show that $\mathrm{J}_G(S)$ is the intersection of all maximal graded right ideals of S. Deduce that $\mathrm{J}_G(S)$ is the largest graded ideal of S whose identity component is right quasi-regular. Finally prove that $\mathrm{J}_G(S)$ is the largest graded ideal of S whose identity component is left quasi-regular, and conclude that $\mathrm{J}_G(S)$ is equal to its left analog.

8. Let S be a G-graded ring and let $W \subseteq V$ be graded S-modules. Suppose $V = W \oplus U$ where U is an S-module which is not necessarily graded. For each $x \in G$ show that $V_x = W_x \oplus [(W_x' \oplus U) \cap V_x]$ where $W_x' = \sum_{y \neq x} W_y$. Deduce that $V = W \oplus U'$ for some graded module U'.

2   Delta Methods and Semiprime Rings

5. Delta Methods

It's time to move on and consider crossed products $R*G$ of infinite groups G. Our goal is to discover when these rings are semiprime or prime. Of course this is not completely settled even when G is finite. Thus our answer will be a reduction to $R*N$ for certain finite subgroups N of G. Specifically, the subgroups N to be checked depend upon the action of G on the ideals of R.

Various aspects of this problem have been studied for over 25 years. The first results concerned ordinary group algebras and appeared in [155] and [44]. These papers handled the semiprime and prime problems respectively using a coset counting technique known as the Δ-*method* to effectively reduce these questions to the finite normal subgroups of G. Furthermore, the same techniques handled twisted group algebras with little additional difficulty [157]. However, the crossed product situation is considerably more complicated. Thus it is appropriate to begin by briefly discussing the Δ-method in the case of ordinary group algebras $K[G]$.

A *linear identity* is an equation in $K[G]$ of the form

$$\alpha_1 x \beta_1 + \alpha_2 x \beta_2 + \cdots + \alpha_n x \beta_n = 0$$

which holds for all $x \in G$. Such identities arise in a number of different contexts, most notably in studying when $K[G]$ is semiprime, prime or satisfies a polynomial identity. Perhaps the simplest example here is the equation $\alpha x - x \alpha = 0$ for all $x \in G$ which of course merely says that α is central in $K[G]$. Amazingly, it turns out that all linear identities are intimately related to $\mathbf{Z}(K[G])$, the center of $K[G]$.

Now it is easy to see (Exercise 1) that $\mathbf{Z}(K[G])$ has as a K-basis the set of all *finite class sums* of G. In other words, these basis elements are of the form $\hat{\mathcal{K}} = \sum_{x \in \mathcal{K}} x \in K[G]$ where $\mathcal{K}$ is a conjugacy class in G of finite size. It follows that the elements of G which appear in the support of elements of $\mathbf{Z}(K[G])$ are precisely those which belong to

$$\Delta = \Delta(G) = \left\{ x \in G \,\middle|\, |G : \mathbf{C}_G(x)| < \infty \right\}.$$

$\Delta(G)$ is called the *f.c. center* (finite conjugate center) of G and any group G with $G = \Delta(G)$ is called an *f.c. group*. It turns out that the elements x of finite order, $o(x)$, in $\Delta(G)$ are of particular importance. Thus we define

$$\Delta^+ = \Delta^+(G) = \left\{ x \in G \,\middle|\, |G : \mathbf{C}_G(x)| < \infty \text{ and } o(x) < \infty \right\}.$$

Key properties of these subsets are listed below (see [**161**, Section 4.1]).

Lemma 5.1. *Let G be any group.*
   i. *Δ and Δ^+ are characteristic subgroups of G.*
   ii. *Δ/Δ^+ is a torsion-free abelian group.*
   iii. *Δ^+ is generated by the finite normal subgroups of G.*

This explains the first half of the name "Δ-method". The second half is apparent in the following proof.

Lemma 5.2. *Let*

$$\alpha_1 x \beta_1 + \alpha_2 x \beta_2 + \cdots + \alpha_n x \beta_n = 0 \qquad \text{for all } x \in G.$$

be a linear identity in $K[G]$. *Then*

$$\pi_\Delta(\alpha_1)\beta_1 + \pi_\Delta(\alpha_2)\beta_2 + \cdots + \pi_\Delta(\alpha_n)\beta_n = 0$$

and

$$\pi_\Delta(\alpha_1)\pi_\Delta(\beta_1) + \pi_\Delta(\alpha_2)\pi_\Delta(\beta_2) + \cdots + \pi_\Delta(\alpha_n)\pi_\Delta(\beta_n) = 0.$$

Proof. We show first that $\sum_1^n \pi_\Delta(\alpha_i)\beta_i = 0$. To this end, suppose the expression is not zero, fix v in its support and, for each i, write $\alpha_i = \pi_\Delta(\alpha_i) + \bar\alpha_i$ with Supp $\bar\alpha_i \cap \Delta = \emptyset$. Let W denote the intersection of the centralizers $\mathbf{C}_G(y)$ over all $y \in$ Supp $\pi_\Delta(\alpha_i)$ for all i. Then, by definition of Δ, we have $|G : W| < \infty$ and we will restrict our attention to group elements $x \in W$. Note that each such x centralizes each $\pi_\Delta(\alpha_i)$.

Let $x \in W$ and multiply the given identity for x on the left by x^{-1} to obtain $\sum_i (\alpha_i)^x \beta_i = 0$. This yields $\sum_i \pi_\Delta(\alpha_i)\beta_i = -\sum_i \bar\alpha_i^x \beta_i$, and since v is in the support of the left-hand term, it must occur on the right. Thus there exist $a \in$ Supp $\bar\alpha_i$, $b \in$ Supp β_i, for some i, with $v = a^x b$. Hence $a^x = vb^{-1}$ and $x \in \mathbf{C}_W(a)w_{a,b}$, a fixed right coset of $\mathbf{C}_W(a)$ depending only on a, b and v. We conclude that W is the finite union $W = \bigcup_{a,b} \mathbf{C}_W(a)w_{a,b}$. As we will see in the next section (Lemma 6.2), this implies that $|W : \mathbf{C}_W(a)| < \infty$ for some a and we have a contradiction since $|G : W| < \infty$ and $a \notin \Delta$.

Thus $\sum_1^n \pi_\Delta(\alpha_i)\beta_i = 0$ and by applying π_Δ to this expression we obtain $\sum_1^n \pi_\Delta(\alpha_i)\pi_\Delta(\beta_i) = 0$. ∎

The second conclusion above is from [155]. The first is a particularly useful formulation contained in [194].

The relevance to the primeness or semiprimeness of $K[G]$ is as follows. Suppose A and B are nonzero ideals of $K[G]$ with $AB = 0$ and choose elements $\alpha \in A$ and $\beta \in B$. If x is any element of G, then $\alpha x \in A$ so $\alpha x \beta \in AB = 0$ and we obtain the linear identity

$$\alpha x \beta = 0 \qquad \text{for all } x \in G.$$

We conclude from the above that $\pi_\Delta(\alpha) \cdot \pi_\Delta(\beta) = 0$ and therefore that $\pi_\Delta(A) \cdot \pi_\Delta(B) = 0$. This reduces the various questions to $K[\Delta]$ where they are easily solved. The results are:

Theorem 5.3. *If* char $K = 0$, *then* $K[G]$ *is semiprime.*

Theorem 5.4. [155] *If* char $K = p > 0$, *then* $K[G]$ *is semiprime if and only if* $\Delta^p(G) = \langle 1 \rangle$.

Here $\Delta^p(G)$ is the subgroup of $\Delta^+(G)$ generated by all elements of order a power of p.

Theorem 5.5. [44] $K[G]$ *is prime if and only if* $\Delta^+(G) = \langle 1 \rangle$.

Now let us move on to crossed products $R*G$. As will be apparent, when G acts nontrivially on the ring R, a new dimension is added to the problem. Here the first result, due to [9], asserts that if R is a simple ring and G is outer on R, then $R*G$ is a simple ring (see Exercise 6). This was extended in [127] where it was shown that if R is prime (or semiprime) and if G is X-outer on R, then $R*G$ is prime (or semiprime). Furthermore, as we observed earlier, [58] essentially settled the semiprime question for G finite. The final attack on infinite groups began in [132] where the Δ-method and the techniques of [58] combined to handle the case where R is a prime ring. This was extended in [162] to semiprime coefficients and then the problem was completely solved in [166] and [167]. Somewhat later, the results were generalized in [180], using duality, to component regular group-graded rings.

In this and the next three sections, we will discuss the results of [167]. Following that paper, we will work in the context of strongly group-graded rings. The proofs are no harder in that generality and in fact they are sometimes more natural. We begin by introducing some notation which will enable us to state the main result and prove its easy direction.

Lemma 5.6. *If* S *is strongly* G-graded, *then it is component regular. Furthermore if* α_y *is a nonzero element of* S_y *and* $xyz = 1$, *then* $S_x \alpha_y S_z$ *is a nonzero ideal of* $R = S_1$.

Proof. This is clear since $1 \in S_1 = S_x S_{x^{-1}}$ and since $S_x \alpha_y S_z$ is a nonzero (R, R)-subbimodule of $R = S_1$. ∎

We will use this lemma, and its obvious generalizations, freely throughout the remainder of this chapter.

Now let $S = R(G)$ be G-graded and let $N \lhd H \subseteq G$. For $x \in H$ and I an ideal of $R(N)$, we define $I^x = S_{x^{-1}} I S_x$. In this way H acts on the ideals of $R(N)$ and basic properties are as follows.

Lemma 5.7. *Assume that* $S = R(G)$ *is strongly graded and let* $N \lhd H \subseteq G$. *If* $x, y \in H$ *and* I, J *are ideals of* $R(N)$, *then*
  i. I^x *is an ideal of* $R(N)$,
  ii. $(I^x)^y = I^{xy}$ *and* $I^1 = I$,
  iii. $I \subseteq J$ *implies that* $I^x \subseteq J^x$,
  iv. $(IJ)^x = I^x J^x$.

Proof. Since $N \lhd H$ and $x \in H$, it follows easily that $R(N)S_x = S_x R(N)$. From this we conclude first that $I^x \subseteq R(N)$ and then that I^x is an ideal of $R(N)$. Now for (ii) and (iv) we have

$$(I^x)^y = S_{y^{-1}} I^x S_y = S_{y^{-1}} (S_{x^{-1}} I S_x) S_y = S_{(xy)^{-1}} I S_{xy} = I^{xy}$$

and

$$I^x J^x = (S_{x^{-1}} I S_x)(S_{x^{-1}} J S_x) = S_{x^{-1}} (IRJ) S_x = (IJ)^x$$

since $IR = I$. Finally (iii) is obvious, so the lemma is proved. ∎

If we let $\mathcal{I}$ denote the set of ideals of $R(N)$ in the above situation, then (i) and (ii) assert that the map $x \mapsto {}^x$ is a homomorphism of H into $\mathrm{Sym}(\mathcal{I})$. Furthermore, by (iii), these are inclusion preserving permutations and in particular they preserve the lattice operations of arbitrary intersections and sums. Part (iv) says that finite products are also preserved.

Continuing with this notation, an ideal I of $R(N)$ is said to be *H-invariant* if $I^x = I$ for all $x \in H$. Since $S_x S_{x^{-1}} = R$, this clearly occurs if and only if $I S_x = S_x I$ for all $x \in H$.

Note that if $N = \langle 1 \rangle$, then $R(N) = R$ and, in this way, G acts on the ideals of R. In fact, if $S = R*G$ is a crossed product and I is an ideal of R, then

$$I^x = S_{x^{-1}} I S_x = \bar{x}^{-1}(RIR)\bar{x} = \bar{x}^{-1} I \bar{x}$$

and this is precisely the action we discussed earlier. We can now state the main result for stongly group-graded rings.

Theorem 5.8. [166] [167] *Let $S = R(G)$ be strongly G-graded with base ring R. Then S contains nonzero ideals A, B with $AB = 0$ if and only if there exist*

    i. subgroups $N \triangleleft H \subseteq G$ with N finite,

    ii. an H-invariant ideal I of R with $I^x I = 0$ for all $x \in G \setminus H$,

    iii. nonzero H-invariant ideals $\tilde{A}$, $\tilde{B}$ of $R(N)$ with $\tilde{A}\tilde{B} = 0$ and $\tilde{A}, \tilde{B} \subseteq I \cdot R(N)$.

Furthermore $A = B$ if and only if $\tilde{A} = \tilde{B}$.

As we will see in Section 8, this yields appropriate analogs of Theorems 5.3, 5.4 and 5.5 in the case of strongly group-graded rings. It is convenient now to record the following elementary

Lemma 5.9. *Suppose that $S = R(G)$ is strongly graded, H is a subgroup of G and I is an ideal of R with $I^x I = 0$ for all $x \in G \setminus H$. Then*

    i. $I S_x I = 0$ for all $x \in G \setminus H$,

    ii. $I \cdot R(G) \cdot I \subseteq I \cdot R(H) \subseteq R(H)$.

Proof. Part (i) is clear since S is component regular. Moreover

$$I \cdot R(G) \cdot I = \oplus \sum_{x \in G} I S_x I = \oplus \sum_{x \in H} I S_x I \subseteq I \cdot R(H) \subseteq R(H)$$

and (ii) is proved. ∎

We can now offer the

Proof of Theorem 5.8 (Easy Direction). Here we assume that $S = R(G)$ is given and that $H, N, I, \tilde{A}$ and $\tilde{B}$ exist and satisfy the appropriate properties. We set $A = S\tilde{A}S$ and $B = S\tilde{B}S$ so that these are nonzero ideals of the strongly G-graded ring S. The goal is to show that $AB = 0$ or equivalently that $\tilde{A}S_x\tilde{B} = 0$ for all $x \in G$.

If $x \in H$ then, since $\tilde{A}$ is H-invariant, we have $\tilde{A}S_x\tilde{B} = S_x\tilde{A}\tilde{B} = 0$. On the other hand, if $x \in G \setminus H$, then since $\tilde{A}, \tilde{B} \subseteq I \cdot R(N)$ we have

$$\tilde{A}S_x\tilde{B} \subseteq I \cdot R(N)S_x \cdot I \cdot R(N).$$

But $R(N)S_x = \sum_{y \in Nx} S_y$ and $Nx \subseteq G \setminus H$ so Lemma 5.9(i) implies that $I \cdot R(N)S_x \cdot I = 0$. Therefore $\tilde{A}S_x\tilde{B} = 0$ in this case also, and we have shown that $AB = 0$. Since $\tilde{A} = \tilde{B}$ implies $A = B$, this direction is proved. ∎

The converse proof is considerably more difficult because of problems with the Δ-method. As an indication of this, suppose that $S = R*G$ is a crossed product and that we are given the identity

$$\alpha\bar{x}\beta = 0 \qquad \text{for all } x \in G.$$

Following our earlier group algebra argument, we begin by defining $W = \bigcap_y \mathbf{C}_G(y)$ where the intersection is over all $y \in \text{Supp } \pi_\Delta(\alpha)$. Then we have $|G : W| < \infty$ and if $x \in W$ we might suspect that $\bar{x}$ centralizes $\pi_\Delta(\alpha)$. However, this fails for two distinct reasons. First, although x commutes with all elements $y \in \text{Supp } \pi_\Delta(\alpha)$, we cannot conclude that $\bar{x}$ centralizes each $\bar{y}$ because of the twisting. Moreover, even if there were no twisting, $\bar{x}$ would still not necessarily commute with the coefficients in $\pi_\Delta(\alpha)$ because of the action of $\bar{x}$ on R.

In some sense, we avoid the first difficulty by assuming that $\pi_\Delta(\alpha) = r\bar{1}$ and $\pi_\Delta(\beta) = s\bar{1}$ for some $r, s \in R$. We then obtain an interesting relationship between cosets of centralizers and the action of G on R in the following way.

Write $\alpha = \pi_\Delta(\alpha) + \bar{\alpha} = r\bar{1} + \bar{\alpha}$ and $\beta = \pi_\Delta(\beta) + \bar{\beta} = s\bar{1} + \bar{\beta}$ as before and let $x \in G = W$ (in this case). If $r^x s \neq 0$, then the identity coefficient $r^x s \neq 0$ in $\alpha^{\bar{x}}\beta = 0$ must be cancelled. We conclude from this that $x \in \bigcup_{a,b} \mathbf{C}_G(a)w_{a,b}$ where the union is over suitable $a \in \text{Supp } \bar{\alpha}$ and $b \in \text{Supp } \bar{\beta}$. In other words we have shown that

$$r^x s = 0 \qquad \text{for all } x \in G \setminus \bigcup_{a,b} \mathbf{C}_G(a)w_{a,b}.$$

This relation is probably hopeless to study, but with a little more care, r and s can be replaced by arbitrary elements in some nonzero ideal of R. That condition then becomes quite manageable and indeed the next section shows how to translate it to an extremely useful form.

The proof of Theorem 5.8 is finally completed in Section 7 where most of the work involves finding an appropriate α and β with $\pi_\Delta(\alpha)$ and $\pi_\Delta(\beta)$ relatively well behaved. The formal Δ-method argument shows up in Lemma 7.5.

We close this section with two elementary observations which will be needed at the end of the proof. They require the following graded-ring definitions which are the obvious analogs of their crossed product counterparts. Suppose $S = R(G)$ is G-graded and let $\alpha = \sum_{x \in G} \alpha_x$. Then,

$$\operatorname{Supp} \alpha = \{ x \in G \mid \alpha_x \neq 0 \}.$$

Similarly if H is a subgroup of G, then the map $\pi_H \colon R(G) \to R(H)$ given by

$$\pi_H(\alpha) = \pi_H \left(\sum_{x \in G} \alpha_x \right) = \sum_{x \in H} \alpha_x$$

is an $(R(H), R(H))$-bimodule homomorphism.

Furthermore, suppose A is a nonzero ideal of the strongly W-graded ring $R(W)$ and let $N \lhd W$. Then we denote by $\min_N A$ the additive span of all elements $\alpha \neq 0$ of A whose support meets the minimum number of cosets of N.

Lemma 5.10. *Let $S = R(G)$ be given, let $H \subseteq G$ and suppose $N \lhd W$ are subgroups of G normalized by H. If A is a nonzero H-invariant ideal of $R(W)$, then*
    i. $\min_N A$ is a nonzero H-invariant ideal of $R(W)$,
    ii. $\pi_N(A)$ is a nonzero H-invariant ideal of $R(N)$.

Proof. (i) By definition, $\min_N A$ is nonzero and let $\alpha \in A$ be any generator of this set. If $w \in W$, then $\operatorname{Supp} \alpha S_w \subseteq (\operatorname{Supp} \alpha)w$ and $\operatorname{Supp} S_w \alpha \subseteq w(\operatorname{Supp} \alpha)$ so, since $N \lhd W$, it follows that $\min_N A$

is an ideal of $R(W)$. Finally if $h \in H$, then Supp $S_{h^{-1}}\alpha S_h \subseteq$
$h^{-1}(\text{Supp } \alpha)h$ so, since A is H-invariant and N is normalized by
H, we see that $\min_N A$ is H-invariant.

(ii) We know that π_N is an $(R(N), R(N))$-bimodule homomorphism so $\pi_N(A)$ is an ideal of $R(N)$. Next if $\alpha \in A$ and $h \in H$ then,
since H normalizes N, we have $\pi_N(S_{h^{-1}}\alpha S_h) = S_{h^{-1}}\pi_N(\alpha)S_h$ and
therefore $\pi_N(A)$ is H-invariant. Finally choose $0 \neq \alpha = \sum_{w \in W} \alpha_w \in$
A. If $x \in \text{Supp } \alpha$, then $\alpha S_{x^{-1}} \subseteq A$ and clearly $\pi_N(\alpha S_{x^{-1}}) \neq 0$. ∎

Lemma 5.11. *Let $R(W)$ be given, let $N \triangleleft W$ and assume that W/N
is a unique product group (for example, an ordered group). If A and
B are nonzero ideals of $R(W)$ with $AB = 0$, then*

$$\pi_N(\min_N A) \cdot \pi_N(\min_N B) = 0.$$

Proof. In view of the graded analog of Lemma 1.3, we may assume
that $N = \langle 1 \rangle$. In particular $\pi_N\left(\sum_x \alpha_x\right) = \alpha_1$.

Let $\alpha = \sum \alpha_x$ and $\beta = \sum \beta_y$ be generators of $\min_N A$ and
$\min_N B$, respectively. Since W is a unique product group, we can
let $x_0 y_0$ be a unique product element in $(\text{Supp } \alpha)(\text{Supp } \beta)$. From
$\alpha\beta = 0$ we deduce that $\alpha_{x_0}\beta_{y_0} = 0$. Then $\alpha\beta_{y_0} \in A$ has smaller
support size than that of α. Hence by the minimal nature of α, we
have $\alpha\beta_{y_0} = 0$ and therefore $\alpha_x\beta_{y_0} = 0$ for all $x \in W$. Similarly
we now see that $\alpha_x\beta \in B$ has smaller support size than that of
β. Thus $\alpha_x\beta = 0$ so $\alpha_x\beta_y = 0$ for all $x, y \in W$. In particular
$\pi_N(\alpha)\pi_N(\beta) = \alpha_1\beta_1 = 0$. ∎

EXERCISES

1. Show that $\alpha \in \mathbf{Z}(K[G])$ if and only if $\alpha^x = \alpha$ for all $x \in G$
and hence if and only if the coefficients of α are constant on the
conjugacy classes of G. Deduce that $\mathbf{Z}(K[G])$ has as a K-basis the
finite class sums of G. Find an analogous result in the case of twisted
group algebras $K^t[G]$.

2. Let $K[G]$ satisfy the linear identity $\sum_i \alpha_i x \beta_i = 0$ for all $x \in G$. Prove that $\sum_i \pi_\Delta(\alpha_i) x \beta_i = 0$ and $\sum_i \pi_\Delta(\alpha_i) x \pi_\Delta(\beta_i) = 0$ for all $x \in G$. This generalizes Lemma 5.2.

3. If $\langle 1 \rangle \neq N$ is a finite normal subgroup of G, set $\alpha = \hat{N} = \sum_{x \in N} x \in K[G]$. Show that α is a central element of $K[G]$ satifying $\alpha(\alpha - |N|) = 0$. Conclude that $K[G]$ is not prime and that $K[G]$ is not semiprime if $|N| = 0$ in K. This yields the easy direction of Theorems 5.4 and 5.5.

4. Complete the hard direction of the proofs of Theorems 5.3, 5.4 and 5.5 by using Lemma 5.2 to reduce the problem to $K[\Delta]$ and then Lemma 5.11 to further reduce it to $K[\Delta^+]$. Note that Δ/Δ^+ is a torsion-free abelian group and hence an ordered group.

5. Suppose R is a simple ring, σ is an automorphism of R and $0 \neq u \in R$. If $ur = r^\sigma u$ for all $r \in R$, show that u is a unit of R and σ is the inner automorphism of R induced by the unit u^{-1}.

6. Let $R*G$ be given with R a simple ring and assume that for all $1 \neq x \in G$ the automorphism ^{x} is outer on R. Use a minimal length argument to show that $R*G$ is simple ([9]). To this end, let I be a nonzero ideal of $R*G$ and choose $0 \neq \alpha \in I$ of minimal support size. Show that we can assume that $1 \in \text{Supp } \alpha$ and then, since R is simple, that the coefficient of $\bar{1}$ is $1 \in R$. Deduce that α centralizes R and then use the result of the preceding exercise to conclude that $\alpha = 1$.

6. Coset Calculus

Let H be a group and let R be a ring. We say that H *strongly permutes* the ideals of R if for all $x, y \in H$ and ideals I, J of R we have

   i. I^x is an ideal of R,
   ii. $(I^x)^y = I^{xy}$ and $I^1 = I$,
   iii. $I \subseteq J$ implies that $I^x \subseteq J^x$,
   iv. $(IJ)^x = I^x J^x$.

Thus, for example, if we are given the strongly H-graded ring $R(H)$, then Lemma 5.2 asserts precisely that H strongly permutes the ideals of R. As we observed previously, the first three conditions above imply that this action comes from a homomorphism of H into the group of inclusion preserving permutations on the set of ideals of R. In particular, each ^{x} also preserves the lattice operations of arbitrary sums and intersections. It follows that $0^x = 0$ and $R^x = R$ for all $x \in H$.

Now let J be a nonzero ideal of R. We are interested in studying those $h \in H$ with $J^h J = 0$. As we have seen, the Δ-method usually implies that something occurs for all elements of H except for those in a finite union of cosets of varying subgroups. In this section we show how to reformulate such a conclusion so that it in fact holds for all elements in $H \setminus L$ where L is a single subgroup of H suitably determined by the situation.

Suppose A and B are subgroups of H and that the left cosets xA and yB are not disjoint. If $z \in xA \cap yB$, then $xA = zA$, $yB = zB$ and hence

$$xA \cap yB = zA \cap zB = z(A \cap B).$$

In other words, the intersection of two left cosets is either empty or a coset of the intersection. Property (iv) above will be crucial in the next few lemmas.

Lemma 6.1. *Let H strongly permute the ideals of R and let J be a nonzero ideal of R such that*

$$J^h J = 0 \qquad \text{for all } h \in H \setminus \bigcup_1^n h_k H_k.$$

Here $\bigcup_1^n h_k H_k$ is a fixed finite union of left cosets of the subgroups H_k of H. Then there exists a subgroup L of H and a nonzero product $0 \neq K = J^{y_1} J^{y_2} \cdots J^{y_r}$ of H-conjugates of J, with some $y_i = 1$, such that $K^h K = 0$ for all $h \in H \setminus L$. Furthermore $|L : L \cap H_k| < \infty$ for some k.

Proof. In the course of the proof we will replace $\{ H_1, H_2, \ldots, H_n \}$ by certain other finite sets $\mathcal{A}$ of subgroups of H with the property that $A \in \mathcal{A}$ implies that $A \subseteq H_k$ for some k. We note that if the result is proved for such a set $\mathcal{A}$, then from $|L : L \cap A| < \infty$ for some A and $L \cap A \subseteq L \cap H_k$ we obtain $|L : L \cap H_k| < \infty$. In other words, the result will then follow for the original subgroups $H_1, H_2, \ldots, H_n$.

If $\mathcal{A}$ is the set of all proper (that is, nonempty) intersections of the H_k's, then $\mathcal{A}$ is finite and closed under intersections. Thus without loss of generality we can now assume that the H_k's are contained in a finite set $\mathcal{A}$ closed under intersections and we prove the result by induction on $|\mathcal{A}| \geq 0$. If $|\mathcal{A}| = 0$ then the hypothesis and conclusion both assert that $J^h J = 0$ for all $h \in H$.

Assume now that $|\mathcal{A}| \geq 1$, let A be a maximal member of $\mathcal{A}$ and set $\mathcal{A}' = \mathcal{A} \setminus \{A\}$. Then $|\mathcal{A}'| < |\mathcal{A}|$ and $\mathcal{A}'$ is closed under intersections. We will be concerned with finite unions of left cosets $S = \bigcup h_{i,j} A_i$ with $A_i \in \mathcal{A}$. By the support of S we mean those A_i's which occur in this representation. Suppose $K = J^{y_1} J^{y_2} \cdots J^{y_r} \neq 0$, some $y_i = 1$, and $K^h K = 0$ for all $h \in H \setminus S$. If $A \notin \operatorname{Supp} S$, then $\operatorname{Supp} S \subseteq \mathcal{A}'$ and induction applies. Thus there exists a finite product $I = K^{u_1} K^{u_2} \cdots K^{u_s} \neq 0$ with some $u_j = 1$ and $I^h I = 0$ for all $h \in H \setminus L$. Since I is a suitable product of conjugates of J and since $|L : L \cap A_t| < \infty$ for some $A_t \in \mathcal{A}'$, the result follows in this case.

Thus we can assume, for all such pairs K and S as above, that $A \in \operatorname{Supp} S$. Of course there is at least one such pair by hypothesis and now we choose K and S so that S has the smallest number, say $m \geq 1$, of cosets of A occurring in its representation. Then

$$ S = z_1 A \cup z_2 A \cup \cdots \cup z_m A \cup T $$

where T is a finite union of cosets of groups in $\mathcal{A}'$ and we define L by

$$ L = \left\{ h \in H \ \middle| \ h \cdot \left(\bigcup_1^m z_i A \right) = \bigcup_1^m z_i A \right\}. $$

Suppose $K^x K \neq 0$ for some $x \in H$. Then $K^x K$ is a nonzero finite product of H-conjugates of J with some conjugating element equal to 1. Furthermore the symbolic formula $K^{(H \setminus S)} K = 0$ yields $K^{x \cdot x^{-1}(H \setminus S)} K = 0$ so $(K^x K)^h (K^x K) = 0$ for all h with

$$ h \in (H \setminus S) \cup x^{-1}(H \setminus S) = H \setminus (S \cap x^{-1} S). $$

Since $\mathcal{A}$ is closed under intersections, it is clear that $S \cap x^{-1}S$ is also a finite union of left cosets of members of $\mathcal{A}$. Indeed, since clearly Supp $x^{-1}S$ = Supp S, we see that $S \cap x^{-1}S$ is a union of cosets of groups of the form $B \cap C$ with $B, C \in$ Supp S. Furthermore, since A is a maximal member of $\mathcal{A}$, it is clear that $A = B \cap C$ can occur if and only if $B = C = A$. By the definition of m, we note that $S \cap x^{-1}S$ must contain $m' \geq m$ cosets of A.

Since

$$x^{-1}S = x^{-1}z_1 A \cup x^{-1}z_2 A \cup \cdots \cup x^{-1}z_m A \cup x^{-1}T,$$

the A-cosets of $S \cap x^{-1}S$ come from

$$\left(\bigcup_1^m z_i A \right) \cap \left(\bigcup_1^m x^{-1} z_i A \right).$$

But cosets of A are either disjoint or identical so this intersection has $m' \leq m$ terms. Thus we must have $m' = m$, so

$$\left(\bigcup_1^m z_i A \right) = \left(\bigcup_1^m x^{-1} z_i A \right)$$

and hence $x \in L$. In other words, we have shown that $K^{(H \backslash L)} K = 0$.

Now H permutes the set Ω of left cosets of A by left multiplication and L is the stabilizer of the finite set $\Lambda = \{ z_1 A, z_2 A, \ldots, z_m A \}$ of Ω. Thus L is a subgroup of H. Moreover if $\alpha = zA \in \Lambda$, then we have $H_\alpha = \{ h \in H \mid h\alpha = \alpha \} = zAz^{-1}$ and it follows that $|L : L \cap (zAz^{-1})| < \infty$. We can eliminate the conjugating element z^{-1} by conjugating both L and K by z. However in so doing we lose the property of K that some $y_i = 1$. Thus we must take a different approach.

Suppose first that $L \cap z_i A = \emptyset$ for all $i = 1, 2, \ldots, m$. Then $K^h K = 0$ for all h with

$$h \in (H \backslash S) \cup (H \backslash L) = H \backslash (L \cap S)$$

and by the above assumption, $L \cap S = L \cap T$ is a finite union of cosets of the groups $\mathcal{B} = \left\{ L \cap \tilde{A} \mid \tilde{A} \in \mathcal{A}' \right\}$. Since $|\mathcal{B}| \leq |\mathcal{A}'| < |\mathcal{A}|$ and since $\mathcal{B}$ is clearly closed under intersections, induction applies here. Thus, as before, there exists $I = K^{u_1}K^{u_2}\cdots K^{u_s} \neq 0$ with some $u_j = 1$ such that $I^h I = 0$ for all $h \in H \setminus \tilde{L}$ where $|\tilde{L} : \tilde{L} \cap B| < \infty$ for some $B \in \mathcal{B}$. Since I has the appropriate form, the result follows in this case.

Finally, if $L \cap z_i A \neq \emptyset$ for some i, then we may take $z = z_i \in L$. Since $|L : L \cap (zAz^{-1})| < \infty$, conjugating this expression by $z \in L$ then yields $|L : L \cap A| < \infty$ and the lemma is proved. ∎

We remark that the same result holds with $J^h J$ replaced by JJ^h. Indeed, in the proof, merely replace all occurrences of $K^h K$ by KK^h. Furthermore both of these hold if left cosets are replaced by right cosets. Merely note that $J^h J = 0$ is equivalent to $JJ^{h^{-1}} = 0$ and that replacing h by h^{-1} effectively interchanges right and left cosets. The form of the lemma proved here is the one we will use.

Moreover, suppose that $H = \bigcup_1^n h_k H_k$ is given, let H act trivially on the ideals of any ring R with 1 and take $J = R$. Then the hypothesis of the above lemma is vacuously satisfied and the conclusion implies that $R^h R = 0$ for all $h \in H \setminus L$. Thus we must have $L = H$ and $|H : H_k| < \infty$ for some k. In other words, Lemma 6.1 generalizes the well-known result given below concerning the existence of subgroups of finite index.

Lemma 6.2. *Given the set theoretic union*

$$G = \bigcup_{i,j} x_{i,j} H_i$$

where $H_1, H_2, \ldots, H_n$ are finitely many subgroups of the group G. If the union contains finitely many (or countably many) terms, then for some k, the index $|G : H_k|$ is finite (or countable).

Proof. We proceed by induction on n. If a full set of cosets of H_n occurs in the above, in particular if $n = 1$, then the result is clear. So assume that the coset xH_n is missing. Then $xH_n \subseteq G = \bigcup_{i,j} x_{i,j} H_i$ and $xH_n \cap x_{n,j} H_n = \emptyset$, so $xH_n \subseteq \bigcup_{i \neq n, j} x_{i,j} H_i$. Thus

$$x_{n,r} H_n \subseteq \bigcup_{\substack{i \neq n \\ j}} x_{n,r} x^{-1} x_{i,j} H_i$$

and G can be written as a finite (or countable) union of left cosets of the $n - 1$ subgroups $H_1, H_2, \ldots, H_{n-1}$. The result follows by induction. ∎

We will require a slight strengthening of Lemma 6.1.

Lemma 6.3. *Let H strongly permute the ideals of R and let J be a nonzero ideal of R such that*

$$J^h J = 0 \qquad \text{for all } h \in W \setminus \bigcup_1^n w_k H_k.$$

Here W is a subgroup of H of finite index and $\bigcup_1^n w_k H_k$ is a fixed finite union of cosets of the subgroups H_k of W. Then there exists a subgroup L of H and a nonzero product $0 \neq K = J^{y_1} J^{y_2} \cdots J^{y_r}$ of H-conjugates of J, with some $y_i = 1$, such that $K^h K = 0$ for all $h \in H \setminus L$. Furthermore, $|L : L \cap H_k| < \infty$ for some k.

Proof. Since $|H : W| < \infty$ and $J \neq 0$ we can choose the sequence $\Lambda = \{ h_1, h_2, \ldots, h_s \} \subseteq H$ to be of maximal size subject to

  i. $1 \in \Lambda$ and the h_i are in distinct right cosets of W in H,
  ii. $K = J^{h_1} J^{h_2} \cdots J^{h_s} \neq 0$.

Now let $x \in H$ and suppose that $K^x K \neq 0$. Then by considering the right cosets of W, we have $W \Lambda x = W \Lambda$. Indeed, if this were not the case then, for some j, the element $h_j x$ would be in a new right coset of W. We would then have

$$J^{h_j x} J^{h_1} J^{h_2} \cdots J^{h_s} \supseteq K^x K \neq 0$$

contradicting the maximality of the size of Λ. Since $1 \in \Lambda$ and $W \Lambda x = W \Lambda$, this implies that $h_i x \in W$ for some i. Furthermore $J^{h_i x} J \supseteq K^x K \neq 0$ so, by hypothesis, since $h_i x \in W$, we must have $h_i x \in \bigcup_1^n w_k H_k$.

Thus $K^x K \neq 0$ implies that $x \in h_i^{-1} (\bigcup_1^n w_k H_k)$ for some i. Equivalently, we have shown that

$$K^h K = 0 \qquad \text{for all } h \in H \setminus \bigcup_{i,k} h_i^{-1} w_k H_k.$$

We can now apply Lemma 6.1 to this situation to obtain a nonzero product

$$I = K^{y_1} K^{y_2} \cdots K^{y_r} \neq 0$$

with some $y_j = 1$ and a subgroup L of H with $|L : L \cap H_k| < \infty$ for some k and with $I^h I = 0$ for all $h \in H \setminus L$. Since $1 \in \Lambda$, we conclude that I is an appropriate product of H-conjugates of J, thereby completing the proof. ∎

Again there are three other forms of this lemma which are also valid. We close this section with some definitions and minor observations. We assume in the remaining three lemmas that G strongly permutes the ideals of R.

Lemma 6.4. *Let I be an ideal of R.*

    i. $I^G = \sum_{x \in G} I^x$ *is the smallest G-invariant ideal of R containing I.*

    ii. *If I is G-invariant, then so are $r_R(I)$ and $\ell_R(I)$.*

Proof. Part (i) is clear since G preserves arbitrary sums and (ii) follows from the formula $(IJ)^x = I^x J^x$. ∎

Now let I be a G-invariant ideal of R. Then I is said to be *G-nilpotent-free* if I contains no nonzero G-invariant nilpotent ideal of R. Similarly I is said to be *G-annihilator-free* if for all nonzero G-invariant ideals $A, B \subseteq I$ we have $AB \neq 0$. Obviously the latter property implies the former. When $I = R$ these conditions are called *G-semiprime* and *G-prime* respectively.

Lemma 6.5. *Let I be a G-invariant ideal of R.*

    i. *If I is G-nilpotent-free then $r_R(I) = r_R(I^2)$.*

    ii. *If H is a subgroup of G of finite index, then I is G-nilpotent-free if and only if it is H-nilpotent-free.*

Proof. (i) Obviously $r_R(I) \subseteq r_R(I^2) = J$ and these are G-invariant ideals of R. Since $I^2 J = 0$, we see that IJ is a G-invariant ideal contained in I with $(IJ)^2 = 0$. Thus $IJ = 0$ and $J \subseteq r_R(I)$.

(ii) If I is H-nilpotent-free then it is obviously G-nilpotent-free. Conversely suppose I is G-nilpotent-free and let J be an H-invariant nilpotent ideal of R contained in I. Then since $|G : H| < \infty$, we see that J^G is a finite sum of the nilpotent ideals J^x and hence J^G is nilpotent. Thus $J^G = 0$ and $J = 0$. ∎

Lemma 6.6. *Let H be a subgroup of G and let I be an ideal of R. Suppose that $I^x I = 0$ for all $x \in G \setminus H$.*

i. I^H is an H-invariant ideal of R with

$$(I^H)^x (I^H) = 0 \qquad \text{for all } x \in G \setminus H.$$

ii. Let $I \subseteq J$ with I an H-invariant ideal and J a G-invariant ideal. If J is G-nilpotent-free (or G-annihilator-free), then I is H-nilpotent-free (or H-annihilator-free).

Proof. (i) If $a, b \in H$ and $x \in G \setminus H$, then $I^{ax} I^b = (I^{axb^{-1}} I)^b = 0$ since $axb^{-1} \notin H$.

(ii) Let A and B be H-invariant ideals of R contained in I with $AB = 0$. By considering the cases $x \in H$ and $x \in G \setminus H$ separately, we see that $A^x B = 0$ for all $x \in G$. It then follows that A^G and B^G are G-invariant ideals contained in J with $A^G B^G = 0$. If J is G-annihilator-free, then A^G or B^G is zero and hence I is H-annihilator-free. By taking $A = B$ we obtain the analogous result for nilpotent-free ideals. ∎

EXERCISES

1. Find an example of a set S and a sequence of subsets $S_1 \subseteq S_2 \subseteq \ldots$ such that $S = \bigcup_1^\infty S_i$ and with each $S \setminus S_n$ uncountable. Then find an example of a group G and a sequence of subgroups $G_1 \subseteq G_2 \subseteq \ldots$ such that $G = \bigcup_1^\infty G_i$ and with each $|G : G_n|$ uncountable. Compare this with Lemma 6.2.

In the following, let G be a group and let $H_1, H_2, \ldots, H_n$ be finitely many (not necessarily distinct) subgroups of G.

2. If $D = \bigcap_1^n H_i$ show that every left coset of D is an intersection of cosets of the various H_i's. Deduce that $|G : D| \leq \prod_1^n |G : H_i|$.

Now suppose that $G = \bigcup_1^n x_i H_i$.

3. Show that $G = \bigcup' x_i H_i$ where the union is over those H_i of finite index in G. To this end, suppose $H_1, H_2, \ldots, H_k$ are the subgroups of finite index in G and set $D = \bigcap_1^k H_i$. Then $\bigcup_1^k x_i H_i$ is a finite union of left cosets of D. If $G \neq \bigcup_1^k H_i$, then some coset of D is missing.

4. Prove that $|G : H_k| \leq n$ for some k. For this we can assume that all H_j have finite index in G. Then each $x_j H_j$ is a finite union of left cosets of the subgroup $D = \bigcap_1^n H_i$ and one can then count the number of cosets of D which occur.

5. Finally suppose that the union $\bigcup_1^n x_i H_i$ is irredundant so that all H_i have finite index in G. Suppose $D = H_1 \cap H_2 \cap \cdots \cap H_k$ with $k < n$. Then $\bigcup_1^k x_i H_i$ is a finite union of left cosets of D properly smaller than G because of the irredundancy. Thus some coset of D is missing. Deduce that $|D : D \cap H_j| \leq n - k$ for some $j > k$. Conclude by induction that $|G : \bigcap_1^n H_i| \leq n!$.

7. Minimal Forms

The goal here is to complete the proof of Theorem 5.8. We begin with some notation and observations.

First, suppose $S = R(G)$ is G-graded and let $\alpha = \sum_{x \in G} \alpha_x$. Then we recall that

$$\operatorname{Supp} \alpha = \{ x \in G \mid \alpha_x \neq 0 \}.$$

Also if H is a subgroup of G, then the projection map $\pi_H : R(G) \to R(H)$ is an $(R(H), R(H))$-bimodule homomorphism.

Next, if H is a subgroup of an arbitrary group G, then the *almost centralizer* of H in G is defined by

$$\mathbf{D}_G(H) = \left\{ x \in G \,\middle|\, |H : \mathbf{C}_H(x)| < \infty \right\}.$$

It is clear that $\mathbf{D}_G(H)$ is a subgroup of G normalized by H. Furthermore, $H \cap \mathbf{D}_G(H) = \mathbf{D}_H(H) = \Delta(H)$, where $\Delta(H)$ is the f.c. center of H.

Finally, if $S = R(G)$ is strongly G-graded then, by Lemma 5.7, G strongly permutes the ideals of R. Thus the results and definitions of Section 6 apply here and we will freely use this observation without further comment. In particular, recall that R is said to be G-semiprime if it contains no nonzero G-invariant nilpotent ideal.

When dealing with infinite groups, we obviously cannot proceed by induction on the size of the group. Thus it is frequently necessary to produce another numerical parameter to study. In the case of the following crucial result, we introduce a structure called a form. We can then consider forms of minimal size.

Proposition 7.1. *Let* $S = R(G)$ *be a strongly G-graded ring and assume that the base ring R is G-semiprime. Suppose that A and B are nonzero ideals of S with $AB = 0$. Then there exists a subgroup $H \subseteq G$, a nonzero H-invariant ideal I of R and an element $\beta \in B$ such that*

    i. $I^x I = 0$ *for all* $x \in G \setminus H$,
    ii. $I\pi_\Delta(A) \neq 0$ *and* $I\pi_\Delta(\beta) \neq 0$ *where* $\Delta = \Delta(H)$,
    iii. $I\pi_\Delta(A) \cdot I\beta = 0$.

The above conditions motivate the following definition. Suppose A and B are nonzero ideals of S with $AB = 0$. We say that the 4-tuple (H, D, I, β) is a *form* for A, B if

    i. H is a subgroup of G and $D = \mathbf{D}_G(H)$,
    ii. I is an H-invariant ideal of R with $I^x I = 0$ for all $x \in G \setminus H$,
    iii. $\beta \in B$, $I\beta \neq 0$ and $IA \neq 0$.

The proof of Proposition 7.1 proceeds in a sequence of lemmas, the final one being Lemma 7.6. Throughout the argument we assume that the hypotheses of this proposition are satisfied.

Lemma 7.2. *Forms exist.*

Proof. Take $H = G$, $D = \Delta(G)$ and $I = R$. Since $A, B \neq 0$ and $1 \in R$ we have $IA \neq 0$ and $I\beta \neq 0$ for any $\beta \in B \setminus 0$. ∎

We define $n = (H, D, I, \beta)^{\#}$, the *size of the form*, to be the number of right D-cosets meeting Supp β. We now assume for the rest of this argument that (H, D, I, β) is a form whose size n is minimal. In the next lemma we make a slight change in β. Afterwards, no additional changes in the form will be made. Set $\Delta = \Delta(H)$.

Lemma 7.3. *With the above notation we have*
  i. I is H-nilpotent-free and $r_S(I) = r_S(I^2)$,
  ii. $I\pi_\Delta(A) \neq 0$ and we may assume that $I\pi_\Delta(\beta) \neq 0$,
  iii. for all $\gamma \in R(D)$ we have $I\gamma\beta = 0$ if and only if $I\gamma\pi_D(\beta) = 0$.

Proof. (i) By assumption, R is G-semiprime. Since I is H-invariant and $I^x I = 0$ for all $x \in G \setminus H$, it follows from Lemma 6.6(ii) that I is H-nilpotent-free. Hence, by Lemma 6.5(i), $r_R(I) = r_R(I^2)$. Finally $r_S(I) \subseteq r_S(I^2)$ and suppose $\gamma = \sum_x \gamma_x$ is contained in the latter ideal. Then for all x we have $I^2\gamma_x = 0$ so $\gamma_x S_{x^{-1}} \subseteq r_R(I^2) = r_R(I)$. This yields $I\gamma_x S_{x^{-1}} = 0$ so $I\gamma_x = 0$ as required.

(ii) Since $IA \neq 0$, we have $IAS_x \neq 0$ for all x. Thus since $AS_x \subseteq A$ it follows that $I\pi_\Delta(A) \neq 0$. Now write $\beta = \sum_x \beta_x$. Since $I\beta \neq 0$ we have $I\beta_x \neq 0$ for some $x \in G$. Thus $I\beta_x S_{x^{-1}} \neq 0$ and we can choose $\sigma_{x^{-1}} \in S_{x^{-1}}$ with $I\beta_x\sigma_{x^{-1}} \neq 0$. It is now clear that $(H, D, I, \beta\sigma_{x^{-1}})$ is also a form with the additional property that $I\pi_\Delta(\beta\sigma_{x^{-1}}) \neq 0$. Furthermore we have

$$(H, D, I, \beta\sigma_{x^{-1}})^{\#} \leq (H, D, I, \beta)^{\#}$$

so $(H, D, I, \beta\sigma_{x^{-1}})$ also has minimal size n. We now replace β by $\beta\sigma_{x^{-1}}$ for the remainder of the proof of the proposition, or equivalently we can assume that $I\pi_\Delta(\beta) \neq 0$. This implies in particular that $\pi_\Delta(\beta) \neq 0$ and hence that $\pi_D(\beta) \neq 0$ since $D \supseteq \Delta$.

(iii) If $I\gamma\beta = 0$, then applying π_D yields $I\gamma\pi_D(\beta) = 0$. Conversely suppose $I\gamma\pi_D(\beta) = 0$. Then for any $r \in I$ we have $r\gamma\beta \in B$ and Supp $r\gamma\beta$ meets less than n right cosets of D since $\gamma \in R(D)$ and $\pi_D(r\gamma\beta) = 0$ while $\pi_D(\beta) \neq 0$. By definition of n, this implies that $(H, D, I, r\gamma\beta)$ is not a form. Thus $Ir\gamma\beta = 0$ for all $r \in I$ so $I^2\gamma\beta = 0$. We conclude from (i) above that $I\gamma\beta = 0$. ∎

It follows from the above that H, I and β satisfy (i) and (ii) of the conclusion of Proposition 7.1. If in addition they satisfy (iii), then the result is proved. Thus we will assume throughout the remainder of the proof that $I\pi_\Delta(A) \cdot I\beta \neq 0$ and we derive a contradiction.

Lemma 7.4. *With the above assumption, there exist W a subgroup of H of finite index, $\alpha = \sum_x \alpha_x \in A \cap R(H)$ and $d \in$ Supp $\pi_D(\alpha)$ such that*

  i. W centralizes Supp $\pi_D(\alpha)$ and Supp $\pi_D(\beta)$,

  ii. for some $u \in W$

$$I\pi_D(\alpha)\beta \supseteq (I\alpha_d S_{d^{-1}})^u \pi_D(\alpha)\beta \neq 0,$$

  iii. for all $y \in W$

$$I\alpha_d S_{d^{-1}y}\pi_D(\alpha)\pi_D(\beta) = I\pi_D(\alpha)S_{d^{-1}y}\alpha_d\pi_D(\beta).$$

Proof. (i) By assumption $I\pi_\Delta(A) \cdot I\beta \neq 0$ and hence, since $r_S(I) = r_S(I^2)$ we also have

$$I\pi_\Delta(IAI)\beta = I^2\pi_\Delta(A) \cdot I\beta \neq 0.$$

Thus there exists $\alpha \in IAI \subseteq A$ with $I\pi_\Delta(\alpha)\beta \neq 0$. Observe that $\alpha \in ISI \subseteq R(H)$ by Lemma 5.9(ii). Thus since $D \cap H = \Delta$ we have $\pi_D(\alpha) = \pi_\Delta(\alpha)$ and $I\pi_D(\alpha)\beta \neq 0$.

We can now assume that α is chosen so that $|$Supp $\pi_D(\alpha)|$ is minimal subject to $\alpha \in A \cap R(H)$ and $I\pi_D(\alpha)\beta \neq 0$. Let W be the intersection of the centralizers in H of the elements of Supp $\pi_D(\alpha)$ and Supp $\pi_D(\beta)$. Since Supp $\pi_D(\alpha) \cup$ Supp $\pi_D(\beta)$ is a finite subset of $D = \mathbf{D}_G(H)$, it is clear that $|H : W| < \infty$. Note that I is H-nilpotent-free, by Lemma 7.3(i), and hence it is also W-nilpotent-free by Lemma 6.5(ii). This completes the proof of (i).

(ii) This part does not use the minimal nature of Supp $\pi_D(\alpha)$. Set $\gamma = \pi_D(\alpha)\beta$ and write $\alpha = \sum_x \alpha_x$, $\beta = \sum_x \beta_x$ and $\gamma = \sum_x \gamma_x$.

Then $\pi_D(\alpha) = \sum_{d \in D} \alpha_d$ and we let J be the W-invariant ideal of R given by $J = \sum_{d \in D} (R\alpha_d S_{d^{-1}})^W$. Notice that for all $d \in D$, $y \in G$ we have

$$\alpha_d \beta_y S_{y^{-1}d^{-1}} \subseteq R\alpha_d S_{d^{-1}} \cdot S_d \beta_y S_{y^{-1}d^{-1}} \subseteq J \cdot R = J$$

and from this it follows that $\gamma_x S_{x^{-1}} \subseteq J$ for all $x \in G$. Now suppose that $IJ\gamma = 0$. Then $IJ\gamma_x S_{x^{-1}} = 0$ so $\gamma_x S_{x^{-1}} \subseteq \mathrm{r}_R(IJ)$ and hence, by the above,

$$I\gamma_x S_{x^{-1}} \subseteq IJ \cap \mathrm{r}_R(IJ) = 0$$

since the latter is a W-invariant nilpotent ideal contained in I. This yields $I\gamma_x = 0$ and therefore $I\pi_D(\alpha)\beta = I\gamma = 0$, a contradiction by the choice of α. Thus we have $IJ\pi_D(\alpha)\beta \neq 0$ and hence, by definition of J, there exists $d \in \mathrm{Supp}\ \pi_D(\alpha)$ and $u \in W$ with $I(R\alpha_d S_{d^{-1}})^u \pi_D(\alpha)\beta \neq 0$. Since $u \in W \subseteq H$ and I is H-invariant, this part is proved.

(iii) Let $y \in W$ and choose any $a_{y^{-1}} \in S_{y^{-1}}$ and $b_{d^{-1}y} \in S_{d^{-1}y}$. We study the element

$$\delta = a_{y^{-1}} \alpha_d b_{d^{-1}y} \alpha - a_{y^{-1}} \alpha b_{d^{-1}y} \alpha_d.$$

Clearly $\delta \in A \cap R(H)$ and since $\alpha = \sum_{x \in G} \alpha_x$ we can write $\delta = \sum_{x \in G} \sigma(x)$ where

$$\sigma(x) = a_{y^{-1}} \alpha_d b_{d^{-1}y} \alpha_x - a_{y^{-1}} \alpha_x b_{d^{-1}y} \alpha_d.$$

Observe that, since y centralizes d, the summands in $\sigma(x)$ have grades x and $y^{-1}xy$, respectively. In particular, if $x \notin D$, then neither of these grades is in D since $y \in H$ normalizes D. On the other hand, if $x \in D$ then $x \in \mathrm{Supp}\ \pi_D(\alpha)$ so y commutes with x and hence both these summands have grade $x \in D$. It follows that $\pi_D(\delta) = \sum_{x \in D} \sigma(x)$ and that $|\mathrm{Supp}\ \pi_D(\delta)| \leq |\mathrm{Supp}\ \pi_D(\alpha)|$. In fact, this inequality is strict since clearly $\sigma(d) = 0$.

The minimality of $|\mathrm{Supp}\ \pi_D(\alpha)|$ now implies that $I\pi_D(\delta)\beta = 0$ and hence, by applying π_D, that $I\pi_D(\delta)\pi_D(\beta) = 0$. Now as we observed above, $\pi_D(\delta)$ comes precisely from the D-homogeneous components of α so we have

$$\pi_D(\delta) = a_{y^{-1}} \alpha_d b_{d^{-1}y} \pi_D(\alpha) - a_{y^{-1}} \pi_D(\alpha) b_{d^{-1}y} \alpha_d$$

and hence

$$I a_{y^{-1}} \left(\alpha_d b_{d^{-1}y} \pi_D(\alpha) - \pi_D(\alpha) b_{d^{-1}y} \alpha_d \right) \pi_D(\beta) = 0.$$

Notice that this formula holds for all $a_{y^{-1}} \in S_{y^{-1}}$ and that $I S_{y^{-1}} = S_{y^{-1}} I$ since I is H-invariant. We can therefore cancel the $S_{y^{-1}}$ factor and obtain

$$I \alpha_d b_{d^{-1}y} \pi_D(\alpha) \pi_D(\beta) = I \pi_D(\alpha) b_{d^{-1}y} \alpha_d \pi_D(\beta)$$

and since this holds for all $b_{d^{-1}y} \in S_{d^{-1}y}$, the lemma is proved. ∎

The next result is proved by a variant of the Δ-method. Fix $\alpha = \sum_x \alpha_x$, d, W and u as in the preceding lemma for the remainder of the argument.

Lemma 7.5. *With the above notation,*

$$(I \alpha_d S_{d^{-1}})^y \cdot \pi_D(\alpha) \beta = 0$$

for all $y \in W \setminus \bigcup_1^t x_i H_i$. Here $\bigcup_1^t x_i H_i$ is a fixed finite union of left cosets of the subgroups H_i and each H_i is the centralizer in W of some element in $(\mathrm{Supp}\ \beta) \setminus D$.

Proof. We freely use the fact that I is H-invariant and, in particular, that $S_h I = I S_h$ for all $h \in H$.

Let $y \in W$ and suppose that

$$(I \alpha_d S_{d^{-1}})^y \pi_D(\alpha) \beta \neq 0.$$

Then

$$I (R \alpha_d S_{d^{-1}})^y \pi_D(\alpha) \beta \neq 0$$

so Lemma 7.3(iii) yields

$$I (R \alpha_d S_{d^{-1}})^y \pi_D(\alpha) \pi_D(\beta) \neq 0.$$

Thus we have

$$I \alpha_d S_{d^{-1}y} \pi_D(\alpha) \pi_D(\beta) \neq 0$$

so Lemma 7.4(iii) yields

$$I\pi_D(\alpha)S_{d^{-1}y}\alpha_d\pi_D(\beta) \neq 0$$

and therefore finally

$$I\pi_D(\alpha)S_{d^{-1}y}\alpha_d\pi_D(\beta)S_{y^{-1}} \neq 0.$$

Write $\alpha = \pi_D(\alpha) + \tilde{\alpha}$ and $\beta = \pi_D(\beta) + \tilde{\beta}$. Since

$$I\alpha S_{d^{-1}y}\alpha_d\beta S_{y^{-1}} \subseteq AB = 0$$

we have

$$I\big(\pi_D(\alpha) + \tilde{\alpha}\big)S_{d^{-1}y}\alpha_d\big(\pi_D(\beta) + \tilde{\beta}\big)S_{y^{-1}} = 0.$$

We consider the supports of each of the four summands obtained from the above expression to see how cancellation can occur. Observe that $y \in W$ so y normalizes D and centralizes $d \in \text{Supp } \pi_D(\alpha)$. In particular, we have $S_{d^{-1}y}\alpha_d \subseteq S_y$ and from this it follows easily that the sets

$$I\pi_D(\alpha)S_{d^{-1}y}\alpha_d\tilde{\beta}S_{y^{-1}} \qquad \text{and} \qquad I\tilde{\alpha}S_{d^{-1}y}\alpha_d\pi_D(\beta)S_{y^{-1}}$$

have supports disjoint from D. On the other hand,

$$0 \neq I\pi_D(\alpha)S_{d^{-1}y}\alpha_d\pi_D(\beta)S_{y^{-1}} \subseteq R(D)$$

by the work of the preceding paragraph so it follows that this expression must be cancelled by terms from the fourth summand

$$I\tilde{\alpha}S_{d^{-1}y}\alpha_d\tilde{\beta}S_{y^{-1}}.$$

In particular, the latter two summands must have a support element in common.

Thus there exist $f \in \text{Supp } \tilde{\alpha}$, $g \in \text{Supp } \tilde{\beta}$, $a \in \text{Supp } \pi_D(\alpha)$ and $b \in \text{Supp } \pi_D(\beta)$ with $ayby^{-1} = fygy^{-1}$. Since $y \in W$ centralizes $b \in \text{Supp } \pi_D(\beta)$, this yields $g^{y^{-1}} = ygy^{-1} = f^{-1}ab$ so $y \in x\mathbf{C}_W(g)$, some fixed left coset of $\mathbf{C}_W(g)$ depending only on the finitely many parameters f, g, a, b. Since $g \in \operatorname{Supp} \tilde{\beta} = (\operatorname{Supp} \beta) \setminus D$, the lemma is proved. ∎

We remark that the truncation from β to $\pi_D(\beta)$, using Lemma 7.3(iii), in the above argument is crucial. Otherwise the subgroups H_i turn out to be centralizers of elements of $(\operatorname{Supp} \alpha) \setminus D$.

Lemma 7.6. *Contradiction.*

Proof. We use the notation of the preceding two lemmas and we set $\gamma = \pi_D(\alpha)\beta = \sum_{x \in G} \gamma_x$. Then by Lemma 7.4(ii), there exists $z \in \operatorname{Supp} \gamma$ with $(I\alpha_d S_{d-1})^u \gamma_z \neq 0$ and hence

$$J = (I\alpha_d S_{d-1})^u \gamma_z S_{z-1}$$

is a nonzero ideal of R contained in I since I is H-invariant and $u \in W \subseteq H$. Furthermore, since $J \subseteq (I\alpha_d S_{d-1})^u$ we have $J^{u^{-1}} \subseteq I\alpha_d S_{d-1}$ and hence $J^{u^{-1}y} \subseteq (I\alpha_d S_{d-1})^y$ for all $y \in W$.

It follows from the above and Lemma 7.5 that $J^{u^{-1}y} \pi_D(\alpha)\beta = 0$ for all $y \in W \setminus \bigcup_1^t x_i H_i$ or equivalently that $J^y \pi_D(\alpha)\beta = 0$ for all $y \in W \setminus \bigcup_1^t u^{-1}x_i H_i$ since $u \in W$. In particular, $J^y \gamma_z = 0$ for all those y so $J^y(R\gamma_z S_{z-1}) = 0$ and hence $J^y J = 0$. Since $|H : W| < \infty$, Lemma 6.3 applies and there exists a subgroup L of H and a nonzero product $K = J^{v_1} J^{v_2} \cdots J^{v_s}$ such that $K^h K = 0$ for all $h \in H \setminus L$. Furthermore, $v_i = 1$ for some i, so $K \subseteq J \subseteq I$ and $|L : L \cap H_k| < \infty$ for some $1 \leq k \leq t$.

We claim that $(L, \mathbf{D}_G(L), K^L, \pi_D(\alpha)\beta)$ is also a form. To start with, we have $K^h K = 0$ for all $h \in H \setminus L$ and then, since $K \subseteq I$, it follows that $K^g K = 0$ for all $g \in G \setminus L$. Thus, by Lemma 6.6(i), K^L is a nonzero L-invariant ideal of R with $(K^L)^g K^L = 0$ for all $g \in G \setminus L$. Furthermore, since R is G-semiprime, Lemma 6.6(ii) implies that K^L is L-nilpotent-free and in particular $(K^L)^2 \neq 0$. Suppose $K^L \gamma = K^L \pi_D(\alpha)\beta = 0$. Then $K^L \gamma_z = 0$ so $K^L(R\gamma_z S_{z-1}) = 0$ and hence $K^L J = 0$. But $\mathbf{r}_R(K^L)$ is L-invariant, by Lemma 6.4(ii), so this yields $K^L J^L = 0$, a contradiction since $K^L \subseteq J^L$ and $(K^L)^2 \neq 0$. Thus $K^L \pi_D(\alpha)\beta \neq 0$. This then implies that $K^L \pi_D(\alpha) \neq 0$ so, since $\alpha \in A$, we have $K^L A \neq 0$. Finally $\pi_D(\alpha)\beta \in B$ so $(L, \mathbf{D}_G(L), K^L, \pi_D(\alpha)\beta)$ is indeed a form.

It remains to compute the size of this new form. Since $H \supseteq L$ we have $\mathbf{D}_G(L) \supseteq \mathbf{D}_G(H) = D$. Thus since $\pi_D(\alpha) \in R(D)$, it is clear that Supp $\pi_D(\alpha)\beta$ meets at most n right cosets of $\mathbf{D}_G(L)$. But observe that $|L : L \cap H_k| < \infty$ and that $H_k = \mathbf{C}_W(g)$ for some $g \in (\text{Supp } \beta) \setminus D$. Thus $|L : \mathbf{C}_L(g)| < \infty$ so $g \in \mathbf{D}_G(L)$ and in fact $Dg \subseteq \mathbf{D}_G(L)$. Since $\pi_D(\beta) \neq 0$, the two D-cosets D and Dg, which meet elements of Supp β, merge to a single coset of $\mathbf{D}_G(L)$ and therefore Supp $\pi_D(\alpha)\beta$ meets less than n right cosets of $\mathbf{D}_G(L)$. In other words,

$$(L, \mathbf{D}_G(L), K^L, \pi_D(\alpha)\beta)^{\#} < (H, D, I, \beta)^{\#}$$

contradicting the minimal nature of (H, D, I, β). ∎

As we observed previously, the contradiction of Lemma 7.6 is based on the assumption that $I\pi_\Delta(A) \cdot I\beta \neq 0$. Thus $I\pi_\Delta(A) \cdot I\beta = 0$ and Proposition 7.1 is proved.

With this in hand, it is now a simple matter to prove the main result.

Proof of Theorem 5.8 (Hard Direction). Here we assume that $S = R(G)$ is a strongly G-graded ring and that A and B are nonzero ideals of S with $AB = 0$. Suppose first that R is not G-semiprime. Then there exists a nonzero G-invariant ideal $\tilde{A}$ of R with $\tilde{A}^2 = 0$. The result now follows with $H = G$, $I = R$, $N = \langle 1 \rangle$ and $\tilde{B} = \tilde{A}$.

We can therefore assume that R is G-semiprime so Proposition 7.1 applies. Thus there exist a subgroup $H \subseteq G$, a nonzero H-invariant ideal I of R, and an element $\beta \in B$ such that

(i) $I^x I = 0$ for all $x \in G \setminus H$,
(ii) $I\pi_\Delta(A) \neq 0$, $I\pi_\Delta(\beta) \neq 0$ where $\Delta = \Delta(H)$,
(iii) $I\pi_\Delta(A) \cdot I\beta = 0$.

We have therefore found an appropriate H and I. It remains to find N, $\tilde{A}$ and $\tilde{B}$.

Set $A_1 = I\pi_\Delta(A)$ and $B_1 = I \cdot (S\pi_\Delta(\beta)S)^H$. By Lemma 5.9 and (ii) above, A_1 is a nonzero H-invariant ideal of $R(\Delta)$. Since

$I\pi_\Delta(A) \cdot I\beta = 0$ we have $I\pi_\Delta(A) \cdot I\pi_\Delta(\beta) = 0$ and, again by (ii), it follows that B_1 is a nonzero H-invariant ideal of $R(\Delta)$ with $A_1 B_1 = 0$. Note that $A_1, B_1 \subseteq I \cdot R(\Delta)$ and, since $I\pi_\Delta(\beta) \subseteq I\pi_\Delta(B)$, we have $B_1 \subseteq I\pi_\Delta(B)$.

By Lemma 5.1(ii), $\Delta(H)/\Delta^+(H)$ is torsion-free abelian and we set $A_2 = \min_{\Delta^+} A_1$, $A_3 = \pi_{\Delta^+}(A_2)$, $B_2 = \min_{\Delta^+} B_1$ and $B_3 = \pi_{\Delta^+}(B_2)$ in the notation of Lemma 5.10. Then that lemma implies that A_3 and B_3 are nonzero H-invariant ideals of $R(\Delta^+)$ both contained in $I \cdot R(\Delta^+)$. Furthermore, since Δ/Δ^+ is an ordered group and $A_1 B_1 = 0$, Lemma 5.11 implies that $A_3 B_3 = 0$.

Since A_3 and B_3 are nonzero, it follows from Lemma 5.1(iii) that there exists a finite normal subgroup $N \subseteq \Delta^+$ of H with $A_4 = A_3 \cap R(N)$ and $B_4 = B_3 \cap R(N)$ both nonzero. Certainly A_4 and B_4 are H-invariant ideals of $R(N)$ contained in $I \cdot R(N)$ and they satisfy $A_4 B_4 = 0$. For general A and B, the result now follows by taking N as above, $\tilde{A} = A_4$ and $\tilde{B} = B_4$.

Finally if $A = B$, then since $B_1 \subseteq I \cdot \pi_\Delta(B) = I \cdot \pi_\Delta(A) = A_1$ we have $B_1^2 = 0$. It then follows as above that $B_i^2 = 0$ for all i so we can take $\tilde{A} = \tilde{B} = B_4$. This completes the proof. ∎

Using duality, and in particular Proposition 2.6, Theorem 5.8 can be extended to component regular group-graded rings. Suppose $S = R(G)$ is such a ring. If $N \lhd G$ and $I \lhd R(N)$ we can still define $I^x = R(Nx^{-1})IR(xN) \lhd R(N)$ but we no longer have a strong action of G on these ideals. Here we say that I is G-invariant if $I^x \subseteq I$ for all $x \in G$. We state the following result without proof; details of the proof will be considered in the exercises.

Corollary 7.7. [180] *Let $S = R(G)$ be a component regular group-graded ring. Then S contains nonzero ideals A and B with $AB = 0$ if and only if there exist*

  i. *subgroups $N \lhd H \subseteq G$ with N finite,*

  ii. *an H-invariant ideal I of R with $I^x I = 0$ for all $x \in G \setminus H$,*

  iii. *nonzero H-invariant ideals $\tilde{A}$ and $\tilde{B}$ of $R(N)$ both contained in $R(H)IR(H) \cap R(N)$ and with $\tilde{A}\tilde{B} = 0$.*

Furthermore $A = B$ if and only if $\tilde{A} = \tilde{B}$.

Exercises

1. In Section 5 the Δ-method example used right cosets; here we used left. How did this reversal occur?

2. Consider the proof of Lemma 7.5 and suppose we were not able to truncate β to $\pi_D(\beta)$. In other words, assume that we only have the weaker expression $I\pi_D(\alpha)S_{d^{-1}y}\alpha_d\beta S_{y^{-1}} \neq 0$. What conclusion can we obtain from this?

Let $S = R(G)$ be G-graded.

3. If $I \triangleleft R$ and $x \in G$, then by the above $I^x = S_{x^{-1}}IS_x$. Show that the equalities of Lemma 5.7 become inclusions in general. Prove that I is G-invariant if and only if $I = J \cap R$ where $J = SIS \triangleleft S$.

4. If $A \triangleleft S$ and G is finite, define $A^\circ = \bigcap_{x \in G} AS_x$. Show that A° is an ideal of S. Furthermore if A is contained in the graded ideal I, then $A^\circ \subseteq I_1 S$. Finally suppose $x_1, x_2, \ldots, x_n \in G$ are chosen so that the products $y_i = x_i x_{i+1} \cdots x_n$ for $i = 1, 2, \ldots, n$ contain all elements of G. Show that $AS_{x_1} S_{x_2} \cdots S_{x_n} \subseteq A^\circ$ and deduce that $A \neq 0$ and S component regular imply that $A^\circ \neq 0$.

We recall the notation of Proposition 2.6. Thus let H be a subgroup of G and let I be an ideal of $S\{1\}\bar{H} \subseteq M_G(S)$. Then $I^\xi \triangleleft R(H)$ is given by $I^\xi = \{\alpha(1,1) \mid \alpha \in I\}$ so that $I^\xi e_{1,1} = e_{1,1}Ie_{1,1}$.

5. If $g \in G$, prove that $S_{g^{-1}}e_{1,1}\bar{g}$ and $\bar{g}^{-1}e_{1,1}S_g$ are both contained in $S\{1\}$. Next, if I is as above and $g \in N_G(H)$ use the inclusion
$$S_{g^{-1}}e_{1,1}\bar{g} \cdot I^{\bar{g}} \cdot \bar{g}^{-1}e_{1,1}S_g \subseteq I^{\bar{g}}$$
to deduce that $(I^\xi)^g \subseteq (I^{\bar{g}})^\xi$. Conclude that if I is $\bar{g}$-invariant then I^ξ is g-invariant and if $I^{\bar{g}}I = 0$ then $(I^\xi)^g I^\xi = 0$.

6. Suppose $I = J\bar{H} \triangleleft S\{1\}\bar{H}$ where J is an H-invariant ideal of $S\{1\}$. Prove that I^ξ is a graded ideal of $R(H)$ with $(I^\xi)_1 = J^\xi$. Now suppose that $A \triangleleft S\{1\}\bar{H}$ with $A \subseteq I$ and H finite. Conclude from Exercise 4 that $(A^\xi)^\circ \subseteq J^\xi \cdot R(H)$ and hence that $A^\xi \cap [J^\xi \cdot R(H)] \neq 0$ if $R(H)$ is component regular.

7. Prove Corollary 7.7 using Proposition 2.6, Theorem 5.8 (in the case of skew group rings) and the preceding four exercises.

8. Sufficient Conditions

The goal now is to use Theorem 5.8 to obtain sufficient conditions for the strongly graded ring $S = R(G)$ to be either prime or semiprime. In particular, we will obtain results which are clearly analogous to Theorems 5.3, 5.4 and 5.5 on ordinary group algebras. To avoid trivialities we will assume that R is G-semiprime.

We begin with two improvements on Theorem 5.8. The first yields a better understanding of the ideal $I \subseteq R$. The second sharpens the relationship between N, H and I. In particular we show that H can be taken to be the normalizer of N, thereby giving an indication of how close N is to being normal in G. We require some definitions.

Let G strongly permute the ideals of R and let I be an ideal of R. Then we denote by $G_I = \{ x \in G \mid I^x = I \}$ the *stabilizer* of I in G. Clearly G_I is a subgroup of G. The nonzero ideal I is said to be a *trivial intersection ideal* if for all $x \in G$ either $I^x = I$ or $I^x \cap I = 0$. Note that $I^x \cap I = 0$ implies that $I^x I = 0$.

Lemma 8.1. *Let R be a G-semiprime ring, let H be a subgroup of G and let I be a nonzero H-invariant ideal of R. Suppose that $I^x I = 0$ for all $x \in G \setminus H$. Then*
    i. $G_I = H$,
    ii. *I is a trivial intersection ideal,*
    iii. *if X is a right transversal for H in G, then $I^G = \oplus \sum_{x \in X} I^x$.*

Proof. Since R is G-semiprime, Lemma 6.6(ii) implies that I is H-nilpotent-free and hence $I \cap \ell_R(I) = 0$. In particular, $I^2 \neq 0$ so (i) is immediate. Next observe that

$$I \cap \left(\sum_{x \in G \setminus H} I^x \right) \subseteq I \cap \ell_R(I) = 0$$

and this clearly yields the direct sum in (iii). Finally, part (ii) is immediate from (iii). ∎

Corollary 8.2. *Let $S = R(G)$ be a strongly G-graded ring whose base ring R is G-semiprime. Then S contains nonzero ideals A and B with $AB = 0$ if and only if there exist*

    i. *a trivial intersection ideal I of R,*

    ii. *a finite group N with normalizer $\mathbf{N}_G(N) = G_I$,*

    iii. *nonzero G_I-invariant ideals $\tilde{A}$ and $\tilde{B}$ of $R(N)$ with $\tilde{A}$, $\tilde{B} \subseteq I \cdot R(N)$ and $\tilde{A}\tilde{B} = 0$.*

Furthermore, $A = B$ if and only if $\tilde{A} = \tilde{B}$.

Proof. If I, N, $\tilde{A}$ and $\tilde{B}$ exist, then so do A and B by Theorem 5.8. Conversely suppose A and B are given. We apply Theorem 5.8 and use its notation. In particular, N is a finite group and $N \subseteq H \subseteq \mathbf{N}_G(N) = \bar{H}$. Since $I^x I = 0$ for all $x \in G \setminus \bar{H} \subseteq G \setminus H$, Lemma 6.6(i) implies that $\bar{I} = I^{\bar{H}}$ is an $\bar{H}$-invariant ideal of R with $\bar{I}^x \bar{I} = 0$ for all $x \in G \setminus \bar{H}$. By Lemma 8.1, $\bar{I}$ is a trivial intersection ideal with stabilizer $G_{\bar{I}} = \bar{H} = \mathbf{N}_G(N)$.

Now we consider the action of $\bar{H} \supseteq H$ on $R(N)$. Set $\bar{A} = \tilde{A}^{\bar{H}}$ and $\bar{B} = \tilde{B}^{\bar{H}}$ so that these are $\bar{H}$-invariant ideals of $R(N)$ clearly contained in $\bar{I} \cdot R(N)$. Since $I^x I = 0$ for all $x \in \bar{H} \setminus H$ and since $N \subseteq H$, it follows easily that $\left[I \cdot R(N)\right]^x \cdot \left[I \cdot R(N)\right] = 0$ for all $x \in \bar{H} \setminus H$ and hence that $\tilde{A}^x \tilde{B} = 0$ for all such x. But $\tilde{A}$ is H-invariant and $\tilde{A}\tilde{B} = 0$ so this yields $\tilde{A}^{\bar{H}} \tilde{B} = 0$. Hence since the right annihilator of $\tilde{A}^{\bar{H}}$ is $\bar{H}$-invariant, we conclude that $\bar{A}\bar{B} = \tilde{A}^{\bar{H}} \tilde{B}^{\bar{H}} = 0$. ∎

To paraphrase the above, we see that $S = R(G)$ is prime (or semiprime) if and only if for all trivial intersection ideals $I \subseteq R$ and all finite subgroups $N \subseteq G$ with $\mathbf{N}_G(N) = G_I$ we have $I \cdot R(N)$ a G_I-annihilator-free (or G_I-nilpotent-free) ideal of $R(N)$.

The prototype example here is as follows. Let G be a group, H a subgroup and let K be a field. Then G transitively permutes the set Ω of right cosets of H in G by right multiplication and we let $\omega_0 \in \Omega$ correspond to the coset H itself. Define $R = \prod_{\omega \in \Omega} K_\omega$, the complete direct product of copies of K indexed by the elements of Ω, so that R is a ring and in fact a K-algebra. Note that the permutation action of G on Ω extends to a group homomorphism $G \to \mathrm{Aut}_K(R)$ and we form the skew group ring $S = RG$. It follows that R is G-prime, $I = K_{\omega_0}$ is a trivial intersection ideal of R with $G_I = H$ and $I \cdot RH \cong K[H]$. In particular, if $K[H]$ is not prime or semiprime, then neither is S. Moreover, it is easy to see (Exercise 2) that the converse to this is also true, even though R has trivial intersection ideals other than I. Thus Theorems 5.3, 5.4 and 5.5, applied to $K[H]$, completely settle the question.

Under certain circumstances the primeness or semiprimeness of $R(G)$ depends only on the finite normal subgroups of G. Some of these circumstances are obvious; one is slightly less so.

Let $K\langle \zeta_1, \zeta_2, \dots \rangle$ be the polynomial ring over the field K in the noncommuting indeterminates $\zeta_1, \zeta_2, \dots$. An algebra E over K is said to *satisfy a polynomial identity* if there exists a nonzero polynomial $f(\zeta_1, \zeta_2, \dots, \zeta_n) \in K\langle \zeta_1, \zeta_2, \dots \rangle$ with

$$f(\alpha_1, \alpha_2, \dots, \alpha_n) = 0$$

for all $\alpha_1, \alpha_2, \dots, \alpha_n \in E$. For example, any commutative algebra satisfies $f(\zeta_1, \zeta_2) = \zeta_1 \zeta_2 - \zeta_2 \zeta_1$ and note that this is a special case of the *standard identity* defined by

$$s_n(\zeta_1, \zeta_2, \dots, \zeta_n) = \sum_{\sigma \in \mathrm{Sym}_n} (-)^\sigma \zeta_{\sigma(1)} \zeta_{\sigma(2)} \cdots \zeta_{\sigma(n)}.$$

The well known Amitsur-Levitzki theorem [5] asserts that the matrix ring $M_m(K)$ satisfies s_{2m}. A simple linearization argument (see [161, Lemma 5.1.1]) shows that if E satisfies a polynomial identity of degree n, then it satisfies a *multilinear* one of the form

$$f(\zeta_1, \zeta_2, \dots, \zeta_n) = \sum_{\sigma \in \mathrm{Sym}_n} k_\sigma \zeta_{\sigma(1)} \zeta_{\sigma(2)} \cdots \zeta_{\sigma(n)}$$

with all $k_\sigma \in K$ and $k_1 = 1$.

Proposition 8.3. Let $S = R(G)$ be a strongly G-graded ring whose base ring R is G-semiprime and suppose that either
  i. G acts trivially on the ideals of R,

    ii. R is prime,

    iii. R is Noetherian or at least contains no infinite direct sum of nonzero two-sided ideals, or

    iv. S is a K-algebra satisfying a polynomial identity.

Then S is prime (or semiprime) if and only if, for all finite normal subgroups N of G, the subring $R(N)$ is G-prime (or G-semiprime).

Proof. We use Corollary 8.2 and its notation. In particular we let N, I, $\tilde{A}$ and $\tilde{B}$ satisfy (i), (ii) and (iii) of that result. If G acts trivially on the ideals of R then certainly $G_I = G$. Furthermore if R is a prime ring, then $I \neq 0$ implies $I^x I \neq 0$ and again we have $G_I = G$. Thus in either case we must have $N \triangleleft G$.

    It remains to consider assumptions (iii) and (iv). If R is Noetherian or at least contains no infinite direct sum of nonzero two-sided ideals, then it follows from Lemma 8.1(iii) that $|G : G_I| < \infty$. Now suppose S satisfies a polynomial identity of degree n which we can take to be the multilinear polynomial

$$f(\zeta_1, \zeta_2, \ldots, \zeta_n) = \sum_{\sigma \in \mathrm{Sym}_n} k_\sigma \zeta_{\sigma(1)} \zeta_{\sigma(2)} \cdots \zeta_{\sigma(n)}$$

with all $k_\sigma \in K$ and $k_1 = 1$. We claim that $|G : G_I| \leq n$. Indeed, if $x_0, x_1, \ldots, x_n$ are in distinct left cosets of G_I in G and if $\alpha_i \in S_{x_{i-1}^{-1}} I S_{x_i}$, then it follows from Lemma 5.9(i) that if $\sigma \neq 1$ then $\alpha_{\sigma(1)} \alpha_{\sigma(2)} \cdots \alpha_{\sigma(n)} = 0$. Thus the polynomial identity implies that $\alpha_1 \alpha_2 \cdots \alpha_n = 0$ and we conclude that $S_{x_0} I^n S_{x_n} = 0$, a contradiction by Lemma 6.6(ii).

    In other words, assumptions (iii) and (iv) each imply at least that $|G : G_I| < \infty$. Since N is a finite normal subgroup of G_I it follows that $N \subseteq \Delta^+(G)$ and then, by Lemma 5.1(iii), that $N \subseteq M$ where M is a finite normal subgroup of G. Since $\bar{A} = S\tilde{A}S$ and $\bar{B} = S\tilde{B}S$ are ideals of S with $\bar{A}\bar{B} = 0$, we conclude that $\bar{A} \cap R(M)$ and $\bar{B} \cap R(M)$ are nonzero G-invariant ideals of $R(M)$ with product zero. This completes the proof. ∎

    In particular if $S = R*G$ with R prime, then S is prime (or semiprime) if and only if, for all finite normal subgroups N of G, the subring $R*N$ is G-prime (or G-semiprime). The latter result, due to [**132**], will be considered in more detail in the next chapter (see Theorem 12.7 and the remarks following its proof).

Now let us return to a general strongly graded ring $S = R(G)$ with base ring R which is G-semiprime. Then it is clear from Corollary 8.2 that if $R(N)$ is prime, for all relevant N, then S is prime. In Section 17, we will consider when finite crossed products are prime. Here we just finesse the problem by assuming that all such N are trivial. The following is formulated in a manner analogous to Theorem 5.5.

Corollary 8.4. [**167**] *Let $S = R(G)$ be a strongly G-graded ring whose base ring R is G-prime. Suppose that, for every trivial intersection ideal I of R, we have $\Delta^+(G_I) = \langle 1 \rangle$. Then S is prime.*

Proof. We use Corollary 8.2. Suppose I is a trivial intersection ideal of R and N is a finite subgroup of G with $\mathbf{N}_G(N) = G_I$. Then $N \subseteq \Delta^+(G_I) = \langle 1 \rangle$, by assumption, so clearly $N = \langle 1 \rangle$, $R(N) = R$ and $G_I = G$. Since R is G-prime, it follows from Corollary 8.2 that S is prime. ∎

Observe that if G is torsion free, then the hypothesis $\Delta^+(G_I) = \langle 1 \rangle$ is clearly satisfied. Thus we obtain

Corollary 8.5. [**166**] [**167**] *Let $S = R(G)$ be a strongly G-graded ring whose base ring R is G-prime. If G is torsion free, then S is prime.*

We can of course handle the semiprime problem in a similar manner. It follows from the work of Section 4 that the appropriate subrings $R(N)$ will be semiprime if R is semiprime with no $|N|$-torsion. However, this yields a rather weak result which we must strengthen in two ways. First we want to assume that R is G-semiprime rather than semiprime and second we wish to consider the torsion of I rather than that of R. This is achieved via the following variant of Theorem 4.8. We use the proof of Theorem 4.4.

Lemma 8.6. *Let $S = R(H)$ be a strongly H-graded ring, let N be a finite normal subgroup of H and let I be an H-invariant ideal of R. If I is H-nilpotent-free with no $|N|$-torsion, then $I \cdot R(N)$ is an H-nilpotent-free ideal of $R(N)$.*

Proof. We first observe that $I \cdot R(N)$ has no $|N|$-torsion. Indeed suppose $\gamma = \sum_x \gamma_x \in I \cdot R(N)$ with $|N|\gamma = 0$. Then for all $x \in N$ we have $|N|\gamma_x S_{x^{-1}} = 0$ and $\gamma_x S_{x^{-1}} \subseteq I$. Since I has no $|N|$-torsion, we conclude that $\gamma_x S_{x^{-1}} = 0$ and hence $\gamma = 0$.

Now suppose A is an H-invariant ideal of $R(N)$ contained in $I \cdot R(N)$ with $A^2 = 0$. Then $L = \ell_{I \cdot R(N)}(A)$ is a two sided ideal of $R(N)$ which is essential in $I \cdot R(N)$ as a right $R(N)$-submodule. Since $I \cdot R(N)$ has no $|N|$-torsion and $R(N)$ is strongly graded, it follows from Proposition 4.13 that $L_{|R}$ ess $I \cdot R(N)_{|R}$. Therefore $L \cap I = \ell_I(A)$ is essential in I as a right ideal . Now observe that $L' = \ell_I(A)$ is an H-invariant ideal of R contained in I. Thus since I is H-nilpotent-free we have $L' \cap \mathrm{r}_I(L') = 0$ and since L' ess I this yields $\mathrm{r}_I(L') = 0$. Finally if $\alpha = \sum_x \alpha_x \in A \subseteq I \cdot R(N)$, then $L'\alpha = 0$ implies that $L'\alpha_x S_{x^{-1}} = 0$ for all x. Thus since $\alpha_x S_{x^{-1}} \subseteq I$ we have $\alpha_x S_{x^{-1}} = 0$ and $\alpha = 0$. We conclude that $A = 0$ as required. ∎

It is clear that the above proof only needs H to strongly permute the ideals of $R(N)$ and of R in a suitably compatible manner.

If V is an additive abelian group and G is an arbitary group, we say that V has *no $|G|$-torsion* if, for all finite subgroups N of G, V has no $|N|$-torsion. We can now obtain the analog of Theorems 5.3 and 5.4.

Corollary 8.7. [167] *Let $R(G)$ be a strongly G-graded ring whose base ring R is G-semiprime. Suppose, for every trivial intersection ideal I of R, that I has no $|\Delta^+(G_I)|$-torsion. Then $R(G)$ is semiprime.*

Proof. Let I be any trivial intersection ideal of R and set $H = G_I$. Then $I^x I \subseteq I^x \cap I = 0$ for all $x \in G \setminus H$ so Lemma 6.6(ii) implies that I is H-nilpotent-free. Suppose N is a finite normal subgroup of H. Then $N \subseteq \Delta^+(G_I)$ so, by assumption, I has no $|N|$-torsion. We therefore conclude from Lemma 8.6 that $I \cdot R(N)$ is an H-nilpotent-free ideal of $R(N)$ and Corollary 8.2 yields the result. ∎

Observe that if R has no $|G|$-torsion, then certainly I has no $|\Delta^+(G_I)|$-torsion. Thus we obtain

Corollary 8.8. [166] [167] *Let $R(G)$ be a strongly G-graded ring whose base ring R is G-semiprime. If R has no $|G|$-torsion, then $R(G)$ is semiprime.*

This completes our combinatorial approach, via the Δ-methods, to the problem of determining when $R(G)$ is prime or semiprime. As we will see in the next section, the Δ-methods help in characterizing group-graded rings satisfying a polynomial identity. Furthermore, they are of use in computing certain rings of quotients and in describing the prime ideals in ordinary group algebras of polycyclic-by-finite groups.

EXERCISES

1. Verify the details of the skew group ring example $S = RG$ given immediately after Corollary 8.2. Show that S can fail to be prime or semiprime even if G has no nonidentity finite normal subgroup.

2. In the above example, find all trivial intersection ideals of R. If A is a nonzero ideal of RG, show that $0 \neq IAI \subseteq I \cdot RH$. Determine precisely when S is semiprime or prime.

3. The main results of this section apply equally well to component regular group-graded rings. Verify, in particular, the analogs of Corollaries 8.5 and 8.8.

4. Show by example that there exists a G-graded ring $S = R(G)$ with R a field, G a torsion-free group and with S not semiprime. Obviously S cannot be component regular.

5. Let R be an integral domain with field of fractions K and let $a \in R$. Assume that the polynomial $f(x) = x^n - a \in K[x]$ is irreducible. If α is a root of $f(x)$, show that the integral domain

$R[\alpha]$ is graded by the cyclic group of order n. Conclude that a crossed product can be prime even when the group has a nontrivial normal subgroup. Furthermore show that a component regular group-graded ring need not be strongly graded.

9. Polynomial Identities

We close this chapter by briefly considering group-graded rings and crossed products which satisfy a polynomial identity. We start with the case of finite groups. For this we require certain sufficient conditions for an algebra to satisfy a polynomial identity.

Recall that the Amitsur-Levitzki theorem guarantees that the matrix ring $M_m(K)$ satisfies the standard identity s_{2m}. More generally suppose $R = M_m(T)$. If T satisfies a polynomial identity of degree d then, by [**176**], R satisfies $(s_{2md})^\ell$ for some integer $\ell \geq 1$. Moreover if T is semiprime, we can take $\ell = 1$. Notice that $R \supseteq E = M_m(K)$ and that $T = C_R(E)$. Furthermore, E is a finite-dimensional *separable K-algebra*. That is, E remains semiprimitive under all field extensions. This explains the following result which we quote without proof.

Theorem 9.1. [**139**] *Suppose R is a K-algebra and E is a finite-dimensional separable K-subalgebra. If $C_R(E)$ satisfies a polynomial identity of degree d then, for some integer $\ell \geq 1$, R satisfies $(s_{dn})^\ell$ with $n \leq \dim_K E$. Furthermore if R is semiprime, then $\ell = 1$.*

Notice that by taking $E = K$ we see that if R satisfies a polynomial identity of degree d, then it satisfies $(s_d)^\ell$ for some $\ell \geq 1$. This is an old result of [**1**].

As a consequence we have

Theorem 9.2. [**14**] *Let $S = R(G)$ be a G-graded K-algebra with G finite and $K \subseteq R$. If R satisfies a polynomial identity of degree d then, for some integer $\ell \geq 1$, S satisfies $(s_{dn})^\ell$ with $n \leq |G|$. Furthermore if S is graded semiprime, then we may take $\ell = 1$.*

Proof. Form $T = S \# G^*$ and let $E \subseteq T$ be the K-subalgebra $E = \oplus \sum_{x \in G} p_x K$. Then E is surely separable with $\dim_K E = |G|$. Furthermore it follows easily (see Exercise 1) that $\mathbf{C}_T(E) = \oplus \sum_{x \in G} p_x R$ is isomorphic to the ring direct sum of $|G|$ copies of R. Thus $\mathbf{C}_T(E)$ also satisfies a polynomial identity of degree d. We conclude from Theorem 9.1 that T satisfies $(s_{dn})^\ell$ and hence so does $S \subseteq T$. Finally if S is graded semiprime, then T is semiprime, by Theorem 4.8, and we can take $\ell = 1$. ∎

In particular, if G is finite, then R satisfies a polynomial identity if and only if $S = R(G)$ does. Now let us move on to infinite groups.

In the case of ordinary group algebras $K[G]$, the polynomial identity condition essentially depends on the structure of G. Indeed if $\operatorname{char} K = 0$, then it was shown in [79] that $K[G]$ satisfies a polynomial identity if and only if G is abelian-by-finite. This was proved, in its more precise version, using the character theory of finite groups and simple reductions from the infinite to the finite case. In characteristic $p > 0$ more ring theoretic methods were required. Here the first result, due to [194], handled prime group rings and obtained the same characterization as above. That proof used localization and brought the Δ-methods into play. These ideas were pushed further in [158] and the problem was finally solved in [160].

Let p be a prime. We say that the group A is *p-abelian* if A', the commutator subgroup of A, is a finite p-group. For convenience, *0-abelian* will mean abelian. Then the result is

Theorem 9.3. [79] [160] *Let K be a field of characteristic $p \geq 0$. Then $K[G]$ satisfies a polynomial identity if and only if G has a p-abelian subgroup A of finite index.*

Furthermore there are bounds relating the degree of the polynomial identity and the product $|G : A| \cdot |A'|$.

Now it turns out that most of the proof in [160] carries over essentially verbatim to the context of component regular group-graded rings. One reason for this is that in the necessary Δ-lemma, namely Lemma 9.5, the various $\pi_\Delta(\alpha_i)$ are either 0 or 1. We will offer the, suitably modified, ring theoretic aspects of the proof here. We will just quote the group theoretic facts which are required.

To start with, let G be a group and let T be a subset of G. We say that T has *finite index* in G if there exist $x_1, x_2, \ldots, x_k \in G$ for some finite k with

$$G = Tx_1 \cup Tx_2 \cup \cdots \cup Tx_k.$$

We then define the *index* $|G : T|$ to be the minimum such k. Of course, if no such k exists, then $|G : T| = \infty$. Observe that if T is a subgroup of G, then this agrees with the usual definition of index.

We remark that this definition is not right–left symmetric. If $G = \bigcup_1^k Tx_i$, then taking inverses yields $G = \bigcup_1^k x_i^{-1}T^{-1}$ and we conclude that the index of T is equal to the left index of T^{-1}. On the other hand, it is quite possible for T to have (right) index equal to 2 and left index equal to ∞ (see Exercise 4).

The following lemma is based on the fact that distinct cosets of subgroups of G are disjoint. Proofs can be found in [**161**, Lemmas 5.2.2 and 5.2.1]. See also Exercise 4 of Section 6.

Lemma 9.4. *Let G be a group, let $H_1, H_2, \ldots, H_k$ be subgroups and let $S = \bigcup_1^k H_i g_i$ for some $g_i \in G$.*

    i. *If $|G : H_i| > k$ for all i, then $S \neq G$.*

    ii. *If $S \neq G$, then there exist $x_1, x_2, \ldots, x_t \in G$ with $t = (k+1)!$ such that $\bigcap_1^t Sx_i = \emptyset$. In particular, if $G = S \cup T$ for some subset T of G, then $|G : T| \leq (k+1)!$.*

For any group G and integer $k \geq 1$ we define

$$\Delta_k = \Delta_k(G) = \{\, x \in G \mid |G : \mathbf{C}_G(x)| \leq k \,\}.$$

Then Δ_k is a normal subset which is closed under taking inverses, but it need not be a subgroup of G. Note that $\Delta_1(G) = \mathbf{Z}(G)$, $\Delta(G) = \bigcup_1^\infty \Delta_k(G)$ and that $\Delta_a \Delta_b \subseteq \Delta_{ab}$. The next result is again a coset counting argument.

Lemma 9.5. *Let $R(G)$ be a component regular G-graded ring, let $\alpha_1, \alpha_2, \ldots, \alpha_t, \beta_1, \beta_2, \ldots, \beta_t, \gamma \in R(G)$ and let k be a fixed positive* *integer. Suppose that* $(\text{Supp }\alpha_i) \cap \Delta_k = \emptyset$ *for all* i *and that*

$$\left| \bigcup_i \text{Supp }\alpha_i \right| = r, \qquad \left| \bigcup_i \text{Supp }\beta_i \right| = s$$

with $rs < k$. *Let* T *be a subset of* G *and suppose that for all* $x \in G \backslash T$
and $X \in R(x)$ *we have*

$$\alpha_1 X \beta_1 + \alpha_2 X \beta_2 + \cdots + \alpha_t X \beta_t = X\gamma.$$

Then either $\gamma = 0$ *or* $|G : T| \le k!$.

Proof. Let

$$\bigcup_i \text{Supp }\alpha_i = \{\, y_1, y_2, \ldots, y_r \,\}$$

$$\bigcup_i \text{Supp }\beta_i = \{\, z_1, z_2, \ldots, z_s \,\}.$$

We assume that $\gamma \ne 0$ and let $v \in \text{Supp }\gamma$ so that $\gamma_v \ne 0$. If
y_i is conjugate to vz_j^{-1} in G for some i, j, choose $h_{i,j} \in G$ with
$h_{i,j}^{-1} y_i h_{i,j} = vz_j^{-1}$.

Let $x \in G \backslash T$ and, since $R(G)$ is component regular, choose
$X \in R(x)$ with $X\gamma_v \ne 0$. Then by hypothesis we have

$$\alpha_1 X \beta_1 + \alpha_2 X \beta_2 + \cdots + \alpha_t X \beta_t = X\gamma.$$

Because $xv \in \text{Supp } X\gamma$, it follows that xv occurs in the support of
the above left-hand side, and thus for some i, j we have $y_i x z_j = xv$.
Therefore, $x^{-1} y_i x = vz_j^{-1}$, so y_i and vz_j^{-1} are conjugate in G and
we have

$$x^{-1} y_i x = vz_j^{-1} = h_{i,j}^{-1} y_i h_{i,j}.$$

It follows that $x \in \mathbf{C}_G(y_i) h_{i,j}$.

We have therefore shown that $G = S \cup T$ where S is the set
$S = \bigcup_{i,j} \mathbf{C}_G(y_i) h_{i,j}$. Now $(\text{Supp }\alpha_\ell) \cap \Delta_k = \emptyset$ by assumption so we
have $|G : \mathbf{C}_G(y_i)| > k$ for all i. Moreover there are at most $rs < k$
cosets in the union for S. Thus we conclude from Lemma 9.4(i) that
$S \ne G$ and hence Lemma 9.4(ii) yields $|G : T| \le (rs + 1)! \le k!$ as
required. ∎

Now let K be a field and consider the noncommutative polynomial ring $K\langle \zeta_1, \zeta_2, \ldots, \zeta_n \rangle$. A *linear monomial* is an element $\mu \in K\langle \zeta_1, \zeta_2, \ldots, \zeta_n \rangle$ of the form $\mu = \zeta_{i_1} \zeta_{i_2} \cdots \zeta_{i_r}$ with all i_j distinct and with $r \geq 1$. Thus μ is linear in each variable that occurs. Note that the number of linear monomials of degree n is $n!$ and that any other linear monomial is an initial segment of one of these. Thus there are at most $n \cdot n! \leq (n+1)!$ linear monomials of all degrees.

The next result is the key ingredient in classifying group rings satisfying a polynomial identity. The main idea of the proof is to treat the given multilinear polynomial identity as a linear identity one variable at a time using the preceding lemma. Unfortunately, it is not always true that $(\operatorname{Supp} \alpha_i) \cap \Delta_k = \emptyset$ and it becomes a bookkeeping problem to keep track of when this occurs. It is for this reason that the linear monomials are introduced.

Theorem 9.6. [160] *Let $S = R(G)$ be a component regular group-graded ring which is an algebra over the field $K \subseteq R$. If S satisfies a polynomial identity over K of degree n, then*

$$|G : \Delta_k(G)| < (k+1)!$$

where $k = (n!)^2$.

Proof. We assume by way of contradiction that $|G : \Delta_k| \geq (k+1)!$. As we mentioned earlier, S must satisfy the multilinear polynomial

$$f(\zeta_1, \zeta_2, \ldots, \zeta_n) = \sum_{\sigma \in \operatorname{Sym}_n} a_\sigma \zeta_{\sigma(1)} \zeta_{\sigma(2)} \cdots \zeta_{\sigma(n)}$$

with $a_\sigma \in K$ and $a_1 = 1$. Clearly $n > 1$.

Now, for $j = 1, 2, \ldots, n$ define $f_j \in K\langle \zeta_j, \zeta_{j+1}, \ldots, \zeta_n \rangle$ by

$$f = \zeta_1 \zeta_2 \cdots \zeta_{j-1} f_j + \text{ terms not starting with } \zeta_1 \zeta_2 \cdots \zeta_{j-1}.$$

Thus clearly $f_1 = f$, $f_n = \zeta_n$ and f_j is a multilinear polynomial of degree $n - j + 1$. In particular, for all j, the variable ζ_j occurs in each monomial of f_j. Furthermore we have

$$f_j = \zeta_j f_{j+1} + \text{ terms not starting with } \zeta_j.$$

For each $j = 2, 3, \ldots, n$ let $\mathcal{M}_j$ denote the set of all linear monomials in $K\langle \zeta_j, \zeta_{j+1}, \ldots, \zeta_n \rangle$ and let $\mathcal{M}_1$ be empty. Then, by the remarks preceding this theorem, we have $|\mathcal{M}_j| \leq |\mathcal{M}_2| \leq n!$ for all j. We show now by induction on $j = 1, 2, \ldots, n$ that for any $x_j, x_{j+1}, \ldots, x_n \in G$ we have either $\mu(x_j, x_{j+1}, \ldots, x_n) \in \Delta_k(G)$ for some $\mu \in \mathcal{M}_j$ or $f_j(X_j, X_{j+1}, \ldots, X_n) = 0$ for all $X_i \in R(x_i)$. Because $f = f_1$, the result for $j = 1$ is given.

Suppose the inductive result holds for some $j < n$. Fix the elements $x_{j+1}, x_{j+2}, \ldots, x_n \in G$ and let x play the role of the jth variable. Let $\mu \in \mathcal{M}_{j+1}$. If $\mu(x_{j+1}, x_{j+2}, \ldots, x_n) \in \Delta_k(G)$, then we are done. Thus we may assume that for all such $\mu \in \mathcal{M}_{j+1}$ we have $\mu(x_{j+1}, x_{j+2}, \ldots, x_n) \notin \Delta_k(G)$. Set $\mathcal{M}_j \setminus \mathcal{M}_{j+1} = \mathcal{T}_j$.

Now let $\mu \in \mathcal{T}_j$ so that μ involves the variable ζ_j. Write $\mu = \mu' \zeta_j \mu''$ where μ' and μ'' are monomials in $K\langle \zeta_{j+1}, \zeta_{j+2}, \ldots, \zeta_n \rangle$. Then we have $\mu(x, x_{j+1}, \ldots, x_n) \in \Delta_k(G)$ if and only if

$$x \in \mu'(x_{j+1}, \ldots, x_n)^{-1} \Delta_k(G) \mu''(x_{j+1}, \ldots, x_n)^{-1} = \Delta_k(G) h_\mu,$$

a fixed right translate of Δ_k since Δ_k is a normal subset of G. It therefore follows that for all $x \notin T = \bigcup_{\mu \in \mathcal{T}_j} \Delta_k h_\mu$ we have $\mu(x, x_{j+1}, \ldots, x_n) \notin \Delta_k(G)$, and this holds for all $\mu \in \mathcal{M}_j$ because $\mathcal{M}_j \subseteq \mathcal{M}_{j+1} \cup \mathcal{T}_j$. Because the inductive result holds for j, we conclude that if $x \notin T$ then $f_j(X, X_{j+1}, \ldots, X_n) = 0$ for all $X \in R(x)$ and $X_i \in R(x_i)$.

Write

$$f_j(\zeta_j, \zeta_{j+1}, \ldots, \zeta_n) = \zeta_j f_{j+1} - \sum_{i=1}^{t} \alpha_i \zeta_j \beta_i$$

where $\alpha_i, \beta_i \in K\langle \zeta_{j+1}, \zeta_{j+2}, \ldots, \zeta_n \rangle$ and α_i is a linear monomial. Hence $\alpha_i \in \mathcal{M}_{j+1}$. Now let $X_{j+1}, X_{j+2}, \ldots, X_n$ be fixed elements of S with $X_i \in R(x_i)$. Then by the above

$$X f_{j+1}(X_{j+1}, \ldots, X_n) = \sum_{i=1}^{t} \alpha_i(X_{j+1}, \ldots, X_n) X \beta_i(X_{j+1}, \ldots, X_n)$$

for all $x \in G \setminus T$ and $X \in R(x)$.

We apply Lemma 9.5 to the above with $\gamma = f_{j+1}(X_{j+1}, \ldots, X_n)$. Now f has at most $n!$ monomials in its expression and thus clearly, in the notation of that lemma, we have $r, s < n!$. Hence $rs < (n!)^2 = k$. Moreover, observe that Supp $\alpha_i(X_{j+1}, \ldots, X_n)$ is disjoint from $\Delta_k(G)$ since $\alpha_i \in \mathcal{M}_{j+1}$ implies that $\alpha_i(x_{j+1}, \ldots, x_n) \in G \setminus \Delta_k(G)$. Thus, since S is component regular, the hypotheses of Lemma 9.5 are satisfied and there are two possible conclusions.

If $|G : T| \leq k!$ then since $T = \bigcup_{\mu \in \mathcal{T}_j} \Delta_k(G)h_\mu$ and $|\mathcal{T}_j| \leq |\mathcal{M}_j| \leq n!$ we see that

$$|G : \Delta_k(G)| \leq |\mathcal{T}_j| \cdot |G : T| \leq (n!)(k!) < (k+1)!$$

a contradiction by assumption. Thus we conclude that

$$0 = \gamma = f_{j+1}(X_{j+1}, X_{j+2}, \ldots, X_n)$$

and the inductive step is proved.

In particular, the inductive result holds for $j = n$. Here $f_n(\zeta_n) = \zeta_n$ and $\mathcal{M}_n = \{\zeta_n\}$. Thus we conclude that for all $x \in G$ either $x \in \Delta_k(G)$ or $R(x) = 0$, a contradiction since $G \neq \Delta_k(G)$ and S is component regular. Therefore the assumption $|G : \Delta_k(G)| \geq (k+1)!$ is false and the theorem is proved. ∎

To obtain a more understandable conclusion from the above, we require some additional group theoretic facts. The first is a simple exercise (see [161, Lemma 5.2.3] or Exercise 6). The second is a more substantial result of [207] (see [161, Theorem 5.2.9]).

Lemma 9.7. Let T be a subset of G with $|G : T| \leq k$ and set $T^* = T \cup \{1\} \cup T^{-1}$. Then

$$(T^*)^{4^k} = T^* \cdot T^* \cdot \cdots \cdot T^* \qquad (4^k \text{ times})$$

is a subgroup of G.

Theorem 9.8. [207] Let G be a group and let k be a positive integer.
    i. If $|G'| \leq k$ then $G = \Delta_k(G)$.
    ii. If $G = \Delta_k(G)$ then $|G'| \leq (k^4)^{k^4}$.

The bound in (ii) above is certainly not sharp. Indeed it is a question of some group theoretic interest to obtain a bound of the correct order of magnitude. As a consequence we have

Corollary 9.9. [160] *Let $S = R(G)$ be a component regular G-graded ring which is an algebra over the field $K \subseteq R$. If S satisfies a polynomial identity over K of degree n, then setting $k = (n!)^2$ we have*

i. $|G : \Delta(G)| < (k+1)!$ *and* $\Delta(G)'$ *is finite,*

ii. G *has a characteristic subgroup A with $|G : A| < (k+1)!$ and $|A'|$ bounded above by a finite function of n.*

Proof. For convenience set $m = (k+1)!$. Then by Theorem 9.6 we have $|G : \Delta_k(G)| < m$ and hence $|G : \Delta(G)| < m$ since $\Delta_k \subseteq \Delta$. Moreover every translate of Δ_k is either contained in Δ of disjoint from it, so we have $|\Delta : \Delta_k| < m$. Say $\Delta = \bigcup_1^t \Delta_k y_i$ with $t < m$. Then there exists an integer ℓ with $y_i \in \Delta_\ell$ for all i, so $\Delta \subseteq \Delta_k \Delta_\ell \subseteq \Delta_{k\ell}$. We conclude that $\Delta(G) = \Delta_{k\ell}(G) = \Delta_{k\ell}((\Delta(G))$ so, by Theorem 9.8(ii), $\Delta(G)'$ is finite and (i) is proved.

For (ii) let A be the characteristic subgroup of G generated by $\Delta_k(G)$. Again $A \supseteq \Delta_k$ implies that $|G : A| < m$. Moreover since $1 \in \Delta_k$ and Δ_k is closed under taking inverses, we see that $\Delta_k = (\Delta_k)^*$ in the notation of Lemma 9.7. Thus, by that lemma, $A = (\Delta_k)^{4^k} \subseteq \Delta_q(G)$ with $q = k^{4^k}$. But then $A = \Delta_q(A)$ so, by Theorem 9.8(ii), $|A'| \leq (q^4)^{q^4}$ and the latter is a finite (though very large) function of n. ∎

We remark that $|\Delta(G)'|$ need not be bounded by a function of n (see Exercise 7).

In particular all this applies to crossed products. But it certainly does not solve the problem. If $S = R*G$ satisfies a polynomial identity, then G has the structure given above. But this, in itself, is not sufficient. We must also know something about the twisting and how G acts on R. This can be achieved by studying prime homomorphic images of $R*G$ and is more appropriately left for a later section (see Section 23).

EXERCISES

1. Let $T = S\#G^*$ be as in Theorem 9.2. If $E = \sum_{x \in G} p_x K$, show that $\mathbf{C}_T(E) = \sum_{x \in G} p_x T p_x = \oplus \sum_{x \in G} p_x R$ and that this ring is isomorphic to a direct sum of $|G|$ copies of R. This can be proved directly or one can use the isomorphism $T \cong S\{1\} \subseteq \mathbf{M}_G(S)$.

2. Let E be an infinite dimensional *exterior algebra* over the field K. Thus by definition there are elements $e_1, e_2, e_3, \ldots \in E$ such that 1 and all finite products $e_{i_1} e_{i_2} \cdots e_{i_s}$ with $i_1 < i_2 < \cdots < i_s$ form a K-basis for E. Furthermore multiplication in E is determined by the relations $e_i^2 = 0$ and $e_i e_j = -e_j e_i$ for all $i \neq j$. Let E_1 be the subspace of E spanned by all basic products of even degree and let E_x be spanned by the odd degree ones. Show that $E = E_1 \oplus E_x$ is graded by the group $G = \{1, x\}$ of order 2. Furthermore if char $K \neq 2$ prove that $E_1 = \mathbf{Z}(E)$.

3. Continuing with the above notation, compute the polynomial expressions $s_n(e_1, e_2, \ldots, e_n)$ and $s_2(e_1 + \cdots + e_\ell, e_{\ell+1} + \cdots + e_{2\ell})^\ell$. Deduce that if char $K = 0$ then E satisfies no standard identity or $(s_2)^\ell$. What does Theorem 9.2 say about the polynomial identities satisfied by E?

4. Let $G = \langle x, y \mid y^2 = 1, x^{-1}yx = y^{-1} \rangle$ be the infinite dihedral group and let $T = \{x^n, x^{-n}y \mid n \geq 0\}$. Notice that $yT = T$ so $x^i y^j T$ fails to contain the elements x^m for all sufficiently small m. Conclude that T has infinite left index in G but that $|G : T| = 2$.

5. Both Theorems 9.2 and 9.6 assume that $K \subseteq R$ and this hypothesis is used differently and subtly in each case. Where is it used and what happens if this assumption is dropped?

6. Let T be a subset of the group G with $1 \in T$ and $T = T^{-1}$. Suppose $|G : T| \leq k$ and say $G = \bigcup_1^k T x_i$ with $x_1 = 1$. If $T^2 \not\subseteq T$ then $T^2 \cap T x_i \neq \emptyset$ for some $i > 1$. Deduce that $T^4 \supseteq T x_1 \cup T x_i$ so $|G : T^4| \leq k - 1$ and hence that T^{4^k} is a subgroup of G.

7. Suppose H is the finite dihedral group $H = \langle A, x \rangle$ where A is an abelian group of odd order, $x^2 = 1$ and $a^x = a^{-1}$ for all $a \in A$. Show that the group algebra $K[H]$ satisfies the standard identity s_4 but that $H' = A$ can be arbitrarily large. Conclude that $|\Delta(G)'|$ is not bounded by a function of n in Theorem 9.9(i).

3   The Symmetric Ring
of Quotients

10. The Martindale Ring of Quotients

To go further with our study of crossed products, it is necessary to localize. Fortunately there is a rather easily constructed and well behaved ring of quotients, the Martindale ring of quotients, which suffices for our purposes. This ring was introduced in [116] as a tool in the study of prime rings satisfying a generalized polynomial identity. The concept was extended to semiprime rings in [4]. As we will see, a good deal of the remainder of this book is a consequence of the existence of this quotient ring. It is, in fact, an absolutely necessary ingredient in numerous aspects of ring theory including crossed products, Galois theory, differential operator rings and normalizing extensions.

    In this section, we define the quotient ring, describe its elementary properties and compute some examples. For simplicity, we will restrict our attention to prime rings. We begin with the definition.

    Let R be a prime ring and consider the set of all left R-module homomorphisms $f: {}_R A \to {}_R R$ where A ranges over all nonzero two-sided ideals of R. Two such functions are said to be equivalent if they agree on their common domain which is a nonzero ideal since R is prime. That this is an equivalence relation follows from

Lemma 10.1. Let $f\colon {}_RA \to {}_RR$ with $A \lhd R$. If $Bf = 0$ for some ideal B with $0 \neq B \subseteq A$, then $Af = 0$.

Proof. Let $a \in A$. Since $Ba \subseteq B$ and f is a left R-module homomorphism, we have $0 = (Ba)f = B \cdot (af)$. Thus since R is prime, $af = 0$. ∎

We let $\hat{f}$ denote the equivalence class of f and let $Q_\ell = Q_\ell(R)$ be the set of all such equivalence classes. The arithmetic in Q_ℓ is defined in a fairly obvious manner. Suppose $f\colon {}_RA \to {}_RR$ and $g\colon {}_RB \to {}_RR$ are given. Then $\hat{f} + \hat{g}$ is the class of $f + g\colon {}_R(A \cap B) \to {}_RR$ and $\hat{f}\hat{g}$ is the class of the composite function $fg\colon {}_R(BA) \to {}_RR$. It is easy to see (Exercise 1) that these definitions make sense and, by Lemma 10.1, that they respect the equivalence relation. Furthermore, the ring axioms are surely satisfied so Q_ℓ is a ring with 1. Finally let $r_\rho\colon {}_RR \to {}_RR$ denote right multiplication by $r \in R$. Then the map $r \mapsto \hat{r}_\rho$ is easily seen to be a ring homomorphism from R into Q_ℓ. Moreover if $r \neq 0$ then $Rr_\rho \neq 0$ and hence $\hat{r}_\rho \neq 0$. We conclude therefore that R is embedded isomorphically in Q_ℓ with the same 1 and we will view Q_ℓ as an overring of R. It is the *left Martindale ring of quotients* of R.

Suppose $f\colon {}_RA \to {}_RR$ and $a \in A$. Then $a_\rho f$ is defined on ${}_RR$ and for all $r \in R$ we have

$$r(a_\rho f) = (ra)f = r(af) = r(af)_\rho.$$

Hence $\hat{a}_\rho \hat{f} = \widehat{(af)}_\rho$ and the map f translates in Q_ℓ to right multiplication by $\hat{f}$. This leads to the following abstract characterization of Q_ℓ.

Proposition 10.2. Let R be a prime ring. Then $Q_\ell = Q_\ell(R)$ satisfies
  i. $Q_\ell(R) \supseteq R$ with the same 1,
  ii. if $q \in Q_\ell$ then there exists $0 \neq A \lhd R$ with $Aq \subseteq R$,

iii. if $q \in Q_\ell$ and $0 \neq A \triangleleft R$ with $Aq = 0$, then $q = 0$,

iv. if $f \colon {}_R A \to {}_R R$ is given with $0 \neq A \triangleleft R$, then there exists $q \in Q_\ell$ with $aq = af$ for all $a \in A$.

Furthermore Q_ℓ is uniquely determined by these properties.

Proof. In view of Lemma 10.1 and the preceding paragraph, it is clear that Q_ℓ satisfies the four properties. We need only show that if Q and Q' satisfy (i)–(iv) then they are R-isomorphic. For this we define a map $\sigma \colon Q \to Q'$ as follows.

Let $q \in Q$ and choose $0 \neq A \triangleleft R$ with $Aq \subseteq R$. Then the map $a \mapsto aq$ is a left R-module homomorphism from A to R so by (iv) there exists $q' \in Q'$ with $aq = aq'$ for all $a \in A$. If follows easily from (iii) that q' is uniquely determined by q independent of the choice of A and we set $q' = q^\sigma$. Clearly $q' = q^\sigma$ is the unique element of Q' with $aq = aq^\sigma$ for all elements a in some nonzero ideal of R. From this we conclude easily that σ is a ring homomorphism which is the identity on R.

Now we reverse the roles and define $\tau \colon Q' \to Q$ in a similar manner. Then $(q')^\tau$ satisfies $aq' = a(q')^\tau$ and therefore we see that $\sigma\tau = \tau\sigma = 1$. ∎

A slight modification of the above shows that if $\eta \colon R \to R_1$ is a ring isomorphism, then η extends to an isomorphism $\eta \colon Q_\ell(R) \to Q_\ell(R_1)$. Furthermore, let S contain the prime ring R with the same 1 and assume that S satisfies (ii) and (iii) above. That is for all $s \in S$ we have $As \subseteq R$ for some nonzero ideal A of R and $As = 0$ implies $s = 0$. Then it follows that S embeds isomorphically in $Q_\ell(R)$.

One can of course define $Q_r(R)$, the *right Martindale ring of quotients* of R in a similar manner. It is obtained from the set of all right R-module homomorphisms $g \colon B_R \to R_R$ with $0 \neq B \triangleleft R$ and it satisfies a result analogous to the above.

We briefly consider two well-known examples.

Lemma 10.3. *Let R be a domain, that is a ring without zero divisors.*

i. *If R is commutative then $Q_\ell(R)$ is the field of fractions of R.*

ii. *If $R = K\langle x, y \rangle$ is the free algebra over the field K in the noncommuting indeterminates x and y, then x and y are zero divisors* *in* $Q_\ell(R)$.

Proof. (i) Let Q be the field of fractions of R. We show that Q satisfies (i)–(iv) of the preceding result. Since (i), (ii) and (iii) are obvious, we need only consider (iv). Thus suppose $f: {}_RA \to {}_RR$ and let a, b be nonzero elements of A. Then

$$a(bf) = (ab)f = (ba)f = b(af)$$

yields $a^{-1}(af) = b^{-1}(bf)$, an equation in Q. This implies that the fraction $a^{-1}(af)$ is a constant for all nonzero elements of A and if $q \in Q$ is this constant value, then $af = aq$ for all $a \in A$.

(ii) Let I be the augmentation ideal of $R = K\langle x, y \rangle$, that is the set of all polynomials in x and y with zero constant term. Then $I = Rx + Ry$ is clearly free as a left R-module with basis $\{x, y\}$. Thus we can define $f: {}_RI \to {}_RR$ by $xf = 1$ and $yf = 0$ and there exists $q \in Q_\ell(R)$ with $xq = 1$ and $yq = 0$. ∎

It is because of (i) that Q_ℓ is a ring of quotients and it is because of (ii) that Q_ℓ is in some sense too big. Fortunately there is a smaller, better behaved ring which suffices for the applications; it is the *symmetric Martindale ring of quotients* $Q_s(R)$. We choose to define it abstractly by its properties and then prove its existence later on.

Proposition 10.4. *Let R be a prime ring. Then the ring $Q_s = Q_s(R)$ is uniquely determined by the properties*
   i. $Q_s(R) \supseteq R$ *with the same* 1,
   ii. *if $q \in Q_s$ then there exist $0 \neq A, B \triangleleft R$ with $Aq, qB \subseteq R$,*
   iii. *if $q \in Q_s$ and $0 \neq I \triangleleft R$, then either $Iq = 0$ or $qI = 0$ implies $q = 0$,*
   iv. *let $f: {}_RA \to {}_RR$ and $g: B_R \to R_R$ be given with $0 \neq A, B \triangleleft R$ and suppose that for all $a \in A$ and $b \in B$ we have $(af)b = a(gb)$, then there exists $q \in Q_s(R)$ with $af = aq$ and $gb = qb$ for all $a \in A$, $b \in B$.*

Proof. We proceed as in the proof of Proposition 10.2. Let Q, Q' both satisfy (i)–(iv) and let $q \in Q$. Choose $0 \neq A, B \triangleleft R$ with $Aq, qB \subseteq R$

and define $f: {}_RA \rightarrow {}_RR$ and $g: B_R \rightarrow R_R$ by $af = aq$ and $gb = qb$. Then

$$(af)b = (aq)b = a(qb) = a(gb)$$

so f and g satisfy the balanced condition of (iv). Thus there exists $q' \in Q'$ with $aq = af = aq'$ and $qb = gb = q'b$ for all $a \in A$ and $b \in B$. With this observation, the earlier proof can now apply. ∎

We view the formula $(af)b = a(gb)$ in (iv) as either a *balanced* or an *associativity condition*. We remark that given (ii) above, part (iii) can be weakened to a one-sided condition.

Lemma 10.5. *In the above context, (iii) is equivalent to either of the conditions*

iii'. *if $q \in Q_s$ and $0 \neq I \lhd R$, then $Iq = 0$ implies $q = 0$,*

iii''. *if $q \in Q_s$ and $0 \neq I \lhd R$, then $qI = 0$ implies $q = 0$.*

Proof. Certainly (iii) implies (iii'). Conversely assume (iii') is satisfied and say $qJ = 0$ with $0 \neq J \lhd R$. Let $0 \neq I \lhd R$ with $Iq \subseteq R$. Since R is prime, $J \neq 0$ and $0 = I(qJ) = (Iq)J$, we have $Iq = 0$. Thus by (iii'), $q = 0$. ∎

We are now ready to prove the existence of $Q_s(R)$. One approach of course is to modify the proof of the existence of Q_ℓ by considering equivalence classes of ordered pairs (f, g) of balanced functions. However, we can avoid this by identifying Q_s as a specific subring of the quotient ring Q_ℓ.

Proposition 10.6. *If R is a prime ring, then $Q_s(R)$ exists. Indeed*

$$Q_s(R) = \{\, q \in Q_\ell(R) \mid qB \subseteq R \text{ for some } 0 \neq B \lhd R \,\}$$

and

$$Q_s(R) = \{\, q \in Q_r(R) \mid Aq \subseteq R \text{ for some } 0 \neq A \lhd R \,\}.$$

Proof. Let

$$S = \{\, q \in Q_\ell(R) \mid qB \subseteq R \text{ for some } 0 \neq B \lhd R \,\}.$$

We will show that S is a ring satisfying the four conditions of Proposition 10.4.

Let $q_1, q_2 \in S$ with $q_1 B_1, q_2 B_2 \subseteq R$. Then $(q_1 + q_2)(B_1 \cap B_2)$ and $q_1 q_2 B_2 B_1$ are both contained in R. Thus S is a ring satisfying condition (i). By definition of S and properties of Q_ℓ, S also satisfies (ii) and (iii') and therefore also (iii).

Finally let $f\colon {}_R A \to {}_R R$ and $g\colon B_R \to R_R$ be balanced maps with $0 \neq A, B \lhd R$. By properties of Q_ℓ applied to f, there exists $q \in Q_\ell$ with $af = aq$ for all $a \in A$. By the balanced condition we then have

$$a(gb) = (af)b = (aq)b = a(qb)$$

so $A(gb - qb) = 0$. Since $gb - qb \in Q_\ell$, this yields $gb = qb$ for all $b \in B$. In particular, $qB = gB \subseteq R$ so $q \in S$. Since $af = aq$ and $gb = qb$, the ring S satisfies (iv). ∎

As we mentioned above, $Q_\ell(R)$ was defined in [**116**]. Its subring $Q_s(R)$, as described in Proposition 10.6, was used in the Galois theoretic studies of [**87**] and [**88**]. The formulations given in Propositions 10.2 and 10.4 are from [**171**].

It follows, as in the remarks after Proposition 10.2, that if R and R_1 are prime rings with $\eta\colon R \to R_1$ a ring isomorphism, then η extends to an isomorphism $\eta\colon Q_s(R) \to Q_s(R_1)$. Furthermore, let S contain the prime ring R with the same 1 and assume that S satisfies (ii) and (iii) of Proposition 10.4. Then S embeds isomorphically in $Q_s(R)$.

We observe now that, in comparison to Lemma 10.3(ii), $Q_s(R)$ is quite close to R.

Lemma 10.7. *Let R be a prime ring.*

i. If R is a domain, then so is $Q_s(R)$.

ii. If $X \subseteq R$ is right (or left) regular in R, then it is right (or left) regular in $Q_s(R)$.

Proof. (i) Let $q_1, q_2 \in Q_s(R)$ with $q_1 q_2 = 0$ and choose nonzero ideals A_1, A_2 of R with $A_1 q_1, q_2 A_2 \subseteq R$. Then $(A_1 q_1)(q_2 A_2) = 0$ and since R is a domain either $A_1 q_1 = 0$ or $q_2 A_2 = 0$. Thus $q_1 = 0$ or $q_2 = 0$.

(ii) Let $Xq = 0$ for some $q \in Q_s(R)$ and choose $0 \neq A \lhd R$ with $qA \subseteq R$. Then $X(qA) = 0$ and, since X is appropriately regular in R, we have $qA = 0$ and therefore $q = 0$. $\blacksquare$

Let us compute a few more examples.

Lemma 10.8. *Let R be a prime ring.*

  i. *If R is simple, then $Q_\ell(R) = Q_s(R) = R$.*

  ii. *$Q_\ell(M_n(R)) = M_n(Q_\ell(R))$ and $Q_s(M_n(R)) = M_n(Q_s(R))$.*

  iii. *Let $R \supseteq S$ with the same 1 and assume that $S \supseteq I$ where I is a nonzero ideal of R. Then $Q_\ell(S) = Q_\ell(R)$ and $Q_s(S) = Q_s(R)$.*

Proof. (i) This is clear since any left R-module homomorphism $_RR \to _RR$ is right multiplication by some element of R.

(ii) If $Q = Q_\ell(R)$, then $M_n(Q) \supseteq M_n(R)$ and it is clear that this overring satisfies (ii) and (iii) of Proposition 10.2. It suffices to verify (iv). To this end, let $f: M_n(A) \to M_n(R)$ be a left $M_n(R)$-module homomorphism with $0 \neq A \lhd R$. For each i, j let $e_{i,j}$ denote the usual matrix unit and let $\pi_{i,j}: M_n(R) \to R$ denote the projection into the $(i,j)^{\text{th}}$-entry. Then the map

$$A \to e_{i,i}A \xrightarrow{\;f\;} M_n(R) \xrightarrow{\;\pi_{i,j}\;} R$$

is a left R-module homomorphism and hence is represented by an element $q_{i,j} \in Q$. It now follows that the matrix $[q_{i,j}]$ represents f since for all subscripts u, v and all $a \in A$

$$(e_{u,v}a)f = e_{u,v} \cdot (e_{v,v}a)f$$
$$= e_{u,v} \cdot \sum_w e_{v,w}(aq_{v,w}) = (e_{u,v}a) \cdot [q_{i,j}].$$

Thus $Q_\ell(M_n(R)) = M_n(Q_\ell(R))$. The result for $Q_s(M_n(R))$ follows from Proposition 10.6.

(iii) We prove the result for the symmetric ring of quotients; the proof for Q_ℓ is similar. Let $0 \neq J \lhd S$. Then $0 \neq IJI \subseteq J$ and IJI is an ideal of both R and S. It follows that S is prime. Note that $S \subseteq R \subseteq Q_s(R)$ and we use the characterization of $Q_s(S)$ given by (i)–(iv) of Proposition 10.4 to show that $Q_s(R) = Q_s(S)$.

Let $q \in Q_s(R)$ and say $0 \neq A, B \triangleleft R$ with $Aq, qB \subseteq R$. Then IA and BI are nonzero ideals of S with $IAq \subseteq I \subseteq S$ and $qBI \subseteq I \subseteq S$. Next suppose $0 \neq J \triangleleft S$ with $Jq = 0$ or $qJ = 0$. By the above, J contains a nonzero ideal of R and therefore $q = 0$. Thus $Q_s(R)$ satisfies (i), (ii) and (iii) for the ring S.

Finally suppose $\bar{A}, \bar{B}$ are nonzero ideals of S and that $f: {}_S\bar{A} \to {}_S S$ and $g: \bar{B}_S \to S_S$ are balanced maps. Let A and B be nonzero ideals of R with $A \subseteq \bar{A}$ and $B \subseteq \bar{B}$. We show that $f: A \to R$ is an R-homomorphism. To this end, let $r \in R$, $i \in I$ and $a \in A$. Since $i, ir \in S$ and $ra \in A$ we have

$$i\big((ra)f\big) = (ira)f = ir(af)$$

so $I\big((ra)f - r(af)\big) = 0$ and hence $(ra)f = r(af)$. Similarly $g: B \to R$ is a right R-module homomorphism and the balanced condition is surely satisfied. Thus there exists $q \in Q_s(R)$ with $af = aq$ and $gb = qb$ for all $a \in A$, $b \in B$. We must show that $\bar{a}f = \bar{a}q$ and $g\bar{b} = q\bar{b}$ for all $\bar{a} \in \bar{A}$ and $\bar{b} \in \bar{B}$. For the first, let $a \in A$. Then $a \in S$ and $a\bar{a} \in A$ so

$$a(\bar{a}f) = (a\bar{a})f = (a\bar{a})q = a(\bar{a}q)$$

and $A(\bar{a}f - \bar{a}q) = 0$. Since A is a nonzero ideal of R we conclude that $\bar{a}f = \bar{a}q$. Similarly $g\bar{b} = q\bar{b}$ so the result follows. ∎

Basic properties of these quotient rings are as follows.

Lemma 10.9. *Let R be a prime ring and set $S = Q_\ell(R)$ or $S = Q_s(R)$.*

i. S is a prime ring; in fact if $q_1, q_2 \in S$ with $q_1 R q_2 = 0$ then $q_1 = 0$ or $q_2 = 0$.

ii. Every automorphism (or derivation) of R extends uniquely to an automorphism (or derivation) of S.

iii. If $C = \mathbf{C}_S(R)$, then C is a field which is the center of both $Q_\ell(R)$ and $Q_s(R)$.

Proof. (i) As usual, $q_1 R q_2 = 0$ yields $(Aq_1 R)q_2 = 0$ so either $Aq_1 R = 0$ or $q_2 = 0$.

(ii) Assume that $S = Q_\ell(R)$. If σ is an automorphism of R, then the isomorphism $\sigma: R \to R$ extends to an isomorphism $\sigma: S \to S$. Alternately, if $f: {}_RA \to {}_RR$ is given, we can define $f^\sigma: {}_RA^\sigma \to {}_RR$ by $a^\sigma f^\sigma = (af)^\sigma$.

Now let $\delta: R \to R$ be a derivation. If $s \in S$, choose $0 \neq A \vartriangleleft R$ with $As \subseteq R$ and define $f_s: A^2 \to R$ by $bf_s = \delta(bs) - \delta(b)s$. Note that $\delta(A^2) \subseteq A$ so $\delta(b)s \in R$ and $bf_s \in R$. Certainly f_s is additive and since δ is a derivation of R we have

$$(rb)f_s = \delta(rbs) - \delta(rb)s$$
$$= r\delta(bs) + \delta(r)bs - r\delta(b)s - \delta(r)bs = r(bf_s).$$

It follows that there exists $q_s \in S$ with $bf_s = bq_s$ for all $b \in A^2$. Furthermore, it is clear that q_s is uniquely determined by s and this formula, independent of the choice of A.

Notice that if $r \in R$, then $bf_r = \delta(br) - \delta(b)r = b\delta(r)$ so $q_r = \delta(r)$ in this case. We can therefore define $\delta(s) = q_s$ for all $s \in S$. By definition of $\delta(s) = q_s$ we then have $\delta(bs) = bf_s + \delta(b)s = b\delta(s) + \delta(b)s$ for all $b \in A^2$. It remains to show that δ is a derivation of S. To this end let $s, t \in S$ and choose $0 \neq A \vartriangleleft R$ with $As, At \subseteq R$. Since $A^3s \subseteq A^2$ and $A^2st \subseteq R$ it follows that for all $b \in A^4$

$$b\delta(st) = \delta(bst) - \delta(b)st = \delta(bs)t + bs\delta(t) - \delta(b)st$$
$$= b\delta(s)t + \delta(b)st + bs\delta(t) - \delta(b)st = b\big(\delta(s)t + s\delta(t)\big)$$

and thus $\delta(st) = \delta(s)t + s\delta(t)$.

It is easy to see that σ and δ both map $Q_s(R)$ to $Q_s(R)$. For uniqueness it suffices to assume that σ is the identity on R and that δ is the zero map. If $s \in S$ with $As \subseteq R$ then for all $a \in A$ we have $as = (as)^\sigma = a^\sigma s^\sigma = as^\sigma$ so $A(s - s^\sigma) = 0$ and σ is the identity on S. Similarly $0 = \delta(as) = \delta(a)s + a\delta(s) = a\delta(s)$ so $A\delta(s) = 0$ and $\delta(s) = 0$ as required.

(iii) Again let $S = Q_\ell(R)$ and fix $q \in \mathbf{C}_S(R)$. If $s \in S$ with $As \subseteq R$, then for all $a \in A$

$$(aq)s = (qa)s = q(as) = (as)q$$

so $A(qs - sq) = 0$. Thus $qs = sq$ and we conclude from Proposition 10.6 that $\mathbf{Z}(Q_\ell) = \mathbf{Z}(Q_s) = \mathbf{C}_S(R)$.

Now if $q \neq 0$ choose $0 \neq B \triangleleft R$ with $Bq \subseteq R$. In this case $0 \neq qB = Bq \triangleleft R$ so $\ell_R(q) = 0$ and we can define $f : Bq \to R$ by $(bq)f = b$. It follows that f represents an element $q' \in S$ with $qq' = q'q = 1$. Thus $q' = q^{-1} \in \mathbf{Z}(S)$ and $\mathbf{Z}(S)$ is a field. ∎

The field C above is called the *extended centroid* of R and the subring RC of $Q_\ell(R)$ is the *central closure* of R. It is a result of [**116**] (see Exercises 7 and 8) that RC is a prime ring equal to its central closure. In other words, this is a closure operation.

On the other hand, none of the quotients Q_ℓ, Q_r or Q_s is a closure operator. Indeed, let K be a field and define

$$R = K[t][x, y \mid xy = tyx].$$

Then it was shown in [**171**, Section 4] that

$$Q_\ell(R) = K(t)[x^{-1}, x, y \mid xy = tyx]$$
$$Q_r(R) = K(t)[x, y, y^{-1} \mid xy = tyx]$$

and

$$Q_s(R) = K(t)[x, y \mid xy = tyx].$$

Furthermore if S denotes any of the above rings then

$$Q_\ell(S) = Q_r(S) = Q_s(S) = K(t)[x^{-1}, x, y, y^{-1} \mid xy = tyx]$$

is properly larger than S.

In the next section we will continue with additional examples and computations. The true importance of the symmetric Martindale ring of quotients will not begin to emerge until Section 12.

EXERCISES

1. Let R be a prime ring. Show that the definitions of $\hat{f} + \hat{g}$ in $Q_\ell(R)$ and $\hat{f}\hat{g}$ make sense and respect the equivalence relation. Furthermore verify that the ring axioms are satisfied. To this end,

note that two functions are equivalent if they agree on any nonzero ideal contained in both of their domains.

2. Let R be any ring with 1 and let $\mathcal{F}$ be a nonempty multiplicatively closed family of two-sided ideals of R each with zero left and right annihilator. Consider the set of all $f: {}_R A \to {}_R R$ with $A \in \mathcal{F}$ and let two such functions be equivalent if they agree on any $C \in \mathcal{F}$ which is contained in both their domains. Show that the set, $R_{\mathcal{F}}$, of equivalence classes $\hat{f}$ of functions is an overring of R and determine its basic properties.

3. If R is semiprime, show that $\mathcal{F} = \{\, A \triangleleft R \mid A_R \text{ ess } R_R \,\}$ satisfies the hypotheses of the preceding exercise. In this case, $R_{\mathcal{F}}$ is the Martindale ring of quotients defined in [**4**].

4. Let K be a field, let $\mathrm{M}_\infty(K)$ denote the set of all infinite matrices over K and let I be the set of finite matrices in $\mathrm{M}_\infty(K)$. If $R = K + I$, show that R is a prime ring with unique nontrivial ideal I. Furthermore prove that

$$Q_\ell(R) = \{\, \text{row finite matrices} \,\},$$

$$Q_r(R) = \{\, \text{column finite matrices} \,\},$$

and

$$Q_s(R) = \{\, \text{row and column finite matrices} \,\}.$$

The proof of Lemma 10.8(ii) should be helpful.

5. In the above example, let $S = Q_s(R)$. Show that $I \triangleleft S$ and determine $Q_\ell(S)$, $Q_r(S)$ and $Q_s(S)$. Is I an ideal of $Q_\ell(R)$?

6. Let R be a prime ring. An element $0 \neq a \in R$ is said to be *normal* if $aR = Ra$. Show that any such a is regular in R, invertible in $Q_s(R)$ and that $r \mapsto a^{-1}ra$ is an automorphism of R.

7. Suppose R is a prime ring and $f: {}_R A \to {}_R R$ is given. Show that $\hat{f}$ is contained in the extended centroid of R if and only if f is a bimodule homomorphism, that is if $f: {}_R A_R \to {}_R R_R$.

8. Again let R be a prime ring and let $S = RC$ be its central closure. Show that S is prime and choose z in its extended centroid C'. If $Az \subseteq S$ for some $0 \neq A \triangleleft S$, prove that $J = \{\, a \in A \cap R \mid az \in R \,\}$ is a nonzero ideal of R and that right multiplication by z yields a bimodule homomorphism ${}_R J_R \to {}_R R_R$. Deduce that there exists

$c \in C$ with $J(z - c) = 0$ and, since $z - c$ belongs to the field C', that $z = c$. Conclude, as in [116], that S is its own central closure.

11. Separated Groups

In our later studies we will discover that the elements between R and $Q_s(R)$ seem to cause the most difficulty. Thus it is of interest to obtain conditions which guarantee that no such elements exist. Specifically, if R is a prime ring, we say that R is *symmetrically closed* if $R = Q_s(R)$. Our goal in this section is to obtain reasonable sufficient conditions for a crossed product $R*G$ to be symmetrically closed.

To start with, we will assume that R is a prime ring which is itself symmetrically closed. Then we will assume that $\Delta^+(G) = \langle 1 \rangle$ to guarantee that $R*G$ is prime. But this is not sufficient. We will require conditions on G which assert that certain types of separation occur via conjugation. Furthermore we will have to avoid anomalous behavior in the ring R.

We begin with the group theoretic aspects.

Lemma 11.1. Let $K[G]$ be a prime group algebra. If $K[G]$ is symmetrically closed, then $\Delta(G) = \langle 1 \rangle$.

Proof. If $K[G]$ is symmetrically closed then, by Lemma 10.9(iii), $Z = \mathbf{Z}(K[G])$ is a field. Now $K[G]$ is prime so we have $\Delta^+(G) = \langle 1 \rangle$ and hence $\Delta(G)$ is torsion free abelian. Thus all units in $K[\Delta]$ are trivial, that is of the form kx for some $0 \neq k \in K$ and $x \in \Delta$. In particular, $Z \subseteq K[\Delta]$ must consist of 0 and trivial units. But Z is close under addition, so this yields $Z = K$. Finally if $x \in \Delta$, then its class sum α is central so $\alpha \in Z = K$ and $x = 1$. ∎

Therefore it is appropriate to assume that $\Delta(G) = \langle 1 \rangle$ so that G has no finite conjugacy classes other than the identity class. As will be apparent, it is the existence of countable conjugacy classes which keep $R*G$ from being symmetrically closed.

Lemma 11.2. *Let G be a group. The following are equivalent.*

> i. *Every nontrivial conjugacy class of G is uncountable.*
>
> ii. *Every nontrivial normal subgroup of G is uncountable.*
>
> iii. *If H is a countable subgroup of G, then $\mathrm{core}_G(H) = \langle 1 \rangle$.*
>
> iv. *If H is a countable subgroup of G and A is a finite subset of G, then there exists $t \in G$ with $A^t \cap H \subseteq \{1\}$.*

Proof. Since any countable normal subgroup is a union of countable conjugacy classes and since any countable conjugacy class generates a countable normal subgroup, it is clear that (i), (ii) and (iii) are equivalent. Here of course $\mathrm{core}_G(H)$ is the largest normal subgroup of G contained in H. Thus if $x \in \mathrm{core}_G(H)$, then $\{x\}^t \subseteq H$ for all $t \in G$, so (iv) implies (iii).

Finally assume (i), let A and H be given as in part (iv) and let $a_1, a_2, \ldots, a_n$ be the nonidentity elements of A. Observe that

$$\{t \in G \mid a_i^t \in H\} = \bigcup_{j=1}^{\infty} \mathbf{C}_G(a_i)x_{i,j}$$

is a (possibly empty) union of countably many right cosets of $\mathbf{C}_G(a_i)$. If

$$\bigcup_{i=1}^{n} \bigcup_{j=1}^{\infty} \mathbf{C}_G(a_i)x_{i,j} = G$$

then we conclude from Lemma 6.2 that $|G : \mathbf{C}_G(a_i)|$ is countable for some i, contradicting part (i). Thus there exists $t \in G$ not in this union and, for this t, we have $A^t \cap H \subseteq \{1\}$. ∎

We say that G is *separated* if it satisfies any of the four equivalent conditions above. It follows from (ii) that every uncountable simple group is separated.

We also consider a somewhat weaker condition. A group G is *weakly separated* if for every finitely generated subgroup H of G and every finite subset A of G there exists $t \in G$ with $A^t \cap H \subseteq \{1\}$. Note that this notation is slightly different from that of [**171**]. Some additional examples not covered by Lemma 11.2 are as follows.

Lemma 11.3. *If G is a free product of infinitely many nontrivial groups, or if G is a locally finite group with $\Delta(G) = 1$, then G is weakly separated.*

Proof. In the first case, we may assume that $G = F_1 * F_2 * \cdots$ is a countably infinite free product. Given H, A we may assume that $H, A \subseteq F_1 * F_2 * \cdots * F_n$. If $t \in F_{n+1}$ with $t \neq 1$, then $A^t \cap H \subseteq \{1\}$.

In the second case, since finitely generated subgroups of G are finite, the same proof as the implication (i) $\Rightarrow$ (iv) of Lemma 11.2 applies. ∎

It follows that if F is free of infinite rank and G is arbitrary, then the free product $F * G$ is weakly separated. Furthermore, if G is weakly separated then $\Delta(G) = \langle 1 \rangle$ since G has no nontrivial finitely generated normal subgroups.

Now let us turn to the ring theoretic aspects. It will be apparent that we must assume a condition for R somewhat stronger than primeness. Since the precise condition needed is not completely clear, we will choose one which is at least familiar. A ring R is said to be (right and left) *strongly prime* if for every $0 \neq a \in R$ there exists a finite subset T of R such that $\ell_R(Ta) = 0 = r_R(aT)$. We call T a (finite) *insulator* for a. A strongly prime ring is, of course, necessarily prime.

Lemma 11.4. *Let $R*G$ be a prime crossed product and let $q \in Q_s(R*G)$. Suppose there exist $0 \neq A, B \triangleleft R$ with $Aq, qB \subseteq R*G$. If R is strongly prime and symmetrically closed, then $q \in R*G$.*

Proof. Fix a group element $x \in G$. We consider the $\bar{x}$-coefficients in Aq and qB. For each $a \in A$, $b \in B$ write $aq = (af)\bar{x} + \cdots$ and $qb = \bar{x}(gb) + \cdots$. Then clearly $f: {}_R A \to {}_R R$ and $g: B_R \to R_R$. Furthermore, $(aq)b = a(qb)$ yields $(af)^\sigma b = a^\sigma(gb)$ where σ is the automorphism $\bar{x}$. Now by Proposition 10.2(iv), there exists $s \in Q_\ell(R)$ with $af = as$ for all $a \in A$. Hence since σ extends to an automorphism of $Q_\ell(R)$ we have

$$a^\sigma(gb) = (af)^\sigma b = (as)^\sigma b = a^\sigma s^\sigma b$$

so $A^\sigma(s^\sigma b - gb) = 0$. Thus $s^\sigma B = gB \subseteq R$ so $s^\sigma \in Q_s(R) = R$, since R is symmetrically closed, and hence $s \in R$.

We have therefore shown that for each $x \in G$ there exists $r_x \in R$ with $aq = \sum_{x \in G} ar_x \bar{x}$. Of course, since the right hand term here is in $R*G$, it follows that for any $a \in A$ the coefficients ar_x are almost all zero. We show that the r_x's are almost all zero. To this end, let $0 \neq a \in A$ and let T be a finite insulator for a. Then $aT \subseteq A$ and, for each $t \in T$, almost all atr_x are zero. Since T is finite this implies that, for almost all $x \in G$, $aTr_x = 0$ and the claim is proved. We can now set $\alpha = \sum_{x \in G} r_x \bar{x} \in R*G$. Then the above yields $aq = a\alpha$ so $A(q - \alpha) = 0$. Since R is prime, it follows from the freeness of $R*G$ over R that $r_{R*G}(A) = 0$. Hence, by Lemma 10.7(ii), A is right regular in $Q_s(R*G)$ and we conclude that $q = \alpha \in R*G$. ∎

We will need the following elementary Δ-lemma for two different applications. The proof is easy since we assume $\Delta(G) = \langle 1 \rangle$.

Lemma 11.5. *Let $R*G$ be given with $G \neq \langle 1 \rangle$ and $\Delta(G) = \langle 1 \rangle$. Suppose S is a subring of R, $0 \neq \alpha, \beta \in R*G$ and F is a finite subset of G. Assume that some (right) coefficient a of α satisfies $r_R(aS) = 0$. Then there exist $s \in S$, $g \in G$ such that $\delta = \alpha(s\bar{g})\beta \neq 0$ and has support disjoint from F.*

Proof. Set $X = \text{Supp } \alpha$ and $Y = \text{Supp } \beta$ and let $g \in G$. Note that $XgY \cap F \neq \emptyset$ implies that $g \in X^{-1}FY^{-1}$ and the latter is a finite set which we view as a finite union of cosets of the identity subgroup. Next if $x_1 gy_1 = x_2 gy_2$ for distinct elements $x_1, x_2 \in X$ and $y_1, y_2 \in Y$, then $g^{-1}(x_2^{-1}x_1)g = y_2 y_1^{-1}$ so g is contained in a fixed right coset of $\mathbf{C}_G(x_2^{-1}x_1)$ depending only on x_1, x_2, y_1 and y_2.

Since $G \neq \langle 1 \rangle$ and $\Delta(G) = \langle 1 \rangle$ it follows that $|G : \langle 1 \rangle| = \infty$ and $|G : \mathbf{C}_G(x_2^{-1}x_1)| = \infty$. Thus, by Lemma 6.2, we can choose $g \in G$ not in any of the above right cosets. For this g, we have $XgY \cap F = \emptyset$ and $x_1 gy_1 = x_2 gy_2$ implies $x_1 = x_2$ and $y_1 = y_2$. It follows that for all $s \in S$, the element $\delta = \alpha s\bar{g}\beta$ has support disjoint from F and that there is no cancellation between terms in the product.

Finally if a is the coefficient of $\bar{x}_1$ in α and b is the (left) coefficient of y_1 in β, then δ contains the term $\bar{x}_1 a \cdot s\bar{g} \cdot b\bar{y}_1 = \bar{x}_1 \cdot asc \cdot \bar{g}\bar{y}_1$, where $c = \bar{g}b\bar{g}^{-1} \in R \setminus 0$, and this term can be chosen to be nonzero since, by assumption, $aSc \neq 0$. ∎

The next lemma explains how the subring S comes into play.

Lemma 11.6. *Let $R*G$ be given with $\Delta(G) = \langle 1 \rangle$ and with R strongly prime. Suppose $q \in Q_s(R*G)$ and $0 \neq \alpha \in R*G$ with $\alpha(R*G)q \subseteq R*G$. Then there exists a countable subring $S \subseteq R$ and a countable subgroup H of G such that*
   i. $S*H$ is a sub-crossed product of $R*G$,
   ii. $\alpha \in S*H$ and $\Delta(H) = \langle 1 \rangle$,
   iii. $\alpha(S*H)q \subseteq S*H$, and
   iv. for some coefficient a of α we have $r_R(aS) = 0$.

Proof. Let a be a nonzero coefficient of α and let T be a finite insulator. Then $r_R(aT) = 0$ and, since S will be chosen to contain T, (iv) follows. Set $H_0 = \langle \text{Supp } \alpha \rangle$ and let $S_0 \subseteq R$ be generated by T, the coefficients of α, all twistings of pairs of elements in H_0 and all conjugates of these elements under the action of H_0. Since H_0 is countable, S_0 is countably generated and hence countable. Certainly S_0*H_0 is a sub-crossed product of $R*G$ and $\alpha \in S_0*H_0$.

We construct an ascending sequence of countable sub-crossed products S_n*H_n so that each element of $H_n^{\#} = H_n \setminus \{1\}$ has infinitely many conjugates in $H_{n+1} \supseteq H_n$ and with $\alpha(S_n*H_n)q \subseteq S_{n+1}*H_{n+1}$. To this end, suppose S_n*H_n is given. Since $\Delta(G) = \langle 1 \rangle$, for each $h \in H^{\#}$ we can find a countably infinite subset W_h of coset representatives for $\mathbf{C}_G(h)$ in G. Now let H_{n+1} be generated by H_n, the sets W_h for all $h \in H_n^{\#}$ and the countably many supports in $\alpha(S_n*H_n)q$. It is clear that H_{n+1} is countable. In the same way, we let S_{n+1} be generated by S_n, the countably many coefficients in $\alpha(S_n*H_n)q$, all twistings of pairs of elements of H_{n+1} and all images of these elements under the action of H_{n+1}. Thus $S_{n+1}*H_{n+1}$ is a countable crossed product.

Finally, let $H = \bigcup_{n=0}^{\infty} H_n$ and $S = \bigcup_{n=0}^{\infty} S_n$. Then it is clear that $S*H$ has the appropriate properties. ∎

It is now a simple matter to prove

Theorem 11.7 [171] *Let $R*G$ be given with R strongly prime and symmetrically closed and with G separated. Then $R*G$ is symmetrically closed.*

Proof. We may assume that $G \neq \langle 1 \rangle$. Since G is separated, $\Delta(G) = \langle 1 \rangle$ so $R*G$ is prime by Proposition 8.3(ii) (or by Lemma 11.5 with $R = S$ and $F = \emptyset$). Let $q \in Q_s(R*G)$ and let $0 \neq I \lhd R*G$ with $qI, Iq \subseteq R*G$. If $\alpha \in I \setminus 0$, then $\alpha(R*G)q \subseteq R*G$ so we can apply Lemma 11.6 to obtain the sub-crossed product $S*H$ with appropriate properties. In particular, $\alpha \in S*H$, $\Delta(H) = \langle 1 \rangle$, $\alpha(S*H)q \subseteq S*H$ and H is countable. Choose $\delta \in I$ with $1 \in$ Supp δ. Since Supp $\bar{t}^{-1}\delta\bar{t} = t^{-1}(\text{Supp } \delta)t$ and G is a separated group, it follows from Lemma 11.2(iv) that there exists $t \in G$ with Supp $\bar{t}^{-1}\delta\bar{t} \cap H = \{1\}$. But $\bar{t}^{-1}\delta\bar{t} \in I$ so we can clearly assume that Supp $\delta \cap H = \{1\}$.

Let $0 \neq b$ be the identity coefficient of δ, let $r \in R$ and note that $qr\delta \subseteq qI \subseteq R*G$. If $\gamma \in \alpha(S*H) \subseteq S*H$ then $\gamma q \in S*H$ and $(\gamma q)r\delta = \gamma(qr\delta)$. Applying the projection map $\pi_H : R*G \to R*H$ yields $(\gamma q)r\pi_H(\delta) = \gamma\pi_H(qr\delta)$. Note that $\pi_H(\delta) = b$ since Supp $\delta \cap H = \{1\}$ and that $\sigma_r = \pi_H(qr\delta) \in R*H$. Thus we have $\gamma qrb = \gamma\sigma_r$ for all $\gamma \in \alpha(S*H)$ and $r \in R$. In particular, for fixed r we have $\alpha(S*H)(qrb - \sigma_r) = 0$.

Now by Lemma 11.5 applied to α and $S*H \subseteq R*H$ we see that $r_{R*H}(\alpha(S*H)) = 0$. This of course uses the fact, by Lemma 11.6(iv), that $r_R(aS) = 0$ for some nonzero coefficient a of α. Thus since $R*G$ is free over $R*H$ we have $r_{R*G}(\alpha(S*H)) = 0$ and hence, by Lemma 10.7(ii), $\alpha(S*H)$ is right regular in $Q_s(R*G)$. We conclude from the above that $qrb = \sigma_r \in R*H$. In other words, if $B = RbR \lhd R$ then $B \neq 0$ and $qB \subseteq R*G$. By symmetry we also obtain a nonzero ideal A of R with $Aq \subseteq R*G$ and Lemma 11.4 yields the result. ∎

In a similar manner we prove

Theorem 11.8. [171] *Let $R*G$ be a strongly prime ring. Suppose R is strongly prime and symmetrically closed and that G is weakly separated. Then $R*G$ is symmetrically closed.*

Proof. Let $q \in Q_s(R*G)$ and choose $0 \neq I \lhd R*G$ with $qI, Iq \subseteq R*G$. Let $0 \neq \alpha \in I$ and let $T = \{\tau_1, \tau_2, \ldots, \tau_n\}$ be an insulator for α.

Then $\alpha \tau_i q \in R*G$ for all i and we set $H = \langle \text{Supp } \alpha \tau_i, \text{Supp } \alpha \tau_i q \mid i = 1, 2, \ldots, n \rangle$. Since H is finitely generated and G is weakly separated, there exists $\delta \in I$ with $\text{Supp } \delta \cap H = \{1\}$.

Let $0 \neq b$ be the identity coefficient of δ and let $r \in R$. By applying the projection map π_H to $(\alpha \tau_i q)r\delta = \alpha \tau_i (qr\delta)$ we obtain $\alpha \tau_i qrb = \alpha \tau_i \sigma_r$ where $\sigma_r = \pi_H(qr\delta)$. Thus $\alpha T(qrb - \sigma_r) = 0$ and since αT is right regular in $Q_s(R*G)$, by Lemma 10.7(ii), we conclude that $qB \subseteq R*G$ where $B = RbR$ is a nonzero ideal of R. Similarly there exists $0 \neq A \lhd R$ with $Aq \subseteq R*G$ and Lemma 11.4 implies that $q \in R*G$. ∎

In particular, suppose R is a symmetrically closed domain and G is a free group of infinite rank. Since G is an ordered group, we see that $R*G$ is also a domain. Hence Lemma 11.3 and Theorem 11.8 imply that $R*G$ is symmetrically closed. We will discuss the case of finitely generated free groups later on.

If we take a close look at the proof of Lemma 11.4 we arrive at a useful, but unfamiliar, definition. Let R be a ring and let G permute its ideals. We say that R is *not G-cohesive* if there exists nonzero G-invariant ideals A, B of R and an infinite sequence $r_1, r_2, \ldots$ of nonzero elements of R such that for all $a \in A$, $b \in B$ there are at most finitely many subscripts i with $ar_i \neq 0$ or $r_i b \neq 0$. If the above situation can never occur then, of course, R is *G-cohesive*. In case $G = \langle 1 \rangle$, all ideals of R are G-invariant, and we use *cohesive* instead of $\langle 1 \rangle$-cohesive. Note that a cohesive ring is necessarily prime (Exercise 5) and that cohesive implies G-cohesive for all G.

It turns out that Lemma 11.4 only requires that R be cohesive and symmetrically closed. Indeed it was essentially shown in the proof that strongly prime implies cohesive. More to the point, we have

Proposition 11.9. *Let $R*G$ be a prime crossed product.*

    i. *If $R*G$ is symmetrically closed and G is infinite, then R is G-cohesive.*

    ii. *Assume that every nonzero ideal of $R*G$ meets R nontrivially. If R is prime, symmetrically closed and G-cohesive, then $R*G$ is symmetrically closed.*

Proof. (i) Suppose R is not G-cohesive. Then there exist nonzero G-invariant ideals A and B of R and a sequence $r_1, r_2, \ldots$ of nonzero elements of R such that, for all $a \in A$, $b \in B$, there are at most finitely many subscripts i with $ar_i \neq 0$ or $r_i b \neq 0$. Since G is infinite, we can let $x_1, x_2, \ldots$ be a sequence of distinct group elements. Let γ be the formal infinite sum $\gamma = \sum_{i=1}^{\infty} r_i s_i \bar{x}_i r_i$ where each s_i is chosen in the prime ring R so that $r_i s_i \bar{x}_i r_i \neq 0$.

Notice that for each $a \in A$ we have $a\gamma \in R*G$. Hence since A is G-invariant, formal right multiplication by γ defines a map $f\colon A*G \to R*G$ which is certainly a left $R*G$-module homomorphism. Similarly, formal left multiplication by γ defines a right $R*G$-module homomorphism $g\colon B*G \to R*G$. Since the balanced condition is surely satisfied, we deduce from Proposition 10.4(iv) that there exists $q \in Q_s(R*G)$ with $aq = a\gamma$ for all $a \in A$. Finally if $q \in R*G$, then Supp $Aq \subseteq$ Supp q is finite. On the other hand, since R is prime, we see that Supp $A\gamma = \{x_1, x_2, \ldots\}$ is infinite. Thus $q \notin R*G$ and $R*G$ is not symmetrically closed.

(ii) Let $q \in Q_s(R*G)$ and choose nonzero ideals $I, J \lhd R*G$ with $Iq, qJ \subseteq R*G$. By assumption, $A = I \cap R \neq 0$ and $B = J \cap R \neq 0$ and clearly these are G-invariant ideals of R. Since $Aq, qB \subseteq R*G$ and R symmetrically closed, we deduce as in Lemma 11.4 that there exist elements $r_x \in R$ such that for all $a \in A$, $b \in B$ we have

$$aq = \sum_{x \in G} ar_x \bar{x}$$

$$qb = \sum_{x \in G} r_x \bar{x} b = \sum_{x \in G} \bar{x} r'_x b$$

where $r'_x = \bar{x}^{-1} r_x \bar{x}$. The goal is to show that $r_x = 0$ for almost all group elements x.

If this were not the case, let $r_1, r_2, \ldots$ be an infinite sequence of nonzero coefficients and choose $s_i \in R$ so that $\delta_i = r_i s_i r'_i \neq 0$. Then $\delta_1, \delta_2, \ldots$ is an infinite sequence of nonzero elements of R with the property that, for all $a \in A$, $b \in B$, there are at most finitely many subscripts i with $a\delta_i \neq 0$ or $\delta_i b \neq 0$. This is a contradiction since R is G-cohesive and we can now proceed as in Lemma 11.4. ∎

Corollary 11.10. [171] *Let $K[G]$ be a prime group algebra with G locally finite. Then $K[G]$ is symmetrically closed if and only if G is separated.*

Proof. If G is separated, then $K[G]$ is symmetrically closed by Theorem 11.7. Conversely suppose G is not separated and let N be a countably infinite normal subgroup of G. Note that $K[G] = R*(G/N)$ where $R = K[N]$.

Since N is locally finite and countable, we can write $N = \bigcup_1^\infty H_i$, an ascending union of finite subgroups, and we let $r_i = \hat{H}_i$ denote the sum of the elements of H_i in $R = K[N]$. If I is the augmentation ideal of $K[N]$ and $\alpha \in I$, then $\alpha \in K[H_m]$ for some m and then $0 = \alpha\hat{H}_k = \hat{H}_k\alpha$ for all $k \geq m$. Thus since I is (G/N)-invariant, we see that R is not (G/N)-cohesive. By Proposition 11.9(i), $K[G] = R*(G/N)$ is not symmetrically closed. ∎

It is tempting to try to modify the proof of Proposition 11.9(i) to show that if R is a symmetrically closed prime ring, then it must be cohesive. However this is not the case. For a counterexample, merely take R to be the ring of row and column finite matrices in $M_\infty(K)$ (see Exercise 4). On the other hand, as we see below, crossed products are not counterexamples.

Theorem 11.11. [171] *Let $R*G$ be a symmetrically closed crossed product with R prime. If $G \neq \langle 1 \rangle$ and $\Delta(G) = \langle 1 \rangle$, then $R*G$ is cohesive.*

Proof. Suppose by way of contradiction that $R*G$ is not cohesive. Then there exist nonzero ideals A and B of $R*G$ and a sequence of nonzero elements $\gamma_1, \gamma_2 \ldots$ of $R*G$ such that, for all $\alpha \in A$ and $\beta \in B$, there are at most finitely many subscripts i with $\alpha\gamma_i \neq 0$ or $\gamma_i\beta \neq 0$. We proceed in a series of steps.

Step 1. *We may assume that there exist $\alpha_1, \alpha_2, \ldots \in A$ such that $\alpha_i\gamma_i \neq 0$ and $\alpha_i\gamma_j = 0$ if $j > i$.*

Proof. Suppose we have already found $\alpha_1, \alpha_2, \ldots, \alpha_{n-1} \in A$ satisfying the above. Since $\gamma_n \neq 0$ and $R*G$ is prime, there exists $\alpha_n \in A$

with $\alpha_n \gamma_n \neq 0$. Now there are at most finitely many i with $\alpha_n \gamma_i \neq 0$ so we can delete those γ_i with $i > n$ and $\alpha_n \gamma_i \neq 0$. We then renumber the remaining γ_i with $i > n$ and continue the process. ∎

Step 2. *We can assume that*

$$|\text{Supp } \alpha_n(\gamma_1 + \gamma_2 + \cdots + \gamma_n)| > n|\text{Supp } \alpha_n|.$$

Proof. We will replace each γ_i in turn by a suitable $\gamma_i \delta_i \gamma_i$ with $\alpha_i(\gamma_i \delta_i \gamma_i) \neq 0$. This will certainly maintain the earlier properties of the gamma sequence. Suppose $\gamma_1, \gamma_2, \ldots, \gamma_{n-1}$ have already been modified. The goal is to find δ so that

$$\alpha_n(\gamma_1 + \gamma_2 + \cdots + \gamma_{n-1} + \gamma_n \delta \gamma_n)$$

has large support. Since $\alpha_n(\gamma_1 + \gamma_2 + \cdots + \gamma_{n-1})$ is fixed, we need only find δ so that $\alpha_n \gamma_n \delta \gamma_n$ has large support. This is achieved using Lemma 11.5 with $S = R$ as follows.

Suppose δ' is given. Then by Lemma 11.5 there exists $r \in R$ and $g \in G$ such that $\alpha_n \gamma_n r \bar{g} \gamma_n \neq 0$ and has support disjoint from that of $\alpha_n \gamma_n \delta' \gamma_n$. Thus setting $\delta = \delta' + r\bar{g}$ we see that

$$|\text{Supp } \alpha_n \gamma_n \delta \gamma_n| > |\text{Supp } \alpha_n \gamma_n \delta' \gamma_n|.$$

In other words, we have an inductive procedure for enlarging the support of such products and thus the required element δ will certainly exist. ∎

Step 3. *The formal sum $s = \sum_{i=1}^{\infty} \gamma_i$ defines an element of $Q_s(R*G)$ not contained in $R*G$. Hence $R*G$ is not symmetrically closed.*

Proof. As usual formal right and left multiplication by s yield balanced module homomorphisms $f: A \to R*G$ and $g: B \to R*G$. Thus there exists an element $q \in Q_s(R*G)$ with $\alpha q = \alpha s$ for all $\alpha \in A$. Suppose by way of contradiction that $q \in R*G$. Then for all n

$$\alpha_n q = \alpha_n(\gamma_1 + \gamma_2 + \cdots + \gamma_n)$$

by Step 1. Since

$$|\text{Supp } \alpha_n||\text{Supp } q| \geq |\text{Supp } \alpha_n q|$$

and

$$|\text{Supp } \alpha_n(\gamma_1 + \gamma_2 + \cdots + \gamma_n)| > n|\text{Supp } \alpha_n|$$

by Step 2, we obtain a contradiction when $n \geq |\text{Supp } q|$. This completes the proof of the theorem. ∎

There are additional group algebra examples computed in [171]. Of particular interest is the following which we state without proof.

Theorem 11.12. [171] *Let $K[G]$ be a prime group algebra with G polycyclic-by-finite. Then $Q_s(K[G]) = K[G]Z^{-1}$ where $Z = \mathbf{Z}(K[G])$. Moreover, this is the central closure of $K[G]$.*

We will consider free rings in Section 13.

EXERCISES

1. Let G be an algebraically closed group and let $a_1, a_2, \ldots, a_n$ and $b_1, b_2, \ldots, b_m$ be nonidentity elements of G. Show that there exist $x, y \in G$ such that x centralizes all b_j, but x centralizes no a_i^y. Deduce that G is weakly separated.

2. Prove that R is strongly prime if and only if every nonzero ideal contains a finite subset F with $\ell_R(F) = 0 = \mathbf{r}_R(F)$.

3. Let $K\langle\zeta_1, \zeta_2, \ldots\rangle$ be the free K-algebra on infinitely many variables and let J be the homogeneous ideal generated by all products $\zeta_i\zeta_j\zeta_k$ with $i > j > k$. Show that $R = K\langle\zeta_1, \zeta_2, \ldots\rangle/J$ is right strongly prime but not left strongly prime. This is an example of paper [70].

4. Let $S = \{\text{row and column finite matrices in } \mathbf{M}_\infty(K)\}$. In Exercise 5 of the preceding section it was shown that S is symmetrically closed and that $I = \{\text{finite matrices}\}$ is an ideal of S. Prove that S is not cohesive.

5. Prove that cohesive implies prime. To this end, if $AB = 0$ first consider a constant sequence with all $r_i = ba$ to conclude that $BA = 0$. Then consider a constant sequence with all $r_i = b$. Show by example that cohesive does not imply strongly prime. Here we can let $R = K[G]$ with G a locally finite, uncountable simple group.

6. Let R be prime but not cohesive and let A, B and $r_1, r_2, \ldots$ be given. Show by induction on n that, by deleting terms if necessary, we have

   i. $s_{n,\epsilon} = \sum_1^n \epsilon_i r_i \neq 0$ for all $\epsilon_i = 0, \pm 1$ which are not all zero,

   ii. there exist $a_{n,\epsilon} \in A$ with $a_{n,\epsilon} s_{n,\epsilon} \neq 0$,

   iii. $a_{n,\epsilon} r_j = 0$ for all $j > n$.

For each set I of positive integers, let $\alpha_I = \sum_{i \in I} r_i$. Prove that the various α_I determine distinct elements of $Q_s(R)$ and conclude that $Q_s(R)$ is uncountable.

7. Let $R*G$ be given with R prime, countable and symmetrically closed. If G is separated, deduce that $R*G$ is symmetrically closed.

12. X-Inner Automorphisms

We now come to the key property of the symmetric ring of quotients. It appears in [**116**] in the special case $\sigma = 1$. The same proof yields this more general observation of [**87**].

Lemma 12.1. Let σ be an automorphism of the prime ring R and let a, b, c, d be fixed nonzero elements of $Q_\ell(R)$. If

$$arb = cr^\sigma d$$

for all $r \in R$, then there exists a unit $q \in Q_s(R)$ with $c = aq$, $d = q^{-1}b$ and $r^\sigma = q^{-1}rq$ for all $r \in R$.

Proof. Choose a nonzero ideal J of R with $Ja, Jc \subseteq R$ and set $A = JaR$ and $C = JcR$. Then A and C are nonzero ideals of R and we define $f: A \to C$ and $g: C \to A$ by

$$f: \sum_i x_i a y_i \mapsto \sum_i x_i c y_i^\sigma$$

and

$$g: \sum_i x_i cy_i \mapsto \sum_i x_i ay_i^{\sigma^{-1}}$$

with $x_i \in J$ and $y_i \in R$. To see that f is well defined, suppose $\sum_i x_i ay_i = 0$. Then for all $r \in R$ the formula $atb = ct^\sigma d$ yields

$$0 = \left(\sum_i x_i ay_i\right) rb = \left(\sum_i x_i cy_i^\sigma\right) r^\sigma d$$

and hence $\sum_i x_i cy_i^\sigma = 0$ since R is prime and $d \neq 0$. Similarly g is well defined and since both are clearly left R-module homomorphisms, there exist $q, q' \in Q_\ell(R)$ which represent f and g respectively. Since $fg = 1$ on A and $gf = 1$ on C, it follows that $q' = q^{-1}$.

Now for all $x \in J$ we have $(xa)q = (xa)f = xc$ so $J(aq - c) = 0$ and $aq = c$. Similarly for all $x \in J$ and $y, r \in R$ we have

$$(xcy)q^{-1}rq = (xay^{\sigma^{-1}})rq$$
$$= \left[xa(y^{\sigma^{-1}}r)\right]q = (xcy)r^\sigma$$

so $C(q^{-1}rq - r^\sigma) = 0$ and $q^{-1}rq = r^\sigma$. Moreover this shows that $q \in Q_s(R)$ since $Iq \subseteq R$ implies $qI^\sigma \subseteq R$. Finally

$$arb = cr^\sigma d = aq \cdot q^{-1}rq \cdot d = ar(qd)$$

so $aR(b - qd) = 0$ and $b = qd$. ∎

A somewhat disguised version of this result is

Lemma 12.2. Let σ be an automorphism of the prime ring R. Suppose A is a nonzero ideal of R and $f: {}_RA \to {}_RR$ is a nonzero map satisfying $(ar)f = (af)r^\sigma$ for all $a \in A$ and $r \in R$. Then there exists a unit $q \in Q_s(R)$ with $af = aq$ for all $a \in A$ and $r^\sigma = q^{-1}rq$ for all $r \in R$.

Proof. We know that there exists $q \in Q_\ell(R)$ with $af = aq$ for all $a \in A$. Hence

$$(ar)q = (ar)f = (af)r^\sigma = (aq)r^\sigma$$

so $A(rq - qr^\sigma) = 0$ and $rq = qr^\sigma$. Since $q \neq 0$, we conclude from the preceding lemma that there exists a unit $u \in Q_s(R)$ with $1 \cdot u = q$ and $u^{-1}ru = r^\sigma$. ∎

It turns out that the above two results guarantee that $Q_s(R)$ is large enough to deal with most problems in crossed products and Galois theory. Thus the automorphisms σ which occur there are of particular importance.

Definition. Let σ be an automorphism of the prime ring R. Then σ is said to be *X-inner* if there exists a unit $q \in Q_s(R)$ with $r^\sigma = q^{-1}rq$ for all $r \in R$. When this occurs, then $s^\sigma = q^{-1}sq$ for all $s \in Q_s(R)$ since any two automorphisms of $Q_s(R)$ which agree on R must be equal.

We let $\mathrm{Xinn}(R)$ be the set of X-inner automorphisms of R. Then it is clear that this is a subgroup of $\mathrm{Aut}(R)$ containing $\mathrm{Inn}(R)$, the group of ordinary inner automorphisms. Furthermore, let $\sigma \in \mathrm{Aut}(R)$ and let q be a unit of $Q_s(R)$ which induces an X-inner automorphism on R. Since σ extends to an automorphism of $Q_s(R)$, we have $(q^{-1}rq)^\sigma = (q^\sigma)^{-1}r^\sigma q^\sigma$ for all $r \in R$. This shows first that q^σ induces an X-inner automorphism of R and then that, as automorphisms, $q\sigma = \sigma q^\sigma$. Thus $\mathrm{Xinn}(R) \triangleleft \mathrm{Aut}(R)$.

Since the center of R remains central in $Q_s(R)$, it follows that $\mathrm{Xinn}(R)$ acts trivially on $\mathbf{Z}(R)$. In particular, if R is a K-algebra, then $\mathrm{Xinn}(R)$ acts as K-automorphisms. Other basic properties are as follows.

Lemma 12.3. *Let R be a prime ring.*

i. *If $\sigma \in \mathrm{Aut}(R)$, then σ is X-inner if and only if there exist nonzero elements $a, b, c, d \in R$ with $arb = cr^\sigma d$ for all $r \in R$.*

ii. *If $\eta: G \to \mathrm{Aut}(R)$ is a group homomorphism, then*

$$G_{\mathrm{inn}} = \{\, x \in G \mid x^\eta \text{ is X-inner on } R \,\}$$

is a normal subgroup of G.

iii. *If $R*G$ is a crossed product, then*

$$G_{\mathrm{inn}} = \{\, x \in G \mid \bar{x} \text{ is X-inner on } R \,\}$$

is a normal subgroup of G.

Proof. (i) If a, b, c, d are given, then σ is X-inner by Lemma 12.1. Conversely suppose $r^\sigma = q^{-1}rq$ for q a unit of $Q_s(R)$. Let $0 \neq A, B \lhd R$ with $Aq, qB \subseteq R$ and choose $a \in A$, $b \in B$ with $aq, qb \neq 0$. Then $qr^\sigma = rq$ yields $(aq)r^\sigma b = ar(qb)$ as required.

(ii) This is clear since $\mathrm{Xinn}(R) \lhd \mathrm{Aut}(R)$.

(iii) Here we recall that conjugation determines a group homomorphism from $\mathcal{G}$, the group of trivial units of $R*G$, to $\mathrm{Aut}(R)$. Thus $\mathcal{G}_{\mathrm{inn}} \lhd \mathcal{G}$ and the result follows since $\mathcal{G}_{\mathrm{inn}} \supseteq U$, the group of units of R, implies that $G_{\mathrm{inn}} = \mathcal{G}_{\mathrm{inn}}/U$. ∎

In the context of (ii) or (iii) above, we say that G is *X-inner* on R if $G_{\mathrm{inn}} = G$. Similarly, G is *X-outer* on R if $G_{\mathrm{inn}} = \langle 1 \rangle$.

We remark that not every unit of $Q_s(R)$ induces an X-inner automorphism of R. For example if $R = \mathrm{M}_n(Z)$ where Z is the ring of integers then, by Lemma 10.8(ii), $Q_s(R) = \mathrm{M}_n(Q)$ where Q is the rationals. Since Z is a principal ideal domain, it is easy to see (Exercise 1) that the only units of $\mathrm{M}_n(Q)$ which normalize R are of the form $Q^\bullet \cdot \mathrm{GL}_n(Z)$ and this is properly smaller than $\mathrm{GL}_n(Q)$.

Let R be a prime ring and let N be the subgroup of units of $Q_s(R)$ which normalize R. Then, following [128], the *normal closure* of R is defined to be RN, the R-linear span of N. It follows (Exercise 3) that RN is a prime subring of $Q_s(R)$ but that, unlike the central closure, it does not yield a closure operation. Now it will be apparent that the ring RN is sufficiently large to handle problems in crossed products and Galois theory. However, once derivations come into play, for example in the study of differential operator rings or enveloping rings (see [15,90,173,174]), one needs the larger ring $Q_s(R)$. Thus for the most part, we will restrict our attention to the symmetric ring of quotients.

Proposition 12.4. [58] *Let $R*G$ be given with R prime and let $S = Q_\ell(R)$ or $Q_s(R)$.*

i. *$R*G$ extends uniquely to a crossed product $S*G$.*

ii. *$S*G_{\mathrm{inn}} = S \otimes_C E$ where $E = \mathbf{C}_{S*G}(S) = \mathbf{C}_{S*G}(R)$ and $C = \mathbf{Z}(S)$ is the extended centroid of R.*

iii. $E \cong C^t[G_{\text{inn}}]$, *some twisted group algebra of G_{inn} over C.*

iv. *Let $\alpha \in S*G$, $\sigma \in \text{Aut}(R)$ and suppose that $r\alpha = \alpha r^\sigma$ for all $r \in R$. If $g \in \text{Supp } \alpha$, then $\bar{g}\sigma^{-1}$ induces an X-inner automorphism on R and $\alpha = \alpha_0 \bar{g}$ with $\alpha_0 \in S*G_{\text{inn}}$.*

Proof. (i) This uses the fact that every automorphism of R extends uniquely to one of S. Thus we can define $S*G$ to be the set of all formal finite sums $\sum_{x \in G} \bar{x}s_x$ with $s_x \in S$ and with the usual addition. Multiplication is defined distributively using

$$\bar{x}\bar{y} = \overline{xy}\tau(x,y),$$

where $\tau: G \times G \to \text{U}(R) \subseteq \text{U}(S)$ is the given twisting of $R*G$, and

$$s\bar{x} = \bar{x}s^{\sigma(x)}$$

where $\sigma(x) \in \text{Aut}(S)$ is the unique extension of $\sigma(x) \in \text{Aut}(R)$. The associativity of $S*G$ now follows immediately from Lemma 1.1 and the uniqueness of extension of automorphisms.

(iv) Here it is convenient to write $\alpha = \sum_x a_x \bar{x}$ with $a_x \in S$. Then for all $r \in R$ we have

$$\sum_x ra_x\bar{x} = r\alpha = \alpha r^\sigma$$
$$= \sum_x a_x\bar{x}r^\sigma = \sum_x a_x r^{\sigma\bar{x}^{-1}} \bar{x} .$$

Comparing coefficients yields $ra_x = a_x r^{\sigma\bar{x}^{-1}}$ for all $r \in R$ and it follows from Lemma 12.1 that if $a_x \neq 0$ then $\sigma\bar{x}^{-1}$ induces an X-inner automorphism on R. In particular, $\text{Supp } \alpha \subseteq (G_{\text{inn}})g$ and $\alpha = \alpha_0 \bar{g}$ as required.

(ii) It follows from (iv) above with $\sigma = 1$ that $E = \mathbf{C}_{S*G}(R) \subseteq S*G_{\text{inn}}$. Now for each $x \in G_{\text{inn}}$ choose a unit $u_x \in Q_s(R) \subseteq S$ such that $r^{\bar{x}} = u_x r u_x^{-1}$ and set $\tilde{x} = \bar{x}u_x$. Then $\tilde{x}$ centralizes R and hence also S by the uniqueness of extension. Furthermore, we have made a diagonal change of basis, so every element of $S*G_{\text{inn}}$ is uniquely of the form $\sum \tilde{x}b_x$. It now follows easily that $\beta = \sum \tilde{x}b_x \in E$ if and only if each $b_x \in C = \mathbf{C}_S(R) = \mathbf{Z}(S)$. Thus E is a C-algebra with basis $\tilde{G}$, E centralizes S and $S*G_{\text{inn}} = S \otimes_C E$.

(iii) If $\tilde{\tau}$ denotes the twisting for $\tilde{G}$, then $\tilde{x}\tilde{y} = \widetilde{xy}\tilde{\tau}(x,y)$ implies that $\tilde{\tau}(x,y) \in E$. Thus $\tilde{\tau}(x,y) \in C$ and we conclude that E is a twisted group algebra of G_{inn} over the field C. ∎

We remark that part (iv) above with $\sigma \neq 1$ is contained in [**123**]. Furthermore, it is easy to see that α can be written as $\alpha = q\beta\bar{g}$ for some unit $q \in S$ and some $\beta \in E$.

The next proposition is the crossed product interpretation of a Galois theory result of [**87**]. If $M \subseteq R*G$ we let Supp $M = \bigcup_{\alpha \in M}$ Supp α.

Proposition 12.5. *Let $R*G$ be a crossed product with R prime, let $S = \mathbf{Q}_{\ell}(R)$ or $\mathbf{Q}_s(R)$ and extend $R*G$ to $S*G$. If M is a nonzero (R,R)-subbimodule of $S*G$ with $1 \in$ Supp M, then there exists an element $a\gamma \in M \cap R*G$ such that $a \in R \setminus 0$ and $\gamma \in \mathbf{C}_{S*G}(R)$ with identity coefficient equal to 1.*

Proof. For convenience we write all coefficients on the left. If $\alpha = \sum s_x\bar{x} \in M$ with $s_1 \neq 0$, then since R is prime, there exists $r \in R$ with $rs_x \in R$ for all $x \in$ Supp α and $rs_1 \neq 0$. Thus $M' = M \cap R*G$ has the same property as M and we may assume that $M \subseteq R*G$.

Next there exists a finite subset X of G of minimal size with $1 \in$ Supp $(M \cap R*X)$. Replacing M by the smaller bimodule $M \cap R*X$, we can now assume that every $\alpha \in M$ is contained in $R*X$ and if $\alpha = \sum_{x \in X} a_x\bar{x}$ satisfies $a_y = 0$ for some y, then $a_1 = 0$. For each $x \in X$ set

$$A_x = \left\{ r \in R \mid \text{there exists } \alpha = \sum a_y\bar{y} \in M \text{ with } a_x = r \right\}.$$

Since M is an (R,R)-bimodule, it follows from the definition of X that each A_x is a nonzero ideal of R.

Fix $x \in X$. If $\alpha = \sum a_y\bar{y} \in M$, we claim that a_x uniquely determines a_1. Indeed if $\alpha' = \sum a'_y\bar{y} \in M$ satisfies $a_x = a'_x$, then $\alpha' - \alpha \in M$ has zero $\bar{x}$-coefficient and hence zero identity coefficient. This means that we have a well defined map $f_x : A_x \to A_1 \subseteq R$ given by $a_x f_x = a_1$ for all $\alpha = \sum a_y \bar{y} \in M$. Since $r\alpha = \sum r a_y \bar{y} \in M$ it follows that f_x is a left R-module homomorphism. Furthermore, from $\alpha r = \sum a_y (\bar{y} r \bar{y}^{-1}) \bar{y} \in M$ we have $(a_x r) f_x = (a_x f_x) r^{\bar{x}}$. We conclude from Lemma 11.2 that either $f_x = 0$ or there exists a unit $u_x \in \mathbf{Q}_s(R) \subseteq S$ with $a_x f_x = a_x u_x^{-1}$ and $r^{\bar{x}} = u_x r u_x^{-1}$. In particular, $\bar{x}$ commutes with u_x. Note that $f_1 = 1$ so $u_1 = 1$.

Finally choose $\alpha \in M$ with $1 \in \text{Supp } \alpha$. Then $\alpha = \sum a_x \bar{x}$ and $a_1 \neq 0$. Since $a_1 = a_x f_x$ we see that $x \in \text{Supp } \alpha$ implies that $f_x \neq 0$ and then that $a_1 = a_x f_x = a_x u_x^{-1}$. Thus $a_x = a_1 u_x$ so

$$\alpha = a_1 \sum u_x \bar{x} = a_1 \sum \bar{x} u_x = a\gamma$$

with $a = a_1$ and $\gamma \in \mathbf{C}_{S*G}(R)$ as required. ∎

Of course, the above applies if M is any nonzero ideal of $R*G$ or $S*G$. Hence since $\text{Supp } a\gamma \in G_{\text{inn}}$ we have

Corollary 12.6. [127] *Let $R*G$ be a crossed product over the prime ring R. If I is any nonzero ideal of $R*G$, then $I \cap R*G_{\text{inn}} \neq 0$. In particular, if $R*G_{\text{inn}}$ is prime (or semiprime), then so is $R*G$.*

Theorem 12.7. [132] *Let $R*G$ be a crossed product over the prime ring R. Then $R*G$ is prime (or semiprime) if and only if $R*N$ is G-prime (or G-semiprime) for every finite normal subgroup N of G with $N \subseteq G_{\text{inn}}$.*

Proof. By Proposition 8.3(ii), $R*G$ is prime (or semiprime) if and only if $R*N$ is G-prime (or G-semiprime) for every finite normal subgroup N of G. Now we need only observe that $N_{\text{inn}} = N \cap G_{\text{inn}} \triangleleft G$ and that, by the previous corollary, if I is a nonzero G-stable ideal of $R*N$, then $I \cap R*N_{\text{inn}}$ is a nonzero G-stable ideal of $R*N_{\text{inn}}$. ∎

We remark that for R and N as above, $R*N$ is G-semiprime if and only if it is semiprime. This follows from Theorem 16.2(iii) since $R*N$ has a unique maximal nilpotent ideal.

Another application of Proposition 12.5 is as follows. Here, as usual, π_Δ denotes the projection from $R*G$ to $R*\Delta(G)$.

Lemma 12.8. *Let* $0 \neq I \lhd R*G$ *with* R *prime and* $G_{\mathrm{inn}} \cap \Delta^+(G) = \langle 1 \rangle$. *Then there exists* $\alpha \in I$ *such that* $\pi_\Delta(\alpha)R$ *is right regular in* $R*G$ *and* $R\pi_\Delta(\alpha)$ *is left regular in* $R*G$.

Proof. Extend $R*G$ to $S*G$ with $S = Q_s(R)$. Since I is a nonzero (R, R)-subbimodule of $S*G$ with $1 \in \mathrm{Supp}\, I$, Proposition 12.5 implies that there exists an element $\alpha = a\gamma = \gamma a \in I$ such that $a \in R \backslash 0$ and $\gamma \in E = \mathbf{C}_{S*G}(R)$ with identity coefficient 1. Since $\gamma \in S*G_{\mathrm{inn}}$ and $1 \in \mathrm{Supp}\, \gamma$ we have $0 \neq \pi_\Delta(\gamma) \in S*(G_{\mathrm{inn}} \cap \Delta)$. Furthermore, $\pi_\Delta(\gamma)$ is also in E.

By assumption, $G_{\mathrm{inn}} \cap \Delta^+ = \langle 1 \rangle$ so $G_{\mathrm{inn}} \cap \Delta(G)$ is torsion free abelian. This implies that $C^t[G_{\mathrm{inn}} \cap \Delta] \subseteq C^t[G_{\mathrm{inn}}] = E$ is a domain and hence $\pi_\Delta(\gamma)$ is regular in $C^t[G_{\mathrm{inn}} \cap \Delta]$. By freeness, we conclude in turn that $\pi_\Delta(\gamma)$ is regular in $E = C^t[G_{\mathrm{inn}}]$, $S*G_{\mathrm{inn}} = S \otimes_C E$ and $S*G$.

Finally $\pi_\Delta(\alpha) = a\pi_\Delta(\gamma) = \pi_\Delta(\gamma)a$, so since $\mathrm{r}_S(aR) = 0 = \ell_S(Ra)$, it follows that $\pi_\Delta(\alpha)R$ is right regular in $S*G$ and $R\pi_\Delta(\alpha)$ is left regular. ∎

Suppose $R*H$ is given and $G \lhd H$ with $R*G$ prime. Then $R*H = (R*G)*(H/G)$ and it is appropriate to consider $(H/G)_{\mathrm{inn}}$. But the action of H/N is special, in that it normalizes both R and the group $\mathcal{G}$ of trivial units of $R*G$. In particular, there is an induced action on $G \cong \mathcal{G}/\mathrm{U}(R)$. This explains the hypothesis of the next result. The proof in [**136**] is amusing, since it requires that one extend $R*G$ to three different rings, namely $S*G$, $Q_s(R*G)$ and $R*\tilde{G}$ where the latter group is an extension of G. All these rings actually live in some large quotient ring, but we will deal with them separately here. In fact, the first extension already occurred in the previous lemma and the group extension is somewhat finessed.

Theorem 12.9. [**136**] *Let* $R*G$ *be a crossed product with* R *prime and* $G_{\mathrm{inn}} \cap \Delta^+(G) = \langle 1 \rangle$. *Then* $R*G$ *is prime. Now suppose* q *is a unit of* $Q_s(R*G)$ *and* σ *is an automorphism of* G *with* $q^{-1}R\bar{x}q = R\overline{x^\sigma}$ *for all* $x \in G$. *Then* $\sigma = \sigma_1 \sigma_2$ *where* σ_1 *centralizes a subgroup of* G *of finite index and* σ_2 *is an inner automorphism of* G.

Proof. If N is a finite normal subgroup of G, then $N \subseteq \Delta^+(G)$. Thus since $G_{\text{inn}} \cap \Delta^+(G) = \langle 1 \rangle$, by hypothesis, it follows from Theorem 12.7 that $R*G$ is prime.

Now let q and σ be as given and choose $0 \neq A, B \lhd R*G$ with $Aq, qB \subseteq R*G$. By Lemma 12.8 there exists $\alpha \in A$ with $\pi_\Delta(\alpha)R$ right regular in $R*G$. Since q is a unit, $\alpha q \neq 0$ and then, as in the proof of Lemma 12.3(i), there exist $0 \neq \beta, \gamma, \delta \in R*G$ with

$$\alpha r \bar{x} \beta = \gamma (r\bar{x})^q \delta$$

for all $r \in R$ and $x \in G$.

Write $\alpha = \pi_\Delta(\alpha) + \alpha'$ where $\text{Supp}\, \alpha' \cap \Delta = \emptyset$ and fix $x \in G$. Since $\pi_\Delta(\alpha)R$ is right regular and $\bar{x}\beta \neq 0$, there exists $r \in R$ with $\pi_\Delta(\alpha)r\bar{x}\beta \neq 0$. Now, by hypothesis,

$$\pi_\Delta(\alpha)r\bar{x}\beta + \alpha' r\bar{x}\beta = \alpha r\bar{x}\beta$$
$$= \gamma(r\bar{x})^q\delta = \gamma s \overline{x^\sigma} \delta$$

for some $s \in R$. Thus by considering a group element in the support of $\pi_\Delta(\alpha)r\bar{x}\beta \neq 0$ we see that either

$$axb = a'xb'$$

for some $a \in \text{Supp}\, \pi_\Delta(\alpha)$, $a' \in \text{Supp}\, \alpha'$ and $b, b' \in \text{Supp}\, \beta$ or

$$axb = cx^\sigma d$$

for some $a \in \text{Supp}\, \pi_\Delta(\alpha)$, $b \in \text{Supp}\, \beta$, $c \in \text{Supp}\, \gamma$ and $d \in \text{Supp}\, \delta$.

To better understand the last equation, form the semidirect product $\tilde{G} = G \rtimes \langle t \rangle$ where t is an element of infinite order acting like σ on G. Then $x^\sigma = t^{-1}xt$ and we conclude from the above that x belongs to a right coset of $\mathbf{C}_G(a^{-1}a')$ or of $\mathbf{C}_G(tc^{-1}a)$ depending on finitely many parameters.

In other words, we see as usual that G is a finite union of cosets of the subgroups $\mathbf{C}_G(a^{-1}a')$ and $\mathbf{C}_G(tc^{-1}a)$ as above. Hence, by Lemma 6.2, one of these subgroups must have finite index in G. But observe that $a \in \text{Supp}\, \pi_\Delta(\alpha) \subseteq \Delta$ and $a' \in \text{Supp}\, \alpha'$ so $a^{-1}a' \notin$

$\Delta(G)$. Thus for a suitable $g \in G$ we see that $\mathbf{C}_G(tg^{-1})$ has finite index in G.

Since $t = tg^{-1} \cdot g$ and tg^{-1} centralizes a subgroup of G of finite index, the result follows. ∎

Paper [136] then goes on to describe precisely what q looks like and to discuss the group of all such X-inner automorphisms. The material becomes technically quite complicated.

Let G be a group and let H be a subgroup of G. Then we recall, from Section 7, that

$$\mathbf{D}_G(H) = \left\{ x \in G \mid |H : \mathbf{C}_H(x)| < \infty \right\}$$

is a subgroup of G normalized by H. The group algebra version of the following result is due to [62].

Corollary 12.10. *Let $R*G$ be given with R prime and let $H \triangleleft G$. Suppose $H_{\mathrm{inn}} \cap \Delta^+(H) = \langle 1 \rangle$. If $0 \neq I \triangleleft R*G$, then $I \cap R*\big(H\mathbf{D}_G(H)\big) \neq 0$. In particular, if $R*\big(H\mathbf{D}_G(H)\big)$ is prime, then so is $R*G$.*

Proof. We have $R*G = (R*H)*(G/H)$ and $R*H$ is prime by Theorem 12.9. Furthermore, if $\bar{x} \in \bar{G}$ induces an X-inner automorphism on $R*H$, then by Theorem 12.9 again, $x \in \mathbf{D}_G(H)H = H\mathbf{D}_G(H)$. In other words, $(G/H)_{\mathrm{inn}} \subseteq H\mathbf{D}_G(H)/H$ and Corollary 12.6 yields the result. ∎

We can now combine this fact with the methods of the last section to obtain some more symmetrically closed crossed products. For example

Corollary 12.11. *Let $R*G$ be given with R prime and let $H \triangleleft G$ with $\mathbf{D}_G(H) = \langle 1 \rangle$. If $R*H$ is symmetrically closed, then so is $R*G$.*

Proof. We may assume that $G \neq \langle 1 \rangle$. Then $\mathbf{D}_G(H) = \langle 1 \rangle$ implies that $H \neq \langle 1 \rangle$ and that $\Delta(H) = \langle 1 \rangle$. Hence, since $R*H$ is symmetrically closed, it is cohesive by Theorem 11.11. Since $H\mathbf{D}_G(H) = H$ and $R*G = (R*H)*(G/H)$, it follows from Corollary 12.10 and Proposition 11.9(ii) that $R*G$ is symmetrically closed. ∎

Proposition 12.12. [171] *Suppose G is a nonabelian free group and R is strongly prime. Then $R*G$ is symmetrically closed.*

Proof. Since R is strongly prime and G is an ordered group, it follows (Exercise 4) that $R*G$ is also strongly prime. If G has infinite rank, the result follows from Theorem 11.8 and Lemma 11.3.

In case G has finite rank, G has a normal subgroup H with G/H infinite cyclic and with H free of infinite rank. Furthermore, $\mathbf{D}_G(H) = \langle 1 \rangle$. Then $R*H$ is symmetrically closed by the above and Corollary 12.11 yields the result. ∎

We close this section with

Theorem 12.13. [123] *Let $R*G$ be a crossed product over the prime ring R. If u is a unit of $R*G$ which normalizes R, then $u = u_0\bar{g}$ for some $g \in G$ and some unit $u_0 \in R*G_{\mathrm{inn}}$ which induces, by conjugation, an X-inner automorphism on R. In particular if all units of $R*G_{\mathrm{inn}}$ are trivial, then u is trivial.*

Proof. If $\sigma \in \mathrm{Aut}(R)$ with $u^{-1}ru = r^\sigma$, then $ru = ur^\sigma$. Thus by Proposition 12.4(iv), $u = u_0\bar{g}$ with u_0 a unit in $R*G_{\mathrm{inn}}$ and with $\sigma\bar{g}^{-1}$ acting in an X-inner fashion on R. Since $u_0 = u\bar{g}^{-1}$ induces $\sigma\bar{g}^{-1}$ on R, the result follows. ∎

As a consequence we have

Corollary 12.14. [123] *Let $R*G$ be given with R prime and $G_{\mathrm{inn}} = \langle 1 \rangle$. If σ is an automorphism of $R*G$ with $R^\sigma = R$, then σ normalizes the group of trivial units of $R*G$.*

Proof. This follows from the preceding theorem since $G_{\mathrm{inn}} = \langle 1 \rangle$ implies that the group of trivial units of $R*G$ is the set of those units of $R*G$ which normalize R. ∎

This result can then we combined with Theorem 12.9 to yield

Corollary 12.15. *Let $R*G$ be a crossed product with R prime and $G_{\mathrm{inn}} = \langle 1 \rangle$. Suppose q is a unit of $Q_s(R*G)$ which normalizes* both $R*G$ and R. Then there exists an automorphism σ of G with $q^{-1}R\bar{x}q = R\overline{x^\sigma}$ for all $x \in G$. Furthermore $\sigma = \sigma_1\sigma_2$ where σ_1 centralizes a subgroup of G of finite index and σ_2 is an inner automorphism of G.

EXERCISES

1. Let R be a commutative unique factorization domain. Prove that $\mathrm{Xinn}(\mathrm{M}_n(R)) = \mathrm{Inn}(\mathrm{M}_n(R))$. To this end, let $A \in \mathrm{GL}_n(K)$, where K is the field of fractions of R, with $A^{-1}\mathrm{M}_n(R)A = \mathrm{M}_n(R)$. We may assume that $A \in \mathrm{M}_n(R)$ with relatively prime entries and we write $A^{-1} = B/d$ where $B \in \mathrm{M}_n(R)$ and $d \in R$ has no prime factor in common with all entries of B. Since d divides all entries of $B\mathrm{M}_n(R)A$, conclude that d is a unit of R.

2. On the other hand, suppose R is a commutative domain having a nonprincipal ideal $I = \alpha R + \beta R$ with $I^2 = dR$ principal. Prove that $\mathrm{M}_2(R)$ has an X-inner automorphism which is not inner. Here let $\gamma, \delta \in I$ with $\alpha\delta - \beta\gamma = d$ and show that $A = \begin{bmatrix} \alpha & \beta \\ \gamma & \delta \end{bmatrix}$ induces the appropriate automorphism.

3. Prove that RN, the normal closure of the prime ring R, is a prime ring. Now suppose $R = K[t][x, y \mid xy = tyx]$ is the example at the end of Section 10. Using the results given there, show that $S = RN = K(t)[x, y \mid xy = tyx]$ and that $SN = K(t)[x^{-1}, x, y, y^{-1} \mid xy = tyx]$. This is an example of [**19**].

4. Suppose G has the *unique product* property so that for any two nonempty finite subsets $A, B \subseteq G$ there is at least one uniquely represented element in the product AB. For example, any ordered group is a unique product group. Let $R*G$ be given. If R is a domain, or prime or strongly prime, show that the same is true of $R*G$.

Let $K[G]$ be a prime group algebra.

5. If σ is an automorphism of G, note that σ extends to an algebra automorphism of $K[G]$. Suppose σ centralizes a subgroup of finite index, let T be a right transversal for $\mathbf{C}_G(\sigma)$ in G and define

$\alpha = \sum_{t \in T} t^{-1}t^\sigma \in K[G]$. If $g \in G$, prove that $g^{-1}\alpha g = \alpha$. Conclude that $0 \neq \alpha$ is a normal element of $K[G]$ (see Exercise 6 of Section 10) and that conjugation by α induces the X-inner automorphism σ on $K[G]$. Following [133], such automorphisms are said to be of *central type*.

6. Let $\lambda: G \to K$ be a linear character of G and note that λ determines an automorphism $\lambda^\#: K[G] \to K[G]$ given by $\lambda^\#(\sum_g a_g g) = \sum_g a_g \lambda(g)g$. Suppose $\mathrm{Ker}(\lambda) \subseteq \mathbf{C}_G(x)$ for some $x \in \Delta(G)$, let T be a right transversal for $\mathbf{C}_G(x)$ in G and set $\beta = \sum_{t \in T} \lambda(t)x^t \in K[G]$. If $g \in G$, prove that $g^{-1}\beta\lambda(g)g = \beta$. Conclude that $0 \neq \beta$ is a normal element of $K[G]$ and that conjugation by β induces the X-inner automorphism $\lambda^\#$ on $K[G]$. Such automorphisms are said to be of *scalar type*.

It is shown in [133] that every X-inner automorphism of $K[G]$ normalizing the group of trivial units is of the form $\sigma = \sigma_1\sigma_2\sigma_3$ where σ_1 is the extension of an inner automorphism of G, σ_2 is an automorphism of central type and σ_3 is an automorphism of scalar type.

7. Let $G = \langle x, y \mid y^2 = 1, y^{-1}xy = x^{-1} \rangle$ be the infinite dihedral group and let $\lambda: G \to K$ be given by $\lambda(x) = -1$ and $\lambda(y) = 1$. If char $K \neq 2$, prove that $\lambda^\#$ is not an X-inner automorphism of $K[G]$. For this, consider the action of $\lambda^\#$ on $\mathbf{Z}(K[G])$.

8. Let $R*G$ and $\tilde{R}*\tilde{G}$ be two crossed products and let $\sigma : R*G \to \tilde{R}*\tilde{G}$ be a ring isomorphism with $R^\sigma = \tilde{R}$. If R is prime, prove that $G/G_{\mathrm{inn}} \cong \tilde{G}/\tilde{G}_{\mathrm{inn}}$. This follows from Theorem 12.13 and is a result of [123].

13. Free Rings

In this section, we continue with our computations, stressing in particular symmetrically closed rings. As we will see, one such example is the free K-algebra $S = K\langle x, y, \dots \rangle$ on at least two generators. Note that $S = K[F]$ is the semigroup algebra of the free semigroup

$F = \langle x, y, \ldots \rangle$. More generally, we begin by considering the semi-group crossed product $R*F$ as defined in Section 1. In this case, we use a variant of our earlier techniques to obtain the appropriate result.

The following is clearly the direct analog of Lemma 11.4 and Proposition 11.9(ii). The proof is the same.

Lemma 13.1. *Let $R*G$ be a prime semigroup crossed product and let $q \in Q_s(R*G)$. Suppose there exist $0 \neq A, B \lhd R$ with $Aq, qB \subseteq R*G$. If R is prime, cohesive and symmetrically closed, then $q \in R*G$.*

Now let $F = \langle x, y, \ldots \rangle$ be the *free semigroup* on the variables $x, y, \ldots$. A subset D of $F^{\#} = F \setminus \{1\}$ is said to be *separated* if for all elements $1 \neq w \in F$, if w is an initial segment of $a \in D$ and a final segment of $b \in D$, then we must have $a = w = b$. We will use $|a|$ to denote the length of a. All this notation will remain in force until Theorem 13.4 is proved.

Lemma 13.2. *Let D be a separated subset of F and let $a, b \in D$ and $f, g \in F$.*
   i. If $af = bg$ then $a = b$ and $f = g$.
   ii. If $fa = gb$ then $a = b$ and $f = g$.
   iii. If $af = gb$ and either $f \neq 1$ or $g \neq 1$, then $f = wb$ and $g = aw$ for some $w \in F$.

Proof. (i) Say $|a| \leq |b|$. Then $af = bg$ implies that $b = ab'$. But then $w = a \neq 1$ is an initial segment of b and a final segment of a. We conclude that $a = w = b$ and then that $f = g$. Part (ii) is similar.

(iii) Suppose $af = gb$. If $|g| \geq |a|$ then $g = aw$ so $af = awb$ and $f = wb$. Now suppose $|g| < |a|$. Then $a = gw$ with $w \neq 1$ and $gb = af = gwf$ so $b = wf$. Thus w is an initial segment of b and a final segment of a and $w \neq 1$. This yields $a = w = b$ and then $f = g = 1$. ∎

Lemma 13.3. *Let D be a finite nonempty subset of F. If F has at least two generators, then there exist $f, g \in F$ with fDg separated.*

Proof. Let F be generated by $x, y, \ldots$ and choose integer n with $n-2$ larger than the lengths of all elements of D. We claim that $x^n y D x y^n$ is separated. Thus suppose $w \neq 1$ is an initial segment of $x^n y a x y^n$ and a final segment of $x^n y b x y^n$ with $a, b \in D$. If $|w| \leq n$ then w must be both a power of x and a power of y, a contradiction. Thus $|w| > n$. From the $x^n y a x y^n$ term we see that w starts with x^n and from the $x^n y b x y^n$ term we see that w ends with y^n. But n is suitably larger than $|a|$ so the only y^n segment in $x^n y a x y^n$ occurs at the end and thus $w = x^n y a x y^n$. Similarly $w = x^n y b x y^n$. ∎

We can now prove the analog of Theorem 11.8.

Theorem 13.4. **[171]** *Let $R*F$ be a crossed product with F the free semigroup on at least two generators. Assume that R is prime, symmetrically closed and cohesive. Then $R*F$ is prime and symmetrically closed.*

Proof. Since F is ordered, it follows that for all $0 \neq \alpha \in R*F$ we have αR right regular in $R*F$. Hence $R*F$ is prime and, by Lemma 10.7(ii), αR is right regular in $Q_s(R*F)$. Let J denote the augmentation ideal of $R*F$ so that J is the set of all elements with identity coefficient zero.

Let $q \in Q_s(R*F)$ and let $0 \neq I \triangleleft R*F$ with $Iq, qI \subseteq R*F$. Replacing I by $JIJ \neq 0$ if necessary, we can assume that $Iq, qI \subseteq J$. Choose $0 \neq \alpha \in I$. If $f, g \in F$, then $\bar{f}\alpha\bar{g} \in I$ and Supp $\bar{f}\alpha\bar{g} = f(\text{Supp } \alpha)g$. Thus by Lemma 13.3 and the fact that F has at least two generators, we can assume that $D = \text{Supp } \alpha$ is separated.

Now fix $r \in R$. Then for all $r' \in R$ we have

$$\alpha r'(q r \alpha) = (\alpha r' q) r \alpha$$

and note that $qr\alpha, \alpha r'q \in J \subset R*F$ since $r\alpha, \alpha r' \in I$. Set $\beta = qr\alpha$ and let $f \in \text{Supp } \beta$. Since R is prime and D is separated, Lemma 13.2(i) implies that $af \in \text{Supp } \alpha r'\beta$ for some $r' \in R$ and $a \in D$. Hence $af \in \text{Supp } (\alpha r'q)r\alpha$ so $af = gb$ for some $g \in \text{Supp } (\alpha r'q)$ and $b \in D = \text{Supp } \alpha$. Note that $f \neq 1$ since $\beta = qr\alpha \in J$. Hence, by Lemma 13.2(iii) and the fact that D is separated, we have $f = f'b$

for some $f' \in F$. We have therefore shown that every $f \in \mathrm{Supp}\,\beta$ has a final segment in D.

We can now write $\alpha = \sum_{a \in D} r_a \bar{a}$ and $\beta = \sum_{a \in D} \beta_a \bar{a}$ with $0 \neq r_a \in R$ and $\beta_a \in R*F$. The equation $\alpha r' \beta = (\alpha r' q) r \alpha$ then yields

$$\sum_{a \in D} \alpha r' \beta_a \bar{a} = \sum_{a \in D} (\alpha r' q) r r_a \bar{a}.$$

Now Lemma 13.2(ii) asserts that if an element of F has a final segment in D, then that segment is unique. We therefore conclude from the above that $\alpha r' \beta_a = (\alpha r' q) r r_a$ for all $a \in D$ so $\alpha R(q r r_a - \beta_a) = 0$. Since αR is right regular in $Q_s(R*F)$, this yields $q r r_a = \beta_a \in R*F$.

In other words, we have $qB \subseteq R*F$ where $B = R r_b R$ is the nonzero ideal of R generated by r_b for some $b \in D$. Similarly we obtain $Aq \subseteq R*F$ for some $0 \neq A \lhd R$ and Lemma 13.1 yields the result. ∎

Of course, if R is not cohesive, then (Exercise 1) $R[F]$ is not symmetrically closed. In view of the above, $\mathrm{Xinn}(R*F) = \mathrm{Inn}(R*F)$. Furthermore, since F is ordered, it is clear that $\mathrm{U}(R*F) = \mathrm{U}(R)$. Thus we have

Corollary 13.5. *Let $R*F$ be a crossed product with F the free semigroup on at least two generators. Assume that R is prime, symmetrically closed and cohesive. Then $\mathrm{Xinn}(R*F) = \mathrm{Inn}(R*F)$ is generated by the action of the units of R.*

In the case of free algebras, the preceding two results are due to [**89**] (see Theorem 13.11).

Corollary 13.6. [**89**] *If S is the free K-algebra $S = K\langle x, y, \ldots \rangle$, then $\mathrm{Xinn}(S) = \langle 1 \rangle$.*

This follows from Corollary 13.5 in case $\langle x, y, \ldots \rangle$ has at least two generators. Otherwise Lemma 10.3(i) applies. Note that, by Lemma 10.3(ii), $Q_\ell(S) > S$.

Observe that free algebras are filtered via total degree. More generally a ring R is *filtered* if $R = \bigcup_{n=0}^{\infty} R_n$ is the ascending union of additive subgroups R_n with $R_n R_m \subseteq R_{n+m}$. Furthermore, $1 \in R_0$ so R_0 is a subring of R. The *associated graded ring* $\bar{R}$ of R is given by

$$\bar{R} = \oplus \sum_{n=0}^{\infty} R_n / R_{n-1}$$

with $R_{-1} = 0$.

Theorem 13.7. [130] *Let R be a filtered ring with associated graded ring $\bar{R}$ a domain. Then R is a domain and if σ is an X-inner automorphism of R, then σ preserves the filtration and hence acts on $\bar{R}$. Furthermore, σ acts trivially on the center of $\bar{R}$.*

Proof. If $a \in R_n \setminus R_{n-1}$, let $\bar{a} \in R_n / R_{n-1} \subseteq \bar{R}$ be its *leading term*. Since $\bar{R}$ is a domain, it follows that $\bar{a}\,\bar{b} = \overline{ab}$. Hence R is a domain with additive degree function. Now let σ be an X-inner automorphism of R. Then, by Lemma 12.3(i), there exist $a, b, c, d \in R \setminus 0$ with $arb = cr^\sigma d$ for all $r \in R$. We therefore have

$$\deg a + \deg r + \deg b = \deg c + \deg r^\sigma + \deg d$$

and since $\deg 1 = \deg 1^\sigma$ we conclude that $\deg r = \deg r^\sigma$. Thus σ preserves the filtration and hence acts on $\bar{R}$. Finally we have $\bar{a}\,\bar{r}\,\bar{b} = \bar{c}\,\overline{r^\sigma}\,\bar{d}$ for all $r \neq 0$. In particular, setting $r = 1$ yields $\bar{a}\,\bar{b} = \bar{c}\,\bar{d}$. Since σ stabilizes $\mathbf{Z}(\bar{R})$ and $\bar{R}$ is a domain, we conclude that $\bar{r} = \overline{r^\sigma} = \bar{r}^\sigma$ for all $\bar{r} \in \mathbf{Z}(\bar{R})$. ∎

This of course does not apply to free algebras, but it does apply to $U = U(L)$, the universal enveloping algebra of a Lie algebra L over K. Indeed we know that U is filtered and that $\bar{U}$ is a commutative polynomial ring. Also if L is finite dimensional, then U is an Ore domain.

Let q be a unit in $Q_s(U)$ which gives rise to the X-inner automorphism σ. Then the above implies that $q^{-1}\ell q = \ell + \lambda(\ell)$ for all $\ell \in L$ where $\lambda \colon L \to K$. Thus $\lambda \in \mathrm{Hom}_K(L, K)$ and $[\ell, q] = \ell q - q\ell = \lambda(\ell)q$ so q is a *semi-invariant* for L with $q \in Q_s(U)_\lambda$. Conversely, suppose $0 \neq q \in Q_s(U)$ with $[\ell, q] = \ell q - q\ell = \lambda(\ell)q$ for all $\ell \in L$. It follows that $U(L)q = qU(L)$ and hence that q is a unit of $Q_s(U)$ (see Exercise 6 of Section 10). Multiplying by q^{-1} then yields $q^{-1}\ell q = \ell + \lambda(\ell)$ and again we have an X-inner automorphism. We conclude that

Corollary 13.8. [130] *Let L be a Lie algebra over the field K and let $U = U(L)$ be its enveloping ring.*

i. The group of X-inner automorphisms of $U(L)$ is isomorphic to the additive subgroup of $\mathrm{Hom}_K(L, K)$ consisting of those λ with $Q_s(U)_\lambda \neq 0$.

ii. The semi-invariants for L in U are precisely the normal elements of U. Hence the semicenter is a characteristic subring of U.

Filtered rings occur naturally in the study of *coproducts*. Let R_1 and R_2 be rings containing a common division ring D. Then $R = R_1 \coprod R_2$, the coproduct over D, is filtered by $F^0 = D$ and $F^n = (R_1 + R_2)^n = \sum R_{i_1} R_{i_2} \cdots R_{i_n}$. The X-inner automorphisms of such rings have been studied in a series of papers, most notably [119,101,118]. The best result is the following, which we offer without proof.

Theorem 13.9. [118] *Assume that each $R_i > D$, at least one of the four dimensions over D is larger than 2, and that one-sided inverses in R_i are two-sided. Then every X-inner automorphism of $R = R_1 \coprod R_2$ is inner unless one of the following occurs.*

i. Each R_i is primary, that is $R_i = D + T_i$ with $T_i^2 = 0$.

ii. One R_i is primary and the other is 2-dimensional.

iii. char $D = 2$, one R_i is not a domain, and one is quadratic.

In the course of the proof, one shows that the X-inner automorphisms σ are strongly bounded, that is there exists an integer $k \geq 0$ with $\deg r^\sigma \leq \deg r + k$ for all $r \in R$.

Now let $R = \bigcup_{n=0}^\infty R_n$ be an arbitrary filtered ring. Then, as above, the *degree* of $r \in R$ is defined to be $\deg r = \min\{n \mid r \in R_n\}$. This function clearly satisfies

    1. $\deg r \geq 0$ for all $r \neq 0$ and, by convention, $\deg 0 = -\infty$,

    2. $\deg(r - s) \leq \max\{\deg r, \deg s\}$,

3. $\deg rs \leq \deg r + \deg s$,
4. $\deg 1 = 0$.

Of course $\deg(r-s) = \max\{\deg r, \deg s\}$ if $\deg r \neq \deg s$. Conversely, by defining $R_n = \{\, r \in R \mid \deg r \leq n \,\}$, any such degree function gives rise to a filtration on R.

Definition. Let R be a filtered ring with degree funtion as above. With respect to this function, we say that $\{\, a_1, a_2, \ldots, a_n \,\} \subseteq R$ is *right dependent* if either some $a_i = 0$ or there exists $b_1, b_2, \ldots, b_n \in R$ not all zero with

$$\deg\left(\sum_1^n a_i b_i\right) < \max\left\{\, \deg a_i + \deg b_i \mid i = 1, 2, \ldots, n \,\right\}.$$

Otherwise the set is *right independent*.

We say that $a \in R$ is *right dependent on* $\{\, a_1, a_2, \ldots, a_n \,\}$ if either $a = 0$ or there exist $b_i \in R$ such that

$$\deg\left(a - \sum_i a_i b_i\right) < \deg a$$

while $\deg a_i + \deg b_i \leq \deg a$ for all i. Otherwise, of course, a is *right independent of* $\{\, a_1, a_2, \ldots, a_n \,\}$.

Finally we say that R satisfies the *n-term weak algorithm* if given any right dependent set $\{\, a_1, a_2, \ldots, a_m \,\}$ with $m \leq n$ and $\deg a_1 \leq \deg a_2 \leq \cdots \leq \deg a_m$, some a_i is right dependent on $\{\, a_1, a_2, \ldots, a_{i-1} \,\}$. Notice that $i = 1$ can occur here only if $a_1 = 0$ since otherwise we have $\deg(a_1 - 0) < \deg a_1$, a contradiction. If R satisfies the n-term weak algorithm for all n, we say that R satisfies the *weak algorithm*.

To familiarize ourselves with these concepts, we prove the following few elementary properties.

Lemma 13.10. *Let R be a filtered ring satisfying the 2-term weak algorithm.*

   i. *For all $a, b \in R$, we have $\deg a + \deg b = \deg ab$ and hence R is a domain.*

  ii. R_0 *is a division ring.*

  iii. *R satisfies the (left) 2-term weak algorithm.*

  iv. *Let* $a_1, a_2 \in R$ *with* $0 \leq \deg a_1 \leq \deg a_2$ *and suppose that* $a_1 b_1 = a_2 b_2$ *for some* b_1, b_2 *with* $b_2 \neq 0$. *Then* $a_2 = a_1 r + s$ *for some* $r, s \in R$ *with* $\deg s < \deg a_1$.

Proof. (i) We may assume that $b \neq 0$. If $\deg ab < \deg a + \deg b$, then $\{a\}$ is a right dependent set. The 1-term weak algorithm yields $a = 0$.

  (ii) Let $0 \neq a \in R_0$. Then $a1 - 1a = 0$ implies that $\{a, 1\}$ is right dependent. Hence since $\deg a = \deg 1 = 0$, we see that 1 is dependent on the set $\{a\}$ so there exists $b \in R$ with $\deg(1 - ab) < \deg 1 = 0$ and $\deg a + \deg b \leq \deg 1$. Thus $ab = 1$ and $b \in R_0$. Since R is a domain, $a^{-1} = b \in R_0$.

  (iii) Assume that $\{b_1, b_2\}$ is a left dependent set with $0 \leq \deg b_1 \leq \deg b_2$. Then there exist a_1, a_2 not both zero with

$$\deg(a_1 b_1 + a_2 b_2) < \max\{\deg a_1 + \deg b_1, \deg a_2 + \deg b_2\}.$$

In particular, this forces

$$\deg a_1 + \deg b_1 = \deg a_1 b_1 = \deg a_2 b_2 = \deg a_2 + \deg b_2$$

so $\deg a_1 \geq \deg a_2 \geq 0$. Observe that $\{a_1, a_2\}$ is right dependent so the 2-term weak algorithm implies that $a_1 = a_2 c + d$ with $\deg d < \deg a_1 = \deg a_2 + \deg c$. Thus setting $e = a_1 b_1 + a_2 b_2$ and substituting in for a_1, we have $a_2(cb_1 + b_2) = e - db_1$ and

$$\deg(e - db_1) < \deg a_1 + \deg b_1 = \deg a_2 + \deg b_2.$$

We conclude from (i) that $\deg(cb_1 + b_2) < \deg b_2$ with $\deg c + \deg b_1 = \deg b_2$ and this fact is proved.

  (iv) We proceed by induction on $\deg a_2$. Since $\{a_1, a_2\}$ is dependent and $0 \leq \deg a_1 \leq \deg a_2$, the 2-term weak algorithm yields r', s' with $a_2 = a_1 r' + s'$ and $\deg s' < \deg a_2$. If $\deg s' < \deg a_1$ we are done, so we may suppose that $\deg a_1 \leq \deg s'$. Moreover

$$a_1 b_1 = a_2 b_2 = (a_1 r' + s')b_2$$

so $a_1(b_1 - r'b_2) = s'b_2$. Since $0 \leq \deg a_1 \leq \deg s' < \deg a_2$, induction implies that $s' = a_1 r'' + s''$ with $\deg s'' < \deg a_1$. Hence we have $a_2 = a_1 r' + s' = a_1(r' + r'') + s''$ as required. ∎

Part (iii) above is a special case of the fact [**43**, Section 2.3] that the n-term weak algorithm is right-left symmetric. Part (iv) is related to the division algorithm.

Theorem 13.11. [**89**] *Suppose $R = \bigcup_{n=0}^{\infty} R_n$ is a filtered ring satisfying the 2-term weak algorithm. Then either $Q_s(R) = R$ or R is a generalized polynomial ring in one variable over the division ring R_0.*

Proof. By the previous lemma, we know that R is a domain and hence, by Lemma 10.7(i), $Q_s(R)$ is also a domain. We assume that $Q_s(R) > R$ and proceed in a series of steps.

Step 1. *There exists $q \in Q_s(R)$ with $q^{-1} = z \in R \setminus R_0$.*

Proof. Among all elements of $Q_s(R) \setminus R$ choose q so that $0 \neq rq \in R$ has smallest possible degree for any $r \in R$. Now let $s \in R$ with $0 \neq qs \in R$ and consider the relation $(rq)s = r(sq)$ in R. By Lemma 13.10(iv), this yields two possibilities. First if $\deg r \leq \deg rq$, then $rq = rt + u$ for some $t, u \in R$ with $\deg u < \deg r \leq \deg rq$. But then $q - t \in Q_s(R) \setminus R$ and $r(q - t) = u \neq 0$ has degree smaller than that of rq, a contradiction. Thus we must have $\deg rq < \deg r$ and then $r = (rq)z + v$ for some $z, v \in R$ with $\deg v < \deg rq$. Now $r(1 - qz) = v$ and $\deg v < \deg rq$, so this implies that $1 - qz \in R$ and we conclude from Lemma 13.10(i) that $1 - qz = 0$ since $\deg v < \deg r$. Finally, $Q_s(R)$ is a domain so $q^{-1} = z \in R$ and, since R_0 is a division ring and $q \notin R$, we have $\deg z \geq 1$. ∎

Step 2. *Every two element subset of R is right dependent.*

Proof. Let $a_1, a_2 \in R$ and choose integer n with $\deg z^n > \deg a_i$ for $i = 1, 2$. Let $0 \neq C \lhd R$ with $q^n C \subseteq R$ and choose $c \in C \setminus 0$. Then $z^n(q^n a_i c) = a_i c$ and $\deg z^n > \deg a_i$ so, by Lemma 13.10(iv),

$z^n = a_i b_i + e_i$ for some $b_i, e_i \in R$ with $\deg e_i < \deg a_i$. Clearly $b_i \neq 0$. Now

$$a_1 b_1 + e_1 = z^n = a_2 b_2 + e_2$$

so $a_1 b_1 - a_2 b_2 = e_2 - e_1$. Since $\deg(e_2 - e_1) < \max\{\deg a_1, \deg a_2\}$, we conclude that $\{a_1, a_2\}$ is right dependent. ∎

Step 3. *R is a generalized polynomial ring over the division ring R_0.*

Proof. Let $D = R_0$ so that D is a division ring. By Step 1, $R > R_0$ so we can choose $x \in R$ to be of minimal positive degree. Let $d \in D$. Then $\{dx, x\}$ is right dependent, by Step 2, and $\deg dx \leq \deg x$ so the 2-term weak algorithm yields

$$dx = xd_1 + d_2$$

for some $d_1, d_2 \in R$ with $\deg d_1, \deg d_2 \leq 0$. Thus $d_1, d_2 \in D$ and, since $\deg x > 0$, it follows easily that d_1 and d_2 are uniquely determined by d. Furthermore, the map $\sigma: D \to D$ given by $\sigma: d \mapsto d_1$ is clearly a monomorphism and $\delta: d \mapsto d_2$ is a σ-derivation. In addition, by Lemma 13.10(iii) and the left analog of Step 2, $\{xd, x\}$ is left dependent. Thus $xd = d_1' x + d_2'$ for suitable $d_1', d_2' \in D$ so it follows that σ is onto and hence is an automorphism of D.

It remains to show that R is generated by x and D and we proceed by induction on the degree. Let $r \in R \setminus R_0$. Then $\{x, r\}$ is right dependent, by Step 2, and $\deg r \geq \deg x$. Thus, by the 2-term weak algorithm, $r = xs + t$ with $\deg t < \deg r$. Since clearly $\deg s < \deg r$, the result follows by induction. We have therefore shown that every element of R is a polynomial in x with coefficients in D and, since $\deg x > 0$, we conclude that $R = D[x; \sigma, \delta]$. ∎

The ring $D[x; \sigma, \delta]$ will be considered in more detail in Exercises 5, 6 and 7. The above theorem applies to free algebras because of the following result. For a proof, see [**43**, Section 2.4] or the graded special case given in Theorem 32.2.

Theorem 13.12. [**41**] *Let $R = \bigcup_{n=0}^{\infty} R_n$ be a filtered ring with $R_0 = K$ a central subfield. Then R is the free associative K-algebra on a* *right independent generating set if and only if R satisfies the weak algorithm.*

We close by briefly discussing a related ring theoretic property. Let R be a ring and let n be a positive integer. Then R is said to be an *n-fir* if all n-generator right ideals of R are free right R-modules of unique rank. A basic property is as follows.

Lemma 13.13. *Let R be an n-fir and let I be a right ideal of R which is free of rank $k < n$. Let $a, b \in R$ with $ab \in I$. Then either $b = 0$ or $I + aR$ is free of rank $\leq k$.*

Proof. Note that $I + aR$ has $k + 1 \leq n$ generators and hence is free of unique rank. Suppose this rank is $> k$. Then R has a right ideal which is free of rank $k + 1$. By uniqueness it follows that every free R-module of rank $k + 1$ has unique rank.

Let $F = I \oplus R$ so that F is free of unique rank $k + 1$ and map F onto $I + aR$ via the homomorphism $\sigma : i \oplus r \mapsto i - ar$. Since $I + aR$ is free, the map splits so $F \cong (I + aR) \oplus L$ where $L = \mathrm{Ker}(\sigma)$. Note that $L \cong \{ r \in R \mid ar \in I \}$, a right ideal of R, and L has at most $k + 1$ generators since it is a homomorphic image of F. Thus L is also free of unique rank and

$$k + 1 = \mathrm{rank}\, F = \mathrm{rank}\,(I + aR) + \mathrm{rank}\, L > k + \mathrm{rank}\, L.$$

Thus $\mathrm{rank}\, L = 0$ so $L = 0$ and $b = 0$. ∎

In particular, by taking $I = 0$ and $k = 0$, we conclude that any n-fir is a domain. In fact, by Exercise 8, R is a 1-fir if and only if it is a domain.

Now let R be any ring. An element $b \in R$ is said to be *bounded* if $bR \cap Rb$ contains a nonzero two-sided ideal of R. The relevance of this is as follows.

Lemma 13.14. *Let R be a prime ring and let $b \in R$. Then $b^{-1} \in Q_s(R)$ if and only if b is bounded.*

Proof. Suppose $b^{-1} = q \in Q_s(R)$ and choose $0 \neq I \triangleleft R$ with $qI, Iq \subseteq R$. Then $I \subseteq bR \cap Rb$.

Conversely suppose b is bounded and let $I \subseteq bR \cap Rb$. Since R is prime, it follows that b is a regular element of R. Thus the maps $f: {}_R I \to {}_R R$ and $g: I_R \to R_R$ given by $(rb)f = r$ and $g(br) = r$ are well defined. Furthermore, these maps are balanced and hence determine an element $q \in Q_s(R)$ which is clearly the inverse of b. ∎

Now by [**43**, Theorems 2.2.4 and 2.2.5], if R is a filtered ring satisfying the $(n + 1)$-term weak algorithm, then R is an n-fir. Furthermore, if R satisfies the weak algorithm, then it is a *fir,* that is all right ideals of R are free of unique rank. In particular, a free algebra is a fir.

Theorem 13.15. [**100**] *Let R be a 2-fir. If $s \in Q_s(R)$, then there exists a bounded element $b \in R$ with $bs \in R$. In particular, R is symmetrically closed if and only if all bounded elements are units of the ring R.*

Proof. Note that R is a domain and hence so is $Q_s(R)$. Let $s \in Q_s(R)$ and choose $0 \neq i, x \in R$ with $is, sx \in R$. Set $sx = y$ and observe that $(is)x = iy$. Now $I = iR$ is a free right ideal of rank 1 and $x \neq 0$. Thus by Lemma 13.13, $I + (is)R$ is free of rank ≤ 1 and hence $iR + (is)R = rR$ for some $r \in R$. Viewing this as an equation in $Q_s(R)$, we see that $r = iq$ for some $q \in Q_s(R)$. Hence since Q_s is a domain, we can cancel the i factor and conclude that $R + sR = qR$. Now $1 \in R \subseteq qR$ so $1 = qb$ for some $b \in R$ and, since Q_s is a domain, we have $q = b^{-1}$. Finally $s \in qR$ so $bs \in R$ and Lemma 13.14 yields the result. ∎

It is also shown in [**100**] that if R is a 2-fir, then $Q_s(R)$ is a symmetrically closed 2-fir.

EXERCISES

1. Let $F \neq \langle 1 \rangle$ be a free semigroup and let R be a prime ring. If R is not cohesive, show that $R[F]$ is not symmetrically closed. If $R * F$

is symmetrically closed, prove that $R*F$ is cohesive by modifying the argument of Theorem 11.11.

In the next three problems we consider an example from [**118**]. Let K be a field of characteristic not 2, and define the rings $R_i = K[x_i \mid (x_i)^2 = a_i \in K]$ for $i = 1, 2$ so that, depending on the choice of a_i, the ring R_i is either primary, a quadratic field extension of K or $R_i \cong K \oplus K$. Let $R = R_1 \coprod R_2$ be the coproduct of R_1 and R_2 as K-algebras.

2. Show that $R = K\langle x_1, x_2 \mid (x_1)^2 = a_1, (x_2)^2 = a_2 \rangle$ has as a K-basis $\{ y^n, x_2 y^n, y^n x_1, x_2 y^n x_1 \mid n = 0, 1, \ldots \}$ where $y = x_1 x_2$. Prove that R is prime.

3. Show that $\sigma: R \to R$ given by $\sigma(x_i) = -x_i$ defines an automorphism of R. Furthermore, prove that σ is not inner by considering the natural homomorphism of R onto the commutative ring $K[x_1, x_2 \mid (x_1)^2 = a_1, (x_2)^2 = a_2]$.

4. Let $\gamma = x_1 x_2 - x_2 x_1 \in R$. Show that $r\gamma = \gamma r^\sigma$ for all $r \in R$. Deduce that γ is a normal element of R and that σ is an X-inner automorphism induced by the unit γ of $Q_s(R)$.

In the next three problems, let $S = D[x; \sigma, \delta]$ be a generalized polynomial ring in the variable x over the division ring D so that S is filtered by the usual degree function.

5. If $f(x), g(x) \in S$ with $g(x) \neq 0$, show that $f(x) = g(x)q(x) + r(x)$ with $\deg r(x) < \deg g(x)$. Deduce that S satisfies the weak algorithm and that every nonzero one-sided ideal of S is generated by a unique monic polynomial of minimal degree.

6. Let $0 \neq I \lhd S$ and let $f(x)$ be the monic generator of I as a left ideal. Show that $fD = Df$ and that $xf - fx = df$ for some $d \in D$. Conclude that f is a normal element of S.

7. Let R be a prime ring, like S above, with the property that every nonzero ideal contains a nonzero normal element. Prove that $Q_\ell(R) = Q_s(R) = Q_r(R) = RN$ is equal to R localized at the multiplicatively closed set of its nonzero normal elements. Conclude

that $Q_s(R)$ is simple and that R is symmetrically closed if and only if it is simple.

8. Prove that R is a 1-fir if and only if it is a domain. One direction has already been noted.

9. Let F be a free semigroup on at least two generators and let R be a domain. Prove that every bounded element of $R*F$ is contained in R. To this end, if $q = b^{-1} \in Q_s(R*F)$, use the argument of Theorem 13.4 to conclude that $qB \subseteq R*F$ and hence that $B \subseteq b(R*F)$.

4 Prime Ideals – The Finite Case

14. G-Prime Coefficients

In previous sections we discussed when $R*G$ is prime or equivalently when 0 is a prime ideal of $R*G$. Now we would like to know, more generally, the nature of all prime ideals of crossed products. This is obviously a more difficult problem and success has been limited to the cases where G is a finite group or where G is a polycyclic-by-finite group with R a right Noetherian ring.

We must, of course, begin by considering ordinary group algebras. Here, if G is finite, then $K[G]$ is a finite dimensional algebra and we will assume that such rings are reasonably well understood. On the other hand, if G is polycyclic-by-finite, then the determination of the primes of $K[G]$ is a major achievement of [185]. Since this is group algebra material which has been adequately expounded in [185] and [168], we will not offer proofs here. However, we will state the necessary results and we will discuss at least one aspect of the proof where crossed products come into play. This will all be considered in later sections.

In this section and the next, we show how primes in crossed products are determined by primes in certain twisted group algebras. For this we will follow the arguments of [109], [111] and [172]. We begin by recalling some basic definitions and observations.

Let $R*G$ be given and, for each $I \triangleleft R$ and $x \in G$, set $I^x = \bar{x}^{-1} I \bar{x}$. Then we know that this yields a permutation action of G on the ideals of R. Furthermore, if I is G-stable we set

$$I*G = \left\{ \sum_{x \in G} \bar{x} a_x \in R*G \;\middle|\; a_x \in I \text{ for all } x \right\}.$$

By Lemma 1.4, if $J \triangleleft R*G$, then $J \cap R$ is a G-stable ideal of R with $J \supseteq (J \cap R)*G$. Conversely, if I is a G-stable ideal of R, then $I*G \triangleleft R*G$ with $(I*G) \cap R = I$ and $R*G/I*G \cong (R/I)*G$.

Let $I \triangleleft R$ be G-stable. We say that I is a G-*prime ideal* of R if for all G-stable ideals $A, B \triangleleft R$, the inclusion $AB \subseteq I$ implies $A \subseteq I$ or $B \subseteq I$. In particular, R is a G-*prime ring* if and only if $I = 0$ is a G-prime ideal.

Lemma 14.1. *Let $R*G$ be given.*

i. If P is a prime ideal of $R*G$, then $P \cap R$ is a G-prime ideal of R.

ii. If I is a G-prime ideal of R, then there exists a prime P of $R*G$ with $P \cap R = I$.

Proof. (i) Let P be given and let A, B be G-stable ideals of R with $AB \subseteq P \cap R$. Then $(A*G)(B*G) \subseteq (P \cap R)*G \subseteq P$ and, since P is prime, one of $A*G$ or $B*G$ is contained in P. But then, for example, if $A*G \subseteq P$ we have $A = (A*G) \cap R \subseteq P \cap R$.

(ii) Since $(I*G) \cap R = I$, we can apply Zorn's lemma to find an ideal P of $R*G$ maximal with $P \cap R = I$. Suppose A, B are ideals of $R*G$ containing P with $AB \subseteq P$. From $(A \cap R)(B \cap R) \subseteq P \cap R = I$ and the fact that I is G-prime, we conclude that one of $A \cap R$ or $B \cap R$ is contained in I. By the maximality of P we have $A = P$ or $B = P$ and P is prime. ∎

If P is a prime ideal of the crossed product $R*G$, then $P \supseteq (P \cap R)*G$ and $P/[(P \cap R)*G]$ is a prime ideal of $R*G/[(P \cap R)*G] \cong$

$(R/P \cap R)*G$. Since the homomorphism $R*G \rightarrow R*G/[(P \cap R)*G]$ is well understood, we see that to describe P it suffices to replace P and $R*G$ by their images under this map. Equivalently, we can assume that $P \cap R = 0$ and hence that R is a G-prime ring by Lemma 14.1(i). In this section, we study the situation of G-prime coefficients rings and show how to reduce the problem to the case of prime coefficient rings. We therefore assume throughout the proof that $R*G$ is given with R a G-prime ring. Furthermore, we suppose that either G is finite or that G is polycyclic-by-finite and R is right Noetherian. In the latter case, $R*G$ is also right Noetherian by Proposition 1.6.

Lemma 14.2. *Let Q be a minimal prime of R.*

 i. $\bigcap_{x \in G} Q^x = 0$ *so R is semiprime.*

 ii. $\{ Q^x \mid x \in G \}$ *is the finite set of minimal primes of R.*

 iii. *Let H denote the stabilizer of Q in G and set $N = \text{ann}_R(Q)$. Then H is a subgroup of G of finite index,*

$$N = \bigcap_{x \notin H} Q^x \neq 0$$

and for all $x \in G \setminus H$

$$0 = N\bar{x}N = N \cap N^x = N \cap Q.$$

 iv. *If A is a nonzero ideal of R with $A \subseteq N$, then $\text{ann}_R(A) = Q$.*

Proof. (i)(ii) Suppose first that G is finite. Then by Zorn's lemma, we can choose $Q \triangleleft R$ maximal with $\bigcap_{x \in G} Q^x = 0$. The goal is to show that Q is prime. To this end, let $A, B \supseteq Q$ with $AB \subseteq Q$. Then

$$\left(\bigcap_{x \in G} A^x \right) \cdot \left(\bigcap_{x \in G} B^x \right) \subseteq \bigcap_{x \in G} Q^x = 0.$$

Since R is G-prime, one of the first two intersections must be zero. But if $\bigcap_{x \in G} A^x = 0$, then the maximality of Q yields $A = Q$.

On the other hand, suppose R is right Noetherian. Then R has finitely many minimal primes, say $Q = Q_1, Q_2, \ldots, Q_n$ and

$\left(\bigcap_1^n Q_i\right)^m = 0$. Clearly G permutes these ideals and we let $C = \bigcap_{x \in G} Q^x$ and D equal to the intersection of the remaining primes if any. Then C and D are G-stable and $(CD)^m = 0$ so either C or D is zero. But $D = 0$ implies $Q \supseteq D$ contradicting the fact that Q is a minimal prime of R. Thus $C = 0$ as required.

Part (ii) is immediate in either case.

(iii) Since R is semiprime, right and left annihilators of two-sided ideals are equal. Thus $N = \text{ann}_R(Q)$ is unambiguously defined and it is clear from (i) that $N \supseteq \bigcap_{x \notin H} Q^x$. On the other hand, if $x \notin H$ then $Q^x \supseteq NQ = 0$ and $Q^x \not\supseteq Q$ so $Q^x \supseteq N$ and we have $N = \bigcap_{x \notin H} Q^x$. It follows from this formula that if $x \notin H$ then $Q \supseteq N^x$ so $N \cap N^x \subseteq N \cap Q = 0$. Finally $0 = N^x N = \bar{x}^{-1} N \bar{x} N$ yields $N \bar{x} N = 0$.

(iv) If $0 \neq A \subseteq N$ then certainly $Q \subseteq \text{ann}_R(N) \subseteq \text{ann}_R(A)$. On the other hand, $Q \supseteq A \cdot \text{ann}_R(A)$ and $Q \not\supseteq A$ so we conclude that $Q \supseteq \text{ann}_R(A)$. ∎

The above notation will be in force for the remainder of this section. Thus Q is a minimal prime of R, $N = \text{ann}_R(Q)$ and H is the stabilizer of Q in G. In addition we set $M = \sum_{x \in G} N^x$ so that M is a nonzero G-stable ideal of R. Part (ii) of the following lemma is a crucial observation; part (i) is needed for its proof.

Lemma 14.3. Let H and N be as above.

i. Let V be a nonzero right R-submodule of $N\bar{G}$ and let T be a finite subset of G. Suppose $V \cap (R*T) \neq 0$ but that $V \cap (R*T') = 0$ for all $T' \subset T$. Then T is contained in a right coset of H.

ii. Let I be an ideal of $R*G$. Then there exists a nonzero G-stable ideal E of R (depending on I) with

$$EI \subseteq \bar{G}N(I \cap R*H)\bar{G}.$$

Proof. (i) Fix $s \in T$. By assumption there exists $0 \neq \alpha = \sum a_t \bar{t} \in V \cap (R*T)$. Since $V \subseteq N\bar{G}$ we have $a_t \in N$ for all $t \in T$ and the minimality condition on T implies that $a_t \neq 0$ for all t. Now for any $q \in Q^s$ we have

$$\alpha q = \sum_t a_t q^{\bar{t}^{-1}} \bar{t} \in V \cap (R*T)$$

and note that $a_s q^{\bar{s}^{-1}} \in NQ = 0$. Therefore by minimality again, since $s \in T$, we have $a_t q^{\bar{t}^{-1}} = 0$ for all t so $(a_t)^{\bar{t}} \in \text{ann}_R(Q^s) = N^s$. Thus $0 \neq (a_t)^{\bar{t}} \in N^s \cap N^t = (N \cap N^{ts^{-1}})^s$ so $ts^{-1} \in H$ by Lemma 14.2(iii).

(ii) We first show that there exists a nonzero ideal B of R with

$$BI \subseteq \bar{G}N(I \cap R*H)\bar{G}$$

and this is clear if $NI = 0$. Thus we may suppose that $V = NI \neq 0$. Note that V satisfies the hypothesis of (i) and that V is a right ideal of $R*G$. Let $\mathcal{T}$ denote the family of all finite subsets T of G such that $V \cap (R*T) \neq 0$ but $V \cap (R*T') = 0$ for all $T' \subset T$. By (i) above, each such $T \in \mathcal{T}$ is contained in a right coset of H. For convenience we choose a canonical element $y = y(T) \in T$ for each $T \in \mathcal{T}$ and we let

$$A_T = \left\{ r \in R \;\middle|\; \text{there exists } \beta = \sum_{t \in T} b_t \bar{t} \in V \text{ with } r = b_y \right\}.$$

Since V is an (R, R)-subbimodule of $N\bar{G}$, it is clear that each A_T is a nonzero ideal of R contained in N.

For convenience we also arbitrarily linearly order the elements $T \in \mathcal{T}$. If S is a finite subset of G we can then define $B_S = N \cdot \prod_{T \subseteq S} A_T$ where the product is taken with the T's in $\mathcal{T}$ in increasing order. Now let S be a finite subset of G with $|S| = m$. We show by induction on m that $B_S(V \cap R*S) \subseteq N(I \cap R*H)\bar{G}$. This is clear for $m = 0$ so assume that $m > 0$ and that the result holds for all smaller size subsets. Let $0 \neq \alpha \in V \cap R*S$. By definition of $\mathcal{T}$, there exists $T \subseteq S$ with $T \in \mathcal{T}$. We assume T is largest possible in the ordering of $\mathcal{T}$ and we set $y = y(T)$.

Let $\alpha = a\bar{y} + \cdots$ and let $d \in A_T$. Then by definition there exists $\beta \in V \cap R*T \subseteq V \cap R*S$ with $\beta = d\bar{y} + \cdots$. Since $T \subseteq S$ it follows that

$$\gamma = d\alpha - \beta a^{\bar{y}} \in V \cap (R*S).$$

But the $\bar{y}$-coefficient of γ is zero so $\gamma \in R*S'$ where $S' = S \setminus \{y\}$ is a proper subset of S. By induction, $B_{S'}\gamma \subseteq N(I \cap R*H)\bar{G}$. Furthermore since $T \subseteq Hy$, it is clear that $\beta = (\beta \bar{y}^{-1})\bar{y} \in (I \cap R*H)\bar{G}$

so since $B_{S'} \subseteq N$ we have $B_{S'}\beta a^{\bar{y}} \subseteq N(I \cap R*H)\bar{G}$. This all implies that $B_{S'}d\alpha \subseteq N(I \cap R*H)\bar{G}$. Note that $y \notin S'$ so $T \not\subseteq S'$. Thus if B' denotes the obvious product with $B_S = B'A_T$, then $B_{S'} \supseteq B'$ so $B'd\alpha \subseteq N(I \cap R*H)\bar{G}$. But this holds for all $d \in A_T$ so $B_S\alpha \subseteq N(I \cap R*H)\bar{G}$ as required.

Set
$$E = \left\{ r \in R \mid rI \subseteq \bar{G}N(I \cap R*H)\bar{G} \right\}.$$

If G is finite, then $0 \neq B_G N \subseteq E$ so $E \neq 0$. On the other hand, if $R*G$ is right Noetherian, then $NI = \sum_1^n \alpha_i R*G$ for suitable $\alpha_i \in V$. Note that $S = \bigcup_1^n \operatorname{Supp} \alpha_i$ is a finite subset of G and $0 \neq B_S N \subseteq E$. Thus again $E \neq 0$. Finally with the additional $\bar{G}$ factor in front, $\bar{G}N(I \cap R*H)\bar{G}$ is an ideal of $R*G$ so it is clear that E must be a G-stable ideal of R and the result follows. ∎

Definition. We continue with the above assumptions. In addition, if $I \triangleleft R*G$ we let
$$I_{|H} = \left\{ \alpha \in R*H \mid N\alpha \subseteq I \right\}$$
and if $L \triangleleft R*H$ we set
$$L^{|G} = \bigcap_{x \in G} (L\bar{G})^{\bar{x}}.$$

Since $_{|H}$ and $^{|G}$ are defined on ideals here, this notation cannot be confused with the restriction and induction of modules $_{|R*H}$ and $^{|R*G}$ previously considered (see Exercise 1). Note also that $L^{|G}$ can be defined for any subgroup H of G and that a number of the basic properties we prove apply to this more general situation.

Recall that $\pi_H : R*G \to R*H$ is an $(R*H, R*H)$-bimodule homomorphism.

Lemma 14.4. *With the above notation we have*
  i. $I_{|H} \triangleleft R*H$ and if $I \cap R = 0$ then $I_{|H} \cap R = Q$,
  ii. $L^{|G}$ is the unique largest two-sided ideal of $R*G$ contained in $L\bar{G}$ and if $L \cap R = Q$ then $L^{|G} \cap R = 0$,
  iii. $L^{|G}$ is the unique largest ideal of $R*G$ with $\pi_H(L^{|G}) \subseteq L$,

iv. $0_{|H} = Q*H$ and $(Q*H)^{|G} = 0$.

Proof. (i) Since $I \lhd R*G$ and $N \lhd R$, it follows that $I_{|H}$ is a left R-module and a right $R*H$-module. Furthermore, both N and I are H-stable under conjugation so $I_{|H}$ is also H-stable. This along with the above implies that $I_{|H} \lhd R*H$. Finally if $I \cap R = 0$ then $r \in I_{|H} \cap R$ if and only if $Nr \subseteq I \cap R = 0$ so $I_{|H} \cap R = \text{ann}_R(N) = Q$.

(ii) If I is an ideal of $R*G$ contained in $L\bar{G}$, then since I is G-invariant under conjugation we have $I \subseteq \bigcap_{x \in G}(L\bar{G})^{\bar{x}} = L^{|G}$. On the other hand, $L^{|G}$ is clearly a right $R*G$-module, a left R-module and it is G-invariant. Thus $L^{|G} \lhd R*G$. Finally if $L \cap R = Q$, then $L^{|G} \cap R \subseteq L \cap R = Q$. But $L^{|G} \cap R$ is G-stable so $L^{|G} \cap R \subseteq \bigcap_{x \in G} Q^x = 0$.

(iii) Let $I \lhd R*G$. If $I \subseteq L\bar{G}$ then clearly $\pi_H(I) \subseteq L$. On the other hand, if $\pi_H(I) \subseteq L$, then $I \subseteq \pi_H(I)\bar{G} \subseteq L\bar{G}$. Thus (ii) above yields this result.

(iv) The first statement follows since $\text{ann}_{R*H}(N) = Q*H$. For the second, set $I = (Q*H)^{|G}$. Then $I \subseteq (Q*H)\bar{G} = Q\bar{G}$ so $\pi_{\langle 1 \rangle}(I) \subseteq Q$. But $\pi_{\langle 1 \rangle}(I)$ is a G-stable ideal of R so $\pi_{\langle 1 \rangle}(I) = 0$ and $I \subseteq \pi_{\langle 1 \rangle}(I)*G = 0$. ∎

Note that part (iii) of the previous lemma asserts that the definition of $L^{|G}$ is right-left symmetric. This however is not the case for $I_{|H}$ (see Exercise 4). Now it is clear that the maps $^{|G}$ and $_{|H}$ are monotone. They also have the following multiplicative properties.

Lemma 14.5. (i) If I_1 and I_2 are ideals of the crossed product $R*G$ then $(I_1)_{|H}(N*H)(I_2)_{|H} \subseteq (I_1 I_2)_{|H}$.

(ii) If L_1 and L_2 are ideals of $R*H$ then $L_1{}^{|G} L_2{}^{|G} \subseteq (L_1 L_2)^{|G}$.

Proof. (i) By definition $N I_{|H} \subseteq I$ so $N(I_1)_{|H} \cdot N(I_2)_{|H} \subseteq I_1 I_2$. Thus $(I_1)_{|H} N(I_2)_{|H} \subseteq (I_1 I_2)_{|H}$ and the result follows since $(I_1)_{|H} \lhd R*H$.

(ii) Since $L_2{}^{|G} \lhd R*G$ we have $\bar{G}L_2{}^{|G} \subseteq L_2{}^{|G}$ and hence

$$L_1{}^{|G} L_2{}^{|G} \subseteq L_1 \bar{G} L_2{}^{|G} \subseteq L_1 L_2{}^{|G} \subseteq L_1 L_2 \bar{G}.$$

Thus since $L_1{}^{|G} L_2{}^{|G} \lhd R*G$, Lemma 14.4(ii) yields the result. ∎

Lemma 14.6. *(i) Let* $L \lhd R*H$ *with* $L \cap R \supseteq Q$. *Then* $\bar{G}NL\bar{G} \subseteq L^{|G} \subseteq L\bar{G}$ *and* $L \subseteq (L^{|G})_{|H}$. *Furthermore* $N(L^{|G})_{|H} \subseteq L$.

   (ii) If $I \lhd R*G$, *then* $M(I_{|H})^{|G} \subseteq I$. *Moreover there exists a nonzero G-stable ideal* E *of* R *with* $EI \subseteq (I_{|H})^{|G}$.

Proof. (i) If $x \in H$, then $\bar{x}NL\bar{G} \subseteq L\bar{G}$ since L is an ideal of $R*H$. If $x \notin H$ then

$$\bar{x}NL\bar{G} = N^{\bar{x}^{-1}} L^{\bar{x}^{-1}} \bar{G} \subseteq Q(R*G) \subseteq L\bar{G}$$

since $N^{\bar{x}^{-1}} \subseteq Q \subseteq L$. Thus $\bar{G}NL\bar{G} \subseteq L\bar{G}$ and since $\bar{G}NL\bar{G} \lhd R*G$ we have $\bar{G}NL\bar{G} \subseteq L^{|G} \subseteq L\bar{G}$. In particular $NL \subseteq L^{|G}$ so $L \subseteq (L^{|G})_{|H}$. In the other direction, $N(L^{|G})_{|H} \subseteq L^{|G} \subseteq L\bar{G}$ so this clearly yields $N(L^{|G})_{|H} \subseteq L$.

   (ii) We have $N(I_{|H})^{|G} \subseteq NI_{|H}\bar{G} \subseteq I\bar{G} = I$, where the second inclusion holds by definition of $I_{|H}$. Since I and $(I_{|H})^{|G}$ are both G-stable it then follows that $M(I_{|H})^{|G} \subseteq I$. In the other direction, we know by Lemma 14.3(ii) that $EI \subseteq \bar{G}N(I \cap R*H)\bar{G}$ for a suitable nonzero G-invariant ideal E of R. Furthermore $I \cap R*H \subseteq I_{|H}$ and $I_{|H} \lhd R*H$ with $I_{|H} \supseteq Q = \text{ann}_R(N)$. Thus by (i) above

$$EI \subseteq \bar{G}N(I \cap R*H)\bar{G} \subseteq \bar{G}N(I_{|H})\bar{G} \subseteq (I_{|H})^{|G}$$

as required. ∎

   We now come to the main result of this section.

Theorem 14.7. [111] [172] *Let* $R*G$ *be given with* R *a* G-*prime ring. Assume that either* G *is finite or* G *is polycyclic-by-finite and* R *is right Noetherian. Let* Q *be a minimal prime of* R *and let* H *be its stabilizer in* G *so that* $|G : H| < \infty$. *Then the maps* $P \mapsto P_{|H}$ *and* $L \mapsto L^{|G}$ *as described above yield a one-to-one correspondence between the prime ideals* P *of* $R*G$ *with* $P \cap R = 0$ *and the primes* L *of* $R*H$ *with* $L \cap R = Q$.

Proof. We start with an observation on a form of cancellation. Let L be a prime ideal of $R*H$ with $L \cap R = Q$ and suppose $EI \subseteq L^{|G}$ where

$I \triangleleft R * G$ and E is a nonzero G-stable ideal of R. Then $EI \subseteq L\bar{G}$ so by applying the projection map π_H we have $E\pi_H(I) \subseteq L$ and hence $(E*H)\pi_H(I) \subseteq L$. But certainly $E*H \not\subseteq L$ since E is G-stable, $L \cap R = Q$ and $\bigcap_x Q^x = 0$. Thus since L is prime we deduce that $\pi_H(I) \subseteq L$ and hence $I \subseteq L^{|G}$ by Lemma 14.4(iii).

Now let P be a prime ideal of $R*G$ with $P \cap R = 0$ and set $L = P_{|H}$. By Lemma 14.4(i), $L \cap R = P_{|H} \cap R = Q$. Let us first observe, by Lemma 14.6(ii), that $P \supseteq M(P_{|H})^{|G} = (M*G)(P_{|H})^{|G}$. Thus since P is prime and $M*G \not\subseteq P$, we see that $P \supseteq (P_{|H})^{|G} = L^{|G}$. Next we show that L is prime. Indeed if L_1 and L_2 are ideals of $R*H$ containing L with $L \supseteq L_1 L_2$, then Lemma 14.4(ii) yields

$$P \supseteq L^{|G} \supseteq (L_1 L_2)^{|G} \supseteq L_1{}^{|G} L_2{}^{|G}.$$

Since P is prime, $P \supseteq L_i{}^{|G}$ for some i and then, by Lemma 14.6(i), since $L_i \cap R \supseteq L \cap R = Q$, we have $L = P_{|H} \supseteq (L_i{}^{|G})_{|H} \supseteq L_i$. Thus L is a prime ideal of $R*H$ with $L \cap R = Q$. Finally by Lemma 14.6(ii) there exists a nonzero G-stable ideal E of R with $EP \subseteq (P_{|H})^{|G} = L^{|G}$. Hence by the cancellation property of $L^{|G}$ mentioned above we have $P \subseteq L^{|G}$ so $P = L^{|G}$ and this half of the correspondence is proved.

In the other direction let L be a prime ideal of $R*H$ with $Q = L \cap R$ and set $P = L^{|G}$. By Lemma 14.4(ii) we have $P \cap R = 0$ and, by Lemma 14.6(i), $L \subseteq (L^{|G})_{|H} = P_{|H}$ and $L \supseteq N(L^{|G})_{|H} = (N*H)P_{|H}$. But L is a prime ideal of $R*H$ and $L \not\supseteq N*H$ since $N \cap Q = 0$, so the latter yields $L \supseteq P_{|H}$ and hence $L = P_{|H}$. Next we show that P is a prime ideal of $R*G$. To this end, let $I_1, I_2 \triangleleft R*G$ with $I_1 I_2 \subseteq P$. Then by Lemma 14.5(i)

$$(I_1)_{|H}(N*H)(I_2)_{|H} \subseteq (I_1 I_2)_{|H} \subseteq P_{|H} = L$$

and thus since L is prime and $L \not\supseteq N*H$ we have $L \supseteq (I_i)_{|H}$ for some i. Now applying Lemma 14.6(ii) to I_i we obtain $EI_i \subseteq (I_i)_{|H}{}^{|G} \subseteq L^{|G}$ for some nonzero G-stable ideal E of R. Hence, by the cancellation property for $L^{|G}$ we have $I_i \subseteq L^{|G} = P$ and P is prime. Since $L = P_{|H}$, the result follows. ∎

We observe, with the above notation, that there is an obvious one-to-one correspondence between the prime ideals L of $R*H$ with $L \cap R = Q$ and the prime ideals $\tilde{L}$ of

$$(R*H)/(Q*H) = (R/Q)*H = \tilde{R}*H$$

with $\tilde{L} \cap \tilde{R} = 0$. Since $\tilde{R}$ is prime, we have therefore reduced the study of prime ideals in $R*G$ with R a G-prime ring to those of $\tilde{R}*H$ with $\tilde{R}$ a prime ring. We will consider the latter situation in the next section. In view of Lemma 14.4(iv) we have

Corollary 14.8. *Let $R*G$ be given with R a G-prime ring. Assume that either G is finite or G is polycyclic-by-finite and R is right Noetherian. Let Q be a minimal prime of R and let H be its stabilizer in G so that $|G : H| < \infty$. Then $R*G$ is a prime ring if and only if $(R*H)/(Q*H) = (R/Q)*H$ is prime.*

We close with a generalization of Theorem 14.7. We begin by proving the *transitivity of induction* using Lemma 14.4(ii)(iii).

Lemma 14.9. *Let $R*G$ be given and let $H \subseteq A$ be subgroups of G. If L is an ideal of $R*H$, then $(L^{|A})^{|G} = L^{|G}$.*

Proof. Since $L^{|A} \subseteq L\bar{A}$, we have $(L^{|A})^{|G} \subseteq L^{|A}\bar{G} \subseteq L\bar{G}$ and hence $(L^{|A})^{|G} \subseteq L^{|G}$. Conversely set $I = \pi_A(L^{|G})$ so that $I \lhd R*A$. Then $\pi_H(I) = \pi_H(L^{|G}) \subseteq L$ so $I \subseteq L^{|A}$ and $L^{|G} \subseteq (L^{|A})^{|G}$. ∎

Corollary 14.10. [114] *Let $R*G$ be given and assume that either G is finite or G is polycyclic-by-finite and R is right Noetherian. Let Q be a prime of R with finitely many G-conjugates and let H be its stabilizer in G. Suppose A is a subgroup of G containing H and set $J = \bigcap_{x \notin A} Q^x$.*

i. The map $L \mapsto L^{|G}$ yields a one-to-one correspondence between the prime ideals L of $R*A$ with $L \cap R = \bigcap_{a \in A} Q^a$ and the primes P of $R*G$ with $P \cap R = \bigcap_{x \in G} Q^x$.

ii. If $P = L^{|G}$ with P and L as above, then $JL \subseteq P \cap (R*A) \subseteq L$ and L is the unique minimal covering prime of $P \cap (R*A)$ not containing J.

Proof. (i) By Theorem 14.7 applied to $R*G$, the map $T \mapsto T^{|G}$ yields a one-to-one correspondence between the primes T of $R*H$ with $T \cap R = Q$ and the primes P of $R*G$ with $P \cap R = \bigcap_{x \in G} Q^x$. Similarly, by Theorem 14.7 applied to $R*A$, the map $T \mapsto T^{|A}$ yields a one-to-one correspondence between the primes T as above and the primes L of $R*A$ with $L \cap R = \bigcap_{a \in A} Q^a$. Since $L^{|G} = (T^{|A})^{|G} = T^{|G} = P$, by transitivity, this fact is proved.

(ii) If $P = L^{|G}$, then $P \subseteq L\bar{G}$ so, by freeness, $P \cap (R*A) \subseteq L$. Furthermore $\bigcap_{a \in A} Q^a \subseteq L$ so $J \subseteq L^{\bar{x}}$ if $x \notin A$. Since $P = L^{|G} = \bigcap_{x \in G} L^{\bar{x}}\bar{G}$, it follows that $JL \subseteq P \cap (R*A)$. Finally if L' is a minimal covering prime of $P \cap (R*A)$ with $L' \cap R \not\supseteq J$, then $L' \supseteq JL$ yields $L' = L$ as required. ∎

EXERCISES

1. Let $R*G$ be given and let H be a subgroup of G. Suppose V is a right $R*H$-module and $L = \operatorname{ann}_{R*H}(V)$. Show that $\operatorname{ann}_{R*G}(V^{|R*G}) = L^{|G}$. This shows that induced ideals and induced modules are intimately related. In particular, deduce that $L^{|G} = \operatorname{ann}_{R*G}([R*H/L]^{|R*G})$.

2. Suppose $R*G$ is given, $H \triangleleft G$ and $L \triangleleft R*H$. Prove that $L^{|G} = (\bigcap_{x \in G} L^{\bar{x}}) \cdot (R*G)$.

3. Let $R = K\langle x_1, x_2, y_1, y_2 \mid x_i y_j = 0 = y_j x_i$ for all $i, j \rangle$ and let $G = \langle \sigma \rangle$, a group of order 2, act on R by $(x_i)^\sigma = y_i$ and $(y_i)^\sigma = x_i$. Form the skew group ring RG and let $Q = (x_1, x_2) = Rx_1 R + Rx_2 R \triangleleft R$. In the notation of this section, show that Q is a minimal prime of R, $N = Q^\sigma = \operatorname{ann}_R(Q) = (y_1, y_2)$ and $H = \langle 1 \rangle$.

4. Continuing with the example of the previous problem, let $I = (x_1 x_2, (x_2)^2, y_1 y_2, (y_2)^2)$ so that I is a G-stable ideal of R and $IG \triangleleft RG$. Prove that

$$y_2 \in (IG)_{|H} = \{\alpha \in RH \mid N\alpha \subseteq IG\}$$

but that

$$y_2 \notin \{\beta \in RH \mid \beta N \subseteq IG\}.$$

This shows that the definition of $_{|H}$ is not right-left symmetric.

5. Let J be a right ideal of $R*G$ and let H be a subgroup of G. Show that
$$(J \cap R*H)\bar{G} \subseteq J \subseteq \pi_H(J)\bar{G}.$$
Conclude that $(J \cap R*H)\bar{G} = J$ if and only if $\pi_H(J) \subseteq J$.

6. With the above notation, prove that there exists a unique minimal subgroup W of G (depending on J) such that $(J \cap R*W)\bar{G} = J$. W is called the *controller* of J.

15. Prime Coefficients

We continue with the task of describing prime ideals in crossed products. Specifically we study primes P in $R*G$ with $P \cap R = 0$ and now we assume that the coefficient ring R is prime. Furthermore we will assume, at some point, that either G is finite or that G is polycyclic-by-finite and R is right Noetherian. In the latter case, of course, $R*G$ is right Noetherian.

We will follow the approach of [172] which is a modification of the finite arguments in [109]. We note that both of these proofs ultimately depend on ideas in [58] and require that R be localized to its symmetric Martindale ring of quotients $Q_s(R)$.

We begin with two technical lemmas. Here we view $S \otimes T$ as the algebra freely generated by its commuting subalgebras S and T.

Lemma 15.1. *Let C be a field, let S and T be C-algebras and let $\Gamma = (S \otimes_C T)*H$ be a crossed product. Assume that $\bar{H}$ normalizes both S and T and let I be an $\bar{H}$-stable ideal of T. Then*
$$\Gamma/I\Gamma = (S \otimes T)*H/(S \otimes I)*H$$
$$\cong (S' \otimes T')*H = \Gamma'$$
where $S' \cong S$ and $T' = T/I$. Furthermore $\mathbf{C}_{\Gamma'}(S') = \mathbf{C}_\Gamma(S)'$, where the latter is the image of $\mathbf{C}_\Gamma(S)$.

Proof. Since $I \lhd T$ is $\bar{H}$-stable, it is clear that $S \otimes I \lhd S \otimes T$ is H-stable. Thus $I\Gamma = (S \otimes I)*H \lhd \Gamma$ and it follows from Lemma 1.4(ii) that
$$\Gamma/I\Gamma = (S \otimes T)*H/(S \otimes I)*H$$
$$= (S \otimes T/S \otimes I)*H.$$

But $(S \otimes T)/(S \otimes I) \cong S \otimes (T/I)$ so we have obtained the appropriate structure of $\Gamma' = \Gamma/I\Gamma$.

It remains to consider the centralizers and certainly $\mathbf{C}_\Gamma(S)' \subseteq \mathbf{C}_{\Gamma'}(S')$, that is the centralizer of S in Γ maps into the centralizer of S' in Γ'. For the other direction, let $\{t_1, t_2, \dots\}$ be a C-basis for a complement for I in T and let $\alpha' \in \mathbf{C}_{\Gamma'}(S')$. Then we may assume that α, an inverse image of α', is of the form

$$\alpha = \sum_{i,x}(a_{i,x} \otimes t_i)\bar{x}$$

with $a_{i,x} \in S$. Let $s \in S$. Since $\bar{H}$ normalizes S we have

$$s\alpha - \alpha s = \sum_{i,x}\left[\left(sa_{i,x} - a_{i,x}s^{\bar{x}^{-1}}\right) \otimes t_i\right]\bar{x}.$$

On the other hand, $s\alpha - \alpha s \in (S \otimes I){*}H$ since $\alpha' \in \mathbf{C}_{\Gamma'}(S')$. By definition of $\{t_i\}$ it therefore follows that $s\alpha - \alpha s = 0$. Since this is true for all $s \in S$ we have $\alpha \in \mathbf{C}_\Gamma(S)$ and hence $\alpha' \in \mathbf{C}_\Gamma(S)'$. ∎

Lemma 15.2. *Let* $\Gamma = (S \otimes_C T){*}H$ *be as in the previous lemma. In addition assume that* R *is a prime ring,* $S = \mathrm{Q}_s(R)$ *and that* $\bar{H}$ *normalizes* R, S *and* T. *If* I *is a nonzero* (R, R)-*subbimodule of* Γ *with* $I\bar{H}^{-1} = I$, *then there exists* $0 \neq \alpha \in \mathbf{C}_\Gamma(S)$ *and* $0 \neq A \triangleleft R$ *with* $A\alpha \subseteq I$.

Proof. Let $\{t_0, t_1, \dots\}$ be a C-basis for T. Then every element $\beta \in \Gamma$ is uniquely writable as

$$\beta = \sum_{i,x}(b_{i,x} \otimes t_i)\bar{x}$$

with $b_{i,x} \in S$. Choose $0 \neq \beta \in I$ with a minimal number, say n, of nonzero coefficients $b_{i,x}$. We may suppose $b_{0,y} \neq 0$. Furthermore, by multiplying β on the left by an appropriate element of R, we may assume that $b_{0,y} \in R$.

Let $r \in R$. Since I is an (R, R)-subbimodule of Γ we have

$$\gamma = b_{0,y}r\beta - \beta r^{\bar{y}}(b_{0,y})^{\bar{y}} \in I.$$

Furthermore

$$\gamma = \sum_{i,x}\left[\left(b_{0,y}rb_{i,x} - b_{i,x}r^{\bar{y}\bar{x}^{-1}}(b_{0,y})^{\bar{y}\bar{x}^{-1}}\right) \otimes t_i\right]\bar{x}$$

and the $(0, y)$-term here is zero. Thus the minimality of n implies that $\gamma = 0$ and hence that

$$b_{0,y}rb_{i,x} = b_{i,x}r^{\bar{y}\bar{x}^{-1}}(b_{0,y})^{\bar{y}\bar{x}^{-1}}$$

for all $r \in R$. It follows from Lemma 12.1 that there exists a unit $q_{i,x} \in S$ with

$$\bar{x}\bar{y}^{-1}r\bar{y}\bar{x}^{-1} = r^{\bar{y}\bar{x}^{-1}} = q_{i,x}^{-1}rq_{i,x}$$

and $b_{i,x} = b_{0,y}q_{i,x}$ for all i, x which occur in the support of β.
Set

$$\alpha = \sum_{i,x}(q_{i,x} \otimes t_i)\bar{x}\bar{y}^{-1}.$$

Since $q_{i,x}\bar{x}\bar{y}^{-1}$ centralizes R and normalizes S, it follows from the uniqueness part of Lemma 10.9(ii) that $q_{i,x}\bar{x}\bar{y}^{-1}$ centralizes S and hence that $\alpha \in \mathbf{C}_\Gamma(S)$. Furthermore since $b_{i,x} = b_{0,y}q_{i,x}$ we have

$$b_{0,y}\alpha = \sum_{i,x}(b_{i,x} \otimes t_i)\bar{x}\bar{y}^{-1} = \beta\bar{y}^{-1} \in I$$

using $I\bar{H}^{-1} = I$.
    Finally since $0 \neq b_{0,y} \in R$, I is an (R, R)-bimodule and α commutes with R we have

$$(Rb_{0,y}R)\alpha = R(b_{0,y}\alpha)R \subseteq I$$

and the lemma is proved. ∎

    We fix some notation for the remainder of this section. Let $\Gamma = R*G$ be a crossed product with R a prime ring. By Proposition 12.4(i), if $S = Q_s(R)$, then Γ extends uniquely to a crossed product $\Gamma' = S*G$. Set $C = \mathbf{Z}(S) = \mathbf{C}_S(R)$ so that C, the extended centroid of R, is a field. Recall that an automorphism σ of R is said to be X-inner if its unique extension to S becomes inner. By Lemma 12.3(iii)

$$G_{\text{inn}} = \left\{ x \in G \mid \bar{x} \text{ is X-inner on } R \right\}$$

is a normal subgroup of G. The next lemma is essentially all notation.

Lemma 15.3. *If* $T = \mathbf{C}_{\Gamma'}(S)$ *then* $T = C^t[G_{\text{inn}}]$, *some twisted group algebra of* G_{inn} *over* C, *and* G *acts on* T *normalizing its group of trivial units. Furthermore,* $S*G_{\text{inn}} = S \otimes T$ *and* $\Gamma' = (S*G_{\text{inn}})*H = (S \otimes_C T)*H$, *a suitable crossed product of* $H = G/G_{\text{inn}}$ *over* $S \otimes T$.

Proof. Most of this follows from Proposition 12.4 and Lemma 1.3. All that remains is to observe that G acts on T. For this we note that $\mathcal{G}$, the group of trivial units of $R*G$, normalizes S and hence acts by conjugation on $T = \mathbf{C}_{\Gamma'}(S)$. Furthermore $U = \mathrm{U}(R)$ centralizes T so we obtain an action of $G \cong \mathcal{G}/U$ on T. It is clear that G normalizes the group of trivial units of T. ∎

For the remainder of this section we write $T = \mathbf{C}_{\Gamma'}(S)$. In addition, if I is a (Γ, Γ)-subbimodule of $\Gamma' = S*G$ we set

$$\tilde{I} = \{ \alpha \in T \mid A\alpha \subseteq I \text{ for some } 0 \neq A \lhd R \}.$$

Lemma 15.4. *With the above notation,* $\tilde{I}$ *is a* G-stable ideal of T. *Furthermore if* $\beta \in \tilde{I}\Gamma'$ *then there exists* $0 \neq B \lhd R$ *with* $B\beta \subseteq I$.

Proof. If $\alpha, \beta \in \tilde{I}$ with $A\alpha, B\beta \subseteq I$, then $(A \cap B)(\alpha + \beta) \subseteq I$ so $\alpha + \beta \in \tilde{I}$. Now let $\gamma \in T$ and choose $0 \neq C \lhd R$ with $C\gamma \subseteq \Gamma$. Since α and γ centralize R we have

$$AC\alpha\gamma = (A\alpha)(C\gamma) \subseteq I\Gamma = I$$

and

$$CA\gamma\alpha = (C\gamma)(A\alpha) \subseteq \Gamma I = I$$

so $\alpha\gamma, \gamma\alpha \in \tilde{I}$. Thus $\tilde{I} \lhd T$. Furthermore since I and R are $\bar{G}$-stable, so is $\tilde{I}$.

Finally if $\beta \in \tilde{I}\Gamma'$ then $\beta = \sum_1^n \alpha_i \gamma_i$ with $\alpha_i \in \tilde{I}$ and $\gamma_i \in \Gamma'$. Choose $0 \neq A, C \lhd R$ with $A\alpha_i \subseteq I$ and $C\gamma_i \subseteq \Gamma$ for all $i = 1, 2, \ldots, n$. Then

$$ AC\beta \subseteq \sum_1^n AC\alpha_i \gamma_i = \sum_1^n (A\alpha_i)(C\gamma_i) \subseteq I\Gamma = I $$

and the lemma is proved. ∎

The following result is crucial. It applies when $I \lhd \Gamma$ and $I \lhd \Gamma'$.

Lemma 15.5. *If I is a (Γ, Γ)-subbimodule of Γ', then $\tilde{I}\Gamma' \lhd \Gamma'$ and $I \subseteq \tilde{I}\Gamma'$.*

Proof. By Lemma 15.3, $\Gamma' = (S \otimes_C T)*H$ where $H = G/G_{\mathrm{inn}}$ and, choosing $\bar{H} \subseteq \bar{G}$, we see that R, S and T are $\bar{H}$-invariant. Furthermore $\tilde{I}$ is an $\bar{H}$-stable ideal of T. By Lemma 15.1, $\tilde{I}\Gamma' \lhd \Gamma'$ and

$$ \Gamma'/\tilde{I}\Gamma' = \big(S \otimes (T/\tilde{I})\big)*H = \Gamma''. $$

Moreover $T = \mathbf{C}_{\Gamma'}(S)$ maps onto the centralizer of S in Γ''.

Suppose $I \not\subseteq \tilde{I}\Gamma'$. Then the image I'' of I in Γ'' is a nonzero (R, R)-subbimodule of Γ'' with $I''\bar{H}^{-1} = I''$ and Lemma 15.2 implies that there exists $0 \neq \alpha'' \in \mathbf{C}_{\Gamma''}(S)$ with $A\alpha'' \subseteq I''$ for some $0 \neq A \lhd R$. As we observed, α'' is the image of some element $\alpha \in T$ and certainly $A\alpha \subseteq I + \tilde{I}\Gamma'$. Choose $a \in A$ with $a \neq 0$ and write $a\alpha = \gamma + \beta$ with $\gamma \in I$ and $\beta \in \tilde{I}\Gamma'$. Then, by Lemma 15.4, there exists $0 \neq B \lhd R$ with $B\beta \subseteq I$. Since $B\gamma$ is certainly contained in I, we have $Ba\alpha \subseteq I$ and hence, since α centralizes R, we obtain $(BaR)\alpha = (Ba\alpha)R \subseteq IR = I$. But $0 \neq BaR \lhd R$ implies that $\alpha \in \tilde{I}$, by definition of $\tilde{I}$, and therefore the image α'' of α is zero, a contradiction. We conclude that $I \subseteq \tilde{I}\Gamma'$. ∎

Recall that G acts on T. If $Q \neq T$ is a G-stable ideal of T, then Q is G-prime if for all G-stable ideals I, J of T the inclusion $IJ \subseteq Q$ implies $I \subseteq Q$ or $J \subseteq Q$. We can now set up a correspondence between the primes of $\Gamma = R*G$ disjoint from R and the G-prime ideals of T. With little additional work we can add to this link the primes of $\Gamma' = S*G$ disjoint from S.

Lemma 15.6. *Let Q be a G-stable ideal of T and set $P = Q\Gamma' \cap \Gamma$ and $P' = Q\Gamma'$. Then $P \lhd \Gamma$ and $P' \lhd \Gamma'$ with $P \cap R = P' \cap S = 0$ and $\tilde{P} = \widetilde{P'} = Q$. Furthermore if Q is G-prime then P is a prime ideal of Γ and P' is a prime ideal of Γ'.*

Proof. We apply Lemma 15.3 and write $\Gamma' = (S \otimes_C T)*H$ where $H = G/G_{\text{inn}}$. Since Q is a G-stable ideal of T, it follows that $S \otimes Q$ is an ideal of $S \otimes T$ which is H-stable. Thus by Lemma 1.4 we have

$$P' = Q\Gamma' = (S \otimes Q)*H \lhd \Gamma'$$

and $P' \cap (S \otimes T) = S \otimes Q$. Thus $P = P' \cap \Gamma \lhd \Gamma$ and $P \cap (S \otimes T) \subseteq S \otimes Q$. This implies that $P' \cap S = P \cap R = 0$.

Let $\alpha \in \tilde{P}$ or $\widetilde{P'}$. Then by definition there exists $0 \neq A \lhd R$ with

$$A\alpha \subseteq P' \cap (S \otimes T) = S \otimes Q$$

and hence $\alpha \in Q$. Conversely if $\beta \in Q$ then there exists $0 \neq B \lhd R$ with $B\beta \subseteq \Gamma$ and thus

$$B\beta = \beta B \subseteq Q\Gamma' \cap \Gamma = P \subseteq P'.$$

We conclude therefore that $\tilde{P} = \widetilde{P'} = Q$.

Finally assume that Q is G-prime. We show that P is prime, the proof for P' being identical. Let $I, J \lhd R$ with $IJ \subseteq P$ and let $\alpha \in \tilde{I}, \beta \in \tilde{J}$. If $0 \neq A, B \lhd R$ with $A\alpha \subseteq I$ and $B\beta \subseteq J$ then

$$AB\alpha\beta = (A\alpha)(B\beta) \subseteq IJ \subseteq P$$

so $\alpha\beta \in \tilde{P} = Q$. In other words, $\tilde{I}\tilde{J} \subseteq Q$. But $\tilde{I}$ and $\tilde{J}$ are G-stable ideals of T, by Lemma 15.4, and Q is G-prime. It follows that one of $\tilde{I}$ or $\tilde{J}$ is contained in Q, say $\tilde{I} \subseteq Q$. We conclude from Lemma 15.5 that

$$I \subseteq \tilde{I}\Gamma' \cap \Gamma \subseteq Q\Gamma' \cap \Gamma = P$$

and the lemma is proved. ∎

The last step requires the full assumptions on G. If G is finite, we will be able to restrict our attention to the G-stable ideals of R.

If G is infinite, the ideals of R cannot be taken to be G-stable, so we compensate by using the Noetherian hypothesis.

Lemma 15.7. *Assume in addition that either G is finite or that G is polycyclic-by-finite and R is right Noetherian. Let P be a prime ideal of Γ with $P \cap R = 0$ or let P' be a prime ideal of Γ' with $P' \cap S = 0$. Then $\tilde{P}$ and $\widetilde{P'}$ are G-prime ideals of T with $P = \tilde{P}\Gamma' \cap \Gamma$ and $P' = \widetilde{P'}\Gamma'$.*

Proof. It is convenient to proceed in a series of steps.

Step 1. $P = \tilde{P}\Gamma' \cap \Gamma$.

Proof. Let $W = \tilde{P}\Gamma' \cap \Gamma$ so that $W \lhd \Gamma$ and $W \supseteq P$ by Lemma 15.5. Suppose first that G is finite. If $\alpha \in W$ then by Lemma 15.4 there exists $0 \neq B \lhd R$ with $B\alpha \subseteq P$. Letting $A = \bigcap_{x \in G} B^x$ we see that A is a nonzero G-stable ideal of R with $A\alpha \subseteq P$. Thus $(A*G)\alpha \subseteq P$. Finally P is a prime ideal and $P \cap R = 0$ so $A*G \not\subseteq P$ and hence, for all $\alpha \in W$, we have $\alpha \in P$.

Now assume instead that Γ is right Noetherian and write $W = \sum_1^n \alpha_i \Gamma$. Since $\alpha_i \in W \subseteq \tilde{P}\Gamma'$, it follows from Lemma 15.4 that there exist $0 \neq B_i \lhd R$ with $B_i\alpha_i \subseteq P$. Letting $A = \bigcap_1^n B_i$ we have $0 \neq A \lhd R$ with $AW \subseteq P$. Finally P is prime and $P \cap R = 0$ so $A \not\subseteq P$ and hence $W \subseteq P$ as required. ∎

Step 2. $P' = \widetilde{P'}\Gamma'$.

Proof. Set $W' = \widetilde{P'}\Gamma'$ so that $W' \lhd \Gamma'$ and $W' \supseteq P'$. If either G is finite or polycyclic-by-finite, then T is right Noetherian. Thus $\widetilde{P'}$ is a finitely generated right ideal of T and hence W' is a finitely generated right ideal of Γ'. We can now proceed as in the previous paragraph. In other words, if $W' = \sum_1^n \alpha_i \Gamma'$ then there exists $0 \neq B \lhd R$ with $BW' \subseteq P'$. Since P' is a prime ideal of Γ' and $B \not\subseteq P'$ we conclude that $W' \subseteq P'$. ∎

Step 3. $\tilde{P}$ and $\widetilde{P'}$ are G-prime ideals of T.

Proof. Since the proofs are identical in the two cases, we will only consider $\tilde{P}$. Let I and J be G-stable ideals of T with $IJ \subseteq \tilde{P}$. Since $I\Gamma'$ and $J\Gamma'$ are ideals of Γ', by Lemma 15.6, we have $\Gamma'J \subseteq J\Gamma'$ and hence $(I\Gamma')(J\Gamma') \subseteq IJ\Gamma' \subseteq \tilde{P}\Gamma'$. It follows that

$$(I\Gamma' \cap \Gamma)(J\Gamma' \cap \Gamma) \subseteq \tilde{P}\Gamma' \cap \Gamma = P$$

and thus, since P is prime, one of these factors is in P, say $I\Gamma' \cap \Gamma \subseteq P$. By Lemma 15.6, this yields $I = (I\Gamma' \cap \Gamma)^{\tilde{}} \subseteq \tilde{P}$ so $\tilde{P}$ is G-prime. Similarly $\widetilde{P'}$ is G-prime. This completes the proof. ∎

The above two lemmas now combine to form the main theorem of this section. At this point, it is essentially all notation and requires no additional proof.

Theorem 15.8. **[109] [172]** *Let $\Gamma = R*G$ be a crossed product with R prime. Assume that either G is finite or that G is polycyclic-by-finite and R is right Noetherian. Let $S = Q_s(R)$, $C = \mathbf{Z}(S)$ and let $\Gamma' = S*G$ be the natural extension of Γ. Then $T = \mathbf{C}_{\Gamma'}(S) = C^t[G_{\mathrm{inn}}]$ is a twisted group algebra of the group $G_{\mathrm{inn}} \triangleleft G$ and G acts on T as automorphisms normalizing the group of trivial units. Furthermore there exist one-to-one order preserving correspondences between the primes P of Γ with $P \cap R = 0$, the primes P' of Γ' with $P' \cap S = 0$, and the G-prime ideals Q of T. Specifically these maps are given by*

$$Q \mapsto Q\Gamma' = P', \qquad P' \mapsto P' \cap \Gamma = P, \qquad P \mapsto \tilde{P} = Q$$

where

$$\tilde{P} = \{\, \alpha \in T \mid \text{there exists } 0 \neq A \triangleleft R \text{ with } A\alpha \subseteq P \,\}.$$

We remark that the assumption above that G is finite can be replaced by the weaker hypothesis that any nonzero ideal of R contains a nonzero G-stable ideal (see Exercise 1).

Corollary 15.9. *With the above notation, the following are equivalent.*

    i. $R*G$ is prime.

    ii. $S*G$ is prime.

    iii. $C^t[G_{\mathrm{inn}}]$ is G-prime.

Proof. In view of the preceding theorem, it suffices to show that $\tilde{0} = 0$ and this is immediate from Lemma 15.6 with $Q = 0$. ∎

Finally we remark that the G-prime ideals of T are closely related to the primes of that ring. Indeed, since T is Noetherian, it follows from Lemma 14.2(i)(ii) with G acting on T/Q that

Lemma 15.10. Q is a G-prime ideal of T if and only if $Q = \bigcap_{x \in G} \tilde{Q}^x$ where $\left\{ \tilde{Q}^x \mid x \in G \right\}$ is the finite set of minimal covering primes of Q. In particular, there is a one-to-one correspondence between the G-prime ideals of T and the finite G-orbits of primes of T.

EXERCISES

    1. Show that the assumption of Theorem 15.8 that G is finite can be replaced by the weaker hypothesis that any nonzero ideal of R contains a nonzero G-stable ideal. This requires taking a closer look at Lemma 15.7, proving Step 2 along the lines of Step 1.

    In the remaining three problems let R be a centrally closed prime ring with extended centroid $C = \mathbf{Z}(R)$ and let T be any C-algebra. We obtain some results of [**116**].

    2. If I is any nonzero ideal of $R \otimes_C T$, prove that there exist $0 \neq r \in R$ and $0 \neq t \in T$ with $r \otimes t \in I$. Conclude that if I is prime, then either $I \cap R \neq 0$ or $I \cap T \neq 0$.

    3. Suppose S is a prime C-algebra generated by the commuting subalgebras R and T. Prove that $S \cong R \otimes_C T$ by mapping the tensor product onto S.

    4. Show that the prime ideals P of $R \otimes_C T$ with $P \cap R = 0$ are all of the form $R \otimes Q$ with Q a prime ideal of T.

16. Finite Groups and Incomparability

Let $R \subseteq S$ be a finite integral extension of commutative domains. In algebraic number theory one studies, among other things, the relationship between the prime ideals of these two rings, obtaining in particular (see [34,95]) the classical properties known as *Lying Over*, *Going Up*, *Going Down* and *Incomparability*.

In noncommutative ring theory, we are also interested in the relationships between the prime ideals in certain finite extensions $R \subseteq S$. Here there are various candidates for such extensions. For example we have: (i) crossed products $S = R*G$ with G finite; (ii) group-graded rings $S = R(G)$ with G finite; (iii) fixed rings $R = S^G$ where G is a finite group of automorphisms of S; (iv) finite centralizing, normalizing and intermediate extensions; (v) triangular or finite subnormalizing extensions; and (vi) extensions analogous to (i) and (iii) determined by finite dimensional restricted Lie algebras or more generally by finite dimensional Hopf algebras.

In this section, we will consider crossed products in detail and we will briefly discuss the situation for finite normalizing extensions. In later sections we will show how the theorems obtained here translate to analogous results on group-graded rings and on fixed rings.

We start with crossed products where the main tools are of course the one-to-one correspondences given in the preceding two sections. These ultimately relate the prime ideals of $R*G$ to those of a finite dimensional twisted group algebra. The following lemma lists the basic multiplicative and intersection properties of these correspondences. It uses the specialized notation of the previous two sections which we do not repeat.

Lemma 16.1. *Let $R*G$ be a crossed product satisfying the hypotheses of Section 14 or 15.*

i. In the notation of Section 14, if $I, J \lhd R*H$ then $I^{|G} J^{|G} \subseteq (IJ)^{|G}$ and $I^{|G} \cap J^{|G} = (I \cap J)^{|G}$.

ii. In the notation of Section 15, if I, J are G-stable ideals of $T = C^t[G_{\mathrm{inn}}]$ then we have $(I\Gamma' \cap \Gamma)(J\Gamma' \cap \Gamma) \subseteq (I \cap J)\Gamma' \cap \Gamma$ and $(I\Gamma' \cap \Gamma) \cap (J\Gamma' \cap \Gamma) = (I \cap J)\Gamma' \cap \Gamma$.

Proof. (i) The first part follows from Lemma 14.5(ii) and since $^{|G}$ is order preserving we have $(I \cap J)^{|G} \subseteq I^{|G} \cap J^{|G}$. For the other direction, note that, by Lemma 14.4(iii), $I^{|G} \cap J^{|G}$ is an ideal of $R*G$ with $\pi_H(I^{|G} \cap J^{|G}) \subseteq I \cap J$ so $I^{|G} \cap J^{|G} \subseteq (I \cap J)^{|G}$.

(ii) The first part was noted in the proof of Lemma 15.7, Step 3. For the second, since we have $\Gamma' = (S*G_{\mathrm{inn}})*(G/G_{\mathrm{inn}})$ and $S*G_{\mathrm{inn}} = S \otimes_C C^t[G_{\mathrm{inn}}]$, it follows that Γ' is a free right and left $C^t[G_{\mathrm{inn}}]$-module. Thus the map $I \mapsto I\Gamma'$ preserves intersections. ∎

The next result is clearly an analog of the Wedderburn theorem.

Theorem 16.2. [109] *Let $R*G$ be a crossed product with G finite and with R a G-prime ring.*

i. A prime ideal P of $R*G$ is minimal if and only if $P \cap R = 0$.

ii. There are finitely many minimal primes, say $P_1, P_2, \ldots, P_n$, and in fact $n \leq |G|$.

iii. $N = P_1 \cap P_2 \cap \cdots \cap P_n$ is the unique maximal nilpotent ideal of $R*G$ and $N^{|G|} = 0$.

iv. If Q is a minimal prime ideal of R, then $\{\, Q^x \mid x \in G \,\}$ is the set of all minimal primes of R and $\bigcap_{x \in G} Q^x = 0$.

Proof. Part (iv) is just Lemma 14.2(i)(ii). We show below that $R*G$ has $n \leq |G|$ primes P_i with $P_i \cap R = 0$. Furthermore, these are incomparable and satisfy $\bigcap_1^n P_i = N$ with $N^{|G|} = 0$. Once this is obtained, the result follows immediately. Indeed if P is any prime ideal of $R*G$, then P contains the nilpotent ideal N, so P contains some P_i. Thus since the P_i are incomparable, (i) and (ii) are proved. Finally (iii) follows since every nilpotent ideal is contained in each P_i. There are two cases to consider.

Case 1. *R is prime.*

Proof. We use the notation and conclusion of Theorem 15.8. Thus $T = C^t[G_{\mathrm{inn}}]$ is a finite dimensional C-algebra and hence has at most $\dim_C T = |G_{\mathrm{inn}}|$ prime ideals all of which are minimal. It therefore follows from Lemma 15.10 that T has finitely many G-prime ideals $Q_1, Q_2, \ldots, Q_n$ with $n \leq |G_{\mathrm{inn}}|$. Furthermore, these ideals are incomparable and $\bigcap_1^n Q_i = J$ is the Jacobson radical of T. Thus $J^{|G_{\text{inn}}|} = 0$.

Set $P_i = \Gamma'Q_i \cap \Gamma$ and $N = \Gamma'J \cap \Gamma$. Then it follows from Theorem 15.8 that $P_1, P_2, \ldots, P_n$ are precisely the prime ideals of $\Gamma = R*G$ with $P_i \cap R = 0$. Since the Q_i's are incomparable, so are the P_i's. Furthermore, by Lemma 16.1(ii), we have $\bigcap_1^n P_i = N$ and $N^{|G_{\text{inn}}|} = 0$. ∎

Case 2. *R is G-prime.*

Proof. Here we use the notation and conclusion of Theorem 14.7. By the prime case, $R*H/Q*H = (R/Q)*H$ has $n \leq |H|$ incomparable prime ideals $\tilde{L}_1, \tilde{L}_2, \ldots, \tilde{L}_n$ with $\tilde{L}_i \cap (R/Q) = 0$. Furthermore $\bigcap_1^n \tilde{L}_i = \tilde{J}$ is nilpotent with $\tilde{J}^{|H|} = 0$. Lifting these to ideals of $R*H$, we see that $R*H$ has $n \leq |H|$ incomparable prime ideals $L_1, L_2, \ldots, L_n$ with $L_i \cap R = Q$. Furthermore $\bigcap_1^n L_i = J$ and $J^{|H|} \subseteq Q*H$.

Set $P_i = L_i^{|G}$ and $N = J^{|G}$. Then Theorem 14.7 implies that $P_1, P_2, \ldots, P_n$ are precisely the prime ideals of $R*G$ with $P_i \cap R = 0$. Since the L_i's are incomparable, so are the P_i's. Finally $N^{|H|} \subseteq Q*H$ and $(Q*H)^{|G} = 0$ by Lemma 14.4(iv). Thus by Lemma 16.1(i) we have $\bigcap_1^n P_i = N$ and $N^{|H|} = 0$. Since $|H| \leq |G|$, the result clearly follows. ∎

There is obviously a good deal of additional information contained in the above. Aside from a precise description of the primes, there are sharper bounds on the number of primes in certain special cases. To start with, we have

Lemma 16.3. *Let $K^t[G]$ be a twisted group algebra with G a finite p-group and with K a field of characteristic p.*

i. *$K^t[G]/J(K^t[G])$ is a purely inseparable field extension of K of finite degree.*

ii. *$K^t[G]$ is commutative if and only if G is abelian.*

iii. *If $J(K^t[G]) \neq 0$ then there exists a central elementary abelian subgroup H of G with $J(K^t[H]) \neq 0$.*

Proof. (i) Let $\tilde{K}$ denote the algebraic closure of K and embed $K^t[G]$ in $\tilde{K}^t[G] = \tilde{K} \otimes_K K^t[G]$. Then by [**161**, Lemma 1.2.10], $\tilde{K}^t[G]$ is isomorphic to the ordinary group algebra $\tilde{K}[G]$. Furthermore, by [**161**, Lemma 3.1.6], $J = J(\tilde{K}[G])$ is a nilpotent ideal with $\tilde{K}[G]/J = \tilde{K}$ and, via the isomorphism, the same is true of $\tilde{K}^t[G]$. Thus if $P = J(\tilde{K}^t[G]) \cap K^t[G]$, then P is a nilpotent ideal of $K^t[G]$ with $K^t[G]/P$ a K-subalgebra of $\tilde{K}$. It follows that $K^t[G]/P$ is a finite dimensional field extension of K which is purely inseparable since it is generated by the images of the elements $\bar{x}$ with $x \in G$ and these satisfy $\bar{x}^{p^n} \in K$ for some n. Obviously $P = J(K^t[G])$.

(ii) It is clear that $K^t[G]$ is a commutative ring if and only if $\tilde{K} \otimes_K K^t[G] \cong \tilde{K}[G]$ is commutative and the latter occurs if and only if G is abelian.

(iii) Suppose first that G is nonabelian. By (i), it follows that the group of trivial units of $K^t[G']$ maps to K in $K^t[G]/J(K^t[G])$. Thus if W is any subgroup of G', then $J(K^t[G]) \cap K^t[W]$ is a nilpotent ideal of codimension 1 in $K^t[W]$. Now take W to be a subgroup of order p in $G' \cap \mathbf{Z}(G) \neq \langle 1 \rangle$.

On the other hand, suppose G is abelian so $K^t[G]$ is commutative by (ii). Let $H = \{ x \in G \mid x^p = 1 \}$ so that H is a central elementary abelian subgroup of G. Since $J(K^t[G]) \neq 0$ is nilpotent, we can choose $\alpha \in J(K^t[G])$ with $\alpha \neq 0$ but $\alpha^p = 0$. Furthermore, by commutativity, we can assume that $1 \in \mathrm{Supp}\ \alpha$ and say $\alpha = \sum_x k_x \bar{x}$ with $k_x \in K$. Then by commutativity again $0 = \alpha^p = \sum_x (k_x)^p \bar{x}^p$ and hence if $\beta = \pi_H(\alpha)$ then $\beta \neq 0$ and $\beta^p = 0$. Thus β generates a nonzero nilpotent ideal of $K^t[H]$. ∎

In view of the proof of Theorem 16.2, the preceding lemma yields

Proposition 16.4. *Let $R*G$ be given with G a finite p-group and with R a G-prime ring of characteristic p. Then $R*G$ has a unique minimal prime P which is necessarily nilpotent.*

Now it turns out that Theorem 16.2 contains within it all the basic relations between the prime ideals of R and of $R*G$ except for one aspect of Going Up. The original proof of that latter fact in [**110**]

used integrality methods. However here we use a simpler argument based on the following lemma of [**23**].

Lemma 16.5. *Let* $R \subseteq S$ *be an extension of rings and assume that* $S = \sum_1^n Rx_i$ *with* $Rx_i = x_i R$ *for all* i. *Let* Q *be an ideal of* R *which intersects nontrivially all nonzero ideals of* R. *Then there exists* $I \triangleleft S$ *with* $0 \neq I \cap R \subseteq Q$.

Proof. In this proof we consider the (R, R)-subbimodules of S so the hypothesis asserts that Q ess R. Let M be a maximal complement for R in S so that $R \oplus M$ ess S. It follows that $L = Q \oplus M$ ess S.

For each i, j set $L_{i,j} = \{ s \in S \mid x_i s x_j \in L \}$. It is then easy to see that $L_{i,j}$ is an (R, R)-subbimodule of S. Furthermore, $L_{i,j}$ ess S. Indeed, let U be a nonzero (R, R)-subbimodule of S. If $x_i U x_j = 0$, then $U \subseteq L_{i,j}$ and $L_{i,j} \cap U \neq 0$. On the other hand, if $x_i U x_j \neq 0$, then $x_i U x_j$ is a nonzero (R, R)-bimodule and L ess S so $x_i U x_j \cap L \neq 0$ and again $L_{i,j} \cap U \neq 0$. It now follows that $I = \bigcap_{i,j} L_{i,j}$ is essential in S.

Since L is an (R, R)-bimodule and $S = \sum x_i R = \sum Rx_i$, it is clear that $I = \{ s \in S \mid SsS \subseteq L \}$. Thus I is clearly the largest two-sided ideal of S contained in L. Next, since $I \subseteq L = Q \oplus M \subseteq R \oplus M$, we see that $I \cap R \subseteq Q$. Finally since I ess S, we have $I \cap R \neq 0$. ∎

We can now list the *Krull relations* in $R*G$ and it is convenient to do so diagrammatically. Thus for example the diagram in (iii) below indicates that $P_1 \supseteq P_2$ are primes of $R*G$, $Q_1 \supseteq Q_2$ are primes of R and that P_i lies over Q_i. Furthermore, one reads the first diagram of (iv) as follows. Given $Q_1 \supseteq Q_2$ primes of R and a prime P_2 of $R*G$ which lies over Q_2, there exists a prime P_1 of $R*G$ such that $P_1 \supseteq P_2$ and P_1 lies over Q_1. The other three charts have similar interpretations.

Theorem 16.6. [**109**] [**110**] *Let* $R*G$ *be a crossed product with* G *finite. The following basic relations hold between the prime ideals of* R *and of* $R*G$. *Here* P_1 *and* P_2 *denote primes of* $R*G$ *while* Q_1 *and* Q_2 *are primes of* R.

i. **Cutting Down.** *If P is a prime ideal of $R*G$, then there exists a prime ideal Q of R, unique up to G-conjugation, with Q minimal over $P \cap R$. Indeed, we have $P \cap R = \bigcap_{x \in G} Q^x$. When this occurs, we say that P lies over Q.*

ii. **Lying Over.** *If Q is a prime ideal of R, then there are primes $P_1, P_2, \ldots, P_n$ of $R*G$ with $1 \le n \le |G|$ such that P_i lies over Q.*

iii. **Incomparability.** *If*

and $P_1 \ne P_2$, then $Q_1 \ne Q_2$.

iv. **Going Up.** *With the above notation we have*

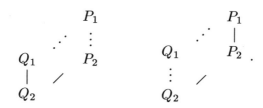

v. **Going Down.** *Similarly we have*

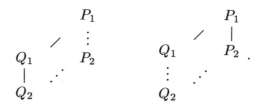

Proof. (i)(ii)(iii) These follow from Theorem 16.2 once we mod out by $(P \cap R)*G$, $\left[\bigcap_{x \in G} Q^x\right]*G$ and $\left[\bigcap_{x \in G}(Q_2)^x\right]*G$ respectively.

(iv) In the first part, note that $P_2 \cap R \subseteq Q_2 \subseteq Q_1$ and thus we can choose $P_1 \lhd R*G$ maximal subject to $P_1 \supseteq P_2$ and $P_1 \cap R \subseteq Q_1$. Since Q_1 is a prime ideal of R, it follows easily that P_1 is a prime of $R*G$. The goal is to show that Q_1 is minimal over $P_1 \cap R$. To this end, let $\tilde{S} = S/P_1$ and let $\tilde{R} = R/(R \cap P_1)$. Then $\tilde{R} \subseteq \tilde{S}$ satisfies the first hypothesis of the preceding lemma by taking $\{x_i\}$ to be the image of $\bar{G}$. Furthermore, by definition of P_1, it is clear that if $\tilde{I}$ is any nonzero ideal of $\tilde{S}$, then $\tilde{I} \cap \tilde{R} \not\subseteq \tilde{Q}_1 = Q_1/(R \cap P_1)$. Thus by Lemma 16.5, $\tilde{Q}_1$ is not an essential ideal of $\tilde{R}$, say $\tilde{Q}_1 \cap \tilde{J} = 0$ with $\tilde{J} \neq 0$. Finally, by (i) above, $\tilde{R}$ is semiprime and if $\tilde{L}$ is a minimal prime of $\tilde{R}$ not containing $\tilde{J}$, then $\tilde{L} \supseteq 0 = \tilde{Q}_1 \tilde{J}$ implies that $\tilde{L} \supseteq \tilde{Q}_1$. Thus equality occurs and $\tilde{Q}_1$ is a minimal prime of $\tilde{R}$ as required.

For the second part, let P_1 lie over Q_1. Then

$$Q_1 \supseteq P_1 \cap R \supseteq P_2 \cap R = \bigcap_{x \in G} (Q_2)^x$$

so $Q_1 \supseteq (Q_2)^x$ for some x. Now replace Q_1 by $(Q_1)^{x^{-1}}$.

(v) In the first part, let $L_1, L_2, \ldots, L_n$ be the primes lying over Q_2. Then by Theorem 16.2, the intersection $L_1 \cap L_2 \cap \cdots \cap L_n$ is nilpotent modulo $\left[\bigcap_{x \in G}(Q_2)^x\right] * G \subseteq P_1$. It follows that P_1 contains some L_i. For the second part, we merely note that $Q_1 \supseteq P_1 \cap R \supseteq P_2 \cap R$. Thus Q_1 contains a minimal covering prime of $P_2 \cap R$. ∎

In particular we have obtained a one-to-one correspondence between certain finite subsets of prime ideals of $R*G$ and the G-orbits of primes of R. It is natural now to see which ring theoretic properties are inherited via this correspondence. One such is primitivity.

Proposition 16.7. *Let P be a prime ideal of $R*G$ which lies over Q. Then P is primitive if and only if Q is primitive.*

Proof. We may assume that $P \cap R = \bigcap_{x \in G} Q^x = 0$. Suppose first that Q is a primitive ideal of R so that $Q = \text{ann}_R(W)$ for some irreducible R-module W. Let V be the induced module $V = W^{|R*G} = \sum_{x \in G} W \otimes \bar{x}$. If $I = \text{ann}_{R*G}(V)$, then it is easy to see that $\pi_{\langle 1 \rangle}(I) \subseteq \text{ann}_R(W \otimes 1) = Q$. Thus since $\pi_{\langle 1 \rangle}(I)$ is a G-stable ideal we have $\pi_{\langle 1 \rangle}(I) = 0$ and hence $I = 0$. Now, by Lemma 3.3, each $W \otimes \bar{x}$ is an irreducible R-module so V has finite length as an R-module and hence as an $R*G$-module. If $L_1, L_2, \ldots, L_t$ are the annihilators of the composition factors of V, then these are primitive ideals of $R*G$ with $L_1 L_2 \cdots L_t \subseteq I = 0$. Since $P \supseteq 0 = L_1 L_2 \cdots L_t$

we conclude that $P \supseteq L_i$ for some i and the result follows from the minimality of P.

In the other direction, let P be primitive with $P = \operatorname{ann}_{R*G}(V')$ for some irreducible $R*G$-module V'. Since $P \cap R = 0$, R acts faithfully on V' and, by Proposition 4.10, $V'|_R$ is completely reducible. Furthermore, since V is cyclic and G is finite, $V'|_R$ is finitely generated and hence has finite length. By taking the annihilators of the composition factors of $V'|_R$, there exist primitive ideals $L_1', L_2', \ldots, L_s'$ of R with $L_1' \cap L_2' \cap \cdots \cap L_s' = 0$. Since Q is a minimal prime of R, we conclude that $Q = L_j'$ for some j. ∎

Now it is obvious that Going Up, Going Down and Incomparability allow us to compare the prime lengths of R and of $R*G$. Recall that the *prime length* of R is the maximal n such that R has a chain of primes $Q_0 \subset Q_1 \subset \cdots \subset Q_n$ of length n. Of course if no such maximum exists then the prime length is infinite. Similarly, the *primitive length* of R is defined to be the maximal n with all Q_i primitive. If Q is a prime of R, then the *height* of Q is the maximal n such that there exists a chain of primes as above with $Q = Q_n$. Similarly, the *depth* of Q is the maximal length of chains of primes with $Q = Q_0$. The following is an immediate consequence of Theorem 16.6 and Proposition 16.7.

Corollary 16.8. [109] [110] *Let $R*G$ be a crossed product with G finite. Then the prime (or primitive) length of R is equal to the corresponding length of $R*G$. Furthermore if P is a prime of $R*G$ which lies over Q, then P and Q have the same height and the same depth.*

We close this section by briefly considering finite normalizing extensions.

Definition. Let $R \subseteq S$ be an extension of rings with the same 1. We say that S is a *finite centralizing extension* or a *liberal extension* of R if $S = \sum_1^n Rx_i$ and each x_i centralizes R. For example we could have $S = R[G]$ with G finite, or $S = \mathrm{M}_m(R)$, or $S = F \otimes_K R$ where R is a K-algebra and F is a finite extension field of K.

More generally we say that S is a *finite normalizing extension* of the ring R if $S = \sum_1^n Rx_i$ with $Rx_i = x_iR$ for all i. For example we could have $S = R*G$ with G finite since $\bar{x}R = R\bar{x}$ for all $x \in G$. Unlike crossed products, these extensions are closed under homomorphic images. In other words, if $I \triangleleft S$ and if $S \supseteq R$ is a finite normalizing (or centralizing) extension, then the same is true of $S/I \supseteq R/(R\cap I)$. Note that the beginning hypothesis of Lemma 16.4 asserts precisely that S is a finite normalizing extension of R.

Finally one also considers *intermediate extensions*. Here S is a finite normalizing extension of R and T is an intermediate ring. For example we could take $R \subseteq S = \mathrm{M}_n(R)$ with T the ring of upper triangular matrices. One can of course study either extension $R \subseteq T$ or $T \subseteq S$.

We will restrict our attention here to normalizing extensions; in particular, we will state the appropriate analog of Theorem 16.5. This result is the product of a series of papers. To start with, [**23**] and [**102**] contain Cutting Down, Lying Over and Going Up. Some of this is also proved in [**164**] in a rather nonstandard manner. However, by far the most difficult relation to prove is Incomparability. Here a special case appears in [**102**]; the general result is in [**72**]. We state the following without proof and, since most of the difficulty concerns Incomparability, we credit it to [**72**]. A proof and many additional details can now be found in the book [**121**].

Theorem 16.9. [**72**] *Let $S = \sum_1^n Rx_i$ be a finite normalizing extension of the ring R with $x_iR = Rx_i$ for all i.*

i. **Cutting Down.** *If P is a prime ideal of S, then $P \cap R$ has only finitely many minimal covering primes. In fact, if these are $Q_1, Q_2, \ldots, Q_t$ then $t \leq n$, $\bigcap_1^t Q_i = P \cap R$ and $R/Q_i \cong R/Q_1$ for all i. We say that P lies over each Q_i.*

ii. **Lying Over.** *If Q is a prime ideal of R, then there are finitely many primes $P_1, P_2, \ldots, P_s$ of S such that P_i lies over Q. Here $1 \leq s \leq n$.*

iii. **Incomparability.** *Let $P \subset I$ be ideals of S with P prime. Then $P \cap R \subset I \cap R$; indeed no prime ideal of R minimal over $P \cap R$ contains $I \cap R$.*

iv. **Going Up and Going Down.** *If P_1, P_2 are primes of S and if Q_1, Q_2 are primes of R, then we have at least*

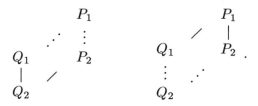

Unlike the crossed product case, the above does not yield a one-to-one correspondence between appropriate finite sets of prime ideals of R and of S. To be precise, if P and P' are prime ideals of S, write $P \sim P'$ if they lie over the same prime of R and extend this transitively to an equivalence relation. Similarly we obtain an equivalence relation on the primes of R and, in either case, equivalent primes are said to be *linked*. The following clever example of [**72**] exhibits some of the pathology which is present here.

Suppose A is a ring and let $\alpha_1, \alpha_2, \ldots, \alpha_n$ be finitely many epimorphisms $\alpha_i \colon A \to A$. Set $S = \mathrm{M}_n(A)$ and define $\rho \colon A \to S$ by $\rho \colon a \mapsto \mathrm{diag}(\alpha_1(a), \alpha_2(a), \ldots, \alpha_n(a))$. If $R = \rho(A)$, then using the matrix units of S, it follows easily (Exercise 5) that S is a finite normalizing extension of R. Let us assume for convenience that $\bigcap_i \mathrm{Ker}(\alpha_i) = 0$ so that ρ is an isomorphism from A to R.

Now every prime $\tilde{P}$ of S is of the form $\tilde{P} = \mathrm{M}_n(P)$ with P a prime ideal of A and we have

$$\rho^{-1}(\tilde{P} \cap R) = \{\, a \in A \mid \alpha_i(a) \in P \text{ for all } i \,\}$$

$$= \bigcap_1^n \alpha_i^{-1}(P).$$

Thus the prime $\tilde{Q}$ of R contains $\tilde{P} \cap R$ if and only if

$$Q = \rho^{-1}(\tilde{Q}) \supseteq \rho^{-1}(\tilde{P} \cap R) = \bigcap_1^n \alpha_i^{-1}(P)$$

and hence if and only if $Q \supseteq \alpha_i^{-1}(P)$ for some i. Thus the minimal covering primes of $\tilde{P} \cap R$ correspond to the minimal members of the set $\{\, \alpha_i^{-1}(P) \,\}$.

Conversely suppose we are given $\tilde{Q}$ and $Q = \rho^{-1}(\tilde{Q})$. If $\tilde{P}$ lies over $\tilde{Q}$, then $Q = \alpha_i^{-1}(P)$ for some i so we must have $Q \supseteq \mathrm{Ker}(\alpha_i)$ and $P = \alpha_i(Q)$ since α_i is onto. This limits the possibilities for P and allows for easy computations.

We specialize now to a concrete situation. Let $A = K[\zeta_1, \zeta_2, \ldots]$ be the polynomial ring in infinitely many variables, let $\alpha_1 = 1$ and let α_2 be the K-algebra epimorphism determined by $\alpha_2(\zeta_{i+1}) = \zeta_i$ and $\alpha_2(\zeta_1) = 0$. We set $S = \mathrm{M}_2(A)$ and note that ρ is indeed an isomorphism. We will consider ideals of A generated by variables and for convenience we write $(i_1, i_2, \ldots, i_m)$ for the ideal I generated by $\zeta_{i_1}, \zeta_{i_2}, \ldots, \zeta_{i_m}$. Furthermore $\tilde{I} = (i_1, i_2, \ldots, i_m)\tilde{}$ denotes the ideal $\mathrm{M}_2(I)$ of S and $I^\rho = (i_1, i_2, \ldots, i_m)^\rho$ denotes the ideal $\rho(I)$ of R.

In this notation we have $\alpha_1^{-1}(1,3) = (1,3)$ and $\alpha_2^{-1}(1,3) = (1,2,4)$. Thus since these ideals are incomparable, $(1,3)\tilde{}$ lies over both $(1,3)^\rho$ and $(1,2,4)^\rho$. Furthermore, it is easy to see (Exercise 5) that the diagram

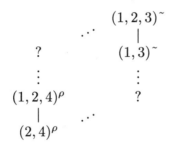

cannot be completed at either corner. Thus the missing Going Up and Going Down relations do indeed fail in general. [34] contains numerous commutative examples.

Finally for each $i \geq 1$ let $P_i = (1, 2, \ldots, i, i+2)$. Since $\alpha_2^{-1}(P_i) = P_{i+1}$ and P_i and P_{i+1} are incomparable, it follows that $\tilde{P}_i$ lies over both P_i^ρ and P_{i+1}^ρ. This yields the infinite diagram

$$\tilde{P}_{i-1} \qquad\qquad \tilde{P}_i \qquad\qquad \tilde{P}_{i+1}$$
$$\diagup \qquad \diagdown \quad \diagup \qquad \diagdown \quad \diagup \qquad \diagdown$$
$$P_i^\rho \qquad\qquad P_{i+1}^\rho$$

where the slanted lines represent Lying Over. Thus we see that both R and S contain infinite equivalence classes of linked primes. It is amusing to note that the above diagram is actually an upwards spiral since $P_{i+2} \supset P_i$ for all i.

1. Let $K = F(\zeta)$ where F is a field of characteristic $p > 0$ and let $K^t[G]$ be the twisted group algebra of the group $G = \langle x \rangle \times \langle y \rangle$ of order p^2 given by $\bar{x}\bar{y} = \bar{y}\bar{x}$, $\bar{x}^p = \zeta$ and $\bar{y}^p = 1 + \zeta$. If z is a nonidentity element of G, show that $\bar{z}^p$ is not a pth power in K and conclude that $K^t[\langle z \rangle]$ is a field. On the other hand, show that $\alpha = 1 + \bar{x} - \bar{y}$ generates a nilpotent ideal of $K^t[G]$. Hence $K^t[G]$ is not semiprime even though $K^t[H]$ is semiprime for all cyclic subgroups H of G.

2. Prove that the two versions of Incomparability given in Theorems 16.6 and 16.9 are equivalent. To this end, if Q is a prime of R and if I is an ideal of $S \supseteq R$ maximal subject to $I \cap R \subseteq Q$, note that I is a prime ideal of S.

3. Let $S = \sum_1^n Rx_i$ be a finite normalizing extension and suppose that $I \neq 0$ is an ideal of R of square 0. Set $I_0 = I$ and define $I_i \lhd R$ inductively for $i \leq n$ as follows. If I_{i-1} is given and $I_{i-1}x_iI_{i-1} = 0$ take $I_i = I_{i-1}$. Otherwise let $I_i = \{\, r \in I_{i-1} \mid x_i r \in I_{i-1}x_i \,\}$. Verify in either case that I_i is an ideal of R contained in I_{i-1}, that $I_i x_j I_i = 0$ for all $j \leq i$ and that $I_i \neq 0$. For the later, use $I_{i-1}x_iI_{i-1} \subseteq I_{i-1}x_i \cap x_iI_{i-1}$. Deduce that SI_nS is a nonzero ideal of S of square 0. This argument is from [102].

4. Let $S \supseteq R$ be a finite normalizing extension. If Q is a prime ideal of R show that there exists a prime P of S with Q minimal over $P \cap R$. For this use the argument in the proof of Theorem 16.6(iv) along with the result of the preceding problem. Now verify the assertions in Theorem 16.9(iv).

5. Consider the example discussed at the end of this section. First show that $S = M_n(A)$ is a finite normalizing extension of $R = \rho(A)$. Then verify that the diagram cannot be completed at either corner. For this, note that $(1, 2, 3)\tilde{}$ only lies over $(1, 2, 3)^\rho$. Furthermore since $(2, 4) \not\supseteq \text{Ker}(\alpha_2)$, $(2, 4)^\rho$ can only lie under $(2, 4)\tilde{}$.

6. If I is an ideal of the ring A, form $S = A \oplus (A/I)$ and map A into S via $\rho: a \mapsto a \oplus \bar{a}$. Show that S is a finite centralizing extension of $R = \rho(A)$. Use this to construct an example with R a commutative domain and with S a commutative ring having no maximal nilpotent ideal.

17. Primeness and Sylow Subgroups

There are numerous consequences of the preceding work, some of which we study in this and the next section. First let $R*G$ be a crossed product with G finite and consider the possibility that $R*G$ is prime or semiprime. We wish to see to what extent these conditions are inherited by subgroups or conversely inherited from analogous subgroup information. Of particular interest of course are the Sylow subgroups of G. Next let $R(G)$ be a G-graded ring with G finite. Then via duality, we expect that certain crossed product results should translate directly to this new context. We exhibit a few more examples of this phenomenon.

We actually start by considering a completely different topic, namely von Neumann regularity. Recall that a ring R is said to be *von Neumann regular* if for every $r \in R$ there exists $r' \in R$ with $rr'r = r$. This is directly equivalent to the fact that every cyclic right or left ideal of R is generated by an idempotent. Furthermore it is equivalent (see Exercises 1 and 2) to the fact that every finitely generated submodule of a free R-module is a direct summand. We will need the following subgroup version of Maschke's theorem.

Lemma 17.1. *Let $W \subseteq V$ be $R*G$ modules and let H be a subgroup of finite index in G. Suppose that W is a direct summand of V as an $R*H$-module and that $|G:H|^{-1} \in R$. Then W is a direct summand of V as an $R*G$-module.*

Proof. By hypothesis there exists an $R*H$-projection $\pi: V \to W$. Let $\{x_1, x_2, \ldots, x_n\}$ be a right transversal for H in G and, since $\frac{1}{n} \in R$, define $\rho: V \to W$ by $\rho(v) = \frac{1}{n} \sum_i \pi(v\bar{x}_i^{-1})\bar{x}_i$. Then ρ is certainly additive and maps to the $R*G$-submodule W. Furthermore let $\alpha \in R\bar{g}$ for some $g \in G$. Then there exists a permutation $i \mapsto i'$ of the subscripts such that $\bar{x}_i \alpha \bar{x}_{i'}^{-1} \in R*H$ for all i. Thus

$$\rho(v)\alpha = \frac{1}{n}\sum_i \pi(v\bar{x}_i^{-1})\bar{x}_i\alpha\bar{x}_{i'}^{-1}\cdot\bar{x}_{i'}$$

$$= \frac{1}{n}\sum_i \pi(v\bar{x}_i^{-1}\cdot\bar{x}_i\alpha\bar{x}_{i'}^{-1})\bar{x}_{i'}$$

$$= \frac{1}{n}\sum_i \pi(v\alpha\bar{x}_{i'}^{-1})\bar{x}_{i'} = \rho(v\alpha)$$

so ρ is an $R*G$-homomorphism. Since ρ is the identity on W, we conclude that $V = W \oplus U$ where $U = \mathrm{Ker}(\rho)$ is an $R*G$-submodule of V. ∎

As a consequence we have

Proposition 17.2. **[122] [182]** *Let $R*G$ be given and let H be a subgroup of G.*

  i. If $R*G$ is von Neumann regular, then so is $R*H$.

  ii. Suppose H has finite index in G and $|G : H|^{-1} \in R$. If $R*H$ is von Neumann regular, then so is $R*G$.

Proof. (i) Let $\pi_H : R*G \to R*H$ be the natural projection and let $\alpha \in R*H$. Since $R*G$ is von Neumann regular, there exists $\alpha' \in R*G$ with $\alpha\alpha'\alpha = \alpha$. Applying π_H we obtain $\alpha\pi_H(\alpha')\alpha = \alpha$ so $R*H$ is von Neumann regular.

(ii) Let W be a finitely generated submodule of the free $R*G$-module V and observe that $V_{|R*H}$ is also free. Furthermore, since $|G : H| < \infty$, it follows that $W_{|R*H}$ is finitely generated. Since $R*H$ is von Neumann regular, $W_{|R*H}$ is a direct summand of $V_{|R*H}$. Thus using $|G : H|^{-1} \in R$ and the preceding lemma, we conclude that W is a direct summand of V and therefore that $R*G$ is von Neumann regular. ∎

We remark that if R is Artinian, then semiprimitive, semiprime and von Neumann regular are equivalent conditions. Thus the above translates to a result on the semiprimeness of the twisted group algebra $K^t[G]$ with G finite. Note that the hypothesis $|G : H|^{-1} \in R$ of (ii) is certainly needed. Indeed for G finite, $K[G]$ is von Neumann regular if and only if $|G|^{-1} \in K$.

Now we turn to the primeness of $R*G$ with G finite. For this we will follow the route of the proof of Theorem 16.2. Namely we first consider twisted group algebras, then prime coefficient rings and then G-prime coefficients.

Let $K^t[G]$ be a twisted group algebra and for the moment suppose that K is an algebraically closed field of characteristic 0. Then the condition that $K^t[G]$ is prime is closely related to what group theorists call *groups of central type*. To be precise, the group H is of central type if it has a faithful irreducible K-representation of degree $|H : \mathbf{Z}(H)|^{\frac{1}{2}}$ and this is equivalent to a certain twisted group algebra $K^t[H/\mathbf{Z}(H)]$ being simple. It is known [46] that H is of central type if and only if (in a certain specified manner) all its Sylow subgroups are and that [78] a group of central type is necessarily solvable. The proof of the latter result requires, at the present time, the classification of all finite simple groups.

In the next two lemmas we use the following notation and observations. Let $K^t[H]$ be a semiprime twisted group algebra of the finite group H. Then

$$K^t[H] = \oplus \sum_i \mathrm{M}_{n_i}(D_i)$$

is a direct sum of full matrix rings over division algebras. As a consequence, if $\rho(H)$ denotes the right regular module of $K^t[H]$, then $\rho(H) = \sum_i n_i V_i$ where V_i is the unique irreducible module of $\mathrm{M}_{n_i}(D_i)$. We call n_i the multiplicity of V_i in $\rho(H)$ and this integer is uniquely determined by the Jordan-Holder theorem. Computing dimensions over K yields

$$|H| = \dim_K K^t[H] = \sum_i n_i \dim_K V_i.$$

Let L be a subgroup of H so that $K^t[L]$ is semiprime by Proposition 17.2(i). Since $K^t[H]$ is a free $K^t[L]$-module of rank $|H : L|$, it follows that $\rho(H)_{|K^t[L]} = |H : L|\rho(L)$. In the other direction, if W is a $K^t[L]$-module, then $W^{|K^t[H]} = W \otimes_{K^t[L]} K^t[H]$ is a right $K^t[H]$-module. Clearly $\dim_K W^{|K^t[H]} = |H : L|\dim_K W$ and $\rho(L)^{|K^t[H]} = \rho(H)$.

Suppose σ is an automorphism of $K^t[H]$ so that σ permutes the $K^t[H]$-modules as in Section 3. Indeed if V is given, then $V^\sigma = \{v^\sigma \mid v \in V\}$ with module action defined by $v^\sigma \alpha^\sigma = (v\alpha)^\sigma$ for all $\alpha \in K^t[H]$. Assume in addition that σ stabilizes $K^t[L]$. Then clearly $(V^\sigma)_{|K^t[L]} = (V_{|K^t[L]})^\sigma$. Conversely if W is a $K^t[L]$-module, then $(W^{|K^t[H]})^\sigma \cong (W^\sigma)^{|K^t[H]}$ via the isomorphism $(w \otimes \alpha)^\sigma \mapsto w^\sigma \otimes \alpha^\sigma$. Observe that if V_i is the unique irreducible module of $M_{n_i}(D_i)$, then V_i^σ is the unique irreducible of $M_{n_i}(D_i)^\sigma$.

Now let A be a finite group of automorphisms of $K^t[H]$ normalizing both K and the group $\mathcal{H}$ of trivial units. Then A acts on $\mathcal{H}/K^\bullet \cong H$. Furthermore, $K^t[H]$ is A-simple if and only if there is precisely one A-orbit of irreducible $K^t[H]$-modules. Note that, in any case, all irreducible modules in a fixed A-orbit occur with the same multiplicity in $\rho(H)$.

Finally let p be a prime and let A_p be a Sylow p-subgroup of A. Then A_p is a p-group which permutes the Sylow p-subgroups of H. Since the number of Sylow subgroups is congruent to 1 modulo p, it follows that A_p must normalize one such subgroup. Let H_p be a fixed Sylow p-subgroup of H normalized by A_p. Thus A_p acts as automorphisms on $K^t[H_p]$. For any positive integer n, we use $|n|_p$ to denote its p-part; when vertical lines are already present, we will not repeat them so for example $|H|_p = |H_p|$.

Lemma 17.3. *With the above notation, if $K^t[H]$ is A-simple, then $K^t[H_p]$ is A_p-simple.*

Proof. Since $K^t[H]$ is A-simple, we have $\rho(H) = aV$ where V is the sum of an A-orbit of irreducible modules. Write $\rho(H_p) = bW + U$ where W is the sum of an A_p-orbit of irreducible modules and U contains the remaining irreducibles. Now

$$aV_{|K^t[H_p]} = \rho(H)_{|K^t[H_p]}$$
$$= |H : H_p|\rho(H_p) = |H : H_p|(bW + U)$$

so $a \mid |H : H_p|b$ and hence $|a|_p \mid b$.

Furthermore, let T be a right transversal for A_p in A. Then $W^{|K^t[H]}$ is A_p-stable, so $\sum_{\tau \in T}(W^{|K^t[H]})^\tau$ is A-stable and hence equals cV for some integer c. Computing dimensions then yields

$$c \dim_K V = \dim_K \sum_{\tau \in T}(W^{|K^t[H]})^\tau$$

$$= |T| \cdot |H : H_p| \cdot \dim_K W$$

so since $|T| = |A : A_p|$ is a p'-number, $|\dim_K V|_p \mid \dim_K W$. Thus

$$|a \dim_K V|_p \mid b \dim_K W.$$

But $a \dim_K V = |H|$ and $b \dim_K W \leq |H_p| = |H|_p$. We conclude therefore that $b \dim_K W = |H_p|$ so $U = 0$ and hence $K^t[H_p]$ is A_p-simple. ■

Lemma 17.4. *With the above notation, write $\rho(H) = aV + U$ where V is the sum of an A-orbit of irreducible modules and U contains the remaining irreducibles. If $K^t[H_p]$ is A_p-simple, then $|H_p| \mid a \dim_K V$. In particular, if $K^t[H_p]$ is A_p-simple for all primes p dividing $|H|$, then $K^t[H]$ is A-simple.*

Proof. By assumption, $\rho(H_p) = bW$ where W is the sum of an A_p-orbit of irreducible modules. Now V is A-stable so $V_{|K^t[H_p]}$ is A_p-stable and hence $V_{|K^t[H_p]} = cW$ for some integer c. Thus $\dim_K W \mid \dim_K V$. Furthermore

$$b(W^{|K^t[H]}) = \rho(H_p)^{|K^t[H]} = \rho(H) = aV + U$$

so clearly $b \mid a$. Combining these yields

$$b \dim_K W \mid a \dim_K V$$

and, since $b \dim_K W = |H_p|$, the first fact follows.

Finally if the above holds for all $p \mid |H|$, then $|H| \mid a \dim_K V$ so $U = 0$ and $K^t[H]$ is A-simple. ■

Note that if $R*G$ is prime, then R must be G-prime. It is now a simple matter to prove

Theorem 17.5. [113] *Let $R*G$ be a crossed product with G finite and with R a G-prime ring. Suppose Q is a minimal prime of R and let H be its stabilizer in G. For each prime number p, let H_p be a Sylow p-subgroup of H, let G_p be a Sylow p-subgroup of G containing H_p and let*

$$R_p*G_p = \left[R/\bigcap\nolimits_{x \in G_p} Q^x\right]*G_p$$

be the naturally obtained crossed product. Then $R*G$ is prime if and only if R_p*G_p is prime for each $p \mid |G|$.

Proof. There are two cases to consider.

Case 1. *R is prime.*

Proof. Here we have $Q = 0$, $H = G$ and $R_p = R$. Thus the goal is to show that $R*G$ is prime if and only if $R*G_p$ is prime for all $p \mid |G|$. In the notation of Section 15, if $T = \mathbf{C}_{S*G}(S) = C^t[G_{\mathrm{inn}}]$, then G acts on T as automorphisms normalizing C and the group of trivial units. Furthermore, by Corollary 15.9, $R*G$ is prime if and only if $C^t[G_{\mathrm{inn}}]$ is G-prime. Let G_p be a Sylow p-subgroup of G so that $(G_{\mathrm{inn}})_p = G_p \cap G_{\mathrm{inn}} = (G_p)_{\mathrm{inn}}$ is a Sylow p-subgroup of $G_{\mathrm{inn}} \lhd G$.

Suppose $R*G$ is a prime ring so that $C^t[G_{\mathrm{inn}}]$ is G-prime. Since $C^t[G_{\mathrm{inn}}]$ is a finite dimensional algebra, it follows immediately that $C^t[G_{\mathrm{inn}}]$ is also semiprime and G-simple. Thus, by Lemma 17.3, we see that $C^t[(G_{\mathrm{inn}})_p]$ is G_p-simple and, by Corollary 15.9 again, $R*G_p$ is prime.

Conversely suppose $R*G_p$ is a prime ring for all $p \mid |G|$. Then by Corollary 15.9, $C^t[(G_{\mathrm{inn}})_p]$ is G_p-prime for all primes p. Again this implies that each $C^t[(G_{\mathrm{inn}})_p]$ is semiprime and G_p-simple and, by Proposition 17.2(ii) applied to $p = \operatorname{char} C$, we see that $C^t[G_{\mathrm{inn}}]$ is semiprime. Now Lemma 17.4 applies and we conclude that $C^t[G_{\mathrm{inn}}]$ is G-simple and hence that $R*G$ is prime. ∎

Case 2. *R is G-prime.*

Proof. Here we use the notation of the statement of the theorem. Then by Corollary 14.8, $R*G$ is prime if and only if $(R/Q)*H$ is prime and hence, by the above, if and only if $(R/Q)*H_p$ is prime for all p. But note that G_p acts on $R_p = R/\left[\bigcap_{x \in G_p} Q^x\right]$ and $H_p = H \cap G_p$ is the stabilizer in G_p of the minimal prime $Q/\left[\bigcap_{x \in G_p} Q^x\right]$. Thus by Corollary 14.8 again, R_p*G_p is prime if and only if $(R/Q)*H_p$ is prime. This completes the proof. ∎

We remark that primeness is not in general inherited by subgroups even in the case of twisted group algebras (see Exercise 5).

We now change topics and consider group-graded rings. The goal is to obtain appropriate analogs of Theorems 16.2 and 16.6. Let $S = R(G)$ be a G-graded ring with G finite. Recall from Section 2 that the smash product $S\#G^*$ is given by

$$S\#G^* = \oplus \sum_{x \in G} p_x S$$

with $1 = \sum_{x \in G} p_x$ a decomposition of 1 into orthogonal idempotents and with

$$p_x s_{x^{-1}y} = p_x s p_y = s_{x^{-1}y} p_y$$

for all $x, y \in G$ and $s \in S$. Furthermore, G acts on $S\#G^*$ via $(p_x s)^g = p_{g^{-1}x} s$ for all $x, g \in G$ and $s \in S$ and duality, Theorem 2.5, asserts that $(S\#G^*)G \cong \mathrm{M}_G(S)$. Here $\mathrm{M}_G(S)$ is the $|G| \times |G|$ matrix ring over S. A precise realization of this isomorphism is given in Lemma 2.4(ii).

A graded ideal I of S is said to be *graded prime* if for all graded ideals A, B of S, the inclusion $AB \subseteq I$ implies that A or B is contained in I. In particular, S is a *graded prime ring* when 0 is a graded prime ideal. If L is any ideal of S, we let L_G denote the largest graded ideal contained in L. Clearly $L_G = \sum_{x \in G}(L \cap S_x)$.

Again if I is a graded ideal of S, then the graded epimorphism $S \to S/I$ gives rise to an epimorphism $S\#G^* \to (S/I)\#G^*$ whose kernel is clearly $(S\#G^*)I = \sum_{x \in G} p_x I$. Thus $I(S\#G^*) = (S\#G^*)I$ is an ideal of $S\#G^*$ which we abbreviate as $I\#G^*$. In view of the homomorphism $S \to S/I$, it is clear that $(I\#G^*) \cap S = I$. Note that $I\#G^*$ is G-stable and conversely, by Lemma 4.7, every G-stable ideal of $S\#G^*$ is of this form.

Lemma 17.6. *Let I be a graded ideal of $S = R(G)$. The following are equivalent.*

    i. I is graded prime.

    ii. $I = P_G$ for some prime ideal P of S.

    iii. $I\#G^*$ is a G-prime ideal of $S\#G^*$.

Proof. (i) $\Leftrightarrow$ (ii) Let I be graded prime. Since $I = I_G$, we can choose, by Zorn's lemma, $P \triangleleft S$ maximal with $P_G = I$. If A, B are ideals of S properly larger than P, then $A_G, B_G \supset I$ so $A_G B_G \not\subseteq I = P_G$ and hence $AB \not\subseteq P$. Thus P is prime.

Conversely let $I = P_G$ with P prime. If A', B' are graded ideals not contained in I, then $A', B' \not\subseteq P$ so $A'B' \not\subseteq P$ and hence $A'B' \not\subseteq P_G = I$.

(i) $\Leftrightarrow$ (iii) Suppose first that I is graded prime and let A, B be G-stable ideals of $S\#G^*$ with $AB \subseteq I\#G^*$. Then by Lemma 4.7, $A \cap S$ and $B \cap S$ are graded ideals of S with $(A \cap S)(B \cap S) \subseteq I\#G^* \cap S = I$. Thus since I is graded prime, say $A \cap S \subseteq I$. By Lemma 4.7 again, $A = (A \cap S)(S\#G^*) \subseteq I(S\#G^*) = I\#G^*$.

Finally let $I\#G^*$ be a G-prime ideal of $S\#G^*$ and let A', B' be graded ideals of S with $A'B' \subseteq I$. Then $A'\#G^*$ and $B'\#G^*$ are G-invariant ideals of $S\#G^*$ with product contained in $I\#G^*$. Since $I\#G^*$ is G-prime, say $A'\#G^* \subseteq I\#G^*$. Then $A' \subseteq I\#G^* \cap S = I$ and I is graded prime. $\blacksquare$

We can now obtain the analog of Theorem 16.2.

Theorem 17.7. [39] *Let $S = R(G)$ be a G-graded ring with G finite and assume that S is graded prime.*

    i. A prime ideal P of S is minimal if and only if $P_G = 0$.

    ii. There are finitely many minimal primes, say $P_1, P_2, \ldots, P_n$ with $n \leq |G|$.

    iii. $N = P_1 \cap P_2 \cap \cdots \cap P_n$ is the unique maximal nilpotent ideal of S and $N^{|G|} = 0$.

iv. *There exists a prime Q of $S\#G^*$ unique up to G-conjugation with $\bigcap_{x \in G} Q^x = 0$.*

Proof. By Lemma 17.6, $S\#G^*$ is G-prime. We consider the skew group ring $(S\#G^*)G \cong M_G(S)$ and we note that (iv) follows from Theorem 16.2(iv). Furthermore, by Theorem 16.2(ii)(iii), there are $n \leq |G|$ minimal primes $\mathcal{P}_1, \mathcal{P}_2, \ldots, \mathcal{P}_n$ of $M_G(S)$ such that $\mathcal{P}_1 \cap \mathcal{P}_2 \cap \cdots \cap \mathcal{P}_n = \mathcal{N}$ satisfies $\mathcal{N}^{|G|} = 0$. Setting $\mathcal{P}_i = M_G(P_i)$ and $\mathcal{N} = M_G(N)$, we see that $P_1, P_2, \ldots, P_n$ are the $n \leq |G|$ minimal primes of S and that $P_1 \cap P_2 \cap \cdots \cap P_n = N$ satisfies $N^{|G|} = 0$. This yields (ii) and (iii).

For (i) we require the precise realization of the isomorphism $M_G(S) \cong (S\#G^*)G$ given by Lemma 2.4(ii). Here

$$S\#G^* = S\{1\}$$
$$= \left\{ \alpha \in M_G(S) \mid \alpha(x, y) \in R(x^{-1}y) \text{ for all } x, y \in G \right\}.$$

Now by Theorem 16.2(i), $\mathcal{P} = M_G(P)$ is a minimal prime if and only if $\mathcal{P} \cap S\{1\} = 0$. Since this is clearly equivalent to $P \cap R(g) = 0$ for all $g \in G$, we see that P is minimal if and only if $P_G = 0$. ∎

We are actually more interested in the relationship between the primes of $S = R(G)$ and those of R. For this we require the following classical result.

Lemma 17.8. *Let e be a nonzero idempotent of the ring R. Then the map $\varphi \colon P \mapsto ePe = P \cap eRe$ determines a one-to-one correspondence between the set $\mathcal{R}_e$ of prime ideals of R not containing e and the set of all prime ideals of eRe. Moreover if P, P_1 and P_2 are in $\mathcal{R}_e$, then $P_1^\varphi \subseteq P_2^\varphi$ if and only if $P_1 \subseteq P_2$ and P^φ is primitive if and only if P is primitive.*

Proof. Observe that if A is an ideal of eRe, then RAR is an ideal of R with $RAR \cap eRe = e(RAR)e = (eRe)A(eRe) = A$.

Suppose that P is a prime ideal of R not containing e. We show that $P^\varphi = ePe$ is prime in eRe. First since $e \notin P$, we have $e \notin P^\varphi$

and hence P^φ is a proper ideal of eRe. Now suppose A_1 and A_2 are ideals of eRe with $A_1 A_2 \subseteq ePe$. Then

$$(RA_1 R)(RA_2 R) = RA_1(eRe)A_2 R = RA_1 A_2 R \subseteq P$$

so $RA_i R \subseteq P$ for some i. Thus $A_i = e(RA_i R)e \subseteq ePe$ and ePe is prime.

Now let $P_1, P_2 \in \mathcal{R}_e$ and assume that $P_1^\varphi \subseteq P_2^\varphi$. Then $eP_1 e \subseteq eP_2 e$ so $(ReR)P_1(ReR) \subseteq P_2$ and since $e \notin P_2$ we have $P_1 \subseteq P_2$. Conversely if $P_1 \subseteq P_2$, then $P_1^\varphi \subseteq P_2^\varphi$. In particular, we see that φ is injective.

Next we show that φ is onto. For this let Q be a prime ideal of eRe and observe that $e(RQR)e = Q$. We can now let P be the unique largest ideal of R with $Q = ePe = P \cap eRe$ and it follows easily that P is prime. Furthermore $e \notin P$ since $e \notin Q$ so $Q = P^\varphi$ as required.

It remains to verify the primitivity assertions. Suppose first that $P \in \mathcal{R}_e$ is a primitive ideal of R and say $P = \text{ann}_R(V)$ with V an irreducible right R-module. Then Ve is a nonzero eRe-submodule of V which is irreducible. Indeed if $0 \neq U \subseteq Ve$ is an eRe-submodule, then from $UR = V$ we have $Ve = URe = U(eRe) = U$. Since ePe is easily seen to be the annihilator of Ve in eRe, it follows that ePe is primitive.

Conversely let Q be a primitive ideal of eRe and say that Q is the annihilator of the irreducible module W. Write $W \cong eRe/X$ where X is a maximal right ideal of eRe. Then $X_1 = XR \oplus (1-e)R$ is a right ideal of R with

$$X_1 \cap eRe = XR \cap eRe = XRe = X$$

and thus X_1 is contained in a maximal right ideal Y of R. Clearly $X = Y \cap eRe$ since X is maximal and $e \notin Y$. Hence, if V is the irreducible R-module R/Y, then $V_{|eRe} \supseteq eRe/X \cong W$. Let $P = \text{ann}_R(V)$. It follows that P is a primitive ideal of R with $P \cap eRe = ePe \subseteq \text{ann}_{eRe}(W) = Q$ and, in particular, $e \notin P$. For the reverse inclusion, observe that $eRQ = (eRe)Q \subseteq X \subseteq Y$ and $(1-e)RQ \subseteq (1-e)R \subseteq Y$. Thus $RQ \subseteq Y$ so RQR is a two-sided ideal of R

contained in Y. This yields $RQR \subseteq \operatorname{ann}_R(V) = P$ and hence $Q \subseteq ePe = P^\varphi$. We have therefore shown that $Q = P^\varphi$ and since P is primitive, the result follows. ∎

Finally we have the analog of Theorem 16.6.

Theorem 17.9. [39] *Let $S = R(G)$ be a group-graded ring with G finite. The following basic relations hold between the prime ideals of R and of S.*

   i. **Cutting Down.** *If P is a prime ideal of S, then there are $n \leq |G|$ primes $Q_1, Q_2, \ldots, Q_n$ of R minimal over $P \cap R$ and we have $P \cap R = Q_1 \cap Q_2 \cap \cdots \cap Q_n$. We say that P lies over each Q_i.*

   ii. **Lying Over.** *If Q is a prime ideal of R, then there exists a prime P of S such that P lies over Q. Indeed there are $m \leq |G|$ such primes P_i which lie over Q and these are precisely the primes satisfying $(P_i)_G = P_G$.*

   iii. **Incomparability.** *Given the lying over diagram*

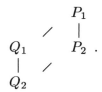

If $P_1 \neq P_2$, then $Q_1 \neq Q_2$.

   iv. **Going Up and Going Down.** *We have at least*

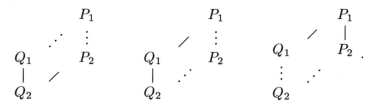

Proof. We will use the smash product $S \# G^* = \sum_{x \in G} p_x S$. Let I be a graded ideal of S. Since $I \# G^* = \sum_{x \in G} p_x I$ we have $p_1(I \# G^*) p_1 = p_1 I p_1 = I_1 p_1$ where $I_1 = I \cap S_1 = I \cap R$. In particular $p_1(S \# G^*) p_1 = R p_1$ and since p_1 centralizes R we see that $R p_1$ is ring isomorphic to R. By the previous lemma there is a one-to-one correspondence between the primes $\mathcal{Q}$ of $S\#G^*$ not containing p_1 and the primes Q of R. Specifically the map $\varphi: \mathcal{Q} \mapsto Q$ is given by $p_1 \mathcal{Q} p_1 = Q p_1 = \mathcal{Q}^\varphi p_1$.

(i)(ii) Let P be a prime ideal of S so that P_G is graded prime by Lemma 17.6. It follows from Theorem 17.7(iv), applied to the graded prime ring S/P_G, that there exists a prime ideal $\mathcal{Q}$ of $S\#G^*$ with $P_G\#G^* = \bigcap_{x\in G} \mathcal{Q}^x$. Since $P_G \cap R = P \cap R$, we conclude that

$$(P \cap R)p_1 = p_1(P_G\#G^*)p_1 = (P_G\#G^*) \cap p_1(S\#G^*)p_1$$

$$= \bigcap_{x\in G} [\mathcal{Q}^x \cap p_1(S\#G^*)p_1] = \bigcap_{x\in G} p_1 \mathcal{Q}^x p_1.$$

Thus, deleting those $\mathcal{Q}^x$ containing p_1, we obtain $P \cap R = \bigcap'(\mathcal{Q}^x)^\varphi$. This shows that $P \cap R$ is a semiprime ideal of R with finitely many minimal covering primes. Furthermore since the $\mathcal{Q}^x$ are incomparable, so are the $(\mathcal{Q}^x)^\varphi$. This proves (i).

Conversely let Q be a prime of R and let $\mathcal{Q}$ be the prime of $S\#G^*$ with $\mathcal{Q}^\varphi = Q$. Set $I = S \cap \bigcap_{x\in G} \mathcal{Q}^x$. Then $\bigcap_{x\in G} \mathcal{Q}^x = I\#G^*$ by Lemma 4.7 and I is graded prime by Lemma 17.6. In view of the work of the preceding paragraph, a prime P of S lies over Q if and only if $P_G = I$. Again by Lemma 17.6, at least one such P exists. Furthermore by Theorem 17.7(ii) applied to S/I, there are at most $|G|$ such primes. Thus (ii) is proved.

(iii)(iv) The third diagram of (iv) is trivial since $Q_1 \supseteq P_1 \cap R \supseteq P_2 \cap R$ implies that Q_1 contains a minimal covering prime of $P_2 \cap R$.

For the rest let $\mathcal{Q}_i$ be the prime of $S\#G^*$ with $(\mathcal{Q}_i)^\varphi = Q_i$ and define the graded ideal I_i of S by $I_i = S \cap \bigcap_{x\in G}(\mathcal{Q}_i)^x$. Then $\mathcal{Q}_1 \supseteq \mathcal{Q}_2$ so $I_1 \supseteq I_2$ and, by (ii) above, P_i lies over Q_i if and only if $(P_i)_G = I_i$.

Suppose P_2 is given. Then $(P_2)_G = I_2 \subseteq I_1$ so, by Zorn's lemma, we can find $P_1 \supseteq P_2 + I_1$ maximal subject to $(P_1)_G = I_1$. Then P_1 is prime and the first diagram is satisfied.

One the other hand suppose P_1 is given. Then $P_1 \supseteq I_1 \supseteq I_2$ so $P_1 \supseteq P_2$ where P_2 is a minimal covering prime of I_2. By Theorem 17.7(i) applied to S/I_2 we have $(P_2)_G = I_2$ and part (iv) is proved.

Finally suppose both P_1 and P_2 are given and that $Q_1 = Q_2$. Then $I_1 = I_2$ so $(P_1)_G = (P_2)_G$. By Theorem 17.7(i) applied to

$S/(P_2)_G$ we conclude that $P_1 = P_2$ and (iii) follows. This completes the proof. ∎

We remark that the above yields a one-to-one correspondence between certain finite sets of primes of R and of S. Furthermore, the missing Going Up result does not hold in general. An example will be offered immediately after Theorem 28.3, a result on primes in fixed rings.

Corollary 17.10. [39] *Let* $S = R(G)$ *be given with* G *finite and let* P *be a prime of* S *lying over* Q. *If* $P \subseteq I \triangleleft S$ *with* $I \cap R \subseteq Q$, *then* $I = P$.

Proof. Choose $P' \supseteq I$ maximal with $P' \cap R \subseteq Q$. Then it follows easily that P' is a prime of S and that P' lies over Q. By Theorem 17.9(iii), $P' = P$ and hence $I = P$. ∎

EXERCISES

1. Let R be a von Neumann regular ring. Show that every cyclic right ideal αR is generated by an idempotent. Now let $\alpha R + \beta R$ be a two generator right ideal and let e be an idempotent with $\alpha R = eR$. Prove first that $\alpha R + \beta R = eR + (1 - e)\beta R$. Now let f be an idempotent with $(1 - e)\beta R = fR$ and set $g = f(1 - e)$. Prove that g is an idempotent orthogonal to e and that $fR = gR$. Conclude that $\alpha R + \beta R = (e + g)R$.

2. Again let R be von Neumann regular. Prove first that every finitely generated right ideal of R is a direct summand of R and hence is projective. Now suppose that W is a finitely generated submodule of a free R-module V. Prove that W is a direct summand of V. For this it suffices to assume that V is finitely generated. Proceed by induction on the number of generators of V by projecting W into the last summand.

3. Let $K^t[G]$ be a twisted group algebra of the finite group G over the field K. For $x \in G$ define the map $\lambda_x \colon \mathbf{C}_G(x) \to K^\bullet$ by $\bar{y}^{-1}\bar{x}\bar{y} = \lambda_x(y)\bar{x}$ for all $y \in \mathbf{C}_G(x)$. Prove that λ_x is a homomorphism

with kernel $\mathbf{C}_G^t(x) = \{\, y \in \mathbf{C}_G(x) \mid \bar{x}\bar{y} = \bar{y}\bar{x} \,\}$ and that $(\lambda_x)^m = 1$ if $x^m = 1$. Show that $\mathbf{Z}(K^t[G])$ has as a K-basis the set of *conjugacy class sums* for all elements $x \in G$ with $\mathbf{C}_G(x) = \mathbf{C}_G^t(x)$.

4. Let $K^t[G]$ be as above and assume when necessary that K is algebraically closed. Prove that $K^t[G]$ is simple if and only if $\mathbf{Z}(K^t[G]) = K$. For this, first show, using the homomorphism λ_x, that if $\mathbf{Z}(K^t[G]) = K$ and if char $K = p > 0$, then G is a p'-group. Observe in addition that if $K^t[G]$ is simple then $|G|$ is a square integer.

5. Assume that the field K contains a nonidentity p^{th} root of unity ϵ for some prime p. Let $G = \langle x_1, \ldots, x_n, y_1, \ldots, y_n \rangle$ be an elementary abelian p-group of order p^{2n} and define $K^t[G]$ by the relations $(\bar{x}_i)^p = (\bar{y}_i)^p = 1$ and $\bar{x}_i \bar{y}_i = \epsilon \bar{y}_i \bar{x}_i$ for $1 \le i \le n$. All other pairs of generators are assumed to commute. Prove that $K^t[G]$ is simple. Find a subgroup H of G with $K^t[H]$ not simple.

6. Let $S = R(G)$ be G-graded with G a finite group. Discuss the analog of Corollary 16.8 in this context. In addition, discuss the one-to-one correspondence between finite sets of prime ideals of R and of S.

18. Semiprimeness and Sylow Subgroups

Now let us move on to consider $R*G$ and the semiprime condition. As we will see, this property is inherited by all subgroups and inherited from at least certain subgroups. In addition, the results translate quite readily to group-graded rings via duality. We start interestingly enough by considering twisted group rings over commutative von Neumann regular rings.

If C is a central subring of a ring R and M is a maximal ideal of C, then we denote by R_M the central localization by the multiplicatively closed set $C \setminus M$. Note that $C \setminus M$ need not consist of regular elements here, so the homomorphism from R to R_M is not necessarily an embedding.

Lemma 18.1. *Let C be a central subring of R and let M be a maximal ideal of C.*

   i. If C is von Neumann regular, then C_M is a field.

   ii. If R is semiprime and a finitely generated C-module, then R_M is semiprime.

   iii. R is embedded in $\prod_M R_M$.

Proof. (i) If $c \in M$, then there is an idempotent $e \in Cc \subseteq M$ with $ce = c$. Hence $c(1 - e) = 0$ and, since $1 - e \in C \setminus M$, the image of c in C_M is zero. Thus all nonzero elements of C_M have numerators in $C \setminus M$ and are invertible.

   (ii) Let I be an ideal of R_M of square zero and let $\alpha \in R$ map to a numerator in I. By assumption we can write $R = \sum_i r_i C$, a finite sum. Now for each i, we have $\alpha r_i \alpha = 0$ in R_M so there exists $c_i \in C \setminus M$ with $\alpha r_i \alpha c_i = 0$. Set $c = \prod_i c_i \in C \setminus M$. Then $(\alpha c) r_i (\alpha c) = 0$ for all i so, since $R = \sum_i r_i C$, we have $(\alpha c) R (\alpha c) = 0$. But R is semiprime so $\alpha c = 0$ and thus the image of α in R_M is zero. In other words, $I = 0$.

   (iii) If $0 \neq r \in R$ then $r_C(r) \neq C$ and the result follows. ∎

   This of course applies to twisted group rings $C^t[G]$ with C commutative and G finite. Observe that if M is a maximal ideal of C, then the freeness of $C^t[G]$ over C implies that $C^t[G]_M = (C_M)^t[G]$. Moreover if H is a subgroup of G, then $C^t[H]_M = (C_M)^t[H]$ is the naturally embedded subring of $(C_M)^t[G]$.

Lemma 18.2. Let $C^t[G]$ be a twisted group ring with C a commutative von Neumann regular ring and with G finite.

   i. If $C^t[G]$ is semiprime and H is a subgroup of G, then $C^t[H]$ is semiprime.

   ii. If $C^t[G]$ is not semiprime, then there exists a prime p and an elementary abelian p-subgroup P of G such that $C^t[P]$ is not semiprime and C has p-torsion.

Proof. (i) Let M be a maximal ideal of C. Since $C^t[G]$ is semiprime, Lemma 18.1(ii) implies that $C^t[G]_M = (C_M)^t[G]$ is semiprime. Thus since C_M is a field, Proposition 17.2(i) implies that $(C_M)^t[H] = C^t[H]_M$ is semiprime. Lemma 18.1(iii) now yields the result.

(ii) Suppose I is a nonzero nilpotent ideal of $C^t[G]$. Then by Lemma 18.1(iii) there exists a maximal ideal M of C such that the image of I in $C^t[G]_M$ is nonzero; thus $(C_M)^t[G]$ is not semiprime. Since C_M is a field, it follows that C_M must have characteristic p for some prime $p > 0$ and that, by Proposition 17.2(ii) applied to a Sylow p-subgroup of G and Lemma 16.3(iii), there exists an elementary abelian p-subgroup P of G with $(C_M)^t[P] = C^t[P]_M$ not semiprime. We conclude from Lemma 18.1(ii) that $C^t[P]$ is not semiprime. Moreover, since the image of $p \in C$ is zero in C_M, it follows that some element of $C \setminus M$ has p-torsion. ∎

We now discuss certain crossed product reductions. Let $R*G$ be given with G finite. Then G permutes the ideals of R and we recall that $0 \neq A \triangleleft R$ is a trivial intersection ideal if for all $x \in G$ either $A^x = A$ or $A^x \cap A = 0$. These occur as follows.

Lemma 18.3. *Let $0 \neq A$ be an ideal of R. Then A contains a trivial intersection ideal $\tilde{A}$.*

Proof. Since $A \neq 0$ and G is finite, we can choose a subset $X \subseteq G$ maximal with $\tilde{A} = \bigcap_{x \in X} A^x \neq 0$. Then $\tilde{A}^g = \bigcap_{x \in Xg} A^x$ so $\tilde{A}^g \cap \tilde{A} \neq 0$ implies that $Xg \subseteq X$. Thus $Xg = X$ and $\tilde{A}^g = \tilde{A}$. ∎

Lemma 18.4. *Let $R*G$ be semiprime and let G be finite.*
  i. If $N \triangleleft G$, then $R*N$ is semiprime.
  ii. Suppose A is a trivial intersection ideal of R with stabilizer $H \subseteq G$. Then $A*H$ contains no nonzero nilpotent ideal of $R*H$.
  iii. Let A and H be as above and set $B = \mathrm{r}_R(A)$. Then $(R/B)*H = (R*H)/(B*H)$ is semiprime.

Proof. (i) If I is a nilpotent ideal of $R*N$, then the finite sum $L = \sum_{x \in G} I^x$ is a G-stable ideal of $R*N$ which is also nilpotent. Hence $L(R*G) = L*G$ is a nilpotent ideal of $R*G$ and therefore $L*G = 0$.

(ii) Let $I \triangleleft R*H$ with $I \subseteq A*H$ and $I^2 = 0$. If $g \in H$, then $I \bar{g} I \subseteq I^2 = 0$. On the other hand, if $g \in G \setminus H$ then from $A^g A \subseteq A^g \cap A = 0$ it follows that $I \bar{g} I \subseteq (A*H)\bar{g}(A*H) = 0$. Thus $I(R*G)I = 0$ so $I = 0$ since $R*G$ is semiprime.

(iii) Observe that R is semiprime by (i) so we have $B \cap A = 0$ and $(A + B)/B$ is essential in R/B. If $(R/B)*H$ has a nonzero nilpotent ideal, then this ideal meets $((A + B)/B)*H$ nontrivially. Since $A \cap B = 0$, the inverse image of this intersection in $A*H$ is a nilpotent ideal of $R*H$, contradicting (ii). Thus $(R/B)*H$ is semiprime. ∎

We remark that (iii) above applies when $A \neq 0$ is a G-invariant ideal of R and $H = G$. The main results of this section are proved by techniques which originated in [58] and were further developed in the work of [109] and [172] as described in Section 15. Therefore some of the preliminary lemmas here are just variants of earlier work and we will be somewhat skimpy with their proofs. For example, the following is essentially contained in Proposition 12.5.

Lemma 18.5. *Let I be an ideal of $R*G$ and let Λ be a subset of G of minimal size with $1 \in \Lambda$ and $I \cap (R*\Lambda) \neq 0$. For each $x \in \Lambda$ define*

$$A_x = \left\{ r \in R \;\middle|\; \sum_{y \in \Lambda} r_y \bar{y} \in I \text{ and } r = r_x \right\}.$$

i. *Each A_x is a nonzero ideal of R.*
ii. *For each $x \in \Lambda$ there is an additive bijection $f_x : A_1 \to A_x$ with $(ras)f_x = r(af_x)s^{\bar{x}^{-1}}$ for all $r, s \in R$ and $a \in A_1$. Here $f_1 = 1$.*
iii. *$I \cap (R*\Lambda) = \left\{ \sum_{x \in \Lambda}(af_x)\bar{x} \mid a \in A_1 \right\}$.*

Proof. Part (i) is clear since $I \cap (R*\Lambda)$ is an (R, R)-bimodule. Now we note that the 1-coefficient of $\alpha \in I \cap (R*\Lambda)$ uniquely determines α. Indeed if $\alpha, \alpha' \in I \cap (R*\Lambda)$ have the same 1-coefficient, then $\alpha - \alpha' \in I \cap (R*\Lambda)$ has smaller support and hence must be zero. This allows us to define $f_x : A_1 \to A_x$ with $f_1 = 1$ so that (iii) is satisfied. Furthermore, f_x is onto by definition of A_x and it is one-to-one by the minimality of Λ. Finally

$$\sum_x [(ras)f_x]\bar{x} = ras = r \sum_x (af_x)\bar{x}s = \sum_x r(af_x)s^{\bar{x}^{-1}}\bar{x}$$

and (ii) follows. ∎

We will also need the symmetric Martindale ring of quotients, this time applied to semiprime rings. As we mentioned in Section 10, the construction and basic properties of $Q_s(R)$ in this context is a routine generalization. Of course it requires that we restrict our attention to essential ideals of R, that is ideals A with $\ell_R(A) = r_R(A) = 0$. With this understanding, $Q_s(R)$ exists and is characterized as in Propositions 10.4 and 10.6. Furthermore, most of the proof of Lemma 10.9(ii)(iii) goes through so that if $S = Q_\ell(R)$ or $Q_s(R)$, then every automorphism of R extends uniquely to one of S. Also if $C = \mathbf{C}_S(R)$, then C is the center of both $Q_\ell(R)$ and $Q_s(R)$. The key difference here is that C, the extended centroid of R, is no longer a field in general. Indeed we have

Lemma 18.6. [4] *If R is a semiprime ring, then its extended centroid C is a commutative von Neumann regular ring.*

Proof. Let $c \in C$ and let A' be an essential ideal of R with $A'c \subseteq R$. If $B = \ell_{A'}(c)$ then $B \lhd R$ and we can choose $A \lhd R$ so that $A \oplus B$ is essential in A' and hence in R. Notice that $Ac \lhd R$ and choose $D \lhd R$ so that $Ac \oplus D$ is essential in R. Since multiplication by c is one-to-one on A, we can define a left R-module homomorphism $f \colon Ac \oplus D \to R$ by $(ac + d)f = a$. Of course f determines an element $c' \in Q_\ell(R)$ satisfying $(ac + d)c' = a$ so, for all $a \in A$ and $b \in B$, we have

$$(a + b)cc'c = (ac)c'c = ac = (a + b)c$$

and thus $cc'c = c$. Finally if $r \in R$ then, for all $a \in A$ and $d \in D$, we have

$$(ac + d)rc' = (arc + dr)c' = ar = (ac + d)c'r$$

so $rc' = c'r$ and $c' \in \mathbf{Z}(Q_s(R)) = C$ as required. ∎

This explains our concern with commutative von Neumann regular rings in Lemmas 18.1 and 18.2. The next result is a variant of Lemma 12.2.

Lemma 18.7. *Let R be a semiprime ring, σ an automorphism of R and A and B ideals of R. Suppose $f \colon A \to B$ is an additive bijection satisfying $(ras)f = r(af)s^\sigma$ for all $r, s \in R$ and $a \in A$.*

i. The image $B' = A'f$ of any ideal $A' \subseteq A$ is an ideal contained in B. Furthermore $\ell_R(A') = \ell_R(B')$ is σ-invariant.

ii. If A is an essential ideal, then there exists a unit $q \in Q_s(R)$ such that $af = aq$ for all $a \in A$. Moreover, σ extended to $Q_s(R)$ is the inner automorphism induced by q.

Proof. (i) It is clear that $B' = A'f$ is an ideal of R. Now for all $r \in R$ we have $(rA')f = r(A'f) = rB'$. Hence since f is a bijection, it follows that $\ell_R(A') = \ell_R(B')$. Similarly since $(A'r)f = (A'f)r^\sigma = B'r^\sigma$, it follows that $r_R(A')^\sigma = r_R(B')$. Now use the fact that right and left annihilators are equal in semiprime rings.

(ii) Suppose A is essential so that f determines an element $q \in Q_\ell(R)$ satisfying $af = aq$ for all $a \in A$. Then $(as)q = (as)f = (af)s^\sigma = aqs^\sigma$ implies that $sq = qs^\sigma$ for all $s \in R$. In particular, $qA^\sigma = Aq \subseteq R$ so $q \in Q_s(R)$. Finally since f is a bijection, its inverse g exists and satisfies $g: B \to A$ with $(rbs)g = r(bg)s^{\sigma^{-1}}$. Furthermore, B is essential by (i) and hence g determines an element $q' \in Q_s(R)$. Clearly $q' = q^{-1}$ and $sq = qs^\sigma$ yields $q^{-1}sq = s^\sigma$ for all $s \in R$. But two automorphisms which agree on R must also agree on $Q_s(R)$, so the result follows. ∎

The final lemma follows as in Proposition 12.4 and Lemma 15.4. It requires minimal proof here.

Lemma 18.8. *Let R be a semiprime ring, $S = Q_s(R)$ and let C be the extended centroid of R. Given $R*G$, there exists a uniquely defined crossed product $S*G$ which contains $R*G$. Assume in addition that the action of $\bar{G}$ on R becomes inner when extended to S.*

i. $S*G = S \otimes_C E$ where $E = \mathbf{C}_{S*G}(S) = C^t[G]$, some twisted group ring of G over C.

ii. If $R*G$ is semiprime, then so is $C^t[G]$.

iii. If $H \subseteq G$, then $S*H = S \otimes_C (E \cap (S*H))$ and $E \cap (S*H) = C^t[H]$ where the latter is the natural subring of $C^t[G]$.

iv. If I is an ideal of $R*G$ then

$$\tilde{I} = \{\, \alpha \in E \mid A\alpha \subseteq I \text{ for some essential } A \vartriangleleft R \,\}$$

is an ideal of E. Furthermore, $I^2 = 0$ implies $(\tilde{I})^2 = 0$.

Proof. (ii) If $I \neq 0$ is an ideal of $C^t[G]$ of square 0, then $S \otimes_C I$ is a nonzero ideal of $S*G$ of square 0. Now observe that any nonzero ideal of $S*G$ meets $R*G$ nontrivially.

(iv) Conversely let $I \lhd R*G$ with $I^2 = 0$. Let $\alpha, \beta \in \tilde{I}$ and say $A\alpha, B\beta \subseteq I$. Then $(AB)\alpha\beta = (A\alpha)(B\beta) \subseteq I^2 = 0$ and, since AB is essential in R, we conclude that $\alpha\beta = 0$. Thus $(\tilde{I})^2 = 0$. ∎

We can now obtain the first main result of this section.

Theorem 18.9. [169] *Let $R*G$ be a crossed product with G finite and let H be a subgroup of G. If $R*G$ is semiprime, then so is $R*H$.*

Proof. We proceed by induction on $|G| + |H|$. Observe that R is semiprime by Lemma 18.4(i). Suppose by way of contradiction that I is a nonzero ideal of $R*H$ of square 0. If N denotes the normal closure of H in G, then $H \subseteq N \lhd G$ and $R*N$ is semiprime by Lemma 18.4(i) again. Thus by induction, $G = N$ is the normal closure of H.

We apply Lemma 18.5 (with $G = H$) to the ideal I of $R*H$ and we let $\Lambda \subseteq H$ and the nonzero ideals $A_x \lhd R$ be as in that lemma. If $H_1 = \langle \Lambda \rangle$, then H_1 is a subgroup of H and $I \cap (R*H_1) \neq 0$. By induction, $H = H_1 = \langle \Lambda \rangle$.

By Lemma 18.3, A_1 contains a nonzero trivial intersection ideal D with stabilizer G_1. Since $D \subseteq A_1$ and R is semiprime, we observe that $0 \neq D(I \cap (R*\Lambda))D \subseteq I \cap (R*\Lambda)$. It follows by the minimality of Λ that the supports in $D(I \cap (R*\Lambda))D$ cannot be properly smaller than Λ. Hence for each $x \in \Lambda$ we have $0 \neq DA_x \bar{x}D \subseteq D\bar{x}D$. Thus since D is a trivial intersection ideal, we have $x \in G_1$ and hence $G_1 \supseteq \langle \Lambda \rangle = H$. Furthermore if $B = \ell_R(D)$ then, by Lemma 18.4(iii), $(R*G_1)/(B*G_1) \cong (R/B)*G_1$ is semiprime. On the other hand, since $B \cap D = 0$, the image of I in $(R/B)*G_1$ is a nonzero nilpotent ideal of $(R/B)*H$. Induction now implies that $G_1 = G$.

Let $'$ denote the natural map $R \to R/B$ onto the semiprime ring $R' = R/B$ and for each $x \in \Lambda$ let $D_x = Df_x$ where f_x is the bijection given by Lemma 18.5. Then by Lemma 18.7(i), each D_x is an ideal of R with left annihilator $\ell_R(D_x) = \ell_R(D) = B$. Thus $B \cap D_x = 0$ and it follows that each f_x gives rise to an additive bijection $f'_x : D' \to D'_x$. Clearly each D'_x is essential in R' and $(r'd's')f'_x = r'(d'f'_x)(s')^{\bar{x}^{-1}}$ for all $r', s' \in R'$ and $d' \in D'$.

It follows from Lemma 18.7(ii) that each $\bar{x}$ with $x \in \Lambda$ acts as an inner automorphism on $S = Q_s(R')$. Hence since $\langle \Lambda \rangle = H$ and G is the normal closure of H, we see that $\bar{G}$ is inner on S. Lemma 18.8(i)(ii) now implies that $R'*G \subseteq S*G = S \otimes_C C^t[G]$ where C is the extended centroid of the semiprime ring R'. Furthermore, C is von Neumann regular, by Lemma 18.6, and $C^t[G]$ is semiprime. By Lemma 18.2(i), $C^t[H]$ is also semiprime.

Finally $I \supseteq \{ \sum_{x \in \Lambda} (df_x)\bar{x} \mid d \in D \}$ so I', the image of I in $R'*H$ contains $\{ \sum_{x \in \Lambda} (d'f'_x)\bar{x} \mid d' \in D' \}$. By Lemma 18.7(ii), there is a unit $q_x \in S$ with $d'f'_x = d'q_x$ for all $d' \in D'$ and such that the action of $\bar{x}^{-1}$ on S is the inner automorphism induced by q_x. Thus in $S*H$ we see that $I' \supseteq D'\gamma$ where $\gamma = \sum_{x \in \Lambda} q_x \bar{x} \in \mathbf{C}_{S*H}(S) = C^t[H]$. In other words, in the notation of Lemma 18.8(iv) (with $G = H$), γ is a nonzero element of the ideal $\widetilde{I'}$ of $C^t[H]$. But $(I')^2 = 0$ so $(\widetilde{I'})^2 = 0$ and this contradicts the fact that $C^t[H]$ is semiprime. The result follows. ∎

We now consider the converse direction. For convenience we say that a crossed product $R*G$ with G finite is *weakly semiprime* if $R*P$ is semiprime for $P = \langle 1 \rangle$ and for every elementary abelian p-subgroup P of G such that R has p-torsion. In view of the preceding theorem, if $R*G$ is semiprime then it is weakly semiprime. Furthermore, it is obvious that the weakly semiprime condition is inherited by subgroups. The second main result of this section is

Theorem 18.10. [169] *Let $R*G$ be a crossed product with G finite. If $R*G$ is weakly semiprime, then it is semiprime.*

Proof. We proceed by induction on $|G|$. If G_1 is a proper subgroup of G, then $R*G_1$ is weakly semiprime and hence, by induction, $R*G_1$ is semiprime.

Suppose by way of contradiction that I is a nonzero ideal of $R*G$ of square 0. We apply Lemma 18.5 and use its notation; in addition we set $A = A_1$ and $B = \ell_R(A)$. Since $I \cap (R*\Lambda) \neq 0$, it is clear that

$\langle \Lambda \rangle = G$. Moreover, by Lemma 18.7(i), $B = \ell_R(A_x)$ for all $x \in \Lambda$ and B is x-invariant. Thus B is G-invariant and hence $B = \ell_R(A^g)$ for all $g \in G$. It follows easily (see Exercise 4), since R is semiprime and $B \cap A^g = 0$, that B is the left annihilator of $\prod_{g \in G} A^g$, where the product is taken in any order. Thus $D = \bigcap_{g \in G} A^g$ is a G-stable ideal of R contained in $A = A_1$ with $\ell_R(D) = B$. For all $x \in \Lambda$ set $D_x = Df_x$. Then by Lemma 18.7(i), each D_x is an ideal of R with $\ell_R(D_x) = \ell_R(D) = B$.

Let $'$ denote the natural map $R \to R/B$ onto the semiprime ring $R' = R/B$. Clearly each D'_x is essential in R'. Furthermore, each f_x gives rise to an additive bijection $f'_x: D' \to D'_x$ satisfying $(r'd's')f'_x = r'(d'f'_x)(s')^{\bar{x}^{-1}}$ for all $r', s' \in R'$ and $d' \in D'$. It follows from Lemma 18.7(ii) that each $\bar{x}$ with $x \in \Lambda$ acts as an inner automorphism on $S = Q_s(R')$. Hence, since $G = \langle \Lambda \rangle$, we have $\bar{G}$ inner on S and Lemma 18.8 applies. Thus $R' * G \subseteq S * G = S \otimes_C C^t[G]$ where C is the extended centroid of R'. Furthermore if R' has p-torsion, then so does R and it follows from Lemma 18.4(iii) and the hypothesis that $R' * G$ is weakly semiprime.

In addition, if C has p-torsion then so does R'. Thus we see from Lemma 18.8(ii)(iii) that $C^t[G]$ is weakly semiprime. Since C is a commutative von Neumann regular ring, we conclude from Lemma 18.2(ii) that $C^t[G]$ is semiprime.

Finally $I \supseteq \{ \sum_{x \in \Lambda} (df_x)\bar{x} \mid d \in D \}$ so I', the image of I in $R' * G$ contains $\{ \sum_{x \in \Lambda} (d'f'_x)\bar{x} \mid d' \in D' \}$. By Lemma 18.7(ii), there exists a unit $q_x \in S$ with $d'f'_x = d'q_x$ for all $d' \in D'$ and such that the action of $\bar{x}^{-1}$ on S is the inner automorphism induced by q_x. Thus in $S * G$ we see that $I' \supseteq D'\gamma$ where $\gamma = \sum_{x \in \Lambda} q_x \bar{x} \in \mathbf{C}_{S*H}(S) = C^t[H]$. In other words, in the notation of Lemma 18.8(iv), γ is a nonzero element of the ideal $\tilde{I}'$ of $C^t[G]$. But $(I')^2 = 0$ so $(\tilde{I}')^2 = 0$ and this contradicts the fact that $C^t[G]$ is semiprime. Therefore, the result follows. ∎

The above of course generalizes Theorem 4.4. It has numerous crossed product corollaries, but we mention just two.

Corollary 18.11. [169] *Let $R*G$ be a crossed product with G finite and R semiprime. Then $R*G$ is semiprime if and only if, for each* prime p such that R has p-torsion, we have $R*G_p$ semiprime where G_p is a Sylow p-subgroup of G.

Proof. One implication is immediate from Theorem 18.9. For the other, suppose R has p-torsion and P is an elementary abelian p-subgroup of G. Then P is conjugate to a subgroup of G_p and $R*G_p$ is semiprime by hypothesis. It follows that $R*P$ is semiprime, so $R*G$ is weakly semiprime; Theorem 18.10 yields the result. ∎

Corollary 18.12. [169] Let $R*G$ be a crossed product with G finite and let H be a subgroup of G. If $R*H$ is semiprime and R has no $|G : H|$-torsion, then $R*G$ is semiprime.

Proof. Since $R*H$ is semiprime, so is R. Furthermore, if R has p-torsion, then p is prime to $|G : H|$ so H contains a Sylow p-subgroup G_p of G. By Theorem 18.9, $R*G_p$ is semiprime; Corollary 18.11 now yields the result. ∎

We close this section with the following consequence of duality.

Theorem 18.13. [180] Let $S = R(G)$ be a G-graded ring with G finite.

   i. If $R(G)$ is semiprime and H is a subgroup of G, then $R(H)$ is semiprime.

   ii. Conversely suppose $R(P)$ is semiprime for $P = \langle 1 \rangle$ and for all elementary abelian p-subgroups P of G such that S has p-torsion. If S is component regular, then $S = R(G)$ is semiprime.

Proof. We use Corollary 2.7 and its notation throughout.

   (i) If $R(G)$ is semiprime, then so is $S\{1\}\bar{G}$ by Corollary 2.7(ii). Thus Theorem 18.9 implies that $S\{1\}\bar{H}$ is semiprime and Corollary 2.7(i) yields the result.

   (ii) Note that the torsions of S and of $S\{1\}$ are identical. Hence Corollary 2.7(ii) and the hypothesis imply that $S\{1\}\bar{G}$ is weakly semiprime. Thus $S\{1\}\bar{G}$ is semiprime by Theorem 18.10 and Corollary 2.7(i) again yields the result. ∎

EXERCISES

1. Let C be a central subring of R. Prove that R is von Neumann regular if and only if R_M is for all maximal ideals M of C. Furthermore, if C is von Neumann regular and if R is a finitely generated C-module, show that R is von Neumann regular if and only if it is semiprime.

2. Let P be a Sylow p-subgroup of G and let K have characteristic $p > 0$. If $K^t[G]$ is semiprime, show that P is abelian and has a normal complement in G. For this, observe from Lemma 16.3(i) that $K^t[P]$ is a purely inseparable field extension of K. Thus for all $g \in \mathbf{N}_G(P)$ the automorphism induced by $\bar{g}$ on $K^t[P]$ is trivial. Deduce that P is in the center of its normalizer.

3. Suppose C is a commutative von Neumann regular ring and that G is a finite group. If $C^t[G]$ is semiprime, prove that G', the commutator subgroup of G, is a p'-group for all primes p for which $p^{-1} \notin C$. For this, merely observe that if $p^{-1} \notin C$, then C_M is a field of characteristic p for some maximal ideal M.

4. Let R be a semiprime ring and let $A_1, A_2, \ldots, A_n$ and B be ideals of R. Suppose that $B = \ell_R(A_i)$ for all i. Prove that $B = \ell_R(A_1 A_2 \cdots A_n)$ and then that $B = \ell_R(A_1 \cap A_2 \cap \cdots \cap A_n)$. Furthermore show that $A_1 \oplus B$ is essential in R.

5. Let $R*G$ be a crossed product with G finite and R semiprimitive. Use Theorem 4.2 to show that $R*G$ is semiprime if and only if it is semiprimitive. Now translate Theorems 18.9 and 18.10 into results on semiprimitivity.

6. Find an example of a group graded ring $R(G)$ and two Sylow p-subgroups P_1 and P_2 of G such that $R(P_1)$ is semiprime while $R(P_2)$ is not. Thus show that direct analogs of Corollaries 18.11 and 18.12 fail in this context. Where explicitly is the crossed product structure used in the proofs of those results?

7. Let $S = R(G)$ be G-graded with G finite. Assume that S is component regular and that R is prime. Use Corollary 2.7, Theorem 17.5 and the argument of Theorem 18.13 to show that S is prime if and only if $R(G_p)$ is prime for each Sylow p-subgroup G_p of G. You should first observe that $S\{1\}$ is prime.

5 Prime Ideals –
The Noetherian Case

19. Orbitally Sound Groups

This chapter is concerned with prime ideals in Noetherian crossed products. Its theme is the interplay between the crossed product theory and that of polycyclic group algebras. The story begins with the work of [185] which comes tantalizingly close to completely answering the group algebra question. The remaining finite index problem is then settled in [114] using crossed product methods. Finally these results on $K[G]$ lift to analogous ones on $K^t[G]$ and then, via Theorems 14.7 and 15.8, to results on $R*G$ with R Noetherian.

We devote this section to the work of [185] on group algebras $K[G]$ with G polycyclic-by-finite. The proofs here are interesting but they are quite long and would really take us too far afield. Fortunately this material is already nicely presented in the original paper [185] and in the survey [168]. Thus, with two brief exceptions, we content ourselves here with merely discussing these results. Even this will take some time since it is necessary to first understand the key concepts and definitions.

The goal, of course, is to describe the primes of $K[G]$ and apriori there are two inductive ways to do this. First, suppose $\langle 1 \rangle \neq N \triangleleft G$. Then the homomorphism $G \to G/N$ extends to the natural algebra epimorphism $K[G] \to K[G/N]$. In particular, if Q is a prime of $K[G/N]$, then its complete inverse image P is a prime of $K[G]$. Thus if Q is adequately described in the "smaller" group algebra $K[G/N]$, then P is certainly well understood in $K[G]$. Second, let Z be the center of G so that Z is a finitely generated abelian group and $K[Z]$ is a finitely generated commutative K-algebra. We assume that the commutative theory tells us all we need to know about the primes of this ring. Suppose Q' is such a prime and that $P' = Q'K[G]$, its extension to $K[G]$, is also prime. Then certainly we understand P'. The main result of [185] essentially asserts that these two schemes describe all the primes of $K[G]$. More precisely, one must use the f.c. center $\Delta(G)$ instead of the center $\mathbf{Z}(G)$. Furthermore, this description does not apply to $K[G]$ but rather to $K[G_0]$ where G_0 is a certain characteristic subgroup of finite index in G. We begin our formal arguments by discussing this subgroup.

Definition. In this section, G will always be a polycyclic-by-finite group. Note that his property is inherited by subgroups and factor groups. In particular, all subgroups of G are finitely generated and therefore G satisfies the maximal condition on subgroups. We will use Γ to denote an arbitrary group.

Suppose Γ acts as permutations on a set Ω. An element $\omega \in \Omega$ is said to be *orbital,* or more precisely Γ-*orbital,* if its Γ-orbit is finite. In particular, when Γ acts on itself by conjugation then

$$\Delta(\Gamma) = \left\{ x \in \Gamma \mid |\Gamma : \mathbf{C}_\Gamma(x)| < \infty \right\}$$

is the subset of orbital elements. Furthermore, if V is a group on which Γ acts, we can refer to the Γ-orbital elements and subgroups of V.

Now let the polycyclic-by-finite group G permute its subgroups by conjugation. Then $H \subseteq G$ is an *orbital subgroup* if $|G : \mathbf{N}_G(H)| < \infty$. Furthermore H is said to be an *isolated orbital subgroup* if it is orbital and if for any other orbital subgroup $H_1 \supset H$ we have $|H_1 : H| = \infty$.

Lemma 19.1. *Let H be an orbital subgroup of G and define its isolator $i_G(H)$ by*

$$i_G(H) = \langle L \mid L \supseteq H \text{ is orbital and } |L : H| < \infty \rangle.$$

Then $i_G(H)$ is the unique isolated orbital subgroup of G containing H and having H as a subgroup of finite index.

Proof. We first show that if L_1, L_2 are orbital subgroups of G with $|L_i : H| < \infty$, then $|\langle L_1, L_2 \rangle : H| < \infty$. For this it suffices to assume that $G = \langle L_1, L_2 \rangle$ and that $\mathrm{core}_G H = \bigcap_{x \in G} H^x = \langle 1 \rangle$. Since these subgroups are all orbital and $|L_i : H| < \infty$, we can find a normal subgroup A of finite index in G which normalizes H and satisfies $[A, L_i] \subseteq H$ for $i = 1, 2$. Since $A \triangleleft G$, this yields $[A \cap H, L_i] \subseteq A \cap H$ so $A \cap H$ is normalized by $\langle L_1, L_2 \rangle = G$. Thus $A \cap H \subseteq \mathrm{core}_G H = \langle 1 \rangle$ and H is finite. We conclude that L_1 and L_2 are finite orbital subgroups of G and hence, by Lemma 5.1(iii), $G = \langle L_1, L_2 \rangle$ is finite.

Finally, since G satisfies the maximal condition on subgroups, we can choose L maximal subject to L being orbital and $|L : H| < \infty$. By the above, $L = i_G(H)$ has the appropriate properties. ∎

We note that if $x \in G$ normalizes the orbital subgroup H, then x normalizes $i_G(H)$ so $\mathbf{N}_G(i_G(H)) \supseteq \mathbf{N}_G(H)$. In other words, the isolated orbital subgroups have the largest normalizers among *nearby* orbital subgroups.

Definition. We say that G is *orbitally sound* if all its isolated orbital subgroups are normal. Since G is polycyclic-by-finite, this is equivalent (see Exercise 2) to any of the following three conditions applied to all orbital subgroups H of G.

1. There exist $N_1, N_2 \triangleleft G$ with $N_1 \subseteq H \subseteq N_2$ and $|N_2 : N_1| < \infty$.
2. There exists $N_1 \triangleleft G$ with $N_1 \subseteq H$ and $|H : N_1| < \infty$.
3. There exists $N_2 \triangleleft G$ with $H \subseteq N_2$ and $|N_2 : H| < \infty$.

We note that a group of Hirsch number ≤ 1 is orbitally sound since every subgroup of such a group is either finite or of finite index. Furthermore the orbitally sound condition is inherited by all homomorphic images and by subgroups of finite index. On the other hand, it is not inherited by arbitrary subgroups. For example, let $G = A \rtimes D$ be the semidirect product of A by D where $A = \langle a, b \rangle$ is free abelian of rank 2 and $D = \langle x, y \mid x^{-1}yx = y^{-1}, x^2 = 1 \rangle$ is infinite dihedral. The action here is given by $a^y = a^3 b$, $b^y = a^{-1}$, $a^x = b$ and $b^x = a$. Then G is orbitally sound, but $W = \langle A, x \rangle$ is not (Exercises 3 and 4).

For convenience, we record the following standard fact about finitely generated groups.

Lemma 19.2. *Let H be a subgroup of finite index in G. Then $M \subseteq H \subseteq G$ with M a characteristic subgroup of finite index in G.*

Proof. Let $|G : H| = n < \infty$. Since G is finitely generated, there are only finitely many homomorphisms of G into the symmetric group Sym_n. If M is the intersection of all their kernels, then M is clearly a characteristic subgroup of finite index in G. Finally, the permutation action of G on the right cosets of H yields such a homomorphism, so $H \supseteq M$ as required. ∎

For any polycyclic-by-finite group G we define $\text{nio}(G)$ to be the intersection of the normalizers of all isolated orbital subgroups of G. Thus, by definition, G is orbitally sound if and only if $G = \text{nio}(G)$. More generally we have

Theorem 19.3. [185] *Let G be a polycyclic-by-finite group. Then $\text{nio}(G)$ is a characteristic subgroup of finite index in G. Furthermore, it is the unique largest orbitally sound subgroup of finite index in G.*

The main thrust of this result is that G has some orbitally sound subgroup of finite index. Once this is proved, the subgroup found can be easily related to $\text{nio}(G)$ with the help of Lemma 19.1. Later in this section we will state a useful sufficient condition for G to be orbitally sound.

Definition. If I is an ideal of $K[G]$, we let

$$I^{\dagger} = \{\, x \in G \mid 1 - x \in I \,\}.$$

Notice that $G \subseteq K[G]$ and that $I^\dagger$ is the kernel of the homomorphism of G into the group of units of $K[G]/I$. Thus $I^\dagger \lhd G$. We say that I is *faithful* if $I^\dagger = \langle 1 \rangle$ or equivalently if G is faithfully embedded in $K[G]/I$. Furthermore, I is *almost faithful* if $I^\dagger$ is finite. Note that

$$I \supseteq \omega(K[I^\dagger]) = \left\{ \sum_{x \in I^\dagger} k_x x \ \Big| \ \sum_x k_x = 0 \right\}$$

where the latter is the *augmentation ideal* of $K[I^\dagger]$. Hence $I \supseteq \omega(K[I^\dagger]) \cdot K[G]$, the kernel of the natural epimorphism $K[G] \to K[G/I^\dagger]$, and therefore I is the complete inverse image of an ideal $\tilde{I}$ of $K[G/I^\dagger]$ which is clearly faithful.

Let P be a prime ideal of $K[G]$. Then P is said to be a *standard prime* if $P = (P \cap K[\Delta(G)]) \cdot K[G]$, and $P \cap K[\Delta(G)] = \bigcap_{x \in G} Q^x$ where Q is an almost faithful prime of $K[\Delta(G)]$. Note that G acts like a finite group on $K[\Delta(G)]$ since $\Delta(G)$ being finitely generated implies that $G/\mathbf{C}_G(\Delta(G))$ is finite. Thus $\{\, Q^x \mid x \in G \,\}$ is necessarily the finite set of minimal covering primes of the ideal $P \cap K[\Delta(G)]$ in $K[\Delta(G)]$. It is easy to see that a standard prime is almost faithful.

We say that P is *virtually standard* if the image of P in $K[G/P^\dagger]$ is a standard prime. If P is almost faithful and virtually standard, then it follows easily that P is standard. The main result is then

Theorem 19.4. [185] *If G is an orbitally sound polycyclic-by-finite group, then the almost faithful prime ideals of $K[G]$ are standard. Hence all prime ideals of $K[G]$ are virtually standard.*

Thus all the primes of $K[G]$ can be described via the two reductions mentioned in the introduction to this section. As we will see in Theorem 21.2(ii), the above property of prime ideals actually characterizes orbitally sound groups.

One consequence of the preceding theorem is that the prime and primitive lengths of $K[G]$ can be computed. For this we require some more terminology.

Definition. Let $\mathcal{Q}$ denote the field of rational numbers and $\mathcal{Z}$ the ring of integers. A finite dimensional $\mathcal{Q}[\Gamma]$-module V is said to be a *rational plinth* for Γ if V is an irreducible $\mathcal{Q}[\tilde{\Gamma}]$-module for all subgroups $\tilde{\Gamma}$ of finite index in Γ. Now let Γ act on a finitely generated torsion free abelian group A. Then A is a *plinth* for Γ if, in additive notation, $V = A \otimes_{\mathcal{Z}} \mathcal{Q}$ is a rational plinth. Thus A is a plinth if and only if no proper pure subgroup of A is Γ-orbital. Notice that if A is infinite cyclic, then $|\Gamma/\mathbf{C}_\Gamma(A)| \leq 2$ so in this case A is called a *centric plinth*. Otherwise, when rank $A \geq 2$, A is said to be an *eccentric plinth*.

A normal series

$$\langle 1 \rangle = G_0 \subseteq G_1 \subseteq \cdots \subseteq G_n = G$$

for the polycyclic-by-finite group G is called a *plinth series* for G if each quotient G_i/G_{i-1} is either finite or a plinth for G. It is not necessarily true that every G has a plinth series. However we have (see Exercise 7)

Lemma 19.5. *Any polycyclic-by-finite group G has a normal subgroup H of finite index with a plinth series.*

With G and H as above, let

$$\langle 1 \rangle = H_0 \subseteq H_1 \subseteq \cdots \subseteq H_n = H$$

be a plinth series for H. Then the number of infinite factors H_i/H_{i-1} is called the *plinth length* of G and is denoted by $\mathrm{pl}(G)$. Furthermore the number of infinite factors of rank ≥ 2 is called the *eccentric plinth length* of G and is denoted by $\mathrm{epl}(G)$. It is easy to see, by taking common refinements, that these parameters are independent of the choice of H and of the particular series for H. Note that $\mathrm{epl}(G) \leq \mathrm{pl}(G) \leq \hbar(G)$, where the latter is the Hirsch number of G. Moreover, G is nilpotent-by-finite if and only if either $\mathrm{pl}(G) = \hbar(G)$ or $\mathrm{epl}(G) = 0$.

Finally we distinguish two types of fields according to the nature of their multiplicative group $K^\bullet$. We say K is *absolute* if $K^\bullet$ is periodic or equivalently if K is algebraic over a finite field. Otherwise, K is *nonabsolute*. With this we have

Theorem 19.6. [184] [185] *Let $K[G]$ be a group algebra over the field K with G polycyclic-by-finite.*

i. The prime length of $K[G]$ is equal to $\mathrm{pl}(G)$.

ii. If K is a nonabsolute field, then the primitive length of $K[G]$ is equal to $\mathrm{epl}(G)$.

iii. If K is absolute, then all primitive ideals of $K[G]$ are maximal.

Parts (i) and (ii) are first proved for $K[\mathrm{nio}(G)]$ using Theorem 19.4. Then one can apply Corollary 16.8 since

$$K[G] = K[\mathrm{nio}(G)]*(G/\mathrm{nio}(G)).$$

Actually the equality of the prime lengths of $K[G]$ and of $K[\mathrm{nio}(G)]$ is quite easy in this case since the rings involved are Noetherian. Observe that (iii) asserts that $K[G]$ has primitive length 0 if K is absolute. This is the earlier result of [184].

In addition, it is shown in [185] that if G is orbitally sound then $K[G]$ satisfies the *saturated chain condition*. In other words, if P is a prime ideal of $K[G]$, then all saturated chains of primes with largest prime P have the same length. It is not known whether this property lifts to arbitrary polycyclic-by-finite group algebras.

Theorems 19.3, 19.4 and 19.6 are the key facts we need about prime ideals in $K[G]$, but there are other results which should be mentioned. To start with, we offer a sufficient condition for G to be orbitally sound.

Suppose

$$\langle 1 \rangle = G_0 \subseteq G_1 \subseteq \cdots \subseteq G_n = G$$

is a plinth series for G. Then for each i with G_i/G_{i-1} a plinth we obtain the rational plinth $V_i = (G_i/G_{i-1}) \otimes \mathcal{Q}$. Again, via common refinements, it follows that the collection (with multiplicities) of these rational plinths is independent of the particular series and hence is an invariant of G.

Proposition 19.7. [185] *Let G be a polycyclic-by-finite group and assume that*

i. G has a plinth series with rational plinths $V_1, V_2, \ldots, V_t$,

ii. $G/\mathbf{C}_G(V_i)$ is abelian for each i,

iii. $V_i \cong V_j$ if and only if $V_{i|\mathcal{Q}[H]} \cong V_{j|\mathcal{Q}[H]}$ for any subgroup H of G with $|G : H| < \infty$.

Then G is orbitally sound.

It is not difficult to see that any polycyclic-by-finite group has a normal subgroup of finite index with the above properties. Thus Proposition 19.7 supplies the main ingredient in the proof of Theorem 19.3. An alternate proof of this ingredient can be found in [205] where it is observed that a solvable connected linear group over the integers is orbitally sound.

Now suppose Γ is a *group of operators* (that is automorphisms) on G so that Γ acts on $K[G]$. We assume for convenience that Γ contains the inner automorphisms of G. Then the set of Γ-orbital elements of G is

$$\mathbf{D}_G(\Gamma) = \left\{ x \in G \mid |\Gamma : \mathbf{C}_\Gamma(x)| < \infty \right\}$$

and this is a normal subgroup of G contained in $\Delta(G)$. Moreover, Γ permutes the ideals of $K[G]$ and we can speak about Γ-orbital primes.

We now briefly discuss some operator analogs of the preceding results. To begin with, we have the important

Theorem 19.8. [16] *Let A be a finitely generated free abelian group which is an eccentric plinth for the group Γ. If P is a faithful, Γ-orbital prime of $K[A]$, then $P = 0$.*

This is proved by an extremely clever valuation theoretic argument. Next comes a far reaching generalization which is crucial to the proof of Theorem 19.4. We remark that it is really a result about finitely generated abelian groups; the extension to the f.c. case follows from the fact that such groups are center-by-finite.

Theorem 19.9. [185] *Let A be a finitely generated f.c. group and let Γ be a group of operators on A. If P is an almost faithful, Γ-orbital prime of $K[A]$, then $P = (P \cap K[\mathbf{D}_A(\Gamma)]) \cdot K[A]$.*

In other words, the almost faithful prime P is Γ-orbital if and only if it has Γ-orbital generators. As an indication of the power of this theorem, we note that it quickly yields an affirmative answer to a multiplicative analog of Hilbert's 14th problem (see Exercise 8).

Corollary 19.10. [53] *Let R be a commutative integral domain generated as a K-algebra by a finitely generated group of units A. Suppose Γ is a group of algebra automorphisms of R which stabilizes A. Then the fixed ring R^Γ is a finitely generated K-algebra.*

Finally, a second application of Theorem 19.9 to Theorem 19.4 yields

Corollary 19.11. [185] *Let Γ be an operator group on the polycyclic-by-finite group G and let P be a Γ-orbital almost faithful prime of $K[G]$. If G is orbitally sound and Γ contains the inner automorphisms of G, then $P = (P \cap K[\mathbf{D}_G(\Gamma)]) \cdot K[G]$ and $P \cap K[\mathbf{D}_G(\Gamma)] = \bigcap_{x \in G} Q^x$ where Q is an almost faithful prime of $K[\mathbf{D}_G(\Gamma)]$.*

The interested reader should consult [186] for other approaches to Theorem 19.9. We note that a generalization of that theorem appears in [206]. In the latter result, the hypothesis is weakened so that Γ is merely assumed to act on $K[A]$, stabilizing K and the group of trivial units.

We remark that Theorem 19.4 can be used to characterize the primitive ideals of $K[G]$. Indeed, suppose K is a nonabsolute field and P is a standard prime of $K[G]$ with $P \cap K[\Delta] = \bigcap_{x \in G} Q^x$. Then ([185]) P is primitive if and only if $\dim_K K[\Delta]/Q < \infty$. Furthermore, we say that a $K[G]$-module M is *finite induced,* if there exists a subgroup $H \subseteq G$ and a finite K-dimensional $K[H]$-module V with $M = V^{|K[G]}$. While it is rarely true that all irreducible $K[G]$-modules are finite induced, the finite induced irreducibles are nevertheless *ample* in the following precise sense.

Theorem 19.12. [57] *Let P be a primitive ideal of the group algebra $K[G]$ with G a polycyclic-by-finite group. Then P is the annihilator of an irreducible finite induced $K[G]$-module.*

EXERCISES

1. Let G be a polycyclic-by-finite group. Prove that every sub-group and factor group of G is also polycyclic-by-finite and hence finitely generated. Deduce that $\Delta^+(G)$ and $G/\mathbf{C}_G(\Delta(G))$ are finite.

2. Prove the equivalence of the four definitions of orbitally sound given immediately after Lemma 19.1. For this, use the previous exercise along with Lemma 19.2.

3. Show that the orbitally sound condition is inherited by factor groups and by subgroups of finite index. Furthermore show that any group of Hirsch number ≤ 1 is orbitally sound.

4. Let $G = A \rtimes D$ be the semidirect product given immediately before Theorem 19.3. Prove that $W = \langle A, x \rangle$ is not orbitally sound. Now view A additively and describe the action of D on A via 2×2 matrices. Conclude that A is a plinth for G.

5. Continuing with the above example, let H be an isolated orbital subgroup of G. If $H \cap A = \langle 1 \rangle$, prove that H centralizes A and hence that $H = \langle 1 \rangle$. Otherwise show that $H \supseteq A$. Conclude that G is orbitally sound since $\hbar(G/A) = 1$.

6. Let G_0 be an orbitally sound subgroup of finite index in G and let H be an isolated orbital subgroup of G. Use Lemma 19.1 to conclude that $H_0 = H \cap G_0$ is an isolated orbital subgroup of G_0 and that $H = i_G(H_0)$. Since $H_0 \triangleleft G_0$ conclude that G_0 normalizes H and hence that $G_0 \subseteq \mathrm{nio}(G)$.

7. Let G be an arbitrary polycyclic-by-finite group and consider all subnormal series

$$\langle 1 \rangle = H_0 \subseteq H_1 \subseteq \cdots \subseteq H_n = H \subseteq G$$

with $|G/H| < \infty$, $H_i \triangleleft H$ and H_i/H_{i-1} either torsion free abelian or finite. Prove Lemma 19.5 by choosing such a series with a maximal number of infinite factors. Furthermore, by taking common refinements of two such series, prove that $\mathrm{pl}(G)$ and $\mathrm{epl}(G)$ are well defined.

8. Prove Corollary 19.10. To this end, first observe that there is a natural Γ-epimorphism $\sigma \colon K[A] \to R$ and that $\mathrm{Ker}(\sigma)$ is a Γ-stable faithful prime P. Thus if $D = \mathbf{D}_A(\Gamma)$, then $P = (P \cap K[D]) \cdot K[A]$ and

hence $R = S*(A/D)$ where $S = \sigma(K[D])$. Now show that $R^\Gamma = S^\Gamma$ and that Γ acts like a finite group on S.

20. Polycyclic Group Algebras

In this section we return to our past practice of offering complete proofs. Specifically we consider an application of crossed product theory to polycyclic group algebras.

Let $K[G]$ be a group algebra of the polycyclic-by-finite group G over the field K. If $G_0 = \text{nio}(G)$, then G_0 is a characteristic subgroup of finite index in G and all primes in $K[G_0]$ are well understood. The goal is to describe the primes of $K[G]$. This is what is known as a *finite index problem* because we need to lift information from a subgroup of finite index in G to all of G.

Notice that $K[G] = K[G_0]*(G/G_0)$ and that G/G_0 is finite. Thus Theorems 14.7 and 15.8 must surely apply here. In fact the former result (as Corollary 14.10) does apply directly, but the latter seems to require that a certain group of X-inner automorphisms be computed. Fortunately this computation turns out to be unnecessary. What we actually need is to show that a certain ideal is prime. In view of our previous work on primeness and on computing X-inner automorphisms, it is not surprising that Δ-methods come into play in this part.

For convenience we will assume throughout this section that $K[G]$ is given with G polycyclic-by-finite. Furthermore, we will write $\Delta = \Delta(G)$.

Lemma 20.1. Let I be a G-stable ideal of $K[\Delta]$ which is an intersection of almost faithful primes of that ring. Suppose further that $\Delta \subseteq N \triangleleft G$ with $N/\Delta \subseteq \Delta(G/\Delta)$. If J is a G-stable ideal of $K[N]$ and $J \supset I \cdot K[N]$, then $J \cap K[\Delta] \supset I$.

Proof. Let $Z = \mathbf{Z}(\Delta)$ so that $Z \triangleleft G$. Since Δ/Z is finite and $N/\Delta \subseteq \Delta(G/\Delta)$, it follows easily that $N/Z \subseteq \Delta(G/Z)$. In particular there exists a subgroup H of finite index in G which centralizes both Δ and N/Z.

Since $J \supset I \cdot K[N]$, we can choose $\alpha \in J \setminus (I \cdot K[N])$ with support meeting the minimal number, say n, of cosets of Δ. Thus $\alpha = \sum_1^n \alpha_i x_i$ with $\alpha_i \in K[\Delta]$ and with $x_1, x_2, \ldots, x_n \in N$ in distinct cosets of Δ. Multiplying by a group element if necessary, we can assume that $x_1 = 1$. Clearly $\alpha_i \in K[\Delta] \setminus I$ for all i. The goal is to show that $n = 1$.

Suppose by way of contradiction that $n > 1$ so that $x_n \in N \setminus \Delta$. Then x_n has infinitely many G-conjugates and hence infinitely many H-conjugates since $|G : H| < \infty$. Note that H centralizes both $K[\Delta]$ and N/Z. Thus for any $h \in H$ we have $\alpha^h = h^{-1}\alpha h = \sum_1^n \alpha_i (x_i)^h$ and $(x_i)^h \in Zx_i \subseteq \Delta x_i$. Setting $\beta = \alpha^h - \alpha \in J$, we see that $\beta = \sum_2^n \alpha_i[(x_i)^h x_i^{-1} - 1]x_i$ has support meeting less than n cosets of Δ. It follows that $\beta \in I \cdot K[N]$ and hence that $\alpha_n[(x_n)^h x_n^{-1} - 1] \in I$.

Now write $I = \bigcap_j Q_j$, an intersection of almost faithful primes. Then for each $h \in H$ and subscript j we have $\alpha_n[(x_n)^h x_n^{-1} - 1] \in Q_j$. Since $(Q_j)^\dagger$ is finite and x_n has infinitely many H-conjugates, we can choose h so that $(x_n)^h x_n^{-1} \notin (Q_j)^\dagger$. In other words, $(x_n)^h x_n^{-1} - 1 \notin Q_j$. But $(x_n)^h x_n^{-1} - 1$ is central in $K[\Delta]$ and Q_j is prime, so we conclude that $\alpha_n \in Q_j$. This yields $\alpha_n \in \bigcap_j Q_j = I$, a contradiction. Thus $n = 1$ and $\alpha \in (J \cap K[\Delta]) \setminus I$ as required. ∎

The next result is the necessary replacement for Theorem 15.8. It actually follows easily from Lemma 20.1 and Proposition 8.3(iii). However, instead of quoting the latter proposition, we offer a reasonably brief independent argument.

Proposition 20.2.[108] *Let $K[G]$ be a group algebra of the polycyclic-by-finite group G and let I be a G-stable ideal of $K[\Delta]$. If I is an intersection of almost faithful primes of $K[\Delta]$, then $I \cdot K[G]$ is a semiprime ideal of $K[G]$. If, in addition, I is G-prime, then $I \cdot K[G]$ is a prime ideal of $K[G]$.*

Proof. Let $\Delta \subseteq N \subseteq G$ with $N/\Delta = \Delta(G/\Delta)$ and suppose that A, B are ideals of $K[G]$ containing $I \cdot K[G]$ with $AB \subseteq I \cdot K[G]$. Set $A' = \pi_N(A)$ and $B' = \pi_N(B)$ so that A' and B' are G-stable ideals of $K[N]$ containing $I \cdot K[N]$. Here of course $\pi_N : K[G] \to K[N]$ is the natural projection. We first show that $(A' \cap K[\Delta])(B' \cap K[\Delta]) \subseteq I$.

To this end, let $\alpha' \in A' \cap K[\Delta]$ and $\beta' \in B' \cap K[\Delta]$. Then by definition there exist $\alpha \in A, \beta \in B$ with $\alpha = \alpha' + \sum_1^n \alpha_i x_i$ and $\beta = \beta' + \sum_1^m \beta_i y_i$. Here $\{\,1, x_1, \ldots, x_n\,\}$ and $\{\,1, y_1, \ldots, y_m\,\}$ are sets of distinct coset representatives for N in G and $\alpha_i, \beta_i \in K[N]$.

Now let H be a subgroup of finite index in G which centralizes Δ. Since each $x_i\Delta \notin \Delta(G/\Delta)$, the usual coset counting in G/Δ shows that there exists $h \in H$ such that $(x_i)^h y_j \notin \Delta$ for all i, j. Thus since h centralizes $K[\Delta]$, it follows that $\alpha'\beta' = \pi_\Delta(\alpha^h \beta)$. But $\alpha^h \in A$ and $AB \subseteq I \cdot K[G]$ so

$$\alpha'\beta' \in \pi_\Delta(AB) \subseteq \pi_\Delta(I \cdot K[G]) = I$$

as required.

Note that I is a semiprime ideal of $K[\Delta]$. Thus if $A^2 \subseteq I \cdot K[G]$, then $(A' \cap K[\Delta])^2 \subseteq I$ by the above and we have $A' \cap K[\Delta] \subseteq I$. Lemma 20.1 now yields $A' = I \cdot K[N]$ so

$$A \subseteq \pi_N(A) \cdot K[G] = A' \cdot K[G] = I \cdot K[G]$$

and $I \cdot K[G]$ is a semiprime ideal of $K[G]$.

Finally suppose I is a G-prime ideal. If $AB \subseteq I \cdot K[G]$, then we have $(A' \cap K[\Delta])(B' \cap K[\Delta]) \subseteq I$ so one of these two G-stable factors is contained in I, say $A' \cap K[\Delta] \subseteq I$. As above, this yields $A \subseteq I \cdot K[G]$ and therefore $I \cdot K[G]$ is a prime ideal of $K[G]$. ∎

Recall from Section 14 that if $H \subseteq G$ and if $L \triangleleft K[H]$, then the induced ideal $L^{|G}$ is defined by

$$L^{|G} = \bigcap_{x \in G} (L \cdot K[G])^x.$$

It follows from Lemma 14.4 that $L^{|G}$ is the unique largest two-sided ideal of $K[G]$ contained in $L \cdot K[G]$. Furthermore, it is unique largest ideal with $\pi_H(L^{|G}) \subseteq L$. Note that if $H \triangleleft G$, then

$$L^{|G} = \left(\bigcap_{x \in G} L^x \right) \cdot K[G].$$

Now recall from the previous section that a prime ideal P of $K[G]$ is standard if $P = (P \cap K[\Delta]) \cdot K[G]$ and $P \cap K[\Delta] = \bigcap_{x \in G} L^x$ where L is an almost faithful prime of $K[\Delta]$. The following is merely a reformulation of Proposition 20.2 since $\Delta \triangleleft G$ and since the ideal $\bigcap_{x \in G} L^x$ is clearly G-prime.

Lemma 20.3. *Any standard prime P of $K[G]$ can be written as $P = L^{|G}$ with L an almost faithful prime of $K[\Delta]$. Conversely if L is an almost faithful prime of $K[\Delta]$, then $L^{|G}$ is a standard prime ideal of $K[G]$.*

The next lemma describes the behavior of standard primes under restriction to normal subgroups of finite index. Note that $K[G]$ is right and left Noetherian by Proposition 1.6. Thus if $N \triangleleft G$ and P is a prime of $K[G]$, then Lemmas 1.3 and 14.2 imply that $P \cap K[N] = \bigcap_{x \in G} Q^x$ where Q is a G-orbital prime of $K[N]$ unique up to G-conjugacy. Indeed $\{\, Q^x \mid x \in G \,\}$ is the finite set of minimal covering primes of the ideal $P \cap K[N]$.

Lemma 20.4. *Let H be a normal subgroup of finite index in G and let P be a prime ideal of $K[G]$. Write $P \cap K[H] = \bigcap_{x \in G} Q^x$ with Q a prime ideal of $K[H]$.*

  i. P is standard if and only if Q is standard.

  ii. Assume that P is standard and write $P = L^{|G}$ with L an almost faithful prime of $K[\Delta]$. If J is a minimal covering prime of $L \cap K[\Delta(H)]$, then $J^{|H}$ is a minimal covering prime of $P \cap K[H]$ and it is standard.

Proof. We first prove (ii). Set $D = \Delta(H)$ so that $D = H \cap \Delta$ since $|G : H| < \infty$. By assumption, P is standard so $P = (P \cap K[\Delta]) \cdot K[G]$ and hence $\pi_\Delta(P) \subseteq P$. Therefore $D = H \cap \Delta$ implies that

$$\pi_D(P \cap K[H]) = \pi_\Delta(P \cap K[H]) \subseteq P \cap K[H].$$

Setting $I = P \cap K[H]$, it follows from $I \subseteq \pi_D(I) \cdot K[H]$ and the above that $I = (I \cap K[D]) \cdot K[H]$. Moreover, since

$$P \cap K[\Delta] = L^{|G} \cap K[\Delta] = \bigcap_{x \in G} L^x$$

we see that

$$I \cap K[D] = P \cap K[D] = \bigcap_{x \in G}(L^x \cap K[D]) = \bigcap_{x \in G}(L \cap K[D])^x.$$

Since $D \lhd \Delta$ we have $L \cap K[D] = \bigcap_{y \in \Delta} J^y$ where J is any minimal covering prime. Hence $L^\dagger \cap D = \bigcap_{y \in \Delta}(J^\dagger)^y$. In particular, if $Z = \mathbf{Z}(\Delta) \cap D$, then $J^\dagger \cap Z = L^\dagger \cap Z \subseteq L^\dagger$. But L is almost faithful and $|D : Z| < \infty$ so we conclude that J is almost faithful.

It now follows from Lemma 20.3 that $J^{|H}$ is a standard prime of $K[H]$. Moreover, since $D \lhd G$,

$$P \cap K[H] = I = (I \cap K[D]) \cdot K[H] = \bigcap_{x \in G}(L \cap K[D])^x \cdot K[H]$$

$$= \bigcap_{x \in G} J^x \cdot K[H] = \bigcap_{x \in G}\left(\bigcap_{y \in H} J^y \cdot K[H]\right)^x$$

$$= \bigcap_{x \in G}(J^{|H})^x.$$

Thus since $|G : H| < \infty$ we see that $J^{|H}$ is a minimal covering prime of $P \cap K[H]$. This proves (ii) and one implication of (i) since the uniqueness of Q implies that Q is G-conjugate to $J^{|H}$ and hence is clearly standard.

Conversely assume that Q is standard and write $P \cap K[\Delta] = \bigcap_{x \in G} S^x$ where S is a prime ideal of $K[\Delta]$. Since $Q = T^{|H}$ for some almost faithful prime T of $K[D]$, we have $Q \cap K[D] = \bigcap_{y \in H} T^y$ and hence

$$\bigcap_{x \in G}(S \cap K[D])^x = (P \cap K[\Delta]) \cap K[D] = (P \cap K[H]) \cap K[D]$$

$$= \bigcap_{x \in G}(Q^x \cap K[D]) = \bigcap_{x \in G} T^x.$$

Since G acts like a finite group on $K[\Delta]$, these intersections are finite. Thus since T is prime, we have $S \cap K[D] \subseteq T^x$ for some $x \in G$ and it follows that $S^\dagger \cap D \subseteq (T^\dagger)^x$. But the latter group is finite and $|\Delta : D| < \infty$, so S is an almost faithful prime of $K[\Delta]$. By Lemma 20.3,

$$P' = S^{|G} = \left(\bigcap_{x \in G} S^x\right) \cdot K[G] = (P \cap K[\Delta]) \cdot K[G]$$

is a standard prime ideal of $K[G]$ which is clearly contained in P. Finally since $Q = (\bigcap_{y \in H} T^y) \cdot K[H]$ we have

$$(P \cap K[H]) \cdot K[G] = \left(\bigcap_{x \in G} Q^x\right) \cdot K[G] = \left(\bigcap_{x \in G} T^x\right) \cdot K[G]$$
$$\subseteq (P \cap K[\Delta]) \cdot K[G] = P'$$

and thus $(P \cap K[H]) \cdot K[G] \subseteq P' \subseteq P$. Since $K[G] = K[H]*(G/H)$, it now follows from Incomparability, Theorem 16.6(ii), that $P' = P$. Thus P is standard and the lemma is proved. ∎

We require some additional terminology. Let N be a subgroup of G and let I be an ideal of $K[G]$. We say that I is *almost faithful mod N* if $I^\dagger \supseteq N$ and $|I^\dagger : N| < \infty$. In particular if $N \triangleleft G$, then this occurs if and only if I is the complete inverse image of an almost faithful ideal of $K[G/N]$. More importantly, we say I is *almost faithful sub N* if $I^\dagger \subseteq N$ and $|N : I^\dagger| < \infty$. In addition we let $\nabla_G(N)$ denote the complete inverse image in $\mathbf{N}_G(N)$ of $\Delta(\mathbf{N}_G(N)/N)$. We can now obtain the first main result of this section.

Theorem 20.5. [114] *Let G be a polycyclic-by-finite group and let P be a prime ideal of $K[G]$. Suppose H is a normal orbitally sound subgroup of finite index in G and write $P \cap K[H] = \bigcap_{x \in G} Q^x$ with Q a prime of $K[H]$. Define $N = \mathrm{i}_G(Q^\dagger)$, $A = \mathbf{N}_G(N)$ and let $B = \{ x \in G \mid Q^x = Q \}$.*

   i. $H \subseteq B \subseteq A$.

   ii. There exists a unique prime ideal T of $K[A]$ with $P = T^{|G}$ and $T \cap K[H] = \bigcap_{a \in A} Q^a$. Indeed T is the unique minimal covering prime of $P \cap K[A]$ not containing $\bigcap_{x \notin A} Q^x$.

   iii. There exists an almost faithful sub N prime ideal L of $K[\nabla_G(N)]$ with $T = L^{|A}$ and hence with $P = L^{|G}$. Indeed L is a minimal covering prime of $T \cap K[\nabla_G(N)]$.

Proof. We use the notation given in the statement. We note that such a normal orbitally sound subgroup H of finite index in G does exist; indeed we could take $H = \mathrm{nio}(G)$ by Theorem 19.3. Moreover $P \cap K[H]$ has the above structure with Q unique up to G-conjugation.

Also $Q^\dagger \lhd H$ so $Q^\dagger$ is orbital in G and, by Lemma 19.1, $N = i_G(Q^\dagger)$ is an isolated orbital subgroup of G with $|N : Q^\dagger| < \infty$ and $A = \mathbf{N}_G(N) \supseteq \mathbf{N}_G(Q^\dagger) \supseteq H$. Observe that if $Q^x = Q$, then x certainly normalizes $Q^\dagger$ and hence we have $A \supseteq \mathbf{N}_G(Q^\dagger) \supseteq B \supseteq H$. Thus by considering $K[G]$ as the crossed product $K[H]*(G/H)$, we see from Corollary 14.10 that there exists a unique prime ideal T of $K[H]*(A/H) = K[A]$ with $T \cap K[H] = \bigcap_{a \in A} Q^a$ and with $P = T^{|G}$. This proves parts (i) and (ii) since Corollary 14.10(ii) describes T appropriately.

By Lemma 19.2 we can let M be a characteristic subgroup of N of finite index contained in $Q^\dagger$ so that $M \lhd \mathbf{N}_G(N) = A$. We use $\tilde{\ } : K[A] \to K[A/M]$ to denote the natural map. Since $M \lhd H$ and $|Q^\dagger : M| < \infty$, it follows that $\tilde{Q}$, the image of Q is an almost faithful prime of $K[\tilde{H}]$. Furthermore, $\tilde{H}$ is orbitally sound so $\tilde{Q}$ is a standard prime by Theorem 19.4. Note that $T \cap K[H] = \bigcap_{a \in A} Q^a$ and hence $\tilde{T} \cap K[\tilde{H}] = \bigcap_{\tilde{a} \in \tilde{A}} \tilde{Q}^{\tilde{a}}$. We can therefore conclude from Lemma 20.4(i) that $\tilde{T}$ is a standard prime of $K[\tilde{A}]$. In other words there exists an almost faithful prime $\tilde{L}$ of $K[\Delta(\tilde{A})]$ with

$$\tilde{T} = \bigcap_{\tilde{a} \in \tilde{A}} \tilde{L}^{\tilde{a}} \cdot K[\tilde{A}] = \tilde{L}^{|\tilde{A}}.$$

By lifting this expression back to $K[A]$, we see that there exists an almost faithful mod M prime ideal L of $K[\nabla_A(M)]$ with

$$T = \bigcap_{a \in A} L^a \cdot K[A] = L^{|A}.$$

Observe that $N, M \lhd A$ and $|N : M| < \infty$. This implies easily (see Exercise 4) that $\nabla_A(M) = \nabla_A(N) = \nabla_G(N)$ since $A = \mathbf{N}_G(N)$. Also $i_G(M) \supseteq N \supseteq M$ so $i_G(M) = N$. Since $L^\dagger$ is orbital in A, it is orbital in G, and from $L^\dagger \supseteq M$ and $|L^\dagger : M| < \infty$, we conclude that $L^\dagger \subseteq i_G(M) = N$ and that $|N : L^\dagger| < \infty$. Thus we see that L is almost faithful sub N. Finally we have $T = L^{|A}$ and $P = T^{|G}$ so Lemma 14.9 yields $P = (L^{|A})^{|G} = L^{|G}$. Since $T \cap K[\nabla_G(N)] = \bigcap_{a \in A} L^a$, a finite intersection of primes, L is clearly a minimal covering prime of $T \cap K[\nabla_G(N)]$ and the result follows. ∎

Definition. Let P be a prime ideal of $K[G]$, N an isolated orbital subgroup of G, and L an almost faithful sub N prime ideal of $K[\nabla_G(N)]$. If $P = L^{|G}$, then N is said to be a *vertex* of P and we write $N = \mathrm{vx}_G(P)$. Furthermore, for this N, the prime L is said to be a *source* of P.

The previous theorem describes how to find at least one vertex and source for P. This is all the more useful once we prove uniqueness in the next main result.

Theorem 20.6. **[114]** *Let N be an isolated orbital subgroup of G and let L be an almost faithful sub N prime ideal of $K[\nabla_G(N)]$. Then $P = L^{|G}$ is a prime ideal of $K[G]$. Furthermore, let $H = \mathrm{nio}(G)$ and write $P \cap K[H] = Q_1 \cap Q_2 \cap \cdots \cap Q_n$, an intersection of G-conjugate primes.*

i. For some $Q = Q_i$, we have $N = \mathrm{i}_G(Q^\dagger)$ so N is uniquely determined by P up to conjugation in G.

ii. For this N, the ideal L is uniquely determined by P up to conjugation by $A = \mathbf{N}_G(N)$.

iii. If J is a minimal covering prime of $L \cap K[H \cap \nabla_G(N)]$, then $J^{|H} = Q^a$ for some $a \in A$. Furthermore, $H \cap \nabla_G(N) = \nabla_H(Q^\dagger)$.

Proof. Set $A = \mathbf{N}_G(N)$ and let $H = \mathrm{nio}(G)$ so that, by definition, $H \subseteq A$. Since $|N : N \cap H| < \infty$, and $|N : L^\dagger| < \infty$, there exists, by Lemma 19.2, a characteristic subgroup M of finite index in N with $M \subseteq N \cap H$ and $M \subseteq L^\dagger$. Thus $M \triangleleft A$ and therefore also $M \triangleleft H$. Furthermore, since $|N : M| < \infty$, we have $\mathrm{i}_G(M) = N$, so $\mathbf{N}_G(M) = A$ and then clearly $\nabla_G(M) = \nabla_G(N)$ (see Exercise 4 again).

Let $\tilde{} : K[A] \to K[A/M]$ denote the natural map. Then $\tilde{L}$ is an almost faithful prime of $K[\Delta(\tilde{A})]$, since $\nabla_G(M) = \nabla_G(N)$, and we conclude from Lemma 20.3 that

$$\tilde{T} = \tilde{L}^{|\tilde{A}} = \bigcap_{\tilde{a} \in \tilde{A}} \tilde{L}^{\tilde{a}} \cdot K[\tilde{A}]$$

is a standard prime of $K[\tilde{A}]$. Furthermore Lemma 20.4(i) implies that $\tilde{T} \cap K[\tilde{H}] = \bigcap_{\tilde{a} \in \tilde{A}} \tilde{Q}^{\tilde{a}}$ is a finite intersection of $\tilde{A}$-conjugate standard primes of $K[\tilde{H}]$. Lifting this information to $K[A]$, we see immediately that

$$T = L^{|A} = \bigcap_{a \in A} L^a \cdot K[A]$$

is a prime ideal of $K[A]$ and that $T \cap K[H] = \bigcap_{a \in A} Q^a$ is a finite intersection of A-conjugate primes of $K[H]$ which are almost faithful mod M.

Set $B = \{x \in G \mid Q^x = Q\}$ so that $H \subseteq B \subseteq G$ and surely $B \subseteq \mathbf{N}_G(Q^\dagger)$. Observe that $|Q^\dagger : M| < \infty$ and $Q^\dagger$ is orbital in G so $i_G(Q^\dagger) = i_G(M) = N$ and hence $B \subseteq \mathbf{N}_G(Q^\dagger) \subseteq \mathbf{N}_G(N) = A$. We now view $K[G]$ as the crossed product $K[H]*(G/H)$. Then, since $B \subseteq A$, it follows from Corollary 14.10(i), that $P = T^{|G}$ is a prime ideal of $K[G]$ and, since $T^{|G} = (L^{|A})^{|G} = L^{|G}$, by Lemma 14.9, the first assertion is proved. Furthermore, by Corollary 14.10 again, Q is a minimal covering prime of $P \cap K[H] = Q_1 \cap Q_2 \cap \cdots \cap Q_n$ so $Q = Q_i$ for some i. Thus since $N = i_G(Q^\dagger)$, (i) is proved.

Now suppose both N and $P = L^{|G}$ are known. We consider those Q_i with $i_G(Q_i^\dagger) = N$. Since $Q_i = Q^x$ for some $x \in G$, we have

$$N = i_G(Q_i^\dagger) = i_G(Q^\dagger)^x = N^x,$$

and hence this occurs if and only if $x \in A$. In particular, the ideal $\bigcap_{a \in A} Q^a$ is the intersection of all the minimal covering primes Q_i of $P \cap K[H]$ with $i_G(Q_i^\dagger) = N$ and this is surely determined by N and P. It now follows from Corollary 14.10(ii) that the prime ideal T of $K[A]$ is uniquely determined by the conditions $T^{|G} = P$ and $T \cap K[H] = \bigcap_{a \in A} Q^a$. But $T = L^{|A}$ and L is certainly a minimal covering prime of $T \cap K[\nabla_G(N)] = \bigcap_{a \in A} L^a$, so L is unique up to conjugation by A. This proves (ii).

Finally, recall that $\tilde{T} = \tilde{L}^{|\tilde{A}}$ is a standard prime of $K[\tilde{A}]$. Thus Lemma 20.4(ii) asserts that if $\tilde{J}$ is a minimal covering prime of $\tilde{L} \cap K[\Delta(\tilde{H})]$, then $\tilde{J}^{|\tilde{H}}$ is a minimal covering prime of $\tilde{T} \cap K[\tilde{H}] = \bigcap_{\tilde{a} \in \tilde{A}} \tilde{Q}^{\tilde{a}}$ and hence $\tilde{J}^{|\tilde{H}} = \tilde{Q}^{\tilde{a}}$ for some $\tilde{a} \in \tilde{A}$. Lifting this information back to $K[A]$, we see that for any minimal covering prime J of $L \cap K[\nabla_H(M)]$, we have $J^{|H} = Q^a$ for some $a \in A$. Since clearly $\nabla_H(M) = \nabla_H(Q^\dagger) = \nabla_G(N) \cap H$, part (iii) follows. ∎

The preceding two theorems are extremely technical in nature and actually yield more information than we require. Therefore we close this section by offering simpler, more understandable, versions of these results. The formulation divides naturally into three parts.

Theorem 20.7. (Existence) [185] [114] *Let G be a polycyclic-by-finite group and let P be a prime ideal of $K[G]$. Then there exists an isolated orbital subgroup N of G and an almost faithful sub N prime L of $K[\nabla_G(N)]$ with $P = L^{|G}$.*

This is Theorem 20.5(iii). Note that $\nabla_G(N)/N$ is a torsion free abelian group so $\nabla_G(N)/L^\dagger$ is finite-by-abelian and hence an f.c. group. Thus L essentially corresponds to a prime ideal of a commutative group algebra. Recall that N above is a vertex for P and L is a source.

Theorem 20.8. (Uniqueness) [114] *Let G be polycyclic-by-finite and let P be a prime of $K[G]$. Then the vertices of P are unique up to conjugation by G. Furthermore, if N is a vertex, then the sources for this N are unique up to conjugation by $\mathbf{N}_G(N)$.*

This is of course Theorem 20.6(i)(ii). Finally the beginning of Theorem 20.6 yields

Theorem 20.9. (Converse) [114] *Let N be an isolated orbital subgroup of the polycyclic-by-finite group G. If L is an almost faithful sub N prime of $K[\nabla_G(N)]$, then $L^{|G}$ is a prime ideal of $K[G]$.*

In the next section we will consider some corollaries and extensions of these results.

EXERCISES

1. Let I be a G-stable ideal of $K[\Delta]$ which is an intersection of almost faithful primes of that ring. Let $\Delta \subseteq N \triangleleft G$ with

$N/\Delta \subseteq \Delta(N/\Delta)$ and let $\tilde{\alpha}_1, \tilde{\alpha}_2, \ldots, \tilde{\alpha}_n$ and $\tilde{\beta}_1, \tilde{\beta}_2, \ldots, \tilde{\beta}_n$ be elements of $K[N]$. If

$$\sum_i (\tilde{\alpha}_i)^x \beta_i \equiv 0 \qquad \mathrm{mod}\ I \cdot K[N]$$

for all $x \in G$, show that

$$\sum_i \pi_\Delta(\tilde{\alpha}_i)\tilde{\beta}_i \equiv 0 \qquad \mathrm{mod}\ I \cdot K[N]$$

and hence that

$$\sum_i \pi_\Delta(\tilde{\alpha}_i)\pi_\Delta(\tilde{\beta}_i) \equiv 0 \qquad \mathrm{mod}\ I.$$

To this end, let H be a subgroup of finite index in G centralizing both Δ and N/Z where $Z = \mathbf{Z}(\Delta)$. Now proceed by induction on the number of Δ-cosets which meet the supports of the various $\tilde{\alpha}_i$.

2. Let I be as above and let $\alpha_1, \alpha_2, \ldots, \alpha_n$ and $\beta_1, \beta_2, \ldots, \beta_n$ be elements of $K[G]$. If

$$\sum_i (\alpha_i)^x \beta_i \equiv 0 \qquad \mathrm{mod}\ I \cdot K[G]$$

for all $x \in G$, show that

$$\sum_i \pi_\Delta(\alpha_i)\pi_\Delta(\beta_i) \equiv 0 \qquad \mathrm{mod}\ I.$$

For this, let $N/\Delta = \Delta(G/\Delta)$ and set $\tilde{\alpha}_i = \pi_N(\alpha_i)$ and $\tilde{\beta}_i = \pi_N(\beta_i)$. Now use Δ-methods to reduce to the hypothesis of the preceding problem. This is a result of [**108**].

3. If N is a finite normal subgroup of G, show that $\nabla_G(N) = \Delta(G)$. Conclude that a virtually standard, almost faithful prime of $K[G]$ is standard.

4. Let N be an isolated orbital subgroup of G and let M be a characteristic subgroup of finite index in N. Prove that $\mathbf{N}_G(N) = \mathbf{N}_G(M)$ and then that $\nabla_G(N) = \nabla_G(M)$.

5. Let N be a nonidentity torsion free polycyclic-by-finite group, let W be a finite group and set $G = N \wr W$. Thus G is the semidirect product $G = H \rtimes W$ where $H = \prod_{w \in W} N_w$ is the direct product of copies of N indexed by the elements of W. Moreover W permutes the factors of H in the natural manner. Show that each N_w is an isolated orbital subgroup of G and deduce that $\mathrm{nio}(G) \subseteq H$.

21. Polycyclic Crossed Products

We now consider some consequences and extensions of the results of the previous section. We start with group algebras and then move on to twisted group algebras and polycyclic crossed products. Again we let G denote a polycyclic-by-finite group and K will be a field.

We first list some elementary properties of isolators.

Lemma 21.1. Let $N \subseteq H$ be orbital subgroups of G.

i. $i_G(N) \subseteq i_G(H)$ and $i_G(N) \cap H = i_H(N)$.

ii. If $N \triangleleft H$, then $i_H(N)/N = \Delta^+(H/N)$.

iii. If N is an isolated orbital subgroup of G, then $\nabla_G(N)/N$ is torsion free abelian.

Proof. (i) Let $A = \mathbf{N}_H(N) \supseteq N$. Then A is a subgroup of finite index in H and it is G-orbital. In particular, $i_G(A) = i_G(H)$. Since A normalizes N, it normalizes $i_G(N)$ so $A i_G(N)$ is an orbital subgroup of G containing A. Moreover

$$|A i_G(N) : A| = |i_G(N) : A \cap i_G(N)| \leq |i_G(N) : N| < \infty.$$

Thus $A i_G(N) \subseteq i_G(A) = i_G(H)$. The second part follows easily since $i_G(N) \cap H$ is H-orbital and $i_H(N)$ is G-orbital.

(ii) If $M/N = \Delta^+(H/N)$, then we have $M \triangleleft H$ and $|M/N| < \infty$ by Lemma 5.1(iii). Thus $M \subseteq i_H(N)$. Equality follows since every H-orbital subgroup of H/N is contained in $\Delta^+(H/N)$.

(iii) Set $H = \nabla_G(N)$. Then by the above, $\Delta^+(H/N) = \langle 1 \rangle$ so H/N is torsion free abelian by Lemma 5.1(ii). ∎

The following is a converse to Theorem 19.4. It shows that the latter result cannot be extended beyond the class of orbitally sound groups.

Theorem 21.2. **[114]** *Let $K[G]$ be given with G a polycyclic-by-finite group.*

i. Every isolated orbital subgroup N of G is the vertex of a prime ideal of $K[G]$.

ii. If all primes of $K[G]$ are virtually standard, then G is orbitally sound.

Proof. (i) Let N be an isolated orbital subgroup of G and let L denote the kernel of the natural epimorphism $K[\nabla_G(N)] \to K[\nabla_G(N)/N]$. Since $\nabla_G(N)/N$ is torsion free abelian, by Lemma 21.1(iii), its group algebra is a domain and L is a prime ideal of $K[\nabla_G(N)]$. Furthermore, $L^\dagger = N$ so we conclude from Theorem 20.8 that $P = L^{|G}$ is a prime ideal of $K[G]$ with vertex N.

(ii) In view of the above, it suffices to show that every vertex is normal in G. To this end, let P be a prime ideal of $K[G]$. If $N = i_G(P^\dagger)$ then, by Lemma 21.1(iii), $N/P^\dagger = \Delta^+(G/P^\dagger)$ so $N \lhd G$. This also implies that $\nabla_G(N) = \nabla_G(P^\dagger)$. Finally, since P is virtually standard, there exists an almost faithful mod $P^\dagger$ prime L of $K[\nabla_G(P^\dagger)]$ with $P = \left(\bigcap_{x \in G} L^x \right) \cdot K[G]$. Thus $P = L^{|G}$ and since L is almost faithful sub N, we have $\mathrm{vx}_G(P) = N \lhd G$. It follows from Theorem 20.9 that N is the unique vertex of P. ∎

Next we consider an operator version of Theorem 20.7 and for this we need a variant of $\nabla_G(N)$. Thus suppose Γ is a group of operators on G, N is a subgroup of G, and let

$$\Gamma_0 = \mathbf{N}_\Gamma(N) = \{\, y \in \Gamma \mid N^y = N \,\}.$$

Then Γ_0 normalizes N, $\mathbf{N}_G(N)$ and $\nabla_G(N)$ and we define $\nabla_G(N; \Gamma)$ to be the set of $x \in \nabla_G(N)$ having only finitely many Γ_0-conjugates modulo N. It is clear that $\nabla_G(N; \Gamma)$ is a Γ_0-invariant subgroup of $\nabla_G(N)$ containing N. Furthermore if G acts on G by conjugation, then $\nabla_G(N; G) = \nabla_G(N)$.

Proposition 21.3. *Let G be a polycyclic-by-finite group, let Γ be a group of operators on G and let P be a Γ-orbital prime of $K[G]$ with $\text{vx}_G(P) = N$. Then there exists a Γ-orbital almost faithful sub N prime L of $K[\nabla_G(N; \Gamma)]$ with $P = L^{|G}$.*

Proof. We proceed with a series of special cases.

Case 1. *Assume that G is an f.c. group and P is almost faithful.*

Proof. By Lemma 21.1(ii), if $N = \Delta^+(G)$, then N is a finite isolated orbital subgroup of G. Since P is almost faithful sub N and $G = \nabla_G(N)$, it therefore follows that $N = \text{vx}_G(P)$. Moreover since P is Γ-orbital, we conclude from Theorem 19.9 that $P = (P \cap K[\mathbf{D}_G(\Gamma)]) \cdot K[G]$. Now it is clear that $\mathbf{D}_G(\Gamma)$ contains the finite group $\Delta^+(G) = N$ so $\mathbf{D}_G(\Gamma) \triangleleft G$ and it follows easily that $\mathbf{D}_G(\Gamma) = \nabla_G(N; \Gamma)$. Furthermore, $P \cap K[\mathbf{D}_G(\Gamma)] = \bigcap_{x \in G} L^x$ where L is a prime ideal of $K[\mathbf{D}_G(\Gamma)]$ so we have $P = L^{|G}$. Finally if $Z = \mathbf{Z}(G)$, then $|G : Z| < \infty$ and $P^\dagger \cap Z = \bigcap_{x \in G}(L^\dagger)^x \cap Z = L^\dagger \cap Z$ so $L^\dagger \cap Z$ is finite and L is an almost faithful ideal. Certainly L is Γ-orbital. ∎

Case 2. *Assume that $G = \nabla_G(N)$.*

Proof. By definition, P is an almost faithful sub N prime ideal of $K[G]$ and, by Lemma 19.2, we can choose M to be a characteristic subgroup of finite index in N with $M \subseteq P^\dagger$. Since N/M is finite, G/M is an f.c. group. Furthermore, the image of P in $K[G/M]$ is an almost faithful Γ-orbital prime. The preceding case, applied to $K[G/M]$, quickly yields the result. ∎

Case 3. *Assume that G is arbitrary.*

Proof. By Theorem 20.7, $P = J^{|G}$ where J is an almost faithful sub N prime of $K[\nabla_G(N)]$. Since P is Γ-orbital, it follows from uniqueness in Theorem 20.8 that N is Γ-orbital. Thus $|\Gamma : \Gamma_0| < \infty$ where $\Gamma_0 = \mathbf{N}_\Gamma(N)$. Furthermore, by uniqueness again, J is a Γ_0-orbital prime of $K[\nabla_G(N)]$. By the previous case, $J = L^{|\nabla_G(N)}$ with L a

Γ_0-orbital almost faithful sub N prime of $K[\nabla_G(N;\Gamma)]$. Hence, by the transitivity of induction in Lemma 14.9, we have $P = J^{|G} = L^{|G}$. Since L is clearly Γ-orbital, the result follows. $\blacksquare$

At this point it is convenient to consider a few properties of poly-{infinite cyclic} groups. A group G is said to be *poly-Z* if it has a finite subnormal series

$$\langle 1 \rangle = G_0 \triangleleft G_1 \triangleleft \cdots \triangleleft G_n = G$$

with each quotient G_i/G_{i-1} infinite cyclic. It is clear that such groups are polycyclic-by-finite and also torsion free. In addition we have

Lemma 21.4. *i. A polycyclic-by-finite group has a normal poly-Z subgroup of finite index.*

ii. Any poly-Z group is a unique product group.

iii. Let $R*\Gamma$ be a crossed product with Γ a unique product group. If I is a Γ-prime ideal of R, then $I*\Gamma$ is a prime ideal of $R*\Gamma$.

Proof. (i) This is [**161**, Lemma 10.2.5].

(ii) By induction we need only show that if $H \triangleleft G$ is a unique product group and if G/H is infinite cyclic, then G is unique product. For this, let $G = \langle H, x \rangle$ and let A, B be finite nonempty subsets of G. Choose n, m maximal with $A_0 = A \cap Hx^n \neq \emptyset$ and $B_0 = B \cap Hx^m \neq \emptyset$. Then a unique product element in $x^{-n}A_0 \cdot B_0 x^{-m}$ easily determines one in AB.

(iii) We may clearly assume that $I = 0$ so that R is Γ-prime. Suppose $0 \neq A, B \triangleleft R*\Gamma$ with $AB = 0$. Then by Lemma 5.12 we have the product $\pi_{\langle 1 \rangle}(\min_{\langle 1 \rangle} A) \cdot \pi_{\langle 1 \rangle}(\min_{\langle 1 \rangle} B) = 0$. But these factors are nonzero Γ-stable ideals of R, by Lemma 5.11, so we have a contradiction. $\blacksquare$

Of course (iii) follows directly from Corollary 8.5 since a unique product group is torsion free. The next result offers an affirmative answer to a conjecture of [**185**].

Theorem 21.5. [27] *Let H be an isolated orbital subgroup of the polycyclic-by-finite group G and let L be a G-orbital prime of $K[H]$. Then $L^{|G}$ is a prime ideal of $K[G]$.*

Proof. Let $G_0 = \mathbf{N}_G(H)$ so that $|G : G_0| < \infty$. Then G_0 is a group of operators on $K[H]$ and L is G_0-orbital so the previous proposition applies. In particular if $N = \mathrm{vx}_H(L)$ and $D = \nabla_H(N; G_0)$, then there exists a G_0-orbital almost faithful sub N prime I of $K[D]$ with $L = I^{|H}$. Furthermore, N is G_0-orbital and hence also G-orbital.

By Lemma 21.1(i), $i_G(N) \subseteq i_G(H) = H$ so $i_G(N) = i_H(N) = N$ and N is an isolated orbital subgroup of G. Moreover, $|G : G_0| < \infty$ so it follows easily that $\nabla_G(N) \cap H = \nabla_H(N; G_0) = D$. Note that $\nabla_G(N)/N$ is torsion free abelian, by Lemma 21.1(iii), so $D \lhd \nabla_G(N)$. Since D is G-orbital, we conclude from Lemma 21.1(i) that $i_G(D) \subseteq H$ and then, from Lemma 21.1(ii) applied to $D \subseteq \nabla_G(N)$, that $\nabla_G(N)/D$ is torsion free abelian. But a finitely generated torsion free abelian group is poly-Z, so Lemma 21.4(ii)(iii) implies that

$$J = I^{|\nabla_G(N)} = \left(\bigcap_{x \in \nabla_G(N)} I^x \right) \cdot K[\nabla_G(N)]$$

is prime.

Finally the above formula for J shows first that $J^\dagger \subseteq D$ and then that $J^\dagger = \bigcap_{x \in \nabla_G(N)}(I^\dagger)^x$. Thus since $|N : I^\dagger| < \infty$, Lemma 19.2 yields $|N : J^\dagger| < \infty$. We have therefore shown that J is an almost faithful sub N prime of $K[\nabla_G(N)]$ so, by Theorem 20.9, $J^{|G}$ is prime. But by transitivity

$$J^{|G} = (I^{|\nabla_G(N)})^{|G} = I^{|G} = (I^{|H})^{|G} = L^{|G}$$

so the result follows. ∎

We now move on to twisted group algebras. For this it is necessary to briefly discuss finitely presented groups. We recall that a group Γ is *finitely presented* if it is determined by finitely many generators and relations. We write such a group as

$$\Gamma = \langle x_1, x_2, \ldots, x_n \mid w_1, w_2, \ldots, w_s \rangle$$

where $x_1, x_2, \ldots, x_n$ generate Γ and where all relations among these generators can be derived from $w_1 = 1, w_2 = 1, \ldots, w_s = 1$. Formally this means that the homomorphism from the free group $F = \langle X_1, X_2, \ldots, X_n \rangle$ onto Γ, given by $X_i \mapsto x_i$, has as its kernel the normal subgroup $\langle W_1, W_2, \ldots, W_s \rangle^F$ generated by the finitely many words $W_1, W_2, \ldots, W_s \in F$ with $w_i = W_i(x_1, x_2, \ldots, x_n)$.

The next result follows from [**161**, Lemma 12.3.12(ii)(iv)]. We just briefly sketch its proof.

Lemma 21.6. *Let G be a polycyclic-by-finite group.*

i. G is finitely presented.

ii. If H is a finitely generated group, Z is a central subgroup of H and $H/Z \cong G$, then Z is finitely generated and hence H is polycyclic-by-finite.

Proof. (i) Here, by induction, we need only show that if H is finitely presented and G/H is infinite cyclic or finite, then G is finitely presented. In either case, we start with the generators and relations of H. If G/H is infinite cyclic, then $G = \langle H, x \rangle$ and we add x to the generating set. Furthermore we add the finitely many relations which describe the x-conjugate of each generator of H. On the other hand, if G/H is finite, then $G = \bigcup_1^n Hx_i$ and we add $x_1, x_2, \ldots, x_n$ to the generating set. As above, the action of each x_i on H can be described via a finite number of additional relations. Finally for each i, j we have $x_i x_j \in Hx_k$ for some k and we add one more relation for each of these containments.

(ii) We know that G is finitely presented, say

$$G = \langle x_1, x_2, \ldots, x_n \mid w_1, w_2, \ldots, w_s \rangle$$

where $w_i = W_i(x_1, x_2, \ldots, x_n)$. Let H be finitely generated with $H/Z \cong G$ for some central subgroup Z of H. By adding generators if necessary, we can assume that the generating set includes $h_1, h_2, \ldots, h_n$ with $h_i Z = g_i$. Then $H = \langle h_1, h_2, \ldots, h_n \rangle Z$ so we can assume that the remaining finitely many generators $z_1, z_2, \ldots, z_t$ are in Z. Replacing H by $H/\langle z_1, z_2, \ldots, z_t \rangle$ if necessary, we can suppose that $H = \langle h_1, h_2, \ldots, h_n \rangle$. With this, we claim that Z is generated by

$W_i(h_1, h_2, \ldots, h_n)$ for $1 \leq i \leq s$. To this end, let $z \in Z$ and write z in terms of the generators of H so that $z = W(h_1, h_2, \ldots, h_n)$ where W is a word in the free group $F = \langle X_1, X_2, \ldots, X_n \rangle$. Hence, by mapping z to G, we have $W(x_1, x_2, \ldots, x_n) = 1$ and, by definition, W is in the normal closure of $\langle W_1, W_2, \ldots, W_s \rangle$. Replacing each X_i by h_i and using the fact that $W_i(h_1, h_2, \ldots, h_n) \in Z$ is central in H now yields the result. ∎

As an application we have

Lemma 21.7. *Let $K^t[G]$ be a twisted group algebra with group of trivial units $\mathcal{G}$ and let H be a finitely generated subgroup of $\mathcal{G}$ with $HK^\bullet = \mathcal{G}$. Then $Z = H \cap K^\bullet$ is a central subgroup of H and $H/Z \cong G$. Thus H is a polycyclic-by-finite group and the inclusion map $H \to \mathcal{G}$ extends to a K-algebra epimorphism $K[H] \to K^t[G]$.*

Proof. Since $\mathcal{G}/K^\bullet \cong G$ is finitely generated, we know that H exists, $Z = H \cap K^\bullet$ is central in H and $H/Z \cong G$. By Lemma 21.6, H is polycyclic-by-finite and the remaining facts are clear. ∎

In particular, if P is a prime ideal of $K^t[G]$, then it lifts to a prime of $K[H]$ and the results of the previous section apply. We can then map the information back to $K^t[G]$. For example we have

Proposition 21.8. *If G is polycyclic-by-finite, then the prime length of $K^t[G]$ is at most equal to the plinth length of G.*

Proof. By Lemma 21.4(i), let G_0 be a normal poly-Z subgroup of finite index in G. Since $K^t[G] = K^t[G_0]*(G/G_0)$, the rings $K^t[G]$ and $K^t[G_0]$ have the same prime lengths by Corollary 16.8. Furthermore, it is easy to see that $\mathrm{pl}(G) = \mathrm{pl}(G_0)$. Thus, replacing G by G_0 if necessary, we can assume that G is poly-Z.

Let H and Z be as in Lemma 21.7 and suppose that $h(Z) = n$. Then Z has a series $Z_0 \subset Z_1 \subset \cdots \subset Z_n = Z$ with Z_0 torsion and with Z_i/Z_{i-1} infinite cyclic for $i = 1, 2, \ldots, n$. By Lemma 19.5, G has a normal subgroup of finite index with a plinth series. By lifting this series to H and adjoining it to the above series for Z, we see that $\mathrm{pl}(H) = \mathrm{pl}(G) + n$.

Finally let $P_0 \subset P_1 \subset \cdots \subset P_k$ be a chain of primes in $K^t[G]$ and let $P_0' \subset P_1' \subset \cdots \subset P_k'$ be its inverse image in $K[H]$. Note that under the epimorphism $K[H] \to K^t[G]$, we have $Z \to K^\bullet$. Thus if Q_i denotes the kernel of the homomorphism $K[Z_i] \to K$, then Q_i is an H-stable prime ideal of $K[Z_i]$. Hence, since H/Z_i is poly-Z, Lemma 21.4(ii)(iii) implies that $Q_i' = Q_i \cdot K[H]$ is a prime ideal of $K[H]$. We therefore obtain, in $K[H]$, the chain of primes

$$Q_0' \subset Q_1' \subset \cdots \subset Q_n' \subseteq P_0' \subset P_1' \subset \cdots \subset P_k'$$

of length at least $n + k$. By Theorem 19.6(i), $n + k \le \mathrm{pl}(H) = n + \mathrm{pl}(G)$ so $k \le \mathrm{pl}(G)$ as required. ∎

We remark that equality need not occur above. For example (see Exercise 2) there is a simple twisted group algebra $K^t[G]$ with G free abelian of rank n for any $n \ge 2$. Thus $\mathrm{pl}(G) = n$ but $K^t[G]$ has prime length 0.

We can of course go further with Lemma 21.7 and precisely describe the primes of $K^t[G]$ in general. This is done in [172], but it is too technical to include here. However, one aspect of uniqueness is worth discussing. Notice that the choice of H in Lemma 21.7 is quite arbitrary. Therefore finding a normal subgroup of finite index in G whose inverse image in H is orbitally sound might also appear to be arbitrary. This turns out not to be the case as we see below. The proof of the following result is a simple variant on the proof of Proposition 19.7. It uses the alternate characterizations of the orbitally sound property which are given in Section 19.

Lemma 21.9. *Let G be polycylic-by-finite, let Z be a central subgroup of G and assume that G/Z is orbitally sound. Define $W \lhd G$ to be minimal with $W \supseteq Z$ and G/W an elementary abelian 2-group. Then $|G : W| < \infty$ and any automorphism of G which normalizes Z also normalizes W.*

i. Let N be a W-orbital subgroup of W. If N contains no nonidentity normal subgroup of G, then $|N : N \cap \mathbf{Z}(W)| < \infty$.

ii. W is orbitally sound.

Proof. Since G is finitely generated, its homomorphic images which are elementary abelian 2-groups are of bounded order. It follows that there exists a unique minimal W with $G \supseteq W \supseteq Z$ and G/W an elementary abelian 2-group. Thus $|G : W| < \infty$ and the uniqueness of W yields the result on automorphisms.

(i) We proceed by induction on $\hbar(N)$. Since $|G : W| < \infty$, N is G-orbital so there exists $H \lhd G$ with $|G : H| < \infty$ such that H normalizes N. We may assume that $W \supseteq H \supseteq Z$.

Since G/Z is orbitally sound and ZN/Z is an orbital subgroup, it follows that there exists $C \lhd G$ with $ZN \supseteq C \supseteq Z$ and $|ZN : C| < \infty$. Replacing C by $C \cap H$, we may assume that $C \subseteq H$. Furthermore, $C = Z(N \cap C)$ and $|N : N \cap C| < \infty$. Thus since $N \cap C$ is also orbital in W and normalized by H, we can replace N by $N \cap C$, if necessary, and assume that $C = ZN$.

Since N contains no nontrivial normal subgroup of G, we have $Z \cap N = \langle 1 \rangle$. Also $H \lhd G$ and $N \lhd H$, since $N \subseteq C \subseteq H$, so we have $[C, H] \lhd G$ and $[C, H] = [ZN, H] = [N, H] \subseteq N$. Thus $[C, H] = [N, H] = \langle 1 \rangle$ so N is central in H and in particular N is abelian. If $\hbar(N) \geq 2$ we can write $N = N_1 \times N_2$ with $\hbar(N_1), \hbar(N_2) < \hbar(N)$. Since N_1, N_2 are central in H, they are orbital in W. It follows by induction that $|N_i : N_i \cap \mathbf{Z}(W)| < \infty$ for $i = 1, 2$ so $|N : N \cap \mathbf{Z}(W)| < \infty$ as required.

It remains to consider $\hbar(N) = 1$. Choose an integer n so that both Z^n and N^n are torsion free abelian. Then $C^n = Z^n \times N^n \lhd G$ so replacing N by N^n we can assume that $D = T \times N \lhd G$ with $T = Z^n$ a torsion free central subgroup and N infinite cyclic. Let us now think of D as being additive, so the finite group G/H acts linearly on D. Furthermore, if $\mathcal{Q}$ denotes the field of rational numbers, we can let G/H act on $D \otimes \mathcal{Q}$. Since $\mathcal{Q}[G/H]$-modules are completely reducible and $T \otimes \mathcal{Q}$ is a central subspace of codimension 1, we see that G/H can be diagonalized. In fact each element acts like $\mathrm{diag}(1, 1, \ldots, 1, \lambda)$ where $\lambda : G/H \to \mathcal{Q}^{\bullet}$ is a linear character. This implies that $\lambda^2 = 1$ so, by definition of W, we have $W \subseteq \mathrm{Ker}(\lambda)$ and hence W centralizes $D \otimes \mathcal{Q}$. Thus W centralizes $N \subseteq D \otimes \mathcal{Q}$ and (i) is proved.

(ii) Let $N \subseteq W$ with N orbital in W. Set $M = \bigcap_{g \in G} N^g \lhd G$ and consider $\bar{G} = G/M$. We have $\bar{G} \supseteq \bar{Z} = ZM/M$ so $\bar{Z}$ is central in $\bar{G}$. Also $\bar{G}/\bar{Z} \cong G/ZM$ is a homomorphic image of G/Z so $\bar{G}/\bar{Z}$ is orbitally sound. Since $W \supseteq ZM$, it is clear that $\bar{W}$ is minimal in $\bar{G}$ subject to $\bar{W} \supseteq \bar{Z}$ and $\bar{G}/\bar{W}$ being an elementary abelian 2-group. Furthermore $\bigcap_{\bar{g} \in \bar{G}} \bar{N}^{\bar{g}} = \bar{M} = \langle 1 \rangle$ so (i) applies and we conclude that $|\bar{N} : \bar{N} \cap \mathbf{Z}(\bar{W})| < \infty$. If T denotes the complete inverse image of $\bar{N} \cap \mathbf{Z}(\bar{W})$ in G, then clearly $T \triangleleft W$ and $|N : T| < \infty$. Thus W is orbitally sound. $\blacksquare$

We remark that G above need not be orbitally sound. For example let $G = C \wr C_2$ where C is infinite cyclic and $|C_2| = 2$. If $Z = \mathbf{Z}(G)$, then G/Z has Hirsch number 1 and hence is orbitally sound. But G itself is not orbitally sound.

Let G be polycyclic-by-finite. By Theorem 19.3, $G_0 = \mathrm{nio}(G)$ is a characteristic orbitally sound subgroup of G of finite index. Let $G_1 = \mathrm{nio}^2(G)$ be the subgroup of G generated by the squares of all elements of $\mathrm{nio}(G)$. Then G_0/G_1 is an elementary abelian 2-group and clearly the largest such homomorphic image of G_0.

Proposition 21.10. [172] *If G is a polycyclic-by-finite group, then $\mathrm{nio}^2(G)$ is a characteristic subgroup of finite index. Furthermore let X be any polycyclic-by-finite group such that $X/Z \cong G$ for some central subgroup Z of X. If Y is the complete inverse image of $\mathrm{nio}^2(G)$ in X, then Y is orbitally sound.*

Proof. Write $G_0 = \mathrm{nio}(G)$ and $G_1 = \mathrm{nio}^2(G)$. Then $|G : G_0| < \infty$ and $|G_0 : G_1| < \infty$ so G_1 is clearly a characteristic subgroup of G of finite index. Now let $X/Z \cong G$ and let X_i be the complete inverse image of G_i. Then $X_0/Z \cong G_0$ is orbitally sound so, by Lemma 21.9 and the definition of G_1, we have X_1 orbitally sound. $\blacksquare$

In view of the work of Sections 14 and 15, a description of the primes in twisted group algebras leads to a similar result in Noetherian crossed products. Thus for example we have

Theorem 21.11. [172] *Let $R*G$ be a crossed product with R right Noetherian and G polycyclic-by-finite of Hirsch number n. If $P_0 \subset P_1 \subset \cdots \subset P_{n+1}$ is a chain of primes in $R*G$, then $P_0 \cap R \subset P_{n+1} \cap R$.*

Proof. We may clearly assume that $P_0 \cap R = 0$ so that R is G-prime. Suppose by way of contradiction that $P_{n+1} \cap R = P_0 \cap R = 0$. Then $P_i \cap R = 0$ for all i. We proceed with the usual reductions.

Case 1. *R is G-prime.*

Proof. We use the notation and results of Theorem 14.7. Thus Q is a minimal prime of R with stabilizer H and there is a one-to-one correspondence between the primes of $R*G$ disjoint from R and certain primes of $R*H$. Moreover, $|G:H| < \infty$ so $\hbar(H) = \hbar(G) = n$. Now let L_i be the prime ideal of $R*H$ corresponding to P_i. Then $L_i \cap R = Q$ and $L_0 \subset L_1 \subset \cdots \subset L_{n+1}$. Furthermore, if $\tilde{\ }: R*H \to (R*H)/(Q*H) = (R/Q)*H$ denotes the natural map, then we obtain a chain of primes $\tilde{L}_0 \subset \tilde{L}_1 \subset \cdots \subset \tilde{L}_{n+1}$ with $\tilde{L}_i \cap \tilde{R} = 0$ and $\tilde{R}$ prime. Thus we have reduced the problem to the case of prime coefficient rings. ∎

Case 2. *R is prime.*

Proof. Here we use the notation and results of Theorem 15.8. In particular, we have $G_{\mathrm{inn}} \lhd G$ and a correspondence between the primes of $R*G$ disjoint from R and the G-prime ideals of a certain twisted group algebra $C^t[G_{\mathrm{inn}}]$. If $\tilde{P}_i$ is the G-prime corresponding to P_i, then we obtain the chain $\tilde{P}_0 \subset \tilde{P}_1 \subset \cdots \subset \tilde{P}_{n+1}$. In addition it follows easily from Lemma 15.10 that there exists a chain $Q_0 \subset Q_1 \subset \cdots \subset Q_{n+1}$ of primes in $C^t[G_{\mathrm{inn}}]$ with each Q_i minimal over $\tilde{P}_i$. But $\mathrm{pl}(G_{\mathrm{inn}}) \leq \hbar(G_{\mathrm{inn}}) \leq \hbar(G) = n$, so this contradicts Proposition 21.8 and the result follows. ∎

We remark that if $H \lhd G$ then we do not have $\mathrm{pl}(H) \leq \mathrm{pl}(G)$ in general. Thus the last line of the above proof prevents us from replacing $n = \hbar(G)$ by the smaller parameter $\mathrm{pl}(G)$. On the other hand, the result is in fact true with $n = \mathrm{pl}(G)$. This is proved in [**206**] using the extension of Theorem 19.9 mentioned at the end of Section 19.

1. Let $A = \langle a_1, a_2, \ldots, a_n \rangle$ and $B = \langle b_1, b_2, \ldots, b_n \rangle$ be free abelian groups of rank n and set $R = K[A]$. If λ is an element of infinite multiplicative order in $K^\bullet$, define the K-action of B on R by $(a_i)^{b_j} = a_i$ for $i \neq j$ and $(a_i)^{b_i} = \lambda a_i$. Show that B acts faithfully on R and that R is B-simple.

2. Continuing with the above, suppose $\langle \lambda, \mu_1, \mu_2, \ldots, \mu_k \rangle$ is a free abelian subgroup of $K^\bullet$. If $C = \langle c_1, c_2, \ldots, c_k \rangle$ is a free abelian group, define the action of C on R by $(a_i)^{c_j} = \mu_j a_i$. Show that $B \times C$ acts faithfully on R and hence is X-outer since R is commutative. Conclude that the skew group ring $R(B \times C)$ is simple and observe that $R(B \times C) \cong K^t[A \times B \times C]$.

3. Let $\mathcal{F}$ be a finite field extension of the rationals $\mathcal{Q}$ with $(\mathcal{F} : \mathcal{Q}) = n$ and let $\mathcal{O}$ be the ring of algebraic integers in $\mathcal{F}$. Then $\mathcal{O}^+$ is a free abelian group of rank n and if α is a unit in $\mathcal{O}$ then, via multiplication, $\mathcal{O}$ is a module for the cyclic group $\langle \alpha \rangle$. Prove that $\mathcal{O}$ is a plinth for $\langle \alpha \rangle$ if and only if $\mathcal{F} = \mathcal{Q}[\alpha^k]$ for all integers $k \geq 1$.

4. Continuing with the above, let p be a prime and let α be a root of the polynomial $x^p + px - 1$. If $\mathcal{F} = \mathcal{Q}[\alpha]$ show that $(\mathcal{F} : \mathcal{Q}) = p$ so that there are no intermediate fields. If $\mathcal{O}' = \mathcal{Z}[\alpha]$, where $\mathcal{Z}$ is the ring of ordinary integers, show that α is a unit of $\mathcal{O}'$ which is not a root of unity. Conclude that the additive group of $\mathcal{O}'$ is a plinth for $\langle \alpha \rangle$ of rank p. Note that $\mathcal{O}' \subseteq \mathcal{O}$ but we do not claim that equality occurs.

5. Let A be a free abelian group of rank n which is a plinth for the infinite cyclic group $\langle x \rangle$ and let G be the semidirect product $G = A \rtimes \langle x \rangle$. Show that $\hbar(G) = n + 1$ but that $\mathrm{pl}(G) = 2$. Use the result of the preceding problem to construct such examples.

6. Let $I \lhd K^t[G]$ and define

$$I^t = \{\, x \in G \mid \bar{x} - k \in I \text{ for some } k \in K \,\}.$$

Prove that I^t is a normal subgroup of G. Furthermore show that there exists a homomorphism $\tilde{\ } : K^t[G] \to K^t[G/I^t]$, where the latter is some twisted group algebra of the group G/I^t, such that $I \supseteq \mathrm{Ker}(\tilde{\ })$ and $\tilde{I}^t = \langle 1 \rangle$.

7. Let $R*G$ be given with R right Noetherian and G polycyclic-by-finite. If $\hbar(G) = n$ and R has prime length m, show that $R*G$ has prime length less than $(n+1)(m+1)$. This follows from Theorem 21.11.

22. Jacobson Rings

A ring R is said to be a *Jacobson ring* if, for every prime ideal P of R, the factor ring R/P is semiprimitive. This is of course equivalent to the assertion that every prime (or semiprime) homomorphic image of R is semiprimitive. Thus, for example, the Hilbert Nullstellensatz implies that $K[x_1, x_2, \ldots, x_n]$ is a Jacobson ring. In fact, every finitely generated commutative K-algebra is a Jacobson ring, since this property is clearly inherited by homomorphic images. The obvious noncommutative analog concerns crossed products $R*G$ with G polycyclic-by-finite. Namely we ask whether the Jacobson property for R implies the same for $R*G$. As we will see, the answer is "yes" if R is right Noetherian, but "no" in general.

In the course of this work we will obviously have to consider $J(R*G)$. Furthermore, suppose P is a prime of $R*G$ and let $J/P = J(R*G/P)$. Since we wish to show that $J = P$, an incomparability result for $J \supset P$ would be most welcome. This is the approach we follow and we keep the Noetherian hypothesis to a minimum here. It is clear, by induction, that we need only consider G infinite cyclic or finite. We start with the latter case.

Lemma 22.1. *Let $R \subseteq S$ be an extension of rings.*

i. If $S = R \oplus U$ where U is a complementary R-submodule of S, then $J(S) \cap R \subseteq J(R)$.

ii. Suppose S is a finitely generated right R-module and that $J(R)S \subseteq SJ(R)$. Then $J(R) \subseteq J(S)$.

iii. Let $S = R(G)$ be a G-graded ring and let H be a subgroup of G. Then $J(R(G)) \cap R(H) \subseteq J(R(H))$.

Proof. (i) Suppose $r \in R$ has an inverse $s \in S$ and write $s = r' + u \in R \oplus U$. Then $1 = sr = r'r + ur$ implies that $ur = 0$ so $u = 0$ and

$s \in R$. It now follows that $J(S) \cap R$ is a quasi-regular ideal of R and hence is contained in $J(R)$.

(ii) Let V be an irreducible S-module. Then $V_{|S}$ is cyclic and $S_{|R}$ is finitely generated so V is a finitely generated R-module. By Nakayama's Lemma, $VJ(R) \subset V$. In addition, $J(R)S \subseteq SJ(R)$ implies that $VJ(R)$ is an S-submodule of V. Thus since V is irreducible, we have $VJ(R) = 0$ and hence $J(R) \subseteq J(S)$.

(iii) Here we need only observe that $U = R(G \setminus H)$ is a complementary $R(H)$-submodule of $R(G)$ so (i) applies. ∎

Lemma 22.2. [68] *Let $S = R(G)$ be a G-graded ring with G finite and let $I \lhd R(G)$. Set $\bar{R} = R/(R \cap I) \subseteq \bar{S} = S/I$.*

    i. If $\bar{r} \in \bar{R}$ has an inverse in $\bar{S}$, then its inverse is in $\bar{R}$.

    ii. $J(\bar{S}) \cap \bar{R} = J(\bar{R})$.

Proof. (i) Let $r \in R$ with $\bar{r} = r + (R \cap I)$. We show by inverse induction on $|A|$ that if A is a subset of G with $1 \in A$, then there exists $s \in R(A)$ with $rs \equiv 1 \pmod{I}$. By hypothesis, s exists if $A = G$. Now suppose $1 \in A \subset G$ and that the result holds for all subsets of G of larger size. Choose $g \in G \setminus A$. Then $A \cup \{g\}$ is a larger subset of G and $R(A \cup \{g\}) = R(A) + R(g)$. Thus by induction there exists $s = s(A) + s(g) \in R(A) + R(g)$ with $rs \equiv 1 \pmod{I}$. Furthermore $g^{-1}A \cup \{1\}$ is also a subset of G of larger size so again, by induction, there exists $t = t(g^{-1}A) + t(1) \in R(g^{-1}A) + R(1)$ with $rt \equiv 1 \pmod{I}$. Now $1 - r \cdot s(A) \equiv r \cdot s(g)$ and $1 - r \cdot t(1) \equiv r \cdot t(g^{-1}A)$ so multiplying these equations yields

$$1 - r[s(A) + t(1) - s(A) \cdot r \cdot t(1)]$$
$$\equiv r[s(g) \cdot r \cdot t(g^{-1}A)] \pmod{I}.$$

But the factors of r in the square brackets both belong to $R(A)$ since $r \in R = R(1)$ and $1 \in A$. Thus the inductive statement is proved. The result follows by taking $A = \{1\}$.

(ii) By (i) above, $J(\bar{S}) \cap \bar{R}$ is a quasi-regular ideal of $\bar{R}$ and thus $J(\bar{S}) \cap \bar{R} \subseteq J(\bar{R})$. In the other direction, let V be an irreducible $\bar{S}$-module. Then V is an irreducible S-module and hence a completely reducible R-module, by Proposition 4.10. Thus V is a completely reducible $\bar{R}$-module so $VJ(\bar{R}) = 0$ and hence $J(\bar{R}) \subseteq J(\bar{S}) \cap \bar{R}$. ∎

We remark that the above is false in the case of infinite groups (see Exercise 1). We can now prove

Theorem 22.3. *Let $S = R(G)$ be a G-graded ring with G finite. Then R is a Jacobson ring if and only if S is.*

Proof. Suppose first that R is a Jacobson ring and let P be a prime ideal of $S = R(G)$. By moding out by the graded ideal of S generated by $P \cap R$, we may assume that $P \cap R = 0$. It follows from Theorem 17.9(i) that R is semiprime and hence semiprimitive. Now define $J \lhd S$ by $J/P = \mathrm{J}(S/P)$. Then $J \cap R \subseteq \mathrm{J}(R) = 0$, by Lemma 22.2(ii), and we conclude from Corollary 17.10 that $J = P$.

Conversely suppose that S is a Jacobson ring and let Q be a prime of R. By Theorem 17.9(ii), there exists a prime P of S with Q minimal over $P \cap R$. Again we can assume that $P \cap R = 0$. Since S/P is semiprimitive, it follows from Lemma 22.2(ii) that R is semiprimitive. Finally we define $J' \lhd R$ with $J'/Q = \mathrm{J}(R/Q)$ and we let N be the intersection of the other minimal primes of R. Then it follows from Theorem 17.9(i) that $N \neq 0$ and $N \cap Q = 0$. In particular $J'N \subseteq N$ is an ideal of R which embeds isomorphically in J'/Q. Hence $J'N$ is quasi-regular and therefore contained in $\mathrm{J}(R) = 0$. Finally $Q \supseteq J'N = 0$ and $Q \not\supseteq N$, so we conclude that $Q \supseteq J'$. ∎

This completes the finite group argument. Now suppose S is a G-graded ring with G infinite cyclic. In this case, we normally think of G as the additive group of integers Z and say that S is Z-*graded*. Then S has the structure $S = \oplus \sum_{i \in Z} S_i$ with $S_i S_j \subseteq S_{i+j}$. As usual, if $0 \neq s \in S_i$ for some i, then s is said to be *homogeneous* and $\deg s = i$. Furthermore if $s = \sum_{i=m}^{n} s_i$ is a nonzero element of S written in terms of its homogeneous components with $s_m, s_n \neq 0$, then we let $s_+ = s_n$ be the *leading term* of s and $s_- = s_m$ be its *trailing term*. Moreover $\deg s = \deg s_+ = n$ and we let $\mathrm{br}(s) = n - m \geq 0$ denote the *breadth* of s. Notice that $\mathrm{br}(s) = 0$ if and only if s is homogeneous. For convenience we set $\mathrm{br}(0) = -\infty$.

The next goal is an incomparability result for such rings.

Lemma 22.4. *Let S be a Z-graded ring and assume that S is prime. Let $0 \neq I \lhd S$ and let u be an element of minimal breadth in $I \setminus 0$. If*

$0 \neq w \in I$, then there exist $0 \neq t \in S$ and homogeneous $h \in S$ such that

$$tsu = whsu_+ \qquad \text{for all } s \in S.$$

Proof. We first consider u. Note that for any homogeneous $j \in S$ the element $uju_+ - u_+ ju \in I$ has smaller breadth than u. Thus this expression must be zero and by linearity we have

$$usu_+ = u_+ su \qquad \text{for all } s \in S. \tag{$*$}$$

Now we study $0 \neq w \in I$ and we proceed by induction on $\mathrm{br}(w)$. Let j be any homogeneous element of S and form $wju_+ - w_+ ju = w' \in I$. Since $\mathrm{br}(w) \geq \mathrm{br}(u)$ and since the term $w_+ ju_+$ cancels, we see that $\mathrm{br}(w') < \mathrm{br}(w)$ and that $\deg w' < \deg w + \deg j + \deg u$. If $w' = 0$ for all j, then by linearity $wsu_+ = w_+ su$ holds for all $s \in S$ and the result follows with $t = w_+$ and $h = 1$.

Now suppose $w' \neq 0$ for some j. Then we conclude by induction that there exist $0 \neq t' \in S$ and homogeneous $h' \in S$ with

$$t'su = w'h'su_+ \qquad \text{for all } s \in S. \tag{$**$}$$

Notice that $(*)$ yields $w_+ j[uh'su_+] = w_+ j[u_+ h'su]$ and thus by substituting the definition of w' into $(**)$ we have

$$[t' + w_+ ju_+ h']su = w[ju_+ h']su_+ \qquad \text{for all } s \in S.$$

Thus the result follows with $t = t' + w_+ ju_+ h'$ and $h = ju_+ h'$ provided we show that $t \neq 0$. We do this by computing degrees.

If $w_+ ju_+ h' = 0$, then it is clear that $t = t' \neq 0$. Thus we may assume that $w_+ ju_+ h'$ is not zero so

$$\deg(w_+ ju_+ h') = \deg w + \deg j + \deg u + \deg h'.$$

Since S is prime, there exists a homogenous element $s \in S$ with $t'_+ su_+ \neq 0$. Using this s in $(**)$ yields

$$\deg t' + \deg s + \deg u = \deg t'su$$
$$= \deg w'h'su_+$$
$$\leq \deg w' + \deg h' + \deg s + \deg u$$

so $\deg t' \leq \deg w' + \deg h'$. Finally plugging in the upper bound for $\deg w'$ yields

$$\deg t' < \deg w + \deg j + \deg u + \deg h' = \deg(w_+ j u_+ h')$$

so clearly $t = t' + w_+ j u_+ h' \neq 0$. ∎

We can now prove

Theorem 22.5. [18] *Let S be a Z-graded ring and let P be a nonzero prime ideal. If I is an ideal of S properly containing P, then I has a nonzero homogeneous element. In particular, if S is component regular, then $I \cap S_0 \neq 0$.*

Proof. Suppose first that S is not prime and let $a, b \in S \setminus 0$ with $aSb = 0$. Then clearly $a_+ S b_+ = 0 \subset P$ so at least one of a_+ or b_+ is contained in P. Thus P contains a nonzero homogeneous element and hence so does I. We can now assume that S is prime and that P itself has no nonzero homogeneous elements.

We are given $0 \subset P \subset I$ and hence we choose w to be an element of minimal breadth in $P \setminus 0$ and u an element of minimal breadth in $I \setminus P$. Since w_+ is homogeneous, $w_+ \notin P$ so $w_+ S u \nsubseteq P$ and there exists a homogeneous element $j \in S$ with $w_+ j u \notin P$. It follows that $w' = w_+ j u - w j u_+$ is an element of $I \setminus P$ and clearly $\mathrm{br}(w') < \max\{\mathrm{br}(w), \mathrm{br}(u)\}$. But $\mathrm{br}(u) \leq \mathrm{br}(w')$, by definition of u, so $\mathrm{br}(u) < \mathrm{br}(w)$.

It follows that u is an element of minimal breadth in $I \setminus 0$ so the previous lemma applies to yield appropriate $t, h \in S$ with

$$tsu = whsu_+ \qquad \text{for all } s \in S.$$

Since S is prime, we can find homogeneous $a, b \in S$ with $t_+ a u_+ \neq 0$ and $t_- b u_- \neq 0$. Letting $s = a$ or b and computing degrees as before yields

$$\deg t_+ + \deg u_+ \leq \deg w_+ + \deg h + \deg u_+$$
$$\deg t_- + \deg u_- \geq \deg w_- + \deg h + \deg u_+$$

so $\mathrm{br}(t) + \mathrm{br}(u) \le \mathrm{br}(w)$. Moreover $tSu \subseteq P$ and $u \notin P$ so $t \in P$ and $\mathrm{br}(t) \ge \mathrm{br}(w)$ by definition of w. We conclude that $\mathrm{br}(u) = 0$ so that $u \in I$ is homogeneous as required.

Finally if S is component regular and $u \in S_k$, then $0 \ne uS_{-k} \subseteq I \cap S_0$ and the result follows. ∎

Next we consider the Jacobson radical of a Z-graded ring. The first result along this line appeared in [2] where the polynomial ring $R[x]$ was studied. Somewhat later, crossed product algebras were considered in [157]. The techniques used in those two papers are sufficient to prove

Theorem 22.6. *Let S be a Z-graded ring. Then $\mathrm{J}(S)$ is a graded ideal and $\mathrm{J}(S) \cap S_m$ is nilpotent for all $m \ne 0$.*

Proof. We first prove that $\mathrm{J}(S)$ is graded. Specifically, the goal is to show that if $s = \sum_i s_i \in \mathrm{J}(S)$, then each $s_i \in \mathrm{J}(S)$. We proceed by induction on the size n of the support of s over all choices of s and S. If $n = 0$ or 1 the result is clear so assume $n \ge 2$. Say $s_a \ne 0$ and fix $s_b \ne 0$ with $b \ne a$.

Choose any rational prime $p > |b-a|$ and form the ring extension $\tilde{S} = S[\zeta]/(1+\zeta+\cdots+\zeta^{p-1})$ of S. If ϵ denotes the image of ζ in $\tilde{S}$, then ϵ commutes with S and $1+\epsilon+\cdots+\epsilon^{p-1} = 0$ so $\epsilon^p = 1$. Furthermore, $\tilde{S} = S \oplus S\epsilon \oplus \cdots \oplus S\epsilon^{p-2}$ so Lemma 22.1(i)(ii) yields $\mathrm{J}(S) = \mathrm{J}(\tilde{S}) \cap S$. Note that $\tilde{S}$ is clearly Z-graded with $\tilde{S}_i = S_i \oplus S_i\epsilon \oplus \cdots \oplus S_i\epsilon^{p-2}$. Thus $s = \sum_i s_i \in \mathrm{J}(\tilde{S})$ and s_i is the $\tilde{S}_i$-component of s.

Now observe that the map $\epsilon^{\#}\colon \tilde{S} \to \tilde{S}$ which multiplies each element of $\tilde{S}_i$ by ϵ^i is an automorphism of $\tilde{S}$. Thus since $\mathrm{J}(\tilde{S})$ is a characteristic ideal of $\tilde{S}$, we have $\sum_i \epsilon^i s_i = \epsilon^{\#}(s) \in \mathrm{J}(\tilde{S})$. Furthermore

$$t = \sum_i \epsilon^i s_i - \epsilon^b \sum_i s_i \in \mathrm{J}(\tilde{S})$$

and this is an element of support size less than n. We conclude by induction that $t_a = (\epsilon^a - \epsilon^b)s_a \in \mathrm{J}(\tilde{S})$ so $(1 - \epsilon^{b-a})s_a \in \mathrm{J}(\tilde{S})$ and note that p does not divide $b - a$ by the choice of p. It follows that

$$ps_a = \prod_{j=1}^{p-1}(1 - \epsilon^j) \cdot s_a \in \mathrm{J}(\tilde{S})$$

so $ps_a \in \mathrm{J}(\tilde{S}) \cap S = \mathrm{J}(S)$. Finally since this is true for at least two distinct p, we conclude that $s_a \in \mathrm{J}(S)$ as required.

For the second part, let $u_m \in \mathrm{J}(S) \cap S_m$ and assume for convenience that $m > 0$. Then $1 - u_m$ is invertible in S with inverse $v = \sum_{i=k} v_i$ and say $v_k \neq 0$. Since the lowest degree term in $1 = (1 - u_m)v$ is $1 \cdot v_k = v_k$, it follows that $k = 0$ and $v_0 = 1$. Furthermore using $u_m v_i = v_{i+m}$, we conclude by induction that $v_{jm} = (u_m)^j$. Since v_{jm} is eventually zero, the result follows. ∎

Corollary 22.7. [11] *i. Let $R*G$ be a crossed product with G a poly-Z group. Then $\mathrm{J}(R*G) \subseteq \mathrm{J}(R)*G$.*

ii. $\mathrm{J}(R[x;\sigma]) = I_0 + I_1 x R[x;\sigma]$ where $I_0 \subseteq I_1$ are σ-invariant ideals of R. In fact, $I_0 = I_1 \cap \mathrm{J}(R)$.

Proof. (i) By induction, it suffices to assume that G is infinite cyclic. In that case, by the previous theorem $\mathrm{J}(R*G)$ is a graded ideal and hence we have $\mathrm{J}(R*G) = I*G$ where $I = \mathrm{J}(R*G) \cap R$. It follows from Lemma 22.1(iii) that $I \subseteq \mathrm{J}(R)$.

(ii) Again $\mathrm{J}(R[x;\sigma])$ is a graded ideal and hence it is equal to $\sum_{n=0}^{\infty} I_n x^n$ with each $I_n \triangleleft R$. Moreover, it is clear that $I_n \subseteq I_{n+1}$. Now let $n \geq 1$ and observe that

$$\left(I_n x R[x;\sigma]\right)^n \subseteq I_n x^n R[x;\sigma] \subseteq \mathrm{J}(R[x;\sigma]).$$

Thus $I_n x R[x;\sigma] \subseteq \mathrm{J}(R[x;\sigma])$ and $I_n \subseteq I_1$. We conclude that all I_n with $n \geq 1$ are equal. Finally note that $I_0 \subseteq I_1 \cap \mathrm{J}(R)$ by Lemma 22.1(i). Furthermore, $[I_1 \cap \mathrm{J}(R)] + I_1 x R[x;\sigma]$ is a right ideal of $R[x;\sigma]$ whose image in $R[x;\sigma]/\mathrm{J}(R[x;\sigma])$ is quasi-regular. Thus the image is zero and $I_1 \cap \mathrm{J}(R) \subseteq I_0$. In addition, $(R[x;\sigma]x I_1)^2 \subseteq \mathrm{J}(R[x;\sigma])$ so $x I_1 \subseteq I_1 x$ and, by symmetry, these sets are equal. Therefore I_1 is σ-stable and the result follows. ∎

We remark that equality need not occur in (i). Furthermore, in (ii) above it is quite possible to have $I_0 \neq I_1$ (see Exercise 5). The next result is a consequence of the division algorithm.

Lemma 22.8. *Let $R*G$ be given with G infinite cyclic, let $0 \neq I \triangleleft R*G$ with $I \cap R = 0$ and let $J/I = \mathrm{J}(R*G/I)$. Then there exist nonzero G-stable ideals A, B of R with $(J \cap R)AB \subseteq \mathrm{J}(R)$.*

Proof. If $G = \langle x \rangle$ then, by replacing $\overline{x^n}$ by $\bar{x}^n$ if necessary, we can assume that $R*G = R\langle x \rangle$ is a skew group ring. As usual, any R-linear combination of those x^n with $n \geq 0$ will be called a polynomial.

Since $I \neq 0$ it is clear that there exists $\gamma = r_0 + r_1 x + \cdots + r_n x^n \in I$ with $r_0 \neq 0$ and we choose γ so that n is minimal. Note that $n \geq 1$ since $I \cap R = 0$. For this n we define

$$A = \{ r \in R \mid c_0 + c_1 x + \cdots + c_n x^n \in I \text{ and } r = c_0 \}$$
$$B = \{ r \in R \mid c_0 + c_1 x + \cdots + c_n x^n \in I \text{ and } r = c_n \}.$$

Then A and B are both nonzero G-stable ideals of R.

Suppose $\eta \in R\langle x \rangle$. We show that there exist integers $u, v \geq 0$ such that for any $c \in A^u B^v$ we have $\eta c \equiv \rho \pmod{I}$ where ρ is a polynomial of degree less than n. To this end write $\eta = \sum_{i=-m}^{} e_i x^i$ with $m \geq 0$. If $a \in A$, then the x^{-m} coefficient of ηa is contained in A. Thus there exists a polynomial α of degree $\leq n$ in I such that

$$\eta a - \alpha x^{-m} = \sum_{i=-m+1}^{} e_i' x^i$$

or equivalently $\eta a \equiv \sum_{i=-m+1}^{} e_i' x^i \pmod{I}$. Continuing in this manner, using various a's we get $\eta a_1 a_2 \cdots a_m \equiv \mu \pmod{I}$ where μ is a polynomial of degree at most $\max\{\deg \eta, n-1\}$.

Write $\mu = \sum_{i=0}^{t} f_i x^i$ with $t \leq \max\{\deg \eta, n-1\}$. If $b \in B$ and $t \geq n$, then there exists a polynomial $\beta \in I$ of degree $\leq n$ such that

$$\mu b - \beta x^{t-n} = \sum_{i=0}^{t-1} f_i' x^i$$

or equivalently $\mu b \equiv \sum_{i=0}^{t-1} f_i' x^i \pmod{I}$. Continuing in this manner, using various b's we get $\mu b_1 b_2 \cdots b_{t-n+1} \equiv \rho \pmod{I}$ where ρ is a polynomial of degree less than n. Thus this observation is proved with $u = \max\{-\deg \eta_-, 0\}$ and $v = \max\{\deg \eta - n + 1, 0\}$.

Finally let M be any maximal right ideal of R. The goal is to show that $(J \cap R)AB \subseteq M$. If $J \cap R \subseteq M$ this is clear, so assume $J \cap R \not\subseteq M$. Then $R = M + (J \cap R)$ so $1 = m + j$ with $m \in M$ and $j \in J \cap R$. By definition of J, $1 - j = m$ is invertible modulo

I, say $m\eta \equiv 1 \pmod{I}$. We apply the observation of the preceding paragraph to η. Thus there exist integers $u, v \geq 0$ such that for any $c \in A^u B^v$ we have $\eta c \equiv \rho \pmod{I}$ where ρ is a polynomial of degree less than n. Thus

$$c \equiv m\eta c \equiv m\rho \pmod{I}.$$

But $m\rho - c$ is a polynomial of degree $< n$ since $n \geq 1$ so, by definition of n, we have $m\rho - c = 0$. By considering the identity coefficient in this expression, we conclude that $c \in mR \subseteq M$.

In other words, we have shown that $A^u B^v \subseteq M$. But the largest two-sided ideal of R contained in M is a primitive and hence a prime ideal. Thus we conclude that either A or B is contained in M and hence $(J \cap R)AB \subseteq M$ as required. Since this holds for all such M, the result follows. ∎

It is now a simple matter to prove

Theorem 22.9. [65] *Let $R*G$ be a crossed product with G infinite cyclic and assume that every G-prime ideal of R is a semiprimitive ideal. Then $R*G$ is a Jacobson ring.*

Proof. Let P be a prime ideal of $R*G$. Moding out by $(P \cap R)*G$ if necessary, we can assume that $P \cap R = 0$. Thus R is a G-prime ring and it is semiprimitive by hypothesis.

Let $J \lhd R*G$ with $J/P = \mathrm{J}(R*G/P)$. The goal is to show that $J = P$. By Corollary 22.7(i), $R*G$ is semiprimitive, so the result follows if $P = 0$. On the other hand, if $P \neq 0$, then by Lemma 22.8 there exist nonzero G-stable ideals A, B of R with $(J \cap R)AB \subseteq \mathrm{J}(R) = 0$. But R is G-prime and $A, B \neq 0$, so this implies that $J \cap R = 0$. Theorem 22.5 now yields the result. ∎

Corollary 22.10. [65] *Let $R*G$ be a crossed product with R right Noetherian and G polycyclic-by-finite. If R is a Jacobson ring, then so is $R*G$.*

Proof. By induction it suffices to assume that G is either infinite cyclic or finite. If G is finite, then Theorem 22.3 yields the result. If

G is infinite cyclic and P is a prime ideal of $R*G$ then, since R is right Noetherian, Lemma 14.2(i) implies that $P \cap R$ is a semiprime ideal of R. Since R is a Jacobson ring, $P \cap R$ is therefore a semiprimitive ideal and the result follows from Theorem 22.9(i). ∎

We remark that certain analogs of these results were known earlier. In particular, it was shown in [**66,96**] that if R is a commutative Jacobson ring then so is $R[x]$. Furthermore, [**185**] considered $R[G]$ with R Noetherian and G polycyclic-by-finite.

We close this section by sketching three examples of interest. First, it is quite possible for $R*G$ to be a Jacobson ring even though R is not. For example, let $R = K[[\zeta]]$ be a power series ring over the field K and let λ be an element of infinite multiplicative order in $K^{\bullet}$. Define the action of the infinite cyclic group $G = \langle g \rangle$ on R via the obvious extension of $\zeta^g = \lambda\zeta$. Note that R is an integral domain with $J(R) = \zeta R \neq 0$ and thus R is not a Jacobson ring. On the other hand, let P be a prime ideal of RG and let $J/P = J(R/P)$. If $P = 0$, then $J = 0$ by Theorem 22.6 since RG is obviously a domain. On the other hand, if $P \neq 0$, then $P \cap R \neq 0$ by Lemma 10.3(i) and Corollary 12.6. Thus $P \cap R = \zeta^n R$ for some $n \geq 1$. But then $(\zeta RG)^n \subseteq P$ so $\zeta RG \subseteq P$ and P corresponds to a prime of the Jacobson ring $RG/\zeta RG \cong K[G]$. We conclude that $J = P$ in this case also. Note that this is a Noetherian example.

Next we observe that Corollary 22.10 fails without the Noetherian hypothesis. The following counterexample is from [**175**]. Let $T = K[y_i \mid i \in Z]$ be a polynomial ring in infinitely many variables and let I be the ideal of T generated by all monomials of the form $y_{i_1} y_{i_2} \cdots y_{i_n}$ with $n \geq 2$ satisfying $i_1 \leq i_2 \leq \cdots \leq i_n$ and $i_n - i_1 \leq n^2$. Then $R = T/I$ is easily seen to be a Jacobson ring. Indeed if $\bar{y}_i$ denotes the image of y_i in R, then $(\bar{y}_i)^2 = 0$ so $N = \sum_i \bar{y}_i R$ is the prime radical of R. It follows that N is the unique prime of R and $R/N \cong K$ is semiprimitive. Now if $G = \langle g \rangle$ is infinite cyclic then G acts on R via the shift operator $(\bar{y}_i)^g = \bar{y}_{i+1}$ and R is G-prime. Indeed if α and β are nonzero elements of R and if the integer m is chosen sufficiently large, then the subscripts of the variables involved in α and β^{g^m} are so far apart that $\alpha\beta^{g^m} \neq 0$. Thus the skew group ring RG is prime by Lemma 21.4(iii). On the other hand, $J(RG) \neq 0$

since $\bar{y}_0 RG \neq 0$ is nil. To see this, let $\gamma = \sum_{-k}^{k} r_i g^i \in RG$ with $k \geq 2$. Then $(\bar{y}_0 \gamma)^k$ is an RG-linear combination of monomials of the form

$$(\bar{y}_0)^{g^{n_1}} (\bar{y}_0)^{g^{n_2}} \cdots (\bar{y}_0)^{g^{n_k}} = \bar{y}_{n_1} \bar{y}_{n_2} \cdots \bar{y}_{n_k}$$

with $n_j = d_1 + d_2 + \cdots + d_j$, $d_1 = 0$ and $|d_j| \leq k$. But $y_{n_1} y_{n_2} \cdots y_{n_k} \in I$ since this monomial has k factors with subscripts at most k^2 apart. Thus $(\bar{y}_0 \gamma)^k = 0$ and we conclude that RG is not a Jacobson ring.

Finally it is clear that the prime radical and the Jacobson radical of any Jacobson ring coincide. In particular, this implies that the Jacobson radical of such a ring is nil. It was thought for awhile that every finitely generated K-algebra S also had $J(S)$ nil. Indeed this was shown to be true in [3] if K is nondenumerable. However the result is false in general ([12]). To see this, let K be a countable field and let D be the polynomial ring $K[y]$ localized at the prime (y). Then D is a countable local domain and $J(D) = yD \neq 0$. Now let $T = \prod_{i \in Z} D_i$ be the complete direct product of copies of D indexed by the integers and let e_i be the idempotent in T with 1 in the ith entry and 0's elsewhere. Furthermore, since D is countable, we can write $D = \{ d_i \mid i \in Z \}$ and define $\delta \in T$ to have i^{th} component equal to d_i. As usual T admits the shift operator g and we let R be the finitely generated subalgebra (or the finitely generated subring) of the skew group ring $T\langle g \rangle$ generated by $1, e_0, \delta, g$ and g^{-1}. Note that

$$e_0 R e_0 \subseteq e_0 T\langle g \rangle e_0 = e_0 T = e_0 D$$

and that $e_0 \delta^{g^i} e_0 = e_0 d_i$. Thus $e_0 R e_0 = e_0 D \cong D$. Since $JD \cong J(e_0 R e_0) \subseteq J(R)$ and $J(D)$ is not nil, we conclude that $J(R)$ is not nil. We remark that R is in fact the skew group ring $(R \cap T)\langle g \rangle$.

EXERCISES

1. Let R be a commutative integral domain which is not a field and let $S = R[\langle x \rangle]$ be the group ring of the infinite cyclic group $\langle x \rangle$ over R. If $0 \neq a$ is a nonunit of R, let P be the ideal of S generated by $x - a$. Show that P is prime and that $\bar{a} = a + P$ is invertible in S/P but not in $\bar{R} \cong R$.

2. Let G be a group. A ring S is called a G-*system* if $S = \sum_{g \in G} S_g$ is a sum, not necessarily direct, of additive subgroups satisfying $S_x S_y \subseteq S_{xy}$. Show that S is a G-system if and only if it is the natural homomorphic image of a G-graded ring.

3. Let S be a Z-graded ring. If u, w are nonzero elements of S and j is homogeneous, prove that

$$\mathrm{br}(u_+ j w - u j w_+) < \max\{\mathrm{br}(u), \mathrm{br}(w)\}.$$

Furthermore if S is prime and $asb = csd$ holds for all $s \in S$, prove that $\mathrm{br}(a) + \mathrm{br}(b) = \mathrm{br}(c) + \mathrm{br}(d)$.

4. Let $S = R(G)$ be given with G a finitely generated free abelian group. Assume that S is an algebra over the infinite field $K \subseteq R$. Prove that every characteristic ideal of S is graded.

5. Suppose $S = K[y_i \mid i \in Z]$ is the polynomial ring over K in infinitely many variables subject to the relations $y_i y_j = 0$ for $i \neq j$ and consider the skew polynomial ring $S[x; \sigma]$ where σ is the shift operator. Show that $\mathrm{J}(S[x; \sigma]) = I_1 x S[x; \sigma]$ where $I_1 = \sum_i y_i S$.

6. Let P be a prime ideal of $R[x; \sigma]$ with $x \notin P$. Prove that $P \cap R$ is σ-stable and then σ-prime. For this, note that $(R[x; \sigma]x)(P \cap R)^\sigma \subseteq P$. Similarly, if

$$A = \{\, r \in R \mid a_0 + a_1 x + \cdots + a_n x^n \in P \text{ and } r = a_0 \,\}$$

prove that A is a σ-stable ideal of R and that $A \neq 0$ if $P \neq 0$.

7. Apply the ideas of the preceding problem to prove the following result of [65]. Let $R[x; \sigma]$ be a skew polynomial ring with R a right Noetherian Jacobson ring. Then $R[x; \sigma]$ is a Jacobson ring. This also uses the fact that every essential ideal of a semiprime right Noetherian ring contains a regular element.

8. Let S be a finitely generated algebra over the nondenumerable field K. If $s \in JS$, use the fact that the uncountably many elements $(1 - ks)^{-1}$ with $k \in K$ are K-linearly dependent to conclude that s is algebraic over K. Then show that s is nilpotent so that $\mathrm{J}(S)$ is a nil ideal. This is the argument of [3].

23. P. I. Algebras

We close this chapter by completing earlier work on two topics of interest, namely incomparability and crossed products satisfying a polynomial identity. We start with incomparability and here the goal is to extend Theorem 22.2 to more general polycyclic-by-finite groups. To proceed, we first need an operator version of the latter result. This could be proved directly by suitably modifying the original argument. Instead we take the opportunity to exhibit a trick first used in [166]. Let Γ act as automorphisms on the Z-graded ring S. We say that Γ *respects the grading* if each component S_i is Γ-stable.

Corollary 23.1. [30] *Let S be a Z-graded ring and let Γ be a group of operators on S which respects the grading. Suppose I is a Γ-stable ideal of S which properly contains the nonzero Γ-prime ideal P. Then I has a nonzero homogeneous element. In particular, if S is component regular then $I \cap S_0 \neq 0$.*

Proof. If $\tilde{\Gamma}$ is a free group which maps onto Γ, then the action of Γ on S lifts to a corresponding action of $\tilde{\Gamma}$ clearly satisfying the same hypotheses. Thus without loss of generality we may assume that $\Gamma = \tilde{\Gamma}$ is a free group and hence an ordered group (see [161, Corollary 13.2.8]).

Form the skew group ring $S\Gamma$. Since each S_i is Γ-stable, it follows that $S\Gamma$ is also Z-graded with $(S\Gamma)_i = (S_i)\Gamma$. Furthermore $0 \subset P\Gamma \subset I\Gamma$ and $P\Gamma$ is a prime ideal of $S\Gamma$ by Lemma 21.4(iii). We can now conclude from Theorem 22.2 that $I\Gamma$ contains a nonzero homogeneous element of $S\Gamma$ and, by considering the Γ coefficients of this element, we see that I contains a nonzero homogeneous element of S. ■

The following is the best crossed product incomparability result to date without a Noetherian assumption. Recall that a finitely generated nilpotent group G is necessarily polycyclic-by-finite. Furthermore $\mathbf{Z}(G)$ is finite if and only if G is finite (see [161, Lemma 11.4.3] with $N = G$).

Theorem 23.2. [30] *Let $R*G$ be a crossed product with G a finitely generated nilpotent group of Hirsch number n. If $P_0 \subset P_1 \subset \cdots \subset P_m$ is a chain of primes of $R*G$ with $m = 2^n$, then $P_0 \cap R \subset P_m \cap R$.*

Proof. Replacing $R*G$ by $R*G/(P_0 \cap R)*G$, if necessary, we may clearly assume that $P_0 \cap R = 0$. We proceed by induction on n. If $n = 0$ then G is finite and Theorem 16.6(iii) yields the result. Now suppose $n > 0$ so that, by the above comment, $\mathbf{Z}(G)$ is infinite and hence contains an infinite cyclic subgroup Z. Then $Z \lhd G$, $R*G = (R*Z)*(G/Z)$ and $\hbar(G/Z) = n - 1$ so induction implies that

$$P_0 \cap (R*Z) \subset P_{m/2} \cap (R*Z) \subset P_m \cap (R*Z).$$

Observe that $S = R*Z$ is a Z-graded ring and, since $Z \subseteq \mathbf{Z}(G)$, the group $\mathcal{G}$ of trivial units of $R*G$ acts on S and respects the grading. Furthermore, each $P_i \cap S$ is a $\mathcal{G}$-prime ideal of S by Lemma 14.1(i). In particular $P_{m/2} \cap S$ is a nonzero $\mathcal{G}$-prime ideal and $P_m \cap S$ is properly larger. Thus, by the previous corollary, $P_m \cap R \neq 0 = P_0 \cap R$ and the result follows. ∎

Finally we transfer this back to group-graded rings via duality.

Corollary 23.3. [31] *Let $S = R(G)$ be a group-graded ring with G a finitely generated nilpotent group of Hirsch number n. If $P_0 \subset P_1 \subset \cdots \subset P_m$ is a chain of primes of S with $m = 2^n$, then $\sum_{x \in G} P_0 \cap S_x \subset \sum_{x \in G} P_m \cap S_x$.*

Proof. By moding out by the graded ideal $\sum_{x \in G} P_0 \cap S_x$ if necessary, we can assume that P_0 contains no nonzero homogeneous elements. Now, by Proposition 2.8, there is a chain of primes $P_0' \subset P_1' \subset \cdots \subset P_m'$ of the skew group ring $S\{1\}\bar{G}$ where $P_i' = M_G(P_i) \cap S\{1\}\bar{G}$. It therefore follows from the previous result that $P_m' \cap S\{1\} \neq 0$. But $P_m' \subseteq M_G(P_m)$ and the entries in the matrices of $S\{1\}$ are all homogeneous elements of S. Thus $\sum_{x \in G} P_m \cap S_x \neq 0$. ∎

This completes our discussion of incomparability. Now we consider crossed products satisfying a polynomial identity. Recall that a

K-algebra S satisfies a polynomial identity of degree n if there exists a polynomial $f(\zeta_1, \zeta_2, \ldots, \zeta_k) \in K\langle \zeta_1, \zeta_2, \ldots \rangle$ of degree n such that $f(s_1, s_2, \ldots, s_k) = 0$ for all $s_i \in S$. It is a standard fact that f can be taken to be multilinear, that is of the form

$$f(\zeta_1, \zeta_2, \ldots, \zeta_n) = \sum_{\sigma \in \mathrm{Sym}_n} a_\sigma \zeta_{\sigma(1)} \zeta_{\sigma(2)} \cdots \zeta_{\sigma(n)}$$

with $a_\sigma \in K$ and $a_1 \neq 0$.

Group algebras satisfying a polynomial identity have been completely classified (see [161, Chapter 5]) and p. i. crossed products are at least reasonably well understood. The latter study began with the work of [71] on skew group rings. More generally [209] considered crossed products with a central twisting. The approach we take here avoids any concern with the nature of the twisting. We start by quoting an omnibus result from the theory of p. i. algebras. We refer the reader to the books [191] or [161] for complete details.

Theorem 23.4. *Let S be a prime K-algebra satisfying a polynomial identity of degree n and let $Z = \mathbf{Z}(S)$.*

  i. The central localization SZ^{-1} is a simple ring with center $F = ZZ^{-1}$.

  ii. $\dim_F SZ^{-1} = m^2$ where $m \leq [n/2]$ and $SZ^{-1} \subseteq \mathrm{M}_m(T)$ for some extension field T of F.

  iii. $\mathrm{Q}_s(S) = SZ^{-1}$ and $\sigma \in \mathrm{Aut}(S)$ is X-inner if and only if σ is trivial on Z.

Proof. (i) This seems to be due to a number of people independently. See for example [191, Theorem 1.7.9] or [161, Theorem 5.4.10].

(ii) By taking a multilinear identity, it follows that SZ^{-1} also satisfies a polynomial identity of degree n. But SZ^{-1} is simple and hence primitive so [191, Theorem 1.5.16] or [161, Theorem 5.3.4] yields the result.

(iii) By Lemma 10.9(iii) we have $\mathrm{Q}_s(S) \supseteq SZ^{-1}$. Conversely let $q \in \mathrm{Q}_s(S)$ and let $0 \neq A \lhd S$ with $qA \subseteq S$. By [191, Theorem 1.5.33] or [161, Theorem 5.4.9] there exists $0 \neq z \in A \cap Z$. But then $qz = s \in S$ so $q = sz^{-1} \in SZ^{-1}$.

Finally if σ is X-inner on S, then σ acts trivially on $\mathbf{Z}(Q_s(S))$ and hence on Z. Conversely if σ is trivial on Z, then σ extends to an automorphism of the finite dimensional simple ring SZ^{-1} which is trivial on its center $F = ZZ^{-1}$. The Skolem-Noether Theorem ([**73**, Theorem 4.3.1]) implies that σ is inner on SZ^{-1} and hence X-inner on S. ∎

The next lemma is the heart of the matter.

Lemma 23.5. *Let $R*G$ be given with R a G-prime ring. Furthermore, suppose $R*G$ satisfies a polynomial identity of degree n over the field $K \subseteq R$ and set $m = [n/2]$.*

i. There exist $k \leq m$ minimal primes $Q_1, Q_2, \ldots, Q_k$ of R. These are G-conjugate and satisfy $\bigcap_i Q_i = 0$.

ii. If Q is any minimal prime of R, there exists a subgroup H of G with $|G : H| \leq m^2$ such that H stabilizes Q and acts as X-inner automorphisms on the prime ring R/Q.

Proof. By Lemma 14.1(ii) there exists a prime P of $R*G$ with $P \cap R = 0$ and we let $\eta : R*G \to R*G/P = S$ be the natural homomorphism. Then S is a prime algebra satisfying a polynomial identity of degree n. If $Z = \mathbf{Z}(S)$, then the previous theorem implies that SZ^{-1} is a simple algebra with center $F = ZZ^{-1}$ and $\dim_F SZ^{-1} \leq m^2$. It follows that $SZ^{-1} = M_j(D)$ where D is an F-division algebra and $j \leq m$.

Note that S is generated by $R^\eta = R/(P \cap R) \cong R$ and $\mathcal{G}^\eta$ where $\mathcal{G}$ is the group of trivial units of $R*G$. Furthermore, $\mathcal{G}^\eta$ normalizes both R^η and F. It follows that $R^\eta F$ is a $\mathcal{G}^\eta$-prime ring. Indeed if A and B are $\mathcal{G}^\eta$-stable ideals of $R^\eta F$ with $AB = 0$, then $A\mathcal{G}^\eta$ and $B\mathcal{G}^\eta$ are ideals of SZ^{-1} with product zero. Thus since SZ^{-1} is simple, one of these factors must be zero.

Since $R^\eta F$ is finite dimensional over F, we conclude that $R^\eta F$ is semisimple and that $\mathcal{G}^\eta$ transitively permutes its simple summands. If e_i denotes the centrally primitive idempotent corresponding to the ith summand, then from $1 = \sum_{i=1}^k e_i$ and the structure of $S = M_j(D)$, it follows that $k \leq j \leq m$. Furthermore, since the simple summands $e_i R^\eta F$ are all F-isomorphic, $\dim_F e_i R^\eta F \leq (\dim_F S)/k \leq m^2/k$.

Let $L_i' = (1 - e_i) R^\eta F$ be the prime ideal of $R^\eta F$ corresponding to the ith summand. Since F is central, it follows that $L_i = L_i' \cap R^\eta$ is a prime of R^η. Moreover, $\mathcal{G}^\eta$ permutes these primes transitively and $\bigcap_i L_i = \bigcap_i (L_i' \cap R^\eta) = 0$. We conclude that $L_1, L_2, \ldots, L_k$ are precisely the minimal primes of $R^\eta \cong R$ and (i) is proved. Note that we do not claim that all L_i are distinct. Thus there may be less than k such minimal prime ideals of R.

Finally let $Q_1, Q_2, \ldots, Q_k$ be the minimal primes of R with $Q_i^\eta = L_i$. If $\mathcal{G}_i$ is the stabilizer of Q_i is $\mathcal{G}$, then $|\mathcal{G} : \mathcal{G}_i| \leq k$ and $\mathcal{G}_i^\eta$ acts on the simple ring $e_i R^\eta F$. Note that $e_i R^\eta F$ is generated by $R_i = e_i R^\eta \cong R/Q_i$ and the central subfield $F_i = e_i F \cong F$ and that $\dim_{F_i} R_i F_i \leq m^2/k$. Thus if Z_i denotes the center of R_i, then $Z_i F_i$ is a central subfield of $R_i F_i$. Moreover $\mathcal{G}_i^\eta$ acts as field automorphisms on $Z_i F_i$ fixing F_i. We conclude from the Galois theory of fields that

$$|\mathcal{G}_i^\eta : \mathbf{C}_{\mathcal{G}_i^\eta}(Z_i)| \leq \dim_{F_i} Z_i F_i \leq m^2/k.$$

In particular, if $\mathcal{H}_i$ denotes the complete inverse image of $\mathbf{C}_{\mathcal{G}_i^\eta}(Z_i)$ in $\mathcal{G}$, then

$$|\mathcal{G} : \mathcal{H}_i| = |\mathcal{G} : \mathcal{G}_i||\mathcal{G}_i : \mathcal{H}_i| \leq k \cdot m^2/k = m^2.$$

Furthermore $\mathcal{H}_i$ acts on R/Q_i fixing its center. Theorem 23.4(iii) now implies that $\mathcal{H}_i$ is X-inner on R/Q_i and (ii) follows with $H_i = \mathcal{H}_i/\mathrm{U}(R)$. ∎

We now require some fairly simple observations.

Lemma 23.6. Let $R*G$ be given.

i. R is G-semiprime if and only if the G-prime ideals of R intersect to zero.

ii. If $R*G$ satisfies a polynomial identity, then R is G-semiprime if and only if it is semiprime.

iii. Let H be a subgroup of G of finite index k and let I be an H-stable ideal of R. Then there is a homomorphism from $R*G$ into $\mathrm{M}_k(R*H/I*H)$ with kernel $\left(\bigcap_{x \in G} I^x \right)*G$.

Proof. Part (i) is standard (see Exercise 2 or 3) and part (ii) follows from (i) and the fact that every G-prime ideal is semiprime by Lemma 23.5. We consider (iii) and for this we may clearly assume that $\bigcap_{x \in G} I^x = 0$. Let $\{\, 1 = x_1, x_2, \ldots, x_k \,\}$ be a right transversal for H in G. Then $V = R*G$ is a free left $R*H$-module with basis $\{\, 1 = \bar{x}_1, \bar{x}_2, \ldots, \bar{x}_k \,\}$. Furthermore V is a faithful right $R*G$-module. Since right and left multiplication commute as operators, it follows that $R*G$ is isomorphic to a subring of $\operatorname{End}_{R*H}(V) \cong \mathrm{M}_k(R*H)$. To be precise, let $\alpha \in R*G$ and for each i write $\bar{x}_i \alpha = \sum_j \alpha_{i,j} \bar{x}_j$ with $\alpha_{i,j} \in R*H$. Then the map $\alpha \mapsto [\alpha_{i,j}] \in \mathrm{M}_k(R*H)$ is the appropriate embedding. Finally consider the composite map

$$\theta \colon R*G \to \mathrm{M}_k(R*H) \to \mathrm{M}_k(R*H/I*H).$$

If $\alpha \in \operatorname{Ker}(\theta)$, then since $\bar{x}_1 = 1$ we have

$$\alpha = \bar{x}_1 \alpha = \sum_j \alpha_{1,j} \bar{x}_j \in \sum_j (I*H)\bar{x}_j = I(R*G).$$

But $\operatorname{Ker}(\theta)$ is an ideal of $R*G$ so $\alpha \in \left(\bigcap_{x \in G} I^x \right)*G = 0$. ∎

We can now offer the necessary and sufficient conditions for $R*G$ to satisfy a polynomial identity. As is apparent, the result is quite tedious to state and not particularly useful; but it is the correct answer. Note that it seems preferable here to use a lower case q to denote a prime ideal.

Theorem 23.7. **[71] [209]** *Let $R*G$ be a crossed product with R a G-semiprime ring. Assume that $R*G$ is a K-algebra with $K \subseteq R$ and that R satisfies a polynomial identity over K. If q is a prime of R, write $S_q = Q_s(R/q)$ and let $C_q = \mathbf{Z}(S_q)$ be the extended centroid of R/q. Note that these are all K-algebras. Then $R*G$ satisfies a polynomial identity over K if and only if there exists a constant k and a multilinear polynomial $f(\zeta_1, \zeta_2, \ldots) \in K\langle \zeta_1, \zeta_2, \ldots \rangle$ such that for each minimal prime q of R there is a subgroup H_q of G with*

    i. $|G : H_q| \leq k$ and H_q stabilizes q,

    ii. $R*H_q/q*H_q \subseteq S_q*H_q = S_q \otimes C_q^t[H_q]$ where $C_q^t[H_q]$ is a twisted group algebra,

    iii. $C_q^t[H_q]$ satisfies the polynomial identity f.

238 5. Prime Ideals – The Noetherian Case

Proof. Suppose first that $R*G$ satisfies a polynomial identity f of degree n which we may assume to be multilinear. Let q be a minimal prime of R. By replacing R by $R/(\bigcap_{x\in G} q^x)$ if necessary, we can clearly assume that $\bigcap_{x\in G} q^x = 0$ so that R is G-prime. By Lemma 23.5(ii), there exists a subgroup H_q of G with $|G : H_q| \leq [n/2]^2$ such that H_q stabilizes q and acts as X-inner automorphisms on the prime ring R/q. We conclude from Proposition 12.4(i)(ii)(iii) that

$$R*H_q/q*H_q \subseteq S_q*H_q = S_q \otimes C_q^t[H_q].$$

Since each $C_q^t[H_q]$ satisfies the polynomial identity f, the result follows in this direction.

Conversely suppose k, f and the subgroups H_q are given and write $m_q = |G : H_q|$. Then for each minimal prime q, it follows from (i), (ii) and Lemma 23.6(iii) that we have a homomorphism

$$\theta_q \colon R*G \to \mathrm{M}_{m_q}(R*H_q/q*H_q) \subseteq \mathrm{M}_k(R*H_q/q*H_q)$$
$$\subseteq \mathrm{M}_k(S_q \otimes C_q^t[H_q]).$$

Note that this combined map need not send the element 1 to 1. By Lemma 23.6(iii) we have $\mathrm{Ker}(\theta_q) = (\bigcap_{x\in G} q^x)*G$. Furthermore, since R satisfies a polynomial identity of degree n, so does S_q and indeed, by Theorem 23.4(ii)(iii), $S_q \subseteq \mathrm{M}_{[n/2]}(F_q)$ for some field $F_q \supseteq C_q$. Thus

$$S_q \otimes C_q^t[H_q] \subseteq \mathrm{M}_{[n/2]}(F_q \otimes C_q^t[H_q])$$

and θ_q extends to a map

$$\theta_q \colon R*G \to \mathrm{M}_m(F_q \otimes C_q^t[H_q])$$

where $m = k[n/2]$.

Finally by combining all of these maps, we obtain a ring homomorphism (not taking 1 to 1)

$$\theta \colon R*G \to \mathrm{M}_m\left(\prod_q F_q \otimes C_q^t[H_q]\right)$$

with

$$\mathrm{Ker}(\theta) = \bigcap_q \mathrm{Ker}(\theta_q) = \bigcap_q (\bigcap_{x\in G} q^x)*G = 0$$

since R is semiprime by Lemma 23.6(i). Moreover, by hypothesis, each $C_q^t[H_q]$ satisfies the multilinear polynomial identity f and hence so does $\prod_q F_q \otimes C_q^t[H_q]$. It now follows from Theorem 9.1 with $E = \mathrm{M}_m(K)$ that the $m \times m$ matrix ring over this ring also satisfies a polynomial identity and hence, since $\mathrm{Ker}(\theta) = 0$, so does $R*G$. ∎

Thus, as with other crossed product problems, the question of when $R*G$ satisfies a polynomial identity is reduced to the case of twisted group algebras. Here the result is a simple analog of the ordinary group algebra situation, so we just state it without proof. In fact, given Corollary 9.9, the proof is not at all difficult and we discuss some aspects of it in the exercises.

Theorem 23.8. [159] *Let $K^t[G]$ be a twisted group algebra over the field K.*

i. If $K^t[G]$ satisfies a polynomial identity of degree n over K, then G has normal subgroups $A \subseteq B$ with $|A|$ and $|G : B|$ bounded by functions of n and with $K^t[B]/\mathrm{J}(K^t[A])K^t[B]$ commutative.

ii. Conversely suppose G has normal subgroups $A \subseteq B$ with $|A| = a < \infty$ and $|G : B| = b < \infty$ and with $K^t[B]/\mathrm{J}(K^t[A])K^t[B]$ commutative. Then $K^t[G]$ satisfies $(s_{2b})^a$ where s_{2b} is the standard identity of degree $2b$.

Thus we see that this result meshes precisely with Theorem 23.7 to settle the polynomial identity problem for $R*G$ with R a G-semiprime ring.

EXERCISES

1. Let G be an infinite elementary abelian 2-group with generators x_i, y_i for $i = 1, 2, 3, \ldots$. Define the twisted group algebra $R = K^t[G]$ so that $\bar{x}_i \bar{y}_i = -\bar{y}_i \bar{x}_i$ for all i but that all other generators commute. If char $K \neq 2$ show that $K^t[\langle x_1, y_1 \rangle] \cong \mathrm{M}_2(K)$ and hence that $R \cong \mathrm{M}_2(R)$.

2. Let $R*G$ be given. Prove that R is G-semiprime if and only if the G-prime ideals of R intersect to zero. To this end suppose R is G-semiprime and let $0 \neq r \in R$. Define a sequence $T = \{r_0, r_1, r_2, \ldots\}$

of nonzero elements of R inductively by $r_0 = r$ and $r_{n+1} \in r_n R r_n^{\bar{g}_n}$ for some $g_n \in G$. If $P \lhd R$ is maximal subject to being G-stable and disjoint from T, show that P is G-prime.

3. Give an alternate proof of the above using the same trick as in Corollary 23.1. Namely, reduce to the case where G is a free group and use the fact that if R is a G-semiprime ring, then $R*G$ is semiprime.

4. Suppose R is prime, G is X-inner on R and $R*G$ satisfies the multilinear polynomial identity f. If $S = Q_s(R)$ and $C = \mathbf{Z}(S)$, prove that C is central in $S*G$ and that $S*G = C \cdot (R*G)$. Deduce that $S*G$ also satisifies f. This is used in the proof of Theorem 23.7.

5. Let $K^t[G]$ and $A \subseteq B$ be given as in Theorem 23.8 with $|A| = a$ and $|G : B| = b$. Show that $K^t[G]/J(K^t[A])K^t[G]$ is isomorphic to a subalgebra of $M_b\big(K^t[B]/J(K^t[A])K^t[B]\big)$. Since the coefficient ring of this matrix ring is assumed to be commutative and since $J(K^t[A])K^t[G]$ is nilpotent of degree $\leq a$, conclude that $K^t[G]$ satisfies $(s_{2b})^a$.

6. Let $K^t[G]$ be given with G an abelian group and define $W = \{\, x \in G \mid \bar{x} \in \mathbf{Z}(K^t[G]) \,\}$. If I is any ideal of $K^t[G]$, prove that $I = (I \cap K^t[W]) \cdot K^t[G]$ and that $K^t[W]$ maps onto the center of $K^t[G]/I$. In particular, if T is a transversal for W in G, conclude that $(T + I)/I$ is linearly independent over $\mathbf{Z}(K^t[G]/I)$.

7. Let $K^t[G]$ and W be as above. If $K^t[G]$ satisfies a polynomial identity of degree n, show that $|G : W| \leq [n/2]^2$. For this, take I to be a maximal ideal of $K^t[G]$ in the preceding exercise. How close does this come to proving Theorem 23.8?

6   Group Actions and Fixed Rings

24. Fixed Points and Traces

Let G act on the ring R so that we have a group homomorphism $G \to \operatorname{Aut}(R)$. Then we can form the skew group ring RG which is an associative ring containing R, G and the *fixed ring*

$$R^G = \{\, r \in R \mid r^g = r \text{ for all } g \in G \,\}.$$

In other words, RG contains all the ingredients of the Galois theory of rings. Thus we might hope to use crossed product results to obtain Galois theoretic information. Indeed there are such applications, but not an overwhelming amount for a number of reasons.

    First, it is necessary to assume that G is finite. Only certain classes of infinite groups are allowed in the theory, and their description does not readily translate into crossed product terms. Second, the structure of RG is best understood when R has no $|G|$-torsion. Thus the route through crossed products usually requires this assumption. Third, certain natural Galois theory questions do not translate into natural questions about RG and vice versa.

In this chapter we consider skew group ring applications to the Galois theory of rings. As indicated above, this represents only a small part of the theory. Thus the material here is necessarily very selective. For many more details the reader should consult the books [129] and [92] and the survey paper [60]. We start by considering the existence of *fixed points*, that is elements of R^G.

If G acts on R, then obviously $1 \in R$ is a fixed point so nonzero fixed points exist. On the other hand, if $I \neq 0$ is a G-stable (right) ideal of R, then it is not so obvious that nonzero fixed points exist in I. For example, let $S = K\langle x, y \rangle$ be the free algebra on x, y over the field K of characteristic $p > 2$, let $R = \mathrm{M}_2(S)$ and let G be the group of units of R generated by the matrices

$$\begin{pmatrix} -1 & 0 \\ 0 & 1 \end{pmatrix}, \quad \begin{pmatrix} 1 & x \\ 0 & 1 \end{pmatrix}, \quad \begin{pmatrix} 1 & y \\ 0 & 1 \end{pmatrix}.$$

Then $|G| = 2p^2$ and G acts on R by conjugation so that $R^G = C_R(G)$. To compute the latter, first note that the centralizer of $\begin{pmatrix} -1 & 0 \\ 0 & 1 \end{pmatrix}$ is precisely the set of diagonal matrices since char $K \neq 2$. It then follows easily that $R^G = K$, embedded as scalars. But there is a natural homomorphism $S \to K$ obtained by mapping $x, y \mapsto 0$ and this extends to a ring homomorphism $R \to \mathrm{M}_2(K)$ with kernel $I \neq 0$. Since $I \lhd R$, we see that I is G-invariant but $I \cap R^G = I \cap K = 0$.

We note three properties of this example. (1) R is a prime ring but not a domain (and hence has nontrivial nilpotent elements). (2) G is inner on R. (3) R has $|G|$-torsion. As we will see in this and later sections, each of these properties is in some sense a necessary ingredient of the example.

One way to construct fixed points is with a *trace map*. If G is finite and acts on R, define $\mathrm{tr}_G \colon R \to R$ by

$$\mathrm{tr}_G(r) = \sum_{g \in G} r^g.$$

Then clearly $\mathrm{tr}_G(r) \in R^G$ and if r is a fixed point then $\mathrm{tr}_G(r) = |G|r$. Thus we have

Lemma 24.1. *If G is finite, then tr_G is an R^G-bimodule homomorphism from R to R^G. Moreover, $|G| \cdot R^G \subseteq \mathrm{tr}_G(R) \triangleleft R^G$.*

Notice that if $I \triangleleft R$, then I is itself a ring (without 1). Thus, in this section and the next, it is convenient to allow these more general rings. Of course a ring without 1 is merely an ideal in a ring with 1. Indeed if R is any ring (without 1), we let $R^{\#} = R \oplus Z$ be its natural extension to a ring with 1. The main result is

Theorem 24.2. [22] *Let G be a finite group of automorphisms of the ring R (without 1) and set $n = |G|$. If $\mathrm{tr}_G(R) = 0$, then nR is nilpotent of degree $\leq 2^{n^2}$.*

Proof. It is convenient, although not necessary, to formulate this argument in terms of the skew group ring RG. Notice that R is naturally embedded in this ring but that, since R need not have a 1, there is no natural embedding of G. As in the case of ordinary group rings, there is an *augmentation map* $\rho: RG \to R$ given by $\rho(\sum_{g \in G} r_g g) = \sum_{g \in G} r_g$. However here this map is certainly not a ring homomorphism; rather it is just a left R-module homomorphism. Define
$$A = \{ \alpha \in RG \mid \rho(\alpha R) = 0 \}.$$
Then A is clearly both a right and left R-submodule of RG. Observe that if $1 \in R$, then it follows immediately that $\rho(A) = 0$. On the other hand, even if R does not contain a 1, we can still show that $\rho(A)$ is nilpotent.

To this end, recall that $n = |G|$ and, for each integer $m \leq n$, define A_m to be the linear span of all elements of A of support size $\leq m$. Then clearly
$$A = A_n \supseteq A_{n-1} \supseteq \cdots \supseteq A_1 \supseteq A_0 = 0$$
and these are both right and left R-submodules of RG. Suppose $X = \{ x_1, x_2, \ldots, x_m \}$ is a subset of G of size m and let $\alpha, \beta \in A_m$ have support in X so that $\alpha = \sum_1^m a_i x_i$ and $\beta = \sum_1^m b_i x_i$ for suitable $a_i, b_i \in R$. For each subscript k we observe that
$$a_k \beta - \alpha(b_k)^{x_k} = \sum_{i=1}^m (a_k b_i - a_i(b_k)^{x_k x_i^{-1}}) x_i \in A_{m-1}$$

since the $i = k$ summand vanishes. Thus

$$\rho(a_k\beta - \alpha(b_k)^{x_k}) \in \rho(A_{m-1}).$$

But $\rho(\alpha(b_k)^{x_k}) = 0$ since $\alpha \in A$ so $a_k\rho(\beta) = \rho(a_k\beta) \in \rho(A_{m-1})$. Summing over $k = 1, 2, \ldots, m$, we obtain $\rho(\alpha)\rho(\beta) \in \rho(A_{m-1})$.

Now let us consider $\left(\rho(A_m)R^{\#}\right)^{2^n}$. This subset of R is clearly spanned by all products of the form

$$\gamma = \rho(\alpha_1)r_1\rho(\alpha_2)r_2 \cdots \rho(\alpha_{2^n})r_{2^n}$$

with $0 \neq \alpha_i \in A_m$ and $r_i \in R^{\#}$. Since there are less than 2^n nonempty subsets of G of size $\leq m$, it follows that in this expression at least two of the α_i terms must have identical support. Thus for some subset X of G of size m we see that

$$\gamma = r_1'\rho(\alpha')r_2'\rho(\beta')r_3'$$

with α', β' both having support in X and with $r_1', r_2', r_3' \in R^{\#}$. Because $\gamma = \rho(r_1'\alpha')\rho(r_2'\beta')r_3' = \rho(\alpha)\rho(\beta)r_3'$, we conclude from the result of the preceding paragraph that $\gamma \in \rho(A_{m-1})R^{\#}$. In other words, we have shown that

$$\left(\rho(A_m)R^{\#}\right)^{2^n} \subseteq \rho(A_{m-1})R^{\#}.$$

It follows that if $t = (2^n)^n$ then

$$\left(\rho(A)R^{\#}\right)^t = \left(\rho(A_n)R^{\#}\right)^t \subseteq \rho(A_0)R^{\#} = 0.$$

Hence $\rho(A)R^{\#}$ is nilpotent. But $\rho(A) \subseteq \rho(A)R^{\#}$ so $\rho(A)^t = 0$.

Finally, let $a \in R$ and form the element $\alpha = \sum_{g \in G} ag \in RG$. If $r \in R$, then we have

$$\rho(\alpha r) = \rho\left(\sum ar^{g^{-1}}g\right) = a\,\mathrm{tr}_G(r) = 0,$$

by assumption. Since $\rho(\alpha r) = 0$ for all $r \in R$, it follows that $\alpha \in A$. In particular, since $\rho(\alpha) = |G|a = na$, we have $\rho(A) \supseteq nR$. Thus $(nR)^t = 0$ and the result follows. ∎

We note that the nilpotence bound given above is not sharp. In fact, a closer look at the proof shows that it could be improved slightly to $\prod_{m=1}^{n}\left[\binom{n}{m}+1\right]$. However this is presumably also much too large since, for G solvable, the bound is n. As we will see, there are other proofs of this result, but none of them yield better information on the bound.

Lemma 24.3. *Let the finite group G act on R.*

i. If $H \triangleleft G$, then G/H acts on R^H. Moreover $R^G = (R^H)^{G/H}$ and $\mathrm{tr}_G = \mathrm{tr}_{G/H} \circ \mathrm{tr}_H$.

ii. Suppose p is a prime, $pR = 0$ and G is a p-group. If $R \neq 0$ then $R^G \neq 0$.

Proof. (i) Most of this is obvious. For the last part, let T be a transversal for H in G. If $r \in R$, then

$$\mathrm{tr}_{G/H} \circ \mathrm{tr}_H(r) = \mathrm{tr}_{G/H}\left(\sum_{h \in H} r^h\right)$$

$$= \sum_{t \in T}\left(\sum_{h \in H} r^h\right)^t = \sum_{g \in G} r^g = \mathrm{tr}_G(r)$$

since $HT = G$.

(ii) In view of (i), it suffices to assume that $G = \langle x \rangle$ is cyclic of order p. In this case, we forget the multiplicative structure of R and just view R as a module for the group algebra $\mathrm{GF}(p)[G]$. Since $x^p = 1$, we have $(1 - x)^p = 0$ so $1 - x$ is a nilpotent linear transformation on $R \neq 0$. It follows that $R^G = \mathrm{ann}_R(1 - x) \neq 0$. ∎

A finite group G is said to be *p-nilpotent* for the prime p if G has a normal p'-subgroup H with G/H a p-group. In particular, any p-group is p-nilpotent.

Corollary 24.4. **[22]** *Let G be a finite group of automorphisms of the ring R (without 1) and assume that $R^G = 0$. If either R has no $|G|$-torsion or G is p-nilpotent and $pR = 0$, then $R^k = 0$ where $k = 2^{n^2}$ and $n = |G|$.*

Proof. We know that $(nR)^k = 0$. Thus if R has no n-torsion, then $R^k = 0$. Now suppose $pR = 0$ and let H be a normal p'-subgroup of G with G/H a p-group. If $R^H \neq 0$ then, since G/H is a p-group, it follows from the preceding lemma that $R^G = (R^H)^{G/H} \neq 0$, a contradiction. Thus $R^H = 0$ and hence, since R has no $|H|$-torsion, we conclude that $R^\ell = 0$ where $\ell = 2^{m^2}$ and $m = |H| \leq n$. ∎

We consider two examples of interest. The first shows that we can have $R \neq 0$ and $R^G = 0$ even when R has no $|G|$-torsion. To this end, let K be a field and let R be the ring of strictly upper triangular $n \times n$ matrices over K. Suppose in addition that K contains n distinct n^{th} roots of unity, say $\epsilon_1, \epsilon_2, \ldots, \epsilon_n$. This implies that one of these elements has order n and also that the characteristic of K does not divide n. Let G be the cyclic group of units of $\mathrm{M}_n(K)$ generated by the diagonal matrix $g = \mathrm{diag}(\epsilon_1, \epsilon_2, \ldots, \epsilon_n)$. Then $|G| = n$, G acts on $\mathrm{M}_n(K)$ by conjugation and $\mathrm{M}_n(K)^G$ is the ring of diagonal matrices. Furthermore, G acts on R and $R^G = 0$. Note that R is nilpotent of degree $n = |G|$.

The second example, from [**22**], is a partial converse to the preceding corollary. We state it as

Lemma 24.5. *Let K be a field of characteristic $p > 0$ and let G be a finite group which is not p-nilpotent. Then there exists a finite dimensional, nonnilpotent K-algebra R (without 1) on which G acts with $R^G = 0$.*

Proof. Let H be the normal subgroup of G generated by all its p'-elements. Then G/H is a p-group so that, by assumption, $p \mid |H|$ and therefore $\tau = \sum_{h \in H} h$ is contained in the augmentation ideal $\omega(K[H])$. Let R be the right $K[G]$-module $R = K[G]/\tau K[G]$. We show first that $R^G = 0$. To this end, let $\alpha \in K[G]$ correspond to an element in the fixed submodule R^G. Then for all $h \in H$ we have $\alpha(h - 1) \in \tau K[G]$ and hence $\alpha(h - 1)^2 = 0$ since τ is central in $K[G]$ and $\tau(h - 1) = 0$. Furthermore char $K = p \geq 2$ so $\alpha(h^p - 1) = \alpha(h - 1)^p = 0$. Now H is generated by p'-elements and thus $H = \langle h^p \mid h \in H \rangle$. We conclude that α is contained in the left annihilator of $\omega(K[H])$ so $\alpha \in \tau K[G]$ and α corresponds to the zero element of R.

Now let $\rho \colon K[G] \to K$ denote the usual augmentation map with kernel $\omega(K[G])$. Since $\tau \in \omega(K[G])$, we see that ρ gives rise to a nonzero linear functional $f \colon R \to K$. Furthermore for all $\alpha \in R$ and $x \in G$ we have $f(\alpha x) = f(\alpha)$. We now define a ring structure on R by setting $\alpha \times \beta = \alpha f(\beta)$. It is trivial to verify that this is distributive with respect to addition and it is associative since

$$(\alpha \times \beta) \times \gamma = \alpha f(\beta) f(\gamma) = \alpha \times (\beta \times \gamma).$$

Thus R is a ring and in fact a finite-dimensional K-algebra (without 1). Moreover $\beta \in R$ is a right identity if and only if $f(\beta) = 1$. Thus R has numerous right identities and is therefore not nilpotent. Finally because $f(\beta x) = f(\beta)$ we see that G acts on R as ring automorphisms and the result follows. ∎

As we indicated earlier, the nilpotence bound in Theorem 24.2 can be appreciably sharpened in case G is solvable. This is based on the following important observation.

Lemma 24.6. *Let G be an abelian group of period n and let K be a field containing a primitive n^{th} root of unity. Suppose G acts as K-automorphisms on the K-algebra R (without 1). For each linear character $\lambda \in \hat{G} = \mathrm{Hom}(G, K^\bullet)$ define*

$$R_\lambda = \{ r \in R \mid r^g = \lambda(g)r \text{ for all } g \in G \}.$$

In particular, $R_1 = R^G$. Then $R = \oplus \sum_{\lambda \in \hat{G}} R_\lambda$ is a $\hat{G}$-graded ring. Specifically, if $r \in R$, then its λ-component r_λ is given by

$$r_\lambda = \frac{1}{|G|} \sum_{g \in G} \lambda(g^{-1}) r^g.$$

Proof. It is clear that each R_λ is an additive subgroup of G and, since G acts as K-automorphisms, we have easily $R_\lambda R_\mu \subseteq R_{\lambda\mu}$. The goal is to show that $R = \oplus \sum_{\lambda \in \hat{G}} R_\lambda$. This is based on the fact that if

$1 \neq \lambda \in \hat{G}$ then $\sum_{g \in G} \lambda(g) = 0$ and the dual fact that $1 \neq g \in G$ yields $\sum_{\lambda \in \hat{G}} \lambda(g) = 0$.

Let $r \in R$ and define r_λ as above. If $x \in G$, then

$$(r_\lambda)^x = \frac{1}{|G|} \sum_{g \in G} \lambda(g^{-1}) r^{gx}$$

$$= \frac{\lambda(x)}{|G|} \sum_{g \in G} \lambda(x^{-1}g^{-1}) r^{gx} = \lambda(x) r_\lambda$$

so $r_\lambda \in R_\lambda$. Moreover

$$\sum_{\lambda \in \hat{G}} r_\lambda = \frac{1}{|G|} \sum_{g \in G} \left(\sum_{\lambda \in \hat{G}} \lambda(g^{-1}) \right) r^g$$

$$= \frac{1}{|G|} \sum_{\lambda \in \hat{G}} \lambda(1) r = r.$$

Finally if $r \in R_\mu$, then $r_\lambda = 0$ for $\lambda \neq \mu$ and $r_\mu = r$. This implies easily that the sum $R = \sum_{\lambda \in \hat{G}} R_\lambda$ is direct and the result follows. ∎

Of course, not every ring is a K-algebra, but sometimes we can work with a *generic model*. Let G be a group, let Λ be an integral domain and form the free Λ-algebra

$$T = \Lambda\langle \zeta_{i,g} \mid g \in G, \, i \in I \rangle$$

where I is some index set. Then G acts as Λ-automorphisms on this ring by permuting the variables in regular orbits. Specifically $(\zeta_{i,g})^x = \zeta_{i,gx}$ for all $x, g \in G$. Write

$$T' = \Lambda\langle \zeta_{i,g} \mid g \in G, \, i \in I \rangle'$$

for the *augmentation ideal* of T, namely the ideal generated by all $\zeta_{i,g}$. Then T' is a Λ-algebra (without 1) and it is G-stable. Now let $\Lambda = Z$ be the ring of integers. If R is any ring (without 1) acted upon by G and if $r_i \in R$ for all $i \in I$, then the map

$$Z\langle \zeta_{i,g} \mid g \in G, \, i \in I \rangle' \to R$$

given by $\zeta_{i,g} \mapsto (r_i)^g$ is a G-homomorphism of rings. Furthermore, for sufficiently large I, this map can be made an epimorphism. It is convenient to record the following elementary observation.

Lemma 24.7. *Let G act on both R and S and let $\theta \colon S \to R$ be a G-homomorphism of rings. Then $\theta(S^G) \subseteq R^G$ and $\theta(\mathrm{tr}_G(S)) \subseteq \mathrm{tr}_G(R)$ if G is finite. Moreover, if θ is onto then $\theta(\mathrm{tr}_G(S)) = \mathrm{tr}_G(R)$.*

As an application we prove

Lemma 24.8. *Let G be a finite abelian group of order n acting on the ring R (without 1). Then for any integer $d \geq 1$ we have*

$$(nR)^{nd} \subseteq R^{\#}(R^G)^d R^{\#}.$$

Proof. Let ϵ be a primitive complex n^{th} root of unity, set $\Lambda = Z[\epsilon]$ and $K = Q[\epsilon]$. We work with the generic models

$$R = Z\langle \zeta_{i,g} \mid g \in G, i \in I \rangle'$$
$$S = \Lambda\langle \zeta_{i,g} \mid g \in G, i \in I \rangle'$$
$$T = K\langle \zeta_{i,g} \mid g \in G, i \in I \rangle'.$$

By Lemma 24.6 we know that $T = \sum_{\lambda \in \hat{G}} T_\lambda$ is $\hat{G}$-graded. Furthermore, by that lemma,

$$nS \subseteq \sum_{\lambda \in \hat{G}} (T_\lambda \cap S).$$

In particular,

$$(nS)^{nd} \subseteq \sum_{(\lambda)} \prod_{k=1}^{nd} (T_{\lambda_k} \cap S)$$

where the latter sum is over all nd-tuples $(\lambda_1, \lambda_2, \ldots, \lambda_{nd})$ of elements of $\hat{G}$.

Fix such an nd-tuple and consider the $nd + 1$ partial products $\mu_i = \lambda_1 \lambda_2 \cdots \lambda_i$ for $i = 0, 1, \ldots, nd$. Since $|\hat{G}| = n$, the pigeonhole principle implies that at least $d+1$ of these μ_i are equal. But $\mu_i = \mu_j$ for $i < j$ implies that

$$\prod_{k=i+1}^{j} (T_{\lambda_k} \cap S) \subseteq T_1 \cap S \subseteq S^G.$$

Thus we conclude that

$$(nS)^{nd} \subseteq S^{\#}(S^G)^d S^{\#}.$$

Next observe that for suitable m

$$S = \oplus \sum_{i=0}^{m} R\epsilon^i \quad \text{and} \quad S^G = \oplus \sum_{i=0}^{m} R^G \epsilon^i.$$

Thus

$$(nR)^{nd} \subseteq (nS)^{nd} \subseteq S^{\#}(S^G)^d S^{\#}$$

$$\subseteq \oplus \sum_{i=0}^{m} R^{\#}(R^G)^d R^{\#} \epsilon^i$$

and by reading off the coefficient of ϵ^0 we have

$$(nR)^{nd} \subseteq R^{\#}(R^G)^d R^{\#}.$$

Finally, by choosing the index set sufficiently large, we can map R onto any ring on which G acts. It therefore follows, by Lemma 24.7, that the above formula holds for all such rings. ∎

Theorem 24.9. [22] *Let G be a finite solvable group of order n acting on the ring R (without 1). Then for any integer $d \geq 1$ we have*

$$(nR)^{nd} \subseteq R^{\#}(R^G)^d R^{\#}.$$

Proof. We proceed by induction on $|G|$. If G is abelian, the result follows from the previous lemma. Otherwise, G has a nontrivial normal subgroup H with $1 < m = |H| < n$. By induction (using $(n/m)d$ for d)

$$(mR)^{nd} \subseteq R^{\#}(R^H)^{(n/m)d} R^{\#}.$$

Furthermore G/H acts on R^H and $(R^H)^{G/H} = R^G$ so induction yields

$$((n/m)R^H)^{(n/m)d} \subseteq (R^H)^\#(R^G)^d(R^H)^\# \subseteq R^\#(R^G)^d R^\#.$$

Finally $n \geq n/m$ so combining these yields

$$(nR)^{nd} \subseteq R^\#((n/m)R^H)^{(n/m)d}R^\# \subseteq R^\#(R^G)^d R^\#$$

and the result follows. ∎

In particular, if $R^G = 0$ and $d = 1$ we obtain $(nR)^n = 0$, the appropriately sharper version of Theorem 24.2 for solvable groups. We close this section with another nice observation from [22] and then a consequence.

Proposition 24.10. [22] *Let $G = \langle x \rangle$ be a cyclic group of prime order p acting on the ring R (without 1). If R has no nonzero nilpotent elements, then either G acts trivially on R or $\mathrm{tr}_G(R) \neq 0$.*

Proof. Assume by way of contradiction that $\mathrm{tr}_G(R) = 0$ and $R^G \neq R$. Then Theorem 24.2 implies that pR is nilpotent so, by assumption, $pR = 0$. We can therefore view R as a right $\mathrm{GF}(p)[G]$-module. With this notation, if $r, s \in R$ then

$$(rs)(1 - x) = rs - (rx)(sx) = (r - rx)s + rx(s - sx)$$
$$= r(1 - x) \cdot s + rx \cdot s(1 - x)$$

and hence by induction

$$(rs)(1 - x)^k = \sum_{i+j=k} \binom{k}{i} r(1 - x)^i x^j \cdot s(1 - x)^j.$$

Since $(1 - x)^{p-1} = 1 + x + \cdots + x^{p-1}$ and $\mathrm{tr}_G(R) = 0$, it follows that $1 - x$ is nilpotent on R of degree $t \leq p-1$. Moreover $t \geq 2$ since $R^G \neq R$. Now put $k = t$, $r \in R(1 - x)^{t-2}$ and $s \in R$ in the formula of the preceding paragraph. Since $R(1 - x)^t = 0$, it follows that

$$0 = \binom{t}{1} r(1 - x)x^{t-1} \cdot s(1 - x)^{t-1}$$

and hence $r(1-x)x^{t-1} \cdot s(1-x)^{t-1} = 0$ since $t < p$. But $r(1-x)x^{t-1}$ and $s(1-x)^{t-1}$ are typical elements of $R(1-x)^{t-1}$ so we conclude that $\left(R(1-x)^{t-1}\right)^2 = 0$. Since R has no nonzero nilpotent elements, it follows that $R(1-x)^{t-1} = 0$ and this contradicts the definition of the degree t. ∎

Let G act on the ring R and let Λ be a nonempty finite subset of G. The map $\mathrm{tr}_\Lambda \colon R \to R$ given by

$$\mathrm{tr}_\Lambda(r) = \sum_{g \in \Lambda} r^g$$

is called a *partial trace* if $\mathrm{tr}_\Lambda(R) \subseteq R^G$. It is *nontrivial* if $0 \neq \mathrm{tr}_\Lambda(R)$. Certainly any partial trace is an R^G-bimodule homomorphism from R to R^G. In particular, $\mathrm{tr}_\Lambda(R)$ is an ideal of the fixed ring.

Corollary 24.11. [38] *Let G be a finite solvable group acting on the ring $R \neq 0$ (without 1). If R has no nonzero nilpotent elements, then a nontrivial partial trace exists.*

Proof. We proceed by induction on $|G|$, the case $|G| = 1$ being clear. Let $H \triangleleft G$ with G/H cyclic of prime order p. By induction, there exists a subset $\Omega \subseteq H$ with $0 \neq \mathrm{tr}_\Omega(R) \subseteq R^H$. Note that G need not act on the ideal $\mathrm{tr}_\Omega(R) = S$, but it does act on $\sum_{x \in G} S^x \subseteq R^H$ and then so does G/H.

If G/H acts trivially on $\sum_{x \in G} S^x$, then $0 \neq \mathrm{tr}_\Omega(R) \subseteq R^G$ and tr_Ω is a nontrivial partial trace. On the other hand, if G/H acts nontrivially, then by the previous proposition $0 \neq \mathrm{tr}_{G/H}(\sum_{x \in G} S^x) \subseteq R^G$. In particular, $0 \neq \mathrm{tr}_{G/H}(S^x) \subseteq R^G$ for some $x \in G$ and the result follows with $\Lambda = \Omega x T$ where T is a transversal for H in G. ∎

We offer some examples in the following exercises to show that the above result is essentially best possible without additional assumptions on the ring.

EXERCISES

1. Let R be a prime ring. Prove that R is a domain if and only if it has no nonzero nilpotent elements.

2. Let $K[G]$ be a group algebra of the finite group G and let H be a subgroup of G. If $\tau = \sum_{h \in H} h \in K[H]$, prove that $r_{K[G]}(\tau) = \omega(K[H])K[G]$ and $\ell_{K[G]}(\omega(K[H])) = K[G]\tau$.

3. Suppose G is a cyclic group of prime order p acting faithfully on the ring R with $pR = 0$. Show that the only possible partial trace here is the ordinary trace. To this end, observe that every element of $GF(p)[G]$ not in the augmentation ideal is a unit.

4. Let $R = M_2(K)$ where K is a field of characteristic $p > 0$ and let $g = \begin{pmatrix} 1 & 1 \\ 0 & 1 \end{pmatrix} \in R$. Then $G = \langle g \rangle$ acts on R by conjugation. Viewing R as a right $GF(p)[G]$-module, show that $R(1-g)$ consists of upper triangular matrices, $R(1-g)^2$ consists of strictly upper triangular matrices and $R(1-g)^3 = 0$. Conclude that for $p \geq 5$ there are no nontrivial partial traces even though R is a simple ring. This example is from [38].

The remaining exercises are based on [69]. Let G be a finite group of order divisible by p, let K be a field of characteristic p and let Ω be the set of all right cosets of subgroups of G of order p. Then G permutes Ω by right multiplication, each point stabilizer G_ω for $\omega \in \Omega$ is a subgroup of G of order p and all such subgroups occur in this manner. Form $R_p = \oplus \sum_{\omega \in \Omega} K_\omega$, the direct sum of copies of K indexed by the elements of Ω, and let G act on the reduced ring R_p by permuting these summands. Suppose $\mathrm{tr}_\Lambda(R_p) \subseteq (R_p)^G$ for some nonempty subset Λ of G.

5. Let $0 \neq r \in K_\omega$. Show that $\mathrm{tr}_\Lambda(r) = 0$ if and only if Λ is a union of right cosets of G_ω or equivalently if and only if $G_\omega \Lambda = \Lambda$. On the other hand, if $\mathrm{tr}_\Lambda(r) \neq 0$ show that Λ is a disjoint union of $1 \leq n \leq p - 1$ right transversals of G_ω and hence that $|\Lambda| = n|G|/p$. To this end, write $\Lambda = \bigcup_i \Lambda_i t_i$ where $\{t_i\}$ is a right transversal for G_ω in G and $\Lambda_i \subseteq G_\omega$. Then note that $\mathrm{tr}_\Lambda(r) = \sum_i |\Lambda_i| r^{t_i}$.

6. Suppose G is generated by its subgroups of order p. Show that $\text{tr}_\Lambda(R_p) = 0$ if and only if $\Lambda = G$ and conclude that $|G|/p$ divides $|\Lambda|$ in general.

7. Suppose p and q are distinct primes and that G is generated by its subgroups of order p and by its subgroups of order q. Show that there is no nontrivial partial trace for the action of G on $R_p \oplus R_q$.

8. Suppose G is generated by its subgroups of order 2 and that G has no subgroup of index 2. Show that there is no nontrivial partial trace for the action of G on R_2. To this end, if $G_\omega = \langle x \rangle$ has order 2, observe that either $x\Lambda = \Lambda$ or Λ is a right transversal for G_ω and hence $x\Lambda = G \setminus \Lambda$. Conclude that G permutes the subsets Λ and $G \setminus \Lambda$ by left multiplication.

25. Integrality

It is possible that Theorem 24.2 is trying to tell us something more. It may be that R is integral, in some sense, over the fixed ring R^G. For this we need some definitions.

Let R be a ring (possibly without 1) and let T be a subring. If $r_1, r_2, \ldots, r_m \in R$, then a T-*monomial* in the r_i's is a product of these elements in some order with elements of T such that at least one factor from T occurs. For example, if $t_1, t_2 \in T$, then $r_1^2 t_1 r_2 t_2 r_1 r_3$ is a T-monomial but $r_1^2 r_2 r_1 r_3$ is not unless $1 \in T$. The *degree* of such a monomial is the total degree in all the r_i's. We say that R is *fully integral* over T of degree m if for any $r_1, r_2, \ldots, r_m \in R$ we have

$$r_1 r_2 \cdots r_m = \varphi(r_1, r_2, \ldots, r_m)$$

where φ is a sum of T-monomials in the r_i's of degree less than m. In particular, setting $r_1 = r_2 = \cdots = r_m = r \in R$ we see that $r^m = \phi(r)$ where ϕ is a sum of T-monomials in r of degree less than m. This says, by definition, that R is *Schelter integral* over T of degree m.

In particular, if R is fully integral over R^G of degree m and if $R^G = 0$, then $R^m = 0$. Thus such an integrality result is an appropriate generalization of Theorem 24.2. As a simple example, let $G = \{1, x\}$ have order 2 and act on R. If $r \in R$, then $t = r + r^x = \mathrm{tr}_G(r) \in R^G$ and $t - r = r^x$ so

$$r(t - r) + (t - r)r = rr^x + r^x r = s = \mathrm{tr}_G(rr^x) \in R^G.$$

In other words, r satisfies $2r^2 - rt - tr + s = 0$ and $2r$ is Schelter integral over R^G.

To see what might be involved in a general proof of this result, we begin with a crucial special case. Let S be an algebra over the field K and set $R = \mathrm{M}_n(S)$. Furthermore, let G be a finite absolutely irreducible subgroup of $\mathrm{GL}_n(K)$ so that, by definition, the K-linear span of G is $\mathrm{M}_n(K)$. For example, if K has characteristic 0, then the symmetric group Sym_{n+1} has such an irreducible representation. Then G acts on R by conjugation and $R^G = S$, embedded as scalars. Thus at the very least we expect $\mathrm{M}_n(S)$ to be integral over S. Fortunately, this is a result of the extremely clever paper [154]. Variants of that theorem were later obtained in [110], [102] and [163] using the same proof. The formulation of the following is from [163].

If A is a ring with 1, we let $\{e_{i,j}\}$ denote the usual set of matrix units of $\mathrm{M}_n(A)$.

Theorem 25.1. [154] *Let A be a ring with 1 and let $R \supseteq T$ be subrings of $\mathrm{M}_n(A)$ (without 1). Assume that for all $k = 1, 2, \ldots, n$ we have*

    i. $\left(\sum_1^k e_{i,i}\right) R \left(\sum_1^k e_{i,i}\right) \subseteq R$,

    ii. T consists of diagonal matrices and $e_{k,k} R e_{k,k} = e_{k,k} T e_{k,k}$.

Then there exists an integer $m = m(n) \geq 1$ such that R is fully integral over T of degree m. Here $m < 2^{2^{n+1}-2}$.

Proof. For ease of notation we first consider Schelter integrality. As we will see at the end of the proof, full integrality is an easy consequence of this. For each $k = 1, 2, \ldots, n$ we embed $\mathrm{M}_k(A)$ into $\mathrm{M}_n(A)$ as the $k \times k$ upper left corner. The goal is to show by induction on k that $R \cap \mathrm{M}_k(A)$ is Schelter integral over T of degree $m(k) < 2^{2^{k+1}-2}$. Furthermore, the monic polynomials involved will all have constant term zero.

First let $k = 1$ and choose $r = \begin{pmatrix} a & 0 \\ 0 & 0 \end{pmatrix} \in R \cap M_1(A)$. Then (ii) implies that there exists $t = \begin{pmatrix} a & 0 \\ 0 & * \end{pmatrix} \in T$. Thus $r^2 = tr$ and $m(1) = 2 < 2^{2^2-2}$.

Now let $k \geq 1$ and for any $r \in R \cap M_{k+1}(A)$ write

$$r = \begin{pmatrix} r' & r'' & 0 \\ * & * & 0 \\ 0 & 0 & 0 \end{pmatrix}$$

where r' is the $k \times k$ corner block. Furthermore, for this r set

$$\tilde{r} = \begin{pmatrix} r' & 0 & 0 \\ 0 & 0 & 0 \\ 0 & 0 & 0 \end{pmatrix}$$

so that $\tilde{r} \in R$ by assumption (i). Given $r_1, r_2 \in R \cap M_{k+1}(A)$ we say that $r_1 \equiv r_2$ if and only if $r_1'' = r_2''$. We note two properties of $\equiv$.

(1) Let $r_1, r_2, r \in R \cap M_{k+1}(A)$ and $t \in T$. If $r_1 \equiv r_2$, then $\tilde{r}r_1 \equiv \tilde{r}r_2$ and $tr_1 \equiv tr_2$.

(2) Let $r_1, r \in R \cap M_{k+1}(A)$. Then there exists $t \in T$ with $\tilde{r}_1 r \equiv r_1 r - r_1 t$. For this, note that there exists $t \in T$ with

$$r - t = \begin{pmatrix} * & r'' & 0 \\ * & 0 & 0 \\ 0 & 0 & * \end{pmatrix}$$

and then

$$r_1(r - t) = \begin{pmatrix} * & r_1' r'' & 0 \\ * & * & 0 \\ 0 & 0 & 0 \end{pmatrix} \equiv \tilde{r}_1 r.$$

By induction on $j > 0$, if $r_1, r_2, \ldots, r_j, r \in R \cap M_{k+1}(A)$, then

$$\tilde{r}_j \tilde{r}_{j-1} \cdots \tilde{r}_1 r \equiv r_j r_{j-1} \cdots r_1 r + \tau$$

where τ is a sum of nonconstant T-monomials in the r_i's of degree smaller than $j + 1$. Indeed if

$$\tilde{r}_{j-1} \tilde{r}_{j-2} \cdots \tilde{r}_1 r \equiv r_{j-1} r_{j-2} \cdots r_1 r + \bar{\tau}$$

then by the above we have

$$\tilde{r}_j\tilde{r}_{j-1}\cdots\tilde{r}_1 r \equiv \tilde{r}_j(r_{j-1}\cdots r_1 r + \bar{\tau})$$
$$\equiv r_j(r_{j-1}\cdots r_1 r + \bar{\tau}) - r_j t$$
$$\equiv r_j r_{j-1}\cdots r_1 r + \tau$$

where t is a suitable element of T and $\tau = r_j\bar{\tau} - r_j t$ is an appropriate sum of T-monomials of degree less than $j+1$.

Now for the inductive part of the proof. Let $r \in R \cap M_{k+1}(A)$. Then by induction, $\tilde{r}$ satisfies a monic polynomial $\tilde{r}^m - \phi(\tilde{r}) = 0$ over T of degree $m = m(k)$. Hence certainly $(\tilde{r}^m - \phi(\tilde{r}))r \equiv 0$. We now apply the observation of the preceding paragraph to each of the monomials in this expression. Since $TM_{k+1}(A), M_{k+1}(A)T \subseteq M_{k+1}(A)$, each nonconstant T-monomial in r is a product of factors all belonging to $M_{k+1}(A)$. With this it follows easily from the above that

$$(r^m - \phi(r))r + \tau(r) \equiv (\tilde{r}^m - \phi(\tilde{r}))r \equiv 0,$$

where $\tau(r)$ is a suitable sum of nonconstant T-monomials in r of degree less than $m+1$. In other words, we have shown that there exists

$$s = r^{m+1} - \phi(r)r + \tau(r) \in R \cap M_{k+1}(A),$$

a monic polynomial over T in r of degree $m+1$, with $s \equiv 0$.

Note that $\tilde{s} \in R \cap M_k(A)$ also satisfies a monic polynomial say $\tilde{s}^m - \theta(\tilde{s}) = 0$ over T. Hence since $s \equiv 0$, we see that

$$v = s^m - \theta(s) = \begin{pmatrix} 0 & 0 & 0 \\ * & a & 0 \\ 0 & 0 & 0 \end{pmatrix}$$

for some $a \in A$. Furthermore if $b \in T$ with $b = \mathrm{diag}(*, a, *)$, then we have $(v^2 - bv)^2 = 0$. Since the latter expression is a monic polynomial in s of degree $4m$ and therefore a monic polynomial in r of degree $4m(m+1)$, the induction step follows with

$$m(k+1) = 4m(k)(m(k)+1) < 2^2 \cdot 2^{2^{k+1}-2} \cdot 2^{2^{k+1}-2} = 2^{2^{k+2}-2}.$$

This proves that R is Schelter integral over T.

Finally form the free ring $\tilde{A} = A\langle \zeta_1, \zeta_2, \dots \rangle$ and let $\tilde{R}$ and $\tilde{T}$ be the subrings of $M_n(\tilde{A}) = M_n(A)\langle \zeta_1, \zeta_2, \dots \rangle$ given by $\tilde{R} = R\langle \zeta_1, \zeta_2, \dots \rangle$ and $\tilde{T} = T\langle \zeta_1, \zeta_2, \dots \rangle$. It follows easily that $\tilde{T} \subseteq \tilde{R} \subseteq M_n(\tilde{A})$ satisfy hypotheses (i) and (ii). Thus $\tilde{R}$ is Schelter integral over $\tilde{T}$ of degree $m = m(n)$. In particular if $r_1, r_2, \dots, r_m \in R$ then $\tilde{r} = r_1\zeta_1 + r_2\zeta_2 + \cdots + r_m\zeta_m$ satisfies an appropriate monic polynomial of degree m over $\tilde{T}$. By considering the coefficient of $\zeta_1\zeta_2 \cdots \zeta_m$, we conclude that R is fully integral of degree m over T. ∎

The above theorem has numerous consequences. To start with, let R be a ring with 1 and let $1 = e_1 + e_2 + \cdots + e_n$ be a decomposition of $1 \in R$ into orthogonal idempotents. If $\mathcal{E} = \{e_1, e_2, \dots, e_n\}$, then it is trivial to see that $\mathbf{C}_R(\mathcal{E}) = e_1Re_1 + e_2Re_2 + \cdots + e_nRe_n$. If $\mathcal{E}$ is central in R, then an $\mathcal{E}$-*transversal* for R is a subring T such that $e_iR = e_iT$ for all $i = 1, 2, \dots, n$. In the following, $m(n)$ denotes the particular function given by Theorem 25.1.

Corollary 25.2. [163] *Let R be a ring, let $1 = e_1 + e_2 + \cdots + e_n$ be a decomposition of $1 \in R$ into orthogonal idempotents, and set $\mathcal{E} = \{e_1, e_2, \dots, e_n\}$. If T is an $\mathcal{E}$-transversal for $\mathbf{C}_R(\mathcal{E})$, then R is fully integral over T of degree $m(n)$.*

Proof. For each $r \in R$, let $r_{i,j} = e_i r e_j$. Then the map $\sigma : R \to M_n(R)$ given by $\sigma(r) = [r_{i,j}]$ is easily seen to be an isomorphism into. We show that $\sigma(T) \subseteq \sigma(R) \subseteq M_n(R)$ satisfy the hypotheses of the preceding theorem. To start with, since $r_{i,j} \in e_iRe_j$ and $\sigma(e_i) = e_i \cdot e_{i,i}$, it follows that $e_{i,i}\sigma(r) = \sigma(e_ir)$, $\sigma(r)e_{i,i} = \sigma(re_i)$ and hence that

$$\left(\sum_{i=1}^{k} e_{i,i} \right) \sigma(R) \left(\sum_{i=1}^{k} e_{i,i} \right) = \sigma\left(\left(\sum_{i=1}^{k} e_i \right) R \left(\sum_{i=1}^{k} e_i \right) \right) \subseteq \sigma(R)$$

so (i) holds.

Next, $T \subseteq \mathbf{C}_R(\mathcal{E}) = \sum_i e_iRe_i$ so $\sigma(T) \subseteq \sum_i e_{i,i}M_n(R)e_{i,i}$ and therefore consists of diagonal matrices. Also, since $r_{i,i} = e_ire_i$ we have

$$e_{i,i}\sigma(r)e_{i,i} = \sigma(e_i)\sigma(r)\sigma(e_i) = \sigma(e_ire_i).$$

But if $r \in R$, then $e_i r e_i \in \mathbf{C}_R(\mathcal{E})$ so, by assumption,

$$e_i r e_i = e_i \cdot e_i r e_i = e_i t = e_i t e_i$$

for some $t \in T$. Hence

$$e_{i,i} \sigma(r) e_{i,i} = \sigma(e_i r e_i) = \sigma(e_i t e_i) = e_{i,i} \sigma(t) e_{i,i}$$

and (ii) holds. We conclude from Theorem 25.1 that $\sigma(R)$ is fully integral of degree $m(n)$ over $\sigma(T)$. Applying σ^{-1} clearly yields the result. ∎

Note that if $R = \mathrm{M}_n(A)$ and if $\mathcal{E} = \{ e_{1,1}, e_{2,2}, \ldots, e_{n,n} \}$, then the scalar matrices A are an $\mathcal{E}$-transversal for $\mathbf{C}_R(\mathcal{E})$, the diagonal matrices. Thus Corollary 25.2 recaptures much of Theorem 25.1. Another consequence of that result is

Corollary 25.3. [21] *Let G be a finite group and let S be a G-graded ring (without 1). Then S is fully integral of degree $m(|G|)$ over the identity component S_1.*

Proof. Write $S = R(G)$ where $R = S_1$ and form $S^{\#} = S \oplus Z$, the natural extension of S to a ring with 1. In the notation of Section 2, let $\mathrm{M}_G(S^{\#})$ be the ring of $|G| \times |G|$ matrices over $S^{\#}$ and let

$$S\{1\} = \left\{ \alpha \in \mathrm{M}_G(S) \subseteq \mathrm{M}_G(S^{\#}) \mid \alpha(x, y) \in R(x^{-1}y) \right\}.$$

If $T\{1\} = R$, embedded as scalar matrices in $\mathrm{M}_G(S^{\#})$, then it follows immediately that $T\{1\} \subseteq S\{1\} \subseteq \mathrm{M}_G(S^{\#})$ satisfy the hypotheses of Theorem 25.1. We conclude that $S\{1\}$ is fully integral over $T\{1\}$ of degree $m(|G|)$.

Finally if $s \in S$, recall that the matrix $\tilde{s} \in \mathrm{M}_G(S)$ is defined by $\tilde{s}(x, y) = s_{x^{-1}y}$ so that $\tilde{s} \in S\{1\}$. Moreover, by Lemma 2.1(ii), the map $s \mapsto \tilde{s}$ is a ring embedding of S into $S\{1\}$. Since the image of R under this map is $T\{1\}$, the result clearly follows from the above. ∎

With this result in hand, it is not surprising that we can obtain full integrality for abelian group actions.

Theorem 25.4. **[163]** *Let G be a finite abelian group of order n acting on a ring R (without 1). Then nR is fully integral over R^G of degree $m(n)$.*

Proof. Let ϵ be a primitive complex n^{th} root of unity, set $\Lambda = Z[\epsilon]$ and $K = Q[\epsilon]$. We work with the generic models

$$R = Z\langle \zeta_{i,g} \mid g \in G,\ i \in I \rangle'$$
$$S = \Lambda\langle \zeta_{i,g} \mid g \in G,\ i \in I \rangle'$$
$$T = K\langle \zeta_{i,g} \mid g \in G,\ i \in I \rangle'$$

as described in the preceding section. By Lemma 24.6 we know that $T = \sum_{\lambda \in \hat{G}} T_\lambda$ is $\hat{G}$-graded. Furthermore, by that lemma,

$$nS \subseteq \sum_{\lambda \in \hat{G}} (T_\lambda \cap S).$$

Note that $\sum_{\lambda \in \hat{G}} (T_\lambda \cap S)$ is $\hat{G}$-graded so, by Corollary 25.3, this ring is fully integral of degree $m(|\hat{G}|) = m(n)$ over the identity component $T_1 \cap S = T^G \cap S = S^G$. Thus nS is fully integral over S^G of degree $m(n)$. Next observe that for a suitable integer q

$$S = \oplus \sum_{i=0}^{q} R\epsilon^i \quad \text{and} \quad S^G = \oplus \sum_{i=0}^{q} R^G \epsilon^i$$

with ϵ^{q+1} a Z-linear combination of $1, \epsilon, \ldots, \epsilon^q$. It then follows easily by reading off the coefficient of ϵ^0, that nR is fully integral over R^G of degree $m(n)$.

Finally, by choosing the index set I sufficiently large, we can map R onto any ring on which G acts. Lemma 24.7 therefore yields the result. ∎

The original proof of the above in **[163]** used skew group rings directly (see Exercises 2–5). Note that the case of solvable group actions does not follow directly from this result. An example of **[20]**, on Schelter integrality, shows that an integral extension of an integral extension need not be integral.

Now we move on to consider group actions by arbitrary finite groups. We remark that if S and T are merely subsets of a ring A, then it still makes sense to consider whether S is fully integral over T. It is with this understanding that we prove the following lemma since the set eTe need not be a subring of A.

Lemma 25.5. *Let $T \subseteq S \subseteq A$ be rings (without 1) and assume that S is fully integral over T of degree m. If e is an idempotent of A with $eSe \subseteq S$, then eSe is fully integral over eTe of degree m.*

Proof. Let $s_1, s_2, \ldots, s_m \in S$. Since $eSe \subseteq S$ and S is fully integral over T of degree m, we have

$$\prod_{i=1}^{m} (es_i e) = \sum_k \alpha_k$$

where each α_k is a T-monomial in the $es_i e$ of total degree less than m. Certainly

$$\prod_{i=1}^{m} (es_i e) = \sum_k e\alpha_k e.$$

Now T is a subring of A, so adjacent T factors in $e\alpha_k e$ can be merged. Once this is done, each T factor t in $e\alpha_k e$ has an e on either side and hence can be replaced by $ete \in eTe$. Thus $e\alpha_k e$ is an eTe-monomial in the $es_i e$ of total degree less than m and the lemma is proved. ∎

We can now obtain

Theorem 25.6. [181] *Let G be a finite group acting on the ring R (without 1) and suppose that $|G| \cdot R = R$. Then R is fully integral of degree $m(|G|)$ over the fixed ring R^G.*

Proof. Let $n = |G|$ and $m = m(n)$. We first consider the generic models

$$\tilde{S} = Z[1/n]\langle \zeta_{i,g} \mid g \in G, 1 \le i \le m \rangle$$
$$S = Z[1/n]\langle \zeta_{i,g} \mid g \in G, 1 \le i \le m \rangle'.$$

Let $B = M_n(S) \subseteq A = M_n(\tilde{S})$ and if $G = \{x_1, x_2, \ldots, x_n\}$ define $C \subseteq B$ by

$$C = \{\operatorname{diag}(s^{x_1}, s^{x_2}, \ldots, s^{x_n}) \mid s \in S\}.$$

It then follows immediately from Theorem 25.1 that B is fully integral over C of degree m.

Now let $e = \frac{1}{n}\alpha \in A$ where α is the matrix with all entries equal to 1. Then e is an idempotent of A and if $\beta = [b_{i,j}] \in A$, we have $e\beta e = \left(\frac{1}{n}\sum_{i,j} b_{i,j}\right)e$. Thus $eBe \subseteq B$ and the previous lemma implies that eBe is fully integral of degree m over eCe. But $eBe \supseteq Se$, with S embedded as scalars, and $eCe = \frac{1}{n}\operatorname{tr}_G(S)e \subseteq S^G e$. Thus $Se \cong S$ is fully integral over $S^G e \cong S^G$ and this case is proved.

Now set

$$T = Z\langle \zeta_{i,g} \mid g \in G, 1 \le i \le m \rangle' \subseteq S.$$

By the above, we have

$$\zeta_{1,1}\zeta_{2,1}\cdots\zeta_{m,1} = \theta$$

where θ is a sum of S^G-monomials in the $\zeta_{i,1}$ of degree less than m. It then follows by clearing denominators that, for some integer $k \ge 0$, $n^k\theta$ is a sum of T^G-monomials in the $\zeta_{i,1}$. Hence we see that

$$n^k\zeta_{1,1} \cdot n^k\zeta_{2,1}\cdots n^k\zeta_{m,1} = \varphi \qquad (*)$$

where φ is a sum of T^G-monomials of degree $< m$ in the variables $n^k\zeta_{i,1}$. Finally if $r_1, r_2, \ldots, r_m \in R$, then there is a G-homomorphism $T \to R$ given by $\zeta_{i,g} \mapsto r_i^g$ and we conclude from Lemma 24.7 and $(*)$ that n^kR is fully integral over R^G of degree m. Thus since $nR = R$, we have $n^kR = R$ and the result follows. ∎

As we mentioned earlier, this yields a far reaching generalization of Theorem 24.2 with however a poorer bound.

We close this section by obtaining an integrality result for normalizing extensions. Let $R \subseteq S = \sum_{i=1}^{n} Rx_i$ be a finite normalizing extension of rings (with 1). An ideal A of R is said to be *normal* if $x_iA = Ax_i$ for all $i = 1, 2, \ldots, n$. It follows that AS is an ideal of S and hence a subring (without 1) containing A.

Theorem 25.7. **[154] [110]** *Let* $R \subseteq S = \sum_{i=1}^{n} Rx_i$ *be a finite normalizing extension of rings (with 1) and let A be a normal ideal of R. Then AS is fully integral of degree $m(n)$ over A.*

Proof. Let $F = R^n$ be the free left R-module of rank n and let $\pi \colon F \to S$ be the left R-module epimorphism given by $(r_1, r_2, \ldots, r_n)^\pi = \sum_{i=1}^{n} r_i x_i$. If $I = \mathrm{Ker}(\pi) \subseteq F$, then $U = \{\, \varphi \in \mathrm{End}(_R F) \mid I^\varphi \subseteq I \,\}$ is a subring of $\mathrm{End}(_R F) = \mathrm{M}_n(R)$. Furthermore, there is a natural homomorphism $\tilde{} \colon U \to \mathrm{End}(_R S)$ such that if $\varphi \in U$ then

$$\tilde{\varphi} \colon \sum_{1}^{n} r_i x_i \mapsto (r_1, r_2, \ldots, r_n)^{\varphi \pi}.$$

Routine verifications show that $\tilde{\varphi}$ is well defined and that $\tilde{}$ is a ring homomorphism. Indeed $\tilde{\varphi} = \pi^{-1} \varphi \pi$ with the understanding that $s^{\pi^{-1}}$ is any inverse image of s.

Let T be the set of all diagonal matrices in $\mathrm{M}_n(R) = \mathrm{End}(_R F)$ of the form $\alpha = \mathrm{diag}(a_1, a_2, \ldots, a_n)$ with $a_i \in A$ and $a_i x_i = x_i a$ for some $a \in A$. Then T is a subring of $\mathrm{M}_n(A)$ and the normality of A implies easily that $e_{i,i} \mathrm{M}_n(A) e_{i,i} = e_{i,i} T e_{i,i}$ for all i. Furthermore if $f = (r_1, r_2, \ldots, r_n) \in F$, then $f^\alpha = (r_1 a_1, r_2 a_2, \ldots, r_n a_n)$ so

$$f^{\alpha \pi} = \sum_{i} r_i a_i x_i = \sum_{i} r_i x_i a = (f^\pi) a.$$

It follows immediately that $\alpha \in U$ and that $\tilde{\alpha}$ is right multiplication by $a \in A$.

Now let $s \in AS$ be given. Since AS is a two-sided ideal of S, we have $x_i s \in AS$ so $x_i s = \sum_j a_{i,j} x_j$ for suitable $a_{i,j} \in A$. Of course the elements $a_{i,j}$ need not be uniquely determined. Let $\sigma = [a_{i,j}] \in \mathrm{M}_n(A)$. Then it follows easily from the above that if $f = (r_1, r_2, \ldots, r_n) \in F$ then $f^{\sigma \pi} = (f^\pi) s$. Hence we see that $\sigma \in U$ and that the endomorphism $\tilde{\sigma}$ is right multiplication by $s \in AS$.

By Theorem 25.1 applied to $T \subseteq \mathrm{M}_n(A) \subseteq \mathrm{M}_n(R)$ we see that $\mathrm{M}_n(A)$ is fully integral of degree $m(n)$ over T. Hence $U \cap \mathrm{M}_n(A)$ is also fully integral over T and therefore $(U \cap \mathrm{M}_n(A))^{\tilde{}}$ is fully integral over $\tilde{T}$. But as we have shown above, $(U \cap \mathrm{M}_n(A))^{\tilde{}} \supseteq AS$, viewed as right multiplication, and $\tilde{T} = A$. Thus the result follows. ∎

EXERCISES

1.   Let $A = Z\langle x, y \rangle$ be the free ring on x, y and let $\alpha = \mathrm{diag}(x, y) \in \mathrm{M}_2(A)$. Show that α is not integral, in the ordinary sense, over the scalar matrices A. Indeed, show that the powers of α are (right or left) independent over A.

In the next four problems, let G be an abelian group of order n and let K be a field containing n distinct n^{th} roots of unity. Set $\hat{G} = \mathrm{Hom}(G, K^{\bullet})$.

2.   For each $1 \neq \lambda \in \hat{G}$ prove that $\sum_{g \in G} \lambda(g) = 0$ and for each $1 \neq g \in G$ prove that $\sum_{\lambda \in \hat{G}} \lambda(g) = 0$. Show that $\hat{G} \cong G$.

3.   For each $\lambda \in \hat{G}$ define $e_\lambda \in K[G]$ by $e_\lambda = \frac{1}{n} \sum_{g \in G} \lambda(g^{-1}) g$. Show that $1 = \sum_\lambda e_\lambda$ is a decomposition of 1 into orthogonal idempotents.

4.   Suppose G acts as K-automorphisms on the K-algebra R with 1 so that $RG \supseteq KG = K[G]$. If $\mathcal{E} = \{ e_\lambda \mid \lambda \in \hat{G} \}$, show that R^G is an $\mathcal{E}$-transversal for $\mathbf{C}_{RG}(\mathcal{E})$. Conclude that RG and hence R is fully integral over R^G of degree $m(n)$.

5.   Now suppose G acts on any ring R (without 1). Use the above and generic models to obtain an alternate proof of Theorem 25.4.

6.   Let G be a group of order n acting on a ring R with $\frac{1}{n} \in R$. If I is a G-stable ideal of R, prove that $(R/I)^G = (R^G + I)/I$. Lemmas 24.7 and 24.1 apply here.

7.   Let G be a group of order n acting on a ring R (with 1) and form the skew group ring RG. As in Lemma 23.6(iii) we have $RG \subseteq \mathrm{End}_R(RG) \cong \mathrm{M}_n(R)$ using the natural left R-basis G. Show that the element $\alpha = \sum_{x \in G} x \in RG$ corresponds to the matrix with all entries equal to 1. In some sense, this underlies the proof of Theorem 25.6.

26. Finiteness Conditions

We now move on to more direct skew group ring applications. For this we will assume throughout that G is finite. Furthermore we return to our usual assumption that all rings have a 1. For the most part we will be concerned in this section with finiteness conditions, in particular the Artinian, Noetherian and Goldie properties of rings. We will consider just a small sample of what is known. The reader should consult [**129**] for the complete story.

Let G act on the ring R with 1. We recall that the trace map $\mathrm{tr}_G \colon R \to R^G$ is defined by $\mathrm{tr}_G(r) = \sum_{x \in G} r^x$ and satisfies part (i) below. In addition if A is a right ideal of R^G, then $A \subseteq (AR)^G$ and $\mathrm{tr}_G(AR) = A \, \mathrm{tr}_G(R) \subseteq A$ so (ii) follows.

Lemma 26.1. *Let G act on R.*

   i. tr_G is an (R^G, R^G)-bimodule homomorphism from R to R^G.

   ii. If A is a right ideal of R^G, then $|G| \cdot A \subseteq |G| \cdot (AR)^G \subseteq \mathrm{tr}_G(AR) \subseteq A$.

As usual, if G acts on R we can form the skew group ring RG. Recall that G is embedded isomorphically in RG so we do not need the overbars. Define $\dot{G} = \sum_{x \in G} x \in RG$ and observe that for all $g \in G$ we have $\dot{G}g = \dot{G} = g\dot{G}$. We remark that the element $\dot{G}$ is sometimes denoted by $\hat{G}$. We have chosen this alternate notation to avoid confusion with the dual group.

Lemma 26.2. *Let $\dot{G} \in RG$ be as above.*

   i. $\dot{G}(RG) = \dot{G}R \cong R$ as right RG-modules. Here R acts on R by right multiplication and G acts in the given manner. Furthermore $\mathrm{End}_{RG}(R) = R^G$, acting by left multiplication.

   ii. If $r \in R$, then $\dot{G}r\dot{G} = \dot{G}\,\mathrm{tr}_G(r)$. Hence

$$\dot{G}(RG)\dot{G} = \dot{G}\,\mathrm{tr}_G(R) \subseteq \dot{G}R^G.$$

   iii. If I and J are G-invariant right ideals of R, then $R\dot{G}I \triangleleft RG$ and $(R\dot{G}I)(R\dot{G}J) = R\dot{G}\,\mathrm{tr}_G(I)J$. Furthermore, for all $n \geq 0$ we have $(R\dot{G}I)^{n+1} = R\dot{G} \cdot \mathrm{tr}_G(I)^n I$.

Proof. (i) Since $RG = GR$, we have $\dot{G}(RG) = \dot{G}(GR) = \dot{G}R \cong R$ where the latter isomorphism is given by $\theta(\dot{G}r) = r$. This is certainly a right R-module map and for $g \in G$ we have $\theta(\dot{G}rg) = \theta(\dot{G}g^{-1}rg) = r^g = \theta(\dot{G}r)^g$. Finally, $\mathrm{End}_R(R) = R$, acting by left multiplication, and therefore we have $\mathrm{End}_{RG}(R) = R^G$, also acting by left multiplication.

(ii) This is clear since

$$\dot{G}r\dot{G} = \sum_{x \in G} \dot{G}rx = \sum_{x \in G} \dot{G}x^{-1}rx = \dot{G}\,\mathrm{tr}_G(r).$$

(iii) $R\dot{G}I$ is surely an (R, R)-bimodule. Furthermore, since both R and I are G-invariant and since $\dot{G}$ absorbs factors from G, we have $R\dot{G}I \lhd RG$. Finally $(R\dot{G}I)(R\dot{G}J) = R\dot{G}\,\mathrm{tr}_G(I)J$ since $\dot{G}IR\dot{G} = \dot{G}\,\mathrm{tr}_G(I)$ by (ii) above. The result on $(R\dot{G}I)^n$ now follows easily by induction. ∎

We can now obtain

Theorem 26.3. [22] [127] *Let G act on R and suppose that the skew group ring RG is semiprime (or prime). This occurs, for example, if R is semiprime with no $|G|$-torsion (or if R is prime and G is X-outer).*

i. If $I \neq 0$ is a G-invariant right or left ideal of R, then $\mathrm{tr}_G(I)$ is not nilpotent and hence $I^G \neq 0$.

ii. R^G is semiprime (prime).

Proof. The sufficient conditions for RG to be semiprime or prime are contained in Theorem 4.4 and Corollary 12.6. Now let $0 \neq I$ be a G-invariant right ideal of R and consider the nonzero ideal $R\dot{G}I$ of RG. By assumption, RG is semiprime so $R\dot{G}I$ is not nilpotent. Thus Lemma 26.2(iii) implies that $\mathrm{tr}_G(I)$ is not nilpotent. In particular, $\mathrm{tr}_G(I) \neq 0$ and (i) is proved. For (ii), let A and B be nonzero right ideals of R^G and set $I = AR$ and $J = BR$. Since $\mathrm{tr}_G(I) \subseteq A$, Lemma 26.2(iii) implies that $IJ \subseteq R\dot{G}AB$. Thus clearly if RG is semiprime (or prime), then so is R^G. ∎

The following is a special case of Theorem 24.2 on the existence of fixed points. We include it to see how close ring theoretic methods can come to proving the general result. Additional comments are contained in Exercise 6.

Theorem 26.4. [22] *Let G act on R, a ring with no $|G|$-torsion, and suppose that $0 \neq I \lhd R$ is G-invariant. If $\mathrm{tr}_G(I)$ is nilpotent, then $I \cap \mathrm{r}_R(I) \neq 0$ and I is nil.*

Proof. By Lemma 26.2(iii), $R\dot{G}I$ is a nilpotent ideal of RG. Thus $J = \mathrm{r}_{RG}(R\dot{G}I)$ is a two-sided ideal of RG which is essential as a left ideal. By the left analog of Proposition 4.3 it follows that $_RJ$ ess $_RRG$ and hence $_R(J \cap R)$ ess $_RR$. But surely $J \cap R = \mathrm{r}_R(I)$ so, since $I \neq 0$, we have $I \cap \mathrm{r}_R(I) \neq 0$.

For the second part, let N be the nil radical of R and set $\bar{R} = R/N$. Then G acts on the semiprime ring $\bar{R}$ and it is easy to see that $\bar{R}$ has no $|G|$-torsion. Furthermore, $\mathrm{tr}_G(\bar{I}) = \overline{\mathrm{tr}_G(I)}$ is nilpotent. It therefore follows from Theorem 26.3(i) that $\bar{I} = 0$ and hence $I \subseteq N$ is nil. ∎

Let V be an R-module. Then the *Goldie rank* of V, written rank V_R, is defined to be the largest integer k such that V contains $V_1 \oplus V_2 \oplus \cdots \oplus V_k$, a direct sum of k nonzero submodules. If no such maximum exists, then rank $V_R = \infty$. Basic properties of this rank can be found in [**64**]. In particular it is clear that $W \subseteq V$ implies that rank $W \leq$ rank V and if W ess V then the ranks are equal. Modules of rank 1 are called *uniform* so that $U \neq 0$ is uniform if and only if for all nonzero submodules $X, Y \subseteq U$ we have $X \cap Y \neq 0$. Furthermore rank $(V \oplus W) =$ rank $V +$ rank W and if $U_1, U_2, \ldots, U_k$ are uniform and $(U_1 \oplus U_2 \oplus \cdots \oplus U_k)$ ess V, then rank $V = k$.

Lemma 26.5. *Let V be a module for the crossed product $R*G$. Then*

$$\mathrm{rank}\, V_{R*G} \leq \mathrm{rank}\, V_R \leq |G| \cdot \mathrm{rank}\, V_{R*G}.$$

Proof. The first inequality, namely rank $V_{R*G} \leq$ rank V_R, is obvious. We consider the second. For each R-submodule A_R of V_R define $\tilde{A} = \bigcap_{x \in G} A\bar{x}$. Then $\tilde{A} = \{\, v \in V \mid v\bar{x}^{-1} \in A \text{ for all } x \in G \,\}$ and it is easy to see that $\tilde{A}$ is an $R*G$-submodule of V. In fact, $\tilde{A}$ is the largest $R*G$-submodule of V contained in A. Since G is finite, Zorn's lemma applies and there exists A maximal with $\tilde{A} = 0$.

Suppose that $(B_1/A) \oplus (B_2/A) \oplus \cdots \oplus (B_k/A)$ is a direct sum of nonzero R-submodules of V/A. Then B_i is properly larger than A, so $\tilde{B}_i \neq 0$ and it is easy to see that $\tilde{B}_1 \oplus \tilde{B}_2 \oplus \cdots \oplus \tilde{B}_k$ is a direct sum of nonzero $R*G$-submodules of V. Thus we see that $\operatorname{rank}(V/A)_R \leq \operatorname{rank} V_{R*G}$. Finally each $A\bar{x}$ is also an R-submodule of V with the same property so $\operatorname{rank}(V/A\bar{x})_R \leq \operatorname{rank} V_{R*G}$. Thus, since $\tilde{A} = 0$, V_R is embedded isomorphically in $\oplus \sum_{x \in G} V/A\bar{x}$ and

$$\operatorname{rank} V_R \leq \sum_{x \in G} \operatorname{rank}(V/A\bar{x})_R \leq |G| \cdot \operatorname{rank} V_{R*G}.$$

This completes the proof. ∎

Now return to the action of G on R. The following lemma shows that there exists a strong relationship between the essential right ideals of R and those of R^G. In part (i) below, the RG-module structure of R is as described in Lemma 26.2(i). Furthermore, the rank of any ring R is its rank as a right R-module.

Lemma 26.6. *Let G act on R, a semiprime ring with no $|G|$-torsion.*
   i. $\operatorname{rank} R^G = \operatorname{rank} R_{RG}$.
   ii. $\operatorname{rank} R^G \leq \operatorname{rank} R_R \leq |G| \cdot \operatorname{rank} R^G$.
   iii. *If E is an essential right ideal of R, then $(E \cap R^G)$ ess R^G.*
   iv. *If A is an essential right ideal of R^G, then AR ess R.*

Proof. We will freely use Lemma 26.1(ii) and Theorem 26.3(i).
   (i) If $I_1 \oplus I_2 \oplus \cdots \oplus I_k$ is a direct sum of nonzero RG-submodules of R, then $I_1^G \oplus I_2^G \oplus \cdots \oplus I_k^G$ is a direct sum of nonzero right ideals of R^G. Conversely if $A_1 \oplus A_2 \oplus \cdots \oplus A_k$ is a direct sum of nonzero right ideals of R^G, then $A_1R \oplus A_2R \oplus \cdots \oplus A_kR$ is a direct sum of nonzero RG-submodules of R. Indeed the latter sum is direct since, for each i, $B_i = (A_1R + A_2R + \cdots + A_{i-1}R) \cap A_iR$ is a G-invariant right ideal of R with $\operatorname{tr}_G(B_i) \subseteq (A_1 + A_2 + \cdots + A_{i-1}) \cap A_i = 0$. Thus $B_i = 0$ and the ranks are clearly equal.

(ii) This is immediate from (i) above and the previous lemma.

(iii) If E ess R then, $\left(\bigcap_{x \in G} E^x\right)$ ess R, since G is finite, and this intersection is G-invariant. Thus we may assume that E is G-invariant. Now suppose X is a nonzero right ideal of R^G. Then $E \cap XR \neq 0$ and hence $0 \neq \mathrm{tr}_G(E \cap XR) \subseteq (E \cap R^G) \cap X$.

(iv) If A ess R^G then it is clear that AR_{RG} ess RG_{RG}. Indeed if X is a nonzero G-invariant right ideal of R, then $X^G \neq 0$ so $A \cap X^G \neq 0$ and hence $AR \cap X \neq 0$. Proposition 4.3 now yields AR_R ess R_R and the result follows. ∎

Note that a ring R is semisimple Artinian if and only if it has no proper essential right ideal. Thus we have

Theorem 26.7. [98] [36] *Let G act on R and suppose that R is semiprime with no $|G|$-torsion. Then R is semisimple Artinian if and only if R^G is.*

Proof. Suppose R^G is semisimple Artinian and let E ess R. Then $(E \cap R^G)$ ess R^G, by Lemma 26.6(iii), so $1 \in E \cap R^G$ and $E = R$.

Conversely if R is semisimple Artinian, then the center of R is a sum of fields and hence $|G|^{-1} \in R$. If A ess R^G then AR ess R, by Lemma 26.6(iv), so $1 \in AR$. Thus $|G| = \mathrm{tr}_G(1) \in \mathrm{tr}_G(AR) \subseteq A$ and $A = R^G$. In view of the above remarks, the theorem is proved. ∎

R is said to be a *Goldie ring* if and only if rank $R_R < \infty$ and R satisfies the maximum condition on right annihilators of subsets of R. Goldie's theorem [**64**, Theorem 1.37] asserts that a ring R has a classical right quotient ring $Q(R)$ which is semisimple Artinian if and only if R is a semiprime Goldie ring. These rings can be characterized in terms of their essential right ideals and therefore a correspondence analogous to the above theorem is to be expected. For convenience, we quote the following lemma which merely isolates the last few steps in the proof of Goldie's theorem (see Exercises 7).

Lemma 26.8. *Let T be a multiplicatively closed subset of regular elements of R and suppose that*
   i. $t \in T$ implies tR ess R,
   ii. E ess R implies $E \cap T \neq \emptyset$.

Then T is a right divisor set in R, the ring RT^{-1} is semisimple Artinian and R is a semiprime Goldie ring with classical quotient ring $Q(R) = RT^{-1}$.

With this we can now prove

Theorem 26.9. [85] *Let G act on the ring R and suppose that R is semiprime with no $|G|$-torsion. Then R is Goldie if and only if R^G is. Furthermore if this occurs, then $Q(R) = RT^{-1}$ where T is the set of regular elements of R^G and $Q(R)^G = Q(R^G)$.*

Proof. Observe that R^G is semiprime by Theorem 26.3(ii).

Suppose first that R is Goldie. Then $R^G \subseteq R \subseteq Q(R)$ and $Q(R)$ is semisimple Artinian, so R^G surely satisfies the maximum condition on right annihilators of subsets. Furthermore rank R^G is finite, by Lemma 26.6(ii), and hence R^G is semiprime Goldie.

Conversely suppose that R^G is Goldie and hence that $Q(R^G) = R^G T^{-1}$ exists, where T is the set of regular elements of R^G. Certainly T is a multiplicatively closed subset of R. Furthermore, if $t \in T$ and if X is either the right or left annihilator of t in R, then X is a G-invariant right or left ideal of R and $X \cap R^G = 0$. Hence, by Theorem 26.3(i), $X = 0$ and T consists of regular elements of R. We show now that T satisfies (i) and (ii) of the previous lemma. Indeed if $t \in T$, then tR^G ess R^G since R^G is a Goldie ring. But then, by Lemma 26.6(iv), $tR = (tR^G)R$ is essential in R. On the other hand, if E ess R, then $(E \cap R^G)$ ess R^G, by Lemma 26.6(iii), and hence $E \cap T = (E \cap R^G) \cap T \neq \emptyset$ since R^G is semiprime Goldie. Lemma 26.8 now asserts that R is a semiprime Goldie ring with $Q(R) = RT^{-1}$. Finally, since $T \subseteq R^G$, we see immediately that $Q(R)^G = (RT^{-1})^G = R^G T^{-1} = Q(R^G)$ and the result follows. ∎

We remark that the above theorem was discovered independently in papers [35] and [59]. To proceed further, we assume that $|G|^{-1} \in R$ and define $e \in RG$ by

$$e = |G|^{-1} \dot{G} = |G|^{-1} \sum_{x \in G} x.$$

Lemma 26.10. *Let G act on R with $|G|^{-1} \in R$. Then $e = |G|^{-1} \dot{G}$ is an idempotent of RG with*

   i. *$e(AG)e = eA^G$ for any G-invariant ideal A of R,*
   ii. *$e(RG)e = eR^G \cong R^G$ where the latter is a ring isomorphism.*

Proof. Since $g\dot{G} = \dot{G}$, it follows that $(\dot{G})^2 = |G| \dot{G}$ and hence that e is an idempotent. Now let A be a G-invariant ideal of R. Then $A^G = \text{tr}_G(A)$ since $|G|^{-1} \in R$ so (i) is immediate from Lemma 26.2(ii). Furthermore, since R^G clearly commutes with e, the isomorphism $eR^G \cong R^G$ does indeed preserve the ring structure. ∎

In view of part (ii) above, the following lemma is surely of interest. Let S be a ring. Recall that if W is an S-module, then $\mathcal{L}(W_S)$ denotes the lattice of S-submodules of W.

Lemma 26.11. *Let W be a right S-module and let f be an idempotent of S. Then there exist inclusion preserving maps*

$$\sigma: \mathcal{L}(Wf_{|fSf}) \to \mathcal{L}(W_S) \qquad \tau: \mathcal{L}(W_S) \to \mathcal{L}(Wf_{|fSf})$$

such that

   i. *$\sigma\tau = 1$ so σ is one-to-one and τ is onto,*
   ii. *τ preserves direct sums.*

Proof. Define $\sigma: \mathcal{L}(Wf_{|fSf}) \to \mathcal{L}(W_S)$ and τ in the other direction by $A^\sigma = AS$ and $B^\tau = Bf$. Then $A^{\sigma\tau} = ASf = AfSf = A$ and the result clearly follows. ∎

Theorem 26.12. [113] *Let G act on the ring R with $|G|^{-1} \in R$ and let V_R be a right R-module. If $W = V^{|RG}$ is the induced RG-module, then there exist inclusion preserving maps*

$$\sigma: \mathcal{L}(V_{|R^G}) \to \mathcal{L}(W_{RG}) \qquad \tau: \mathcal{L}(W_{RG}) \to \mathcal{L}(V_{|R^G})$$

such that

   i. *$\sigma\tau = 1$ so σ is one-to-one and τ is onto,*
   ii. *τ preserves direct sums.*

Proof. Let $S = RG$. In view of the previous two lemmas, it suffices to show that $V_{|R^G} \cong W e_{|eSe}$. But this is clear since $W = V^{|RG} = \oplus \sum_{x \in G} V \otimes x$ implies that the map $v \mapsto \sum_{x \in G} v \otimes x$ gives the necessary R^G-isomorphism. ∎

Corollary 26.13. *Let G act on R with $|G|^{-1} \in R$ and let V_R be a right R-module.*

i. If V_R is Noetherian or Artinian, then so is $V_{|R^G}$.

ii. If R is right Noetherian, then R is a finitely generated R^G-module and R^G is right Noetherian.

iii. If R is right Artinian, then so is R^G.

iv. If V_R is completely reducible of composition length n, then $V_{|R^G}$ is completely reducible of composition length $\leq n \cdot |G|$.

Proof. Let W be the induced RG-module $W = V^{|RG}$. Then $W_{|R} = \oplus \sum_{x \in G} V \otimes x$ with each $V \otimes x$ an R-module conjugate to V (see Lemma 3.3). In particular, the lattice of R-submodules of $V \otimes x$ is isomorphic to $\mathcal{L}(V_R)$.

(i) If V_R is Noetherian or Artinian, then so is each $V \otimes x$ and hence $W_{|R}$. It therefore follows that W_{RG} is Noetherian or Artinian and hence so is $V_{|R^G}$ by Theorem 26.12 since σ is one-to-one.

(ii) By (i) applied to $V = R$, we conclude that R_{R^G} is Noetherian and hence finitely generated. Since $R^G \subseteq R$, it follows that R^G is a right Noetherian ring. Part (iii) is similar.

(iv) If V_R is completely reducible of composition length $\leq n$, then the same is true of each $V \otimes x$. Thus $W_{|R}$ is completely reducible of length $\leq n \cdot |G|$. By Theorem 4.1, W_{RG} is completely reducible and hence so is $V_{|R^G}$ by Theorem 26.12 since τ is onto and preserves direct sums. ∎

Part (ii) above is essentially due to [**56**]. Analogous results hold for (S, R)-bimodules.

Corollary 26.14. [**113**] *Let G act on the ring R with $|G|^{-1} \in R$ and let H act on the ring S with $|H|^{-1} \in S$. Suppose $_SV_R$ is an (S, R)-bimodule.*

i. If $_SV_R$ is Noetherian or Artinian, then so is $_{S^H|}V_{|R^G}$.

ii. If the ideals of R satisfy the ascending or descending chain condition, then the same is true of the ideals of R^G.

iii. If $_S V_R$ is completely reducible of composition length n, then the restriction $_{S^H}|V_{|R^G}$ is completely reducible of length $\leq n \cdot |H| \cdot |G|$.

Proof. We may suppose that $V \neq 0$. Since V is a unitary right module for the ring $S^{\mathrm{op}} \otimes R$, it follows that the latter is a ring with 1. Also $H \times G$ acts on this ring and $|H \times G|^{-1} \in S^{\mathrm{op}} \otimes R$. Now it is easy to see that

$$\mathrm{tr}_{H \times G}(s \otimes r) = \mathrm{tr}_H(s) \otimes \mathrm{tr}_G(r)$$

and thus

$$
\begin{aligned}
(S^{\mathrm{op}} \otimes R)^{H \times G} &= \mathrm{tr}_{G \times H}(S^{\mathrm{op}} \otimes R) \\
&= \mathrm{tr}_H(S^{\mathrm{op}}) \otimes \mathrm{tr}_G(R) = (S^{\mathrm{op}})^H \otimes R^G.
\end{aligned}
$$

With this observation, (i) and (iii) follow from Corollary 26.13(i)(iv). Part (ii) follows by viewing R as an (R, R)-bimodule. ∎

In particular, if R is a direct sum of n simple rings, then (iii) implies that R^G is a direct sum of at most $n \cdot |G|^2$ simple rings. This bound, however, can be sharpened to $n \cdot |G|$, a fact first observed in [**151**]. We have

Theorem 26.15. [**86**] *Let R be a ring which is a direct sum of n simple rings and let G act on R. If R has no $|G|$-torsion, then RG and R^G are direct sums of at most $n \cdot |G|$ simple rings.*

Proof. The structure of R implies immediately that $|G|^{-1} \in R$. Let $V = RG$ and view V as an (RG, RG)-bimodule. Then $_R V_{|R} = \oplus \sum_{x \in G} Rx$ is a direct sum of $|G|$ bimodules, each of which is clearly a direct sum of n simple bimodules. Thus $_R V_{|R}$ is completely reducible of composition length $n \cdot |G|$. By viewing V as a module for the skew group ring $(R^{\mathrm{op}} \otimes R)(G \times G)$, as in the previous result, Theorem 4.1 implies that V is also a direct sum of at most $n \cdot |G|$ simple (RG, RG)-bimodules. In other words, RG is a direct sum of at most $n \cdot |G|$ simple rings. Since $e(RG)e \cong R^G$, Lemma 17.8 now yields the result. ∎

Finally we prove

Theorem 26.16. [56] *Let G act on R and suppose that R is semiprime with no $|G|$-torsion. If R^G is right Noetherian, then so is R.*

Proof. In view of Corollary 26.13(ii), we expect R to be a finitely generated R^G-module. Observe that R^G is Goldie and hence, by Theorem 26.9, R is semiprime Goldie and G acts on the semisimple Artinian ring $Q(R)$. Note that R has no $|G|$-torsion and hence $|G|^{-1} \in Q(R)$. Thus since $Q(R)$ is left Noetherian, the left analog of Corollary 26.13(ii) implies that $Q(R)$ is a finitely generated left $Q(R)^G$-module. We can of course assume that the generators have a common denominator. Thus there exist $r_1, r_2, \ldots, r_n \in R$ and t regular in R with $Q(R) = \sum_1^n Q(R)^G r_i t^{-1}$ so that

$$Rt \subseteq Q(R)t = \sum_1^n Q(R)^G r_i.$$

We now define a map $\theta \colon R \to (R^G)^n$ by $\theta(r) = \oplus \sum_1^n \mathrm{tr}_G(r_i r)$. It is clear that θ is a right R^G-homomorphism into the free right R^G-module $(R^G)^n$. The goal is to show that θ is one-to-one. To this end, suppose $r \in \mathrm{Ker}(\theta)$. Then $\mathrm{tr}_G(r_i r) = 0$ for all i so $\mathrm{tr}_G(Q(R)^G r_i r) = 0$ and thus, by the above, $\mathrm{tr}_G(Rtr) = 0$. If I is the G-invariant left ideal $I = \sum_{x \in G}(Rtr)^x$, it follows that $\mathrm{tr}_G(I) = 0$. But R is semiprime with no $|G|$-torsion, so Theorem 26.3(i) implies that $I = 0$. We conclude that $tr = 0$ and hence, since t is regular, that $r = 0$. Thus θ is one-to-one as required. In other words, R_{R^G} is isomorphic to a submodule of a finitely generated free R^G-module. Since R^G is right Noetherian, we conclude that R_{R^G} is Noetherian and therefore R_R is also Noetherian. ∎

An interesting special case is as follows. Suppose G acts as automorphisms on the polycyclic-by-finite group H. Then G acts on the Noetherian ring $K[H]$ and, by the above, $K[H]^G$ is Noetherian if $|G|^{-1} \in K$. It is an open question as to whether the hypothesis $|G|^{-1} \in K$ is required here.

EXERCISES

1. Find an example of a prime ring R and a group G such that R is not Artinian, Noetherian or Goldie, even through R^G is a field. An appropriate example occurs in Section 24.

2. Let K be a field of characteristic not 2 and let F be an infinite dimensional extension field. Form $R = \begin{pmatrix} K & F \\ 0 & K \end{pmatrix}$ and let $G = \{1, g\}$ act on R with $g = \text{diag}(1, -1)$. Show that $|G|^{-1} \in R$ and that R^G is Artinian, Noetherian and Goldie, but that R is not.

3. Let D be a K-division algebra generated (as a division algebra) by $x_1, x_2, \ldots, x_n$. Assume that D does not satisfy a polynomial identity and that $\text{char}\, K = p > 0$. Let $R = \text{M}_2(D)$ and let G be the group of units generated by $\begin{pmatrix} 1 & 1 \\ 0 & 1 \end{pmatrix}$ and $\begin{pmatrix} 1 & x_i \\ 0 & 1 \end{pmatrix}$ for $i = 1, 2, \ldots, n$. Show that G is a finite p-group and that the fixed ring R^G equals $\begin{pmatrix} Z & D \\ 0 & Z \end{pmatrix}$ where $Z = \mathbf{Z}(D)$. Conclude that R is Artinian, Noetherian and Goldie, but that R^G is not.

4. Construct an appropriate division ring D for the preceding problem starting with the group algebra of a poly-Z group.

5. Let $R = \text{M}_n(K)$ where K is a field of characteristic not 2 and let $G = \{1, g\}$ act on R with $g = \text{diag}(1, -1, -1, \ldots, -1)$. Prove that R_{R^G} requires at least n generators even though $|G| = 2$. To this end, note that $R^G = \text{M}_1(K) \oplus \text{M}_{n-1}(K)$ and that the first column of R is an R^G-submodule annihilated by $\text{M}_{n-1}(K)$.

6. Let G act on R and let I be a G-stable ideal. If R has no $|G|$-torsion and $I^G = 0$, prove that I is nil of bounded degree by applying Theorem 26.4 to a generic model. Then use the action of G on the free ring $R\langle \zeta_1, \zeta_2, \ldots \rangle$ to conclude that I is nilpotent.

7. If E is an essential right ideal of a ring R and $r \in R$, show that $r^{-1}E = \{s \in R \mid rs \in E\}$ is essential in R. Use this to show that (i) and (ii) of Lemma 26.8 imply that T is a right divisor set. Then observe that RT^{-1} has no essential right ideals and complete the proof of that lemma.

27. Rings With No Nilpotent Elements

We return to the question of the existence of fixed points. Recall that the beginning of Section 24 contains an example where fixed points do not exist. In that example, R is a prime ring but not a domain and hence has nontrivial nilpotent elements. Furthermore G is inner on R and R has $|G|$-torsion. We have already seen in Theorem 26.3(i) that the latter two conditions are necessary ingredients for this example. Now we show that the presence of nilpotent elements also matters. Note that a ring with no nilpotent elements, that is a *reduced ring*, is necessarily semiprime.

The proof of this result requires the use of the symmetric Martindale ring of quotients for semiprime rings (see Section 18 for details). In particular, we need the following observations.

Lemma 27.1. *Let R be a semiprime ring.*

    i. $Q_s(R)$ *is semiprime.*

    ii. *If R has no nilpotent elements, then neither does $Q_s(R)$.*

    iii. *If G acts on R, then*

$$G_{\mathrm{inn}} = \{\, x \in G \mid {}^x \text{ is inner on } Q_s(R) \,\}$$

is a normal subgroup of G.

    iv. *If R has no nilpotent elements and B is an annihilator ideal of R, then R/B has no nilpotent elements.*

Proof. (i) This is clear since every nonzero ideal of $Q_s(R)$ meets R nontrivially.

(ii) Let $q \in Q_s(R)$ with $q^2 = 0$ and choose essential ideals A, B of R with $qA, Bq \subseteq R$. Since $qABq$ is a nilpotent subset of R, we have $qABq = 0$. But then qAB is also nilpotent so $qAB = 0$. Since AB is essential in R, it follows that $q = 0$.

(iii) This is the obvious analog of Lemma 12.3(ii).

(iv) Say $B = \ell_R(A)$ with $A \triangleleft R$ and let $x \in R$ with $x^2 \in B$. Then $x^2 A = 0$ so xAx is nilpotent and hence $xAx = 0$. But then xA is nilpotent so $xA = 0$ and $x \in \ell_R(A) = B$ as required. ∎

The next result is a key ingredient.

Theorem 27.2. [51] [87] *Let R be a ring with no nilpotent elements and let G be a finite group acting faithfully on R. If $pR = 0$ for some prime p, then the commutator subgroup $[G_{\mathrm{inn}}, G_{\mathrm{inn}}]$ has no elements of order p.*

Proof. We may assume that $G = G_{\mathrm{inn}}$. Let H be the set of units of $S = \mathrm{Q}_s(R)$ which act by conjugation on R like some element of G. Then H is clearly a subgroup of the unit group of S and, since G acts faithfully on R, there is a well-defined homomorphism $\theta \colon H \to G$ given by $\theta(u) = g$ if and only if $u^{-1}ru = r^g$ for all $r \in R$. Furthermore, θ is onto since $G = G_{\mathrm{inn}}$.

If $Z = \mathrm{Ker}(\theta)$, then Z centralizes R and hence S. Thus Z is central in H and H is center-by-finite. By Theorem 9.8(ii) (or see [**161**, Lemma 4.1.4]) the commutator subgroup $[H, H]$ of H is finite. Finally let $u \in [H, H]$ with $u^p = 1$. Since $pR = 0$, it follows that $pS = 0$ and hence $(u - 1)^p = 0$. But S has no nilpotent elements by Lemma 27.1(ii) so $u = 1$. We conclude that $[H, H]$ is a finite p'-group and, since $\theta([H, H]) = [G, G]$, the same is true of $[G, G]$. ∎

The proof of the existence of fixed points proceeds via a series of reductions. The next lemma handles the final step.

Lemma 27.3. *Let R be a ring with no nilpotent elements and with $pR = 0$ for some prime p. Let G be a finite nonabelian simple group acting on R and let $0 \neq L \lhd R$ be G-stable. Suppose that every nonzero ideal $A \subseteq L$ of R contains a nonzero G-stable ideal. Then $L^G \neq 0$.*

Proof. Assume by way of contradiction that $L^G = 0$. Form the skew group ring RG and let $I = R\dot{G}L$ so that $0 \neq I \lhd RG$, by Lemma 26.2(iii), and $I \subseteq LG$. If $I^2 \neq 0$ then, by that same lemma, $\mathrm{tr}_G(L) \neq 0$ and hence $L^G \neq 0$, a contradiction. Thus $I^2 = 0$ and then clearly $I \cap R = 0$.

We apply Lemma 18.5 to I and use its notation. In particular, there is a subset $1 \in \Lambda \subseteq G$, nonzero ideals $A_x \subseteq L$ for each $x \in \Lambda$

and additive bijections $f_x: A_1 \to A_x$ with $f_1 = 1$ satisfying

$$(ras)f_x = r(af_x)s^{x^{-1}}$$

for all $r, s \in R$ and $a \in A_1$. Note that $\Lambda \neq \{1\}$ since $I \cap R = 0$.

By assumption, A_1 contains a nonzero G-stable ideal B and for each $x \in \Lambda$ we set $B_x = Bf_x$. Then it follows from Lemma 18.7(i) that each B_x is an ideal of R and that $\ell_R(B_x) = \ell_R(B)$. Since B is G-stable, so is $D = \ell_R(B)$. Moreover, since R is semiprime, $B_x \cap D = 0$ and $B_x \oplus D$ is essential in R.

Let $'$ denote the natural map $R \to R/D$ onto the ring $R' = R/D$. Then each B_x' is clearly essential in R'. Furthermore, each f_x gives rise to an additive bijection $f_x': B' \to B_x'$ satisfying $(r'b's')f_x' = r'(b'f_x')(s')^{x^{-1}}$ for all $r', s' \in R'$ and $b' \in B'$. Since R' is semiprime, it follows from Lemma 18.7(ii) that each $x \in \Lambda$ acts as an inner automorphism on $Q_s(R')$. In particular, in its action on R', we have $G_{\mathrm{inn}} \neq \langle 1 \rangle$ since $\Lambda \neq \{1\}$. But G is simple and $G_{\mathrm{inn}} \triangleleft G$ so we have in fact $G_{\mathrm{inn}} = G$. Furthermore, since B embeds in R' and $B \subseteq L$, we see that G acts nontrivially on R'. But again, since G is simple, this implies that G acts faithfully on R'.

Finally, R' has no nilpotent elements by Lemma 27.1(iv) and $pR' = 0$. Thus the preceding theorem implies that $[G, G]$ has no elements of order p. But G is nonabelian simple so $G = [G, G]$ and we conclude that G is a p'-group. Since R is semiprime, Theorem 4.4 now implies that RG is also semiprime and this certainly contradicts $I^2 = 0$. ∎

With this we can prove the first main result of this section.

Theorem 27.4. [87] *Let G be a finite group acting on R a ring with no nilpotent elements. If L is a nonzero G-stable (right or left) ideal of R, then $L^G \neq 0$.*

Proof. We proceed by induction on $|G|$, the case $|G| = 1$ being trivial. We require three reductions.

First we can assume that $pR = 0$ for some prime p dividing $|G|$ and that $L \triangleleft R$. Indeed if L has no $|G|$-torsion, then $L^G \neq 0$ by

Theorem 26.3(i). Thus we may assume that for some prime $p \mid |G|$ we have $L' = \text{ann}_L(p) \neq 0$. Then G acts on L' and hence on the ring $L' \oplus \text{GF}(p) = R'$. Clearly R' has no nilpotent elements, $pR' = 0$ and $L' \vartriangleleft R'$. Since $L' \subseteq L$, it suffices to show that $(L')^G \neq 0$.

Next we can assume that G is a nonabelian simple group. Indeed suppose N is a proper normal subgroup of G. Then, by induction, L^N is a nonzero ideal of R^N. But G/N acts on R^N so, by induction again, $L^G = (L^N)^{G/N} \neq 0$. Thus G is simple. Now if G is abelian, then G must be cyclic of order p and Lemma 24.3(ii) implies that $L^G \neq 0$. Thus we may take G to be nonabelian.

Finally we can assume that every nonzero ideal of R contained in L contains a nonzero G-stable ideal. Indeed let $0 \neq A \vartriangleleft R$ with $A \subseteq L$ and suppose that A contains no nonzero G-stable ideal. By Lemma 18.3, A contains a trivial intersection ideal $\tilde{A} \neq 0$ and, for convenience, we may take $A = \tilde{A}$. By definition this means that for all $x \in G$ either $A^x = A$ or $A^x \cap A = 0$. Let $H = G_A$ be the stabilizer of A in G and let T be a right transversal for H in G. Then $H \neq G$, since A cannot be G-stable, so $A^H \neq 0$ by induction. Furthermore since R is semiprime, Lemma 8.1 implies that the sum $\sum_{t \in T} A^t$ is direct. But then any nonzero $a \in A^H$ gives rise to $\sum_{t \in T} a^t$ a nonzero element of L clearly fixed by G. Thus we may suppose that no such A exists.

In other words, the hypothesis of Lemma 27.3 is now satisfied and it follows from that result that $L^G \neq 0$. ∎

This completes our work on fixed points; however we continue to study rings with no nilpotent elements. Now we take a closer look at the relationship between the nontriviality of the trace map tr_G and the semiprimeness of the skew group ring RG. The main result in this direction is contained in [**128**] which considers group actions on domains. Here we combine the techniques of that paper with those of Section 18 to obtain a slightly more general result.

Let $R[G]$ be an ordinary group ring. Then the map $\rho: R[G] \to R$ given by $\rho(\sum_{g \in G} r_g g) = \sum_{g \in G} r_g$ is a ring homomorphism called the *augmentation map*. In the case of crossed products, ρ is at best an R-module homomorphism, but it is still useful. For example it is a key ingredient in the proof of Theorem 24.2 and it is crucial for Theorem 27.7 below.

Let us recall how the centralizer structure in RG comes about (see Proposition 12.4). Let G be a group, finite or infinite, which acts on the semiprime ring R and suppose that $G = G_{\text{inn}}$. By definition, this means that for each $g \in G$ we can choose a unit $u_g \in S = Q_s(R)$ such that $r^g = u_g^{-1} r u_g$ for all $r \in R$. It is clear that u_g is unique up to a factor of a unit in C, the extended centroid of R. Setting $\bar{g} = u_g^{-1} g \in SG$, it then follows that $E = \mathbf{C}_{SG}(S)$ is a free C-module with the elements $\bar{g}$ as a basis and in fact that $E = C^t[G]$. Note that the C-linear span of the units u_g is a well-defined C-subalgebra of S. It is denoted by $\mathcal{B}(G)$ and called the *algebra of the group*. With this notation we have

Lemma 27.5. *Let G act on the semiprime ring R with $G = G_{\text{inn}}$ and let $S = Q_s(R)$. Then the augmentation map*

$$\rho: \sum_{x \in G} s_x x \mapsto \sum_{x \in G} s_x$$

defined on $E = \mathbf{C}_{SG}(S) \subseteq SG$ is a ring antihomomorphism from E onto $\mathcal{B}(G)$. Furthermore it is a C-homomorphism and if $g \in G$ then $\rho(\bar{g})$ is a unit of S with $r^g = \rho(\bar{g}) r \rho(\bar{g})^{-1}$ for all $r \in R$.

Proof. The map ρ is certainly a C-module homomorphism. For multiplication, we first note that $\sum_{x \in G} s_x x \in E$ if and only if each $s_x x \in E$. Thus we need only consider $\rho\big((sx)(ty)\big)$ with $s, t \in S$, $x, y \in G$ and $sx, ty \in E$. But then sx commutes with t so $(sx)(ty) = t(sx)y = (ts)(xy)$ and we have

$$\rho\big((sx)(ty)\big) = \rho\big((ts)(xy)\big)$$
$$= ts = \rho(ty)\,\rho(sx)$$

as required. Finally since $\bar{g} = u_g^{-1} g \in E$, we have $\rho(\bar{g}) = u_g^{-1}$ and hence $r^g = \rho(\bar{g}) r \rho(\bar{g})^{-1}$ for all $r \in R$. Moreover since E is the C-linear span of the elements $\bar{g}$, we conclude that $\rho(E) = \mathcal{B}(G)$ and the result follows. ∎

If I is an ideal in a semiprime ring R, then $\ell_R(I) = \mathrm{r}_R(I)$. Thus the annihilator of I is unambiguously defined. The following is a special case of Theorem 27.7.

Lemma 27.6. *Let G act on R, a ring with no nilpotent elements, and let $S = Q_s(R)$. Assume that G is an elementary abelian p-group for some prime p and that $G = G_{\mathrm{inn}}$. If $\mathbf{C}_{SG}(S)$ is not semiprime, then there exists a subgroup H of G such that $\mathrm{tr}_H(R)$ annihilates a nonzero ideal of R.*

Proof. We know that $\mathbf{C}_{SG}(S) = C^t[G]$ where C is the extended centroid of R and that, by the previous lemma, there is a ring antihomomorphism $\rho: C^t[G] \to \mathcal{B}$ where $\mathcal{B}$ is the algebra of the group. Furthermore, ρ is a C-homomorphism, C is a commutative von Neumann regular ring by Lemma 18.6 and $\mathcal{B}$ has no nilpotent elements by Lemma 27.1(ii). Since $C^t[G]$ is not semiprime by hypothesis, Lemma 18.1(iii) implies that there exists a maximal ideal M of C such that the central localization $C^t[G]_M$ is not semiprime. Here we are localizing at the multiplicatively closed set $C \setminus M$.

Since $C^t[G]$ is free over C, we have $C^t[G]_M = (C_M)^t[G]$. Furthermore, by Lemma 18.1(i), $C_M = K$ is a field. Now $K^t[G] = C^t[G]_M$ is not semiprime and G is an elementary abelian p-group. Thus it follows from Theorem 4.4 that $\mathrm{char}\, K = p$ and then from Lemma 16.3(ii) that $K^t[G]$ is commutative. Note that the map ρ extends to a K-algebra epimorphism $\rho_M: K^t[G] = C^t[G]_M \to \mathcal{B}_M$. Furthermore $\mathcal{B}_M$ has no nilpotent elements. Indeed suppose $(bt^{-1})^2 = 0$ (in $\mathcal{B}_M$) with $b \in \mathcal{B}$ and $t \in C \setminus M$. Then there exists $t_1 \in C \setminus M$ with $b^2 t_1 = 0$ (in $\mathcal{B}$). But $bt_1 \in \mathcal{B}$ has square zero, so $bt_1 = 0$ and we conclude that $bt^{-1} = 0$ (in $\mathcal{B}_M$).

Let W be a subgroup of G maximal with $K^t[W]$ being a field and say $K^t[W] = F$. Since $K^t[G]$ is not semiprime, we can choose $g \in G \setminus W$ and we set $H = \langle W, g \rangle = W \times \langle g \rangle$. Again since G is elementary abelian, we have $\bar{g}^p = a_1 \in K \subseteq F$ and then clearly $K^t[H] \cong F[\zeta]/(\zeta^p - a_1)$. But $K^t[H]$ is not a field so the polynomial $\zeta^p - a_1 \in F[\zeta]$ is reducible and hence $\zeta^p - a_1 = (\zeta - a)^p$ for some $a \in F$. In other words, $(\bar{g} - a)^p = 0$. Since $\mathcal{B}_M$ has no nilpotent elements, we conclude that $\rho_M(\bar{g} - a) = 0$ so that $\rho_M(\bar{g}) = \rho_M(a)$ in $\mathcal{B}_M$. It follows easily from this that there exists a nonzero element $e \in C$ and an element $\alpha \in C^t[W]$ with $\rho(\bar{g})e = \rho(\alpha)$ in $\mathcal{B}$. Furthermore, since C is von Neumann regular, we can assume for convenience that e is an idempotent.

We can now relate this to the trace map. Suppose $r \in R^W$. Then r is centralized by $\rho(\alpha)$ and hence by $\rho(\bar{g})e$. Since $r^{g^i} = \rho(\bar{g})^i r \rho(\bar{g})^{-i}$ and e is a central idempotent, we see that $er^{g^i} = er$ and hence $e \, \mathrm{tr}_{\langle g \rangle}(r) = p \cdot er = 0$. Note that $\mathrm{tr}_H = \mathrm{tr}_{\langle g \rangle} \circ \mathrm{tr}_W$ and that $\mathrm{tr}_W(R) \subseteq R^W$. We conclude therefore that $e \, \mathrm{tr}_H(R) = 0$. Finally there exists an ideal A of R with $0 \neq Ae \subseteq R$. Then $(Ae) \, \mathrm{tr}_H(R) = 0$ and, since $e \in C$ centralizes R, Ae is a nonzero ideal of R which annihilates $\mathrm{tr}_H(R)$. ∎

Now we come to the second main result of this section.

Theorem 27.7. *Let G be a finite group acting on a ring R with no nilpotent elements. Then the skew group ring RG is semiprime if and only if, for all elementary abelian p-subgroups H of G, $\mathrm{tr}_H(R)$ annihilates no nonzero ideal of R.*

Proof. Suppose first that RG is semiprime and let H be any subgroup of G. Then RH is semiprime by Theorem 18.9. Furthermore if J is any nonzero ideal of R, then $R \dot{H} J$ is a nonzero left ideal of RH and $0 \neq (R\dot{H}J)^2 = R\dot{H} \, \mathrm{tr}_H(J)J$. Thus $0 \neq \mathrm{tr}_H(J)J \subseteq \mathrm{tr}_H(R)J$ and $\mathrm{tr}_H(R)$ annihilates no nonzero ideal of R.

Conversely suppose that, for all elementary abelian p-subgroups H of G, $\mathrm{tr}_H(R)$ annihilates no nonzero ideal of R. The goal is to show that RG is semiprime. By Theorem 18.10 we need only show that RW is semiprime for every elementary abelian p-subgroup W of G. In other words, it suffices to assume that G itself is an elementary abelian p-group. Furthermore, by induction we can assume that RG_0 is semiprime for all properly smaller subgroups G_0.

Suppose by way of contradiction that I is a nonzero ideal of RG with $I^2 = 0$. We apply Lemma 18.5 to I and use its notation. In particular, there is a subset $1 \in \Lambda \subseteq G$, nonzero ideals $A_x \subseteq L$ for each $x \in \Lambda$ and additive bijections $f_x \colon A \to A_x$ satisfying

$$(rat)f_x = r(af_x)t^{x^{-1}}$$

for all $r, t \in R$ and $a \in A$. Here $A = A_1$ and $f_1 = 1$. In addition

$$I \supseteq \left\{ \sum_{x \in \Lambda} (af_x)x \mid a \in A \right\}.$$

Note that $I \cap R\langle \Lambda \rangle \neq 0$ so we must have $\langle \Lambda \rangle = G$ since the skew group rings of all proper subgroups are semiprime.

Let $B = \ell_R(A)$. Then it follows from Lemma 18.7(i) that $B = \ell_R(A_x)$ for all $x \in \Lambda$ and that B is stabilized by $\langle \Lambda \rangle = G$. Moreover, since R is semiprime, $A_x \cap B = 0$ and $A_x \oplus B$ is essential in R.

Let $'$ denote the natural map $R \to R/B$ onto the ring $R' = R/B$. Then each A'_x is clearly essential in R'. Furthermore, each f_x gives rise to an additive bijection $f'_x \colon A' \to A'_x$ satisfying $(r'a't')f'_x = r'(a'f'_x)(t')^{x^{-1}}$ for all $r', t' \in R'$ and $a' \in A'$. Since R' is semiprime, it follows from Lemma 18.7(ii) that each $x \in \Lambda$ acts as an inner automorphism on $Q_s(R')$. Hence since G is generated by Λ, we have $G = G_{\mathrm{inn}}$ in its action on R'.

Now the map $'$ extends to an epimorphism $RG \to R'G$ and I', the image of I, contains $\left\{ \sum_{x \in \Lambda} (a'f'_x)x \mid a \in A \right\}$. By Lemma 18.7(ii) there exists a unit $q_x \in S = Q_s(R')$ with $a'f'_x = a'q_x$ for all $a' \in A'$ and such that the action of x^{-1} on R' is the inner automorphism induced by q_x. Thus in SG we see that $I' \supseteq A'\gamma$ where $\gamma = \sum_{x \in \Lambda} q_x x \in C_{SG}(S) = C^t[G]$. In other words, in the notation of Lemma 18.8(iv), γ is a nonzero element of the ideal $\widetilde{I}'$ of $C^t[G]$. But $(I')^2 = 0$ so $(\widetilde{I}')^2 = 0$ and we conclude that $C^t[G]$ is not semiprime.

Lemma 27.6 now implies that there exists a subgroup H of G and a nonzero ideal J' of R' with $J' \operatorname{tr}_H(R') = 0$. Thus if J is the complete inverse image of J' in R, then $J \supset B$ and $J \operatorname{tr}_H(R) \subseteq B$. It follows that $AJ \operatorname{tr}_H(R) \subseteq AB = 0$. Since $J \supset B = \mathrm{r}_R(A)$, we see that AJ is a nonzero ideal of R. But $\operatorname{tr}_H(R)$ annihilates AJ, so this contradicts the hypothesis and the result follows. $\blacksquare$

The formulation of the original result on domains is much more satisfying.

Corollary 27.8. [128] *Let G be a finite group acting on a domain R. The following are equivalent.*

  i. $\operatorname{tr}_G(R) \neq 0$.
  ii. $\operatorname{tr}_G(I) \neq 0$ for all nonzero right ideals I of R.
  iii. The skew group ring RG is semiprime.

Proof. (iii) $\Rightarrow$ (ii) is immediate from Lemma 26.2(iii) and (ii) $\Rightarrow$ (i) is obvious. We prove that (i) $\Rightarrow$ (iii). Thus suppose that $\mathrm{tr}_G(R) \neq 0$ and let H be any subgroup of G. If T is a right transversal for H in G and if $r \in R$, then $\mathrm{tr}_G(r) = \sum_{t \in T} \mathrm{tr}_H(r)^t$. Thus it follows that $\mathrm{tr}_H(R) \neq 0$. But R is a domain so $\mathrm{tr}_H(R)$ annihilates no nonzero ideal of R and Theorem 27.7 implies that RG is semiprime. ∎

These results cannot be extended beyond rings with no nilpotent elements. For example, let K be a field of characteristic $p > 0$ and let $R = \mathrm{M}_p(K)$. Then R contains an element u of order p which is a full $p \times p$ Jordan block with 1's on the main diagonal and the super diagonal and with 0's elsewhere. The group $G = \langle g \rangle$ of order p then acts on R with g acting as conjugation by u. We first observe that $\mathrm{tr}_G(R) \neq 0$ and indeed that $\mathrm{tr}_G(e_{1,1}) \neq 0$. To this end, note that $u^{-i} e_{1,1} = e_{1,1}$ for all i and hence that $(e_{1,1})^{u^i} = e_{1,1} u^i$. Thus

$$
\mathrm{tr}_G(e_{1,1}) = \sum_{i=0}^{p-1} (e_{1,1})^{u^i} = e_{1,1} \sum_{i=0}^{p-1} u^i
$$
$$
= e_{1,1}(1 - u)^{p-1} = e_{1,1} e_{1,p} = e_{1,p}.
$$

Now R is simple so $\mathrm{tr}_G(R)$ annihilates no nonzero ideal of R. Furthermore, $R = \mathrm{Q}_s(R)$, $C = K$ and $RG = R \otimes_K K^t[G]$ with $\bar{g} = u^{-1}g$. But $\bar{g}^p = 1$ so $K^t[G] = K[G]$ is not semiprime and therefore neither is RG.

EXERCISES

1. Let R be semiprime and let A be a nonzero ideal of R. If $B = \ell_R(A)$ show that $(A \oplus B)/B$ is essential in R/B.

2. Let $f: {}_R A \to {}_R R$ be a left R-module map with $A \lhd R$ and let $I \lhd R$. If $IA = 0$, prove that $Af \subseteq \mathrm{r}_R(I)$. Now suppose R is semiprime with finitely many minimal primes $P_1, P_2, \ldots, P_n$ and set $A_i = \ell_R(P_i)$. Prove that $A_i \neq 0$, $P_i = \ell_R(A_i)$ and that the sum $\sum_{i=1}^{n} A_i$ is direct and yields an essential ideal of R.

3. Suppose R is a semiprime ring with finitely many minimal primes $P_1, P_2, \ldots, P_n$. Prove that

$$
\mathrm{Q}_s(R) = \oplus \sum_{i=1}^{n} \mathrm{Q}_s(R/P_i).
$$

To this end, use the preceding exercise to show that the above right hand side satisfies the semiprime analog of the defining properties given in Proposition 10.4.

In the remaining problems, R is a ring with no nilpotent elements and P is a prime ideal of R.

4. If $a, b \in R$ with $ab = 0$, show that $ba = 0$ and then that $aRb = 0$. Deduce that R is prime if and only if it is a domain.

5. Let $a, b \in R \setminus P$ and suppose $abx = 0$ with $x \in R$. Show that $cx = 0$ for some $c \in R \setminus P$. Conclude that the multiplicatively closed subset A of R generated by $R \setminus P$ does not contain 0.

6. Let Q be an ideal of R maximal with respect to $Q \cap A = \emptyset$. Show that Q is prime and that $Q \subseteq P$. Finally suppose P is a minimal prime of R so that $Q = P$. Deduce that $A = R \setminus P$ and that R/P is a domain. This is a result of [7].

28. Prime Ideals and Fixed Rings

Probably the most successful application of skew group rings to Galois theory concerns the behavior of prime ideals in the ring extension $R \supseteq R^G$. To start with, let us recall some basic properties of prime ideals in crossed products of finite groups.

Let $R*G$ be given with G finite and with R a G-prime ring. Then Theorem 16.2 asserts that

1. A prime ideal P of $R*G$ is minimal if and only if $P \cap R = 0$.

2. There are finitely many such minimal primes $P_1, P_2, \ldots, P_n$ with $n \leq |G|$.

3. If Q is a minimal prime of R, then $\{ Q^x \mid x \in G \}$ is the set of all minimal primes of R and $\bigcap_{x \in G} Q^x = 0$.

Furthermore if R has no $|G|$-torsion, then (3) and Theorem 4.4 imply that $R*G$ is semiprime and thus that $\bigcap_{i=1}^{n} P_i = 0$.

These results then give rise to appropriate Incomparability, Going Up and Going Down properties as described in Theorem 16.6. In addition, Proposition 16.7 asserts that P_i is a primitive ideal of $R*G$ if and only if Q is a primitive ideal of R. As a consequence of all of this we have

Corollary 28.1. *Let $R*G$ be given with $|G|^{-1} \in R$ and let A be a G-prime ideal of R. Then $A*G = P_1 \cap P_2 \cap \cdots \cap P_n$, an intersection of $n \leq |G|$ minimal covering primes. Furthermore if P is a prime ideal of $R*G$, then $P = P_i$ for some i if and only if $P \cap R = A$.*

Proof. Let $\tilde{\ } : R*G \to (R*G)/(A*G) = (R/A)*G$ denote the natural epimorphism. Then $\tilde{R} = R/A$ is a G-prime ring with no $|G|$-torsion since $|G|^{-1} \in R$. By the above observations, we see that $\tilde{R}*G$ has $n \leq |G|$ minimal primes which intersect to zero and hence, if $P_1, P_2, \ldots, P_n$ are their complete inverse images in $R*G$, then each P_i is a prime ideal and $\bigcap_{i=1}^{n} P_i = A*G$. Finally if P is a prime of $R*G$ then, by the above again, $P = P_i$ for some i if and only if $P \supseteq A*G$ and $\tilde{P} \cap \tilde{R} = 0$. Since the latter conditions are easily seen to be equivalent to $P \cap R = A$, the result follows. ∎

Now suppose that G is a finite group acting on R with $|G|^{-1} \in R$. Then we recall from Lemma 26.10(ii) that $e = \frac{1}{|G|} \sum_{x \in G} x$ is an idempotent in the skew group ring RG and that $eRGe = eR^G$ is ring isomorphic to R^G. In this situation, Lemma 17.8 applies to relate the primes of RG to those of R^G. Specifically if I is an ideal of RG, we define $I^\varphi = eIe = I \cap (eRGe)$ so that I^φ is an ideal of $eRGe$. Then Lemma 17.8 asserts that φ yields an inclusion preserving (in both directions) bijection between the primes of RG not containing e and all the primes of $eRGe$. Note that $I^\varphi = eJ$ for a uniquely determined ideal J of R^G.

By combining all of the above observations, it was shown in [113] that R and R^G have equal prime lengths. However that paper did not specifically look at the primes of R^G. It remained for [131] to complete the formalities. The following result uses the function φ as defined above.

Lemma 28.2. *Let G act on the ring R with $|G|^{-1} \in R$ and let A be a G-prime ideal of R. Then $A^G = Q_1 \cap Q_2 \cap \cdots \cap Q_k$, a finite*

intersection of $k \le |G|$ minimal covering primes. Furthermore, if Q is a prime ideal of R^G, then $Q = Q_i$ for some i if and only if $eQ = P^\varphi$ for some prime P of RG with $P \cap R = A$.

Proof. By Corollary 28.1 we have $AG = P_1 \cap P_2 \cap \cdots \cap P_n$ with $n \le |G|$. Thus if $P_i^\varphi = eQ_i$, then Lemma 26.10(i) yields

$$eA^G = (AG)^\varphi = P_1^\varphi \cap P_2^\varphi \cap \cdots \cap P_n^\varphi$$
$$= eQ_1 \cap eQ_2 \cap \cdots \cap eQ_n$$

and hence $A^G = Q_1 \cap Q_2 \cap \cdots \cap Q_n$. We can of course delete those Q_i's equal to R^G or equivalently those P_i's containing e. When this is done and the primes are suitably relabeled, we then have, by Lemma 17.8, $A^G = Q_1 \cap Q_2 \cap \cdots \cap Q_k$ an intersection of $k \le n \le |G|$ primes. Furthermore, since the P_i's are incomparable, so are the Q_i's. Hence $Q_1, Q_2, \ldots, Q_k$ are precisely the minimal covering primes of the ideal A^G of R^G.

Finally if Q is a prime of R^G, write $eQ = P^\varphi$ for some prime P of RG. Then $Q = Q_i$ if and only if $P = P_i$ and hence if and only if $P \cap R = A$. Note that this P necessarily satisfies $e \notin P$. ∎

With this result in hand, we can relate the primes of R^G to those of R. To start with, if T is a prime ideal of R and Q is a prime of R^G, we say that T *lies over* Q or equivalently Q *lies under* T if Q is a minimal covering prime of $T \cap R^G$. As usual we will describe basic relations between the primes diagramatically. Thus for example the middle diagram in (iv) below is read as follow. Suppose $Q_1 \supseteq Q_2$ are primes of R^G and T_2 is a prime of R lying over Q_2. Then there exists a prime T_1 of R such that T_1 lies over Q_1 and $T_1 \supseteq T_2$.

The key result on primes is

Theorem 28.3. [131] Let G act on R and suppose that $|G|^{-1} \in R$. The following basic relations hold between the prime ideals of R and of R^G.

    i. **Cutting Down.** If T is a prime ideal of R, then there are $k \le |G|$ primes $Q_1, Q_2, \ldots, Q_k$ of R^G minimal over $T \cap R^G$ and we have $T \cap R^G = Q_1 \cap Q_2 \cap \cdots \cap Q_k$.

ii. **Lying Over.** If Q is a prime ideal of R^G, then there exists a prime T of R, unique up to G-conjugation, such that T lies over Q. Furthermore, the distinct T^x with $x \in G$ are incomparable.

iii. **Incomparability.** Given the lying over diagram

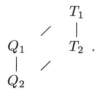

Then $T_1 = T_2$ if and only if $Q_1 = Q_2$.

iv. **Going Up and Going Down.** We have at least

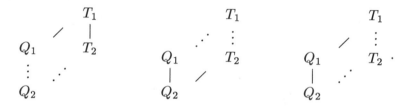

Proof. (i) Let T be a prime of R and observe that for any $x \in G$ we have $T^x \cap R^G = (T \cap R^G)^x = T \cap R^G$. Hence if $A = \bigcap_{x \in G} T^x$, then A is a G-prime ideal of R and $A^G = A \cap R^G = T \cap R^G$. Now apply Lemma 28.2.

(ii) Let Q be a prime ideal of R^G and define the prime P of RG and $A \lhd R$ by $P^\varphi = eQ$ and $A = P \cap R$. Then A is a G-prime ideal of R so, by our earlier observations, $A = \bigcap_{x \in G} T^x$ for some prime T. Since $T \cap R^G = A \cap R^G$, we see from Lemma 28.2 again that T lies over Q. Furthermore that lemma shows that A is the unique G-prime ideal of R with Q minimal over A^G. Since A uniquely determines $\{ T^x \mid x \in G \}$, its set of minimal covering primes, we conclude that T is unique up to G-conjugation. In addition, the distinct T^x are clearly incomparable.

(iii) If $T_1 = T_2$, then since Q_1 is minimal over $T_1 \cap R^G$ we have $Q_1 = Q_2$. Conversely if $Q_1 = Q_2$, then the uniqueness of T in (ii) above, along with the incomparability of the various T^x, implies that $T_1 = T_2$.

(iv) The first relation is obvious. We have $Q_1 \supseteq T_1 \cap R^G \supseteq T_2 \cap R^G$ so Q_1 contains a minimal covering prime Q_2 of $T_2 \cap R^G$. For the second and third, define P_i prime in RG and $A_i \lhd R$ by $P_i^\varphi = eQ_i$ and $A_i = P_i \cap R$. Then $Q_1 \supseteq Q_2$ implies, by Lemma 17.8, that $P_1 \supseteq P_2$ and hence that $A_1 \supseteq A_2$. It is now a simple matter to compare the corresponding T_i's. ∎

We consider two examples of interest. First let $R = \mathrm{M}_n(K)$ where K is a field of characteristic not 2 and, for each $1 \le i \le n-1$, let d_i be the diagonal matrix $d_i = \mathrm{diag}(1,1,\ldots,-1,1,\ldots,1)$ with -1 in the ith entry and 1's elsewhere. Let G be the group of automorphisms of R generated by $g_1, g_2, \ldots, g_{n-1}$ where each g_i acts like conjugation by d_i. Then $|G| = 2^{n-1}$ and G is faithful and inner on R. The latter implies that $RG = R \otimes_K E$ where $E = K^t[G]$. But E is semiprime and is generated by the commuting elements $d_i g_i$ of order 2, so $E \cong \oplus \sum_1^{|G|} K$, a direct sum of $|G|$ copies of K. Thus $RG \cong \oplus \sum_1^{|G|} R$ and therefore RG has $|G| = 2^{n-1}$ minimal primes. On the other hand, R^G is the ring of diagonal matrices, so R^G has precisely n minimal primes. Thus for $n \ge 3$ we see that there are primes of RG lost under the φ map; that is, there are primes containing e. This extends an example of [**131**].

Second, let A be a simple domain over the field K with char $K \ne 2$ and assume that A is not a division ring. Then we can choose $I \ne A$ to be a nonzero left ideal. For example we could take A to be the *Weyl algebra* $A_1(K) = K[x, y \mid xy - yx = 1]$ with char $K = 0$ and $I = Ax$. Now define

$$R = \begin{pmatrix} K+I & A \\ I & A \end{pmatrix} \subseteq \mathrm{M}_2(A).$$

It follows easily that R is a prime ring, so $T_2 = 0$ is a prime ideal of R. Also observe that $T_1 = \begin{pmatrix} I & A \\ I & A \end{pmatrix}$ is a maximal two-sided ideal of R with $R/T_1 = K$. Let G be the group of automorphisms of R generated by conjugation by $\mathrm{diag}(1, -1)$. Then $|G| = 2$ and $R^G = \mathrm{diag}(K+I, A)$. Thus R^G has two minimal primes, one of which is $Q_2 = \mathrm{diag}(K+I, 0)$. But $R^G/Q_2 \cong A$ so Q_2 is also maximal.

Thus we see that there exists no prime Q_1 of R^G which completes
the diagram

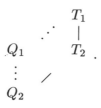

In other words, the missing Going Up result in Theorem 28.3 does
indeed fail. It does not fail in RG, but the prime we get may contain
e and hence not correspond to a prime of R^G. This is from [**138**].

Thus the missing Going Up result in Theorem 17.9 also fails.
Indeed, we need only observe, by Lemma 24.6, that R is a $\hat{G}$-graded
ring with identity component $R_1 = R^G$.

Now it is clear that Theorem 28.3 yields a one-to-one correspondence between the G-conjugacy classes of prime ideals of R and
certain finite subsets of primes of R^G. To be precise, if Q_1 and Q_2
are prime ideals of R^G, write $Q_1 \sim Q_2$ if and only if Q_1 and Q_2 lie
under the same prime T of R. This then defines an equivalence re-
lation whose classes are finite of size $\leq |G|$ and there is a one-to-one
correspondence between these classes and the G-conjugacy classes of
primes of R. It is natural to consider which ring theoretic properties are shared by equivalent primes. Obviously these include those
properties which are inherited by lying over and lying under primes.
Some examples are as follows. Recall that the *height* of a prime P
of R is the largest n such that $P_0 \subset P_1 \subset \cdots \subset P_n = P$ is a chain
of primes in R. If no such maximum n exists, then the height is infinite. The *depth* of P can be defined similarly by looking at primes
containing P. Equivalently, it is the prime length of R/P.

Proposition 28.4. [**113**] [**131**] *Let G act on R with $|G|^{-1} \in R$ and
let T be a prime ideal of R which lies over Q.*

    i. T is primitive if and only if Q is primitive.

    ii. T and Q have the same height.

    *iii. R and R^G have the same prime lengths and the same prim-
itive lengths.*

Proof. (i) This is a consequence of Proposition 16.7, Lemma 17.8 and the argument in the proof of Theorem 28.3(ii).

(ii)(iii) These are immediate from Going Up, Incomparability and (i) above. ∎

In the case of Goldie rings, we are also concerned with the relationship between the ranks of R/T and of the corresponding factor rings R^G/Q_i. For this we require the following special case of the *additivity principle* of [84]. If e and f are idempotents of a ring R, we write $e \sim f$ if and only if $eR_R \cong fR_R$. As is well known (see Exercise 1), this occurs if and only if there exist elements $u, v \in R$ with $e = uv$ and $f = vu$.

Lemma 28.5. *Let $S \subseteq R$ be rings with S semisimple Artinian and write $S = \oplus \sum_1^k S_i$, a direct sum of simple rings. If f_i is a primitive idempotent in S_i, then* rank $R = \sum_{i=1}^k (\text{rank } S_i)(\text{rank } f_i R_R)$.

Proof. We know that $S_i \cong \mathrm{M}_{n_i}(D_i)$, a full matrix ring over a division ring, and we let $\{ f_{i,1}, f_{i,2}, \ldots, f_{i,n_i} \}$ be a family of primitive orthogonal idempotents summing to the identity of S_i. Then $1 = \sum_{i,j} f_{i,j}$ is an orthogonal decomposition of $1 \in S \subseteq R$ and hence $R = \oplus \sum_{i,j} f_{i,j} R$. Observe that $f_{i,j} \sim f_i$ in S and hence in R, by the above comments, so $f_{i,j} R_R \cong f_i R_R$. Thus computing ranks yields rank $R = \sum_1^k n_i \cdot \text{rank } f_i R_R$ and the lemma is proved since $n_i = \text{rank } S_i$. ∎

As a consequence we have the additivity principle applied to fixed rings.

Theorem 28.6. [107] *Let G act on R, a ring with no $|G|$-torsion, and suppose that R is G-prime and Goldie. If T is a minimal prime of R with $H = G_T$, the stabilizer of T in G, and if $Q_1, Q_2, \ldots, Q_k$ are the minimal primes of R^G, then*

$$\text{rank}\,(R/T) = \sum_{i=1}^k z_i \cdot \text{rank}\,(R^G/Q_i)$$

for suitable integers z_i satisfying $1 \le z_i \le |H|$.

Proof. Since R is G-prime, it is semiprime and Theorem 26.9 applies. In particular, R^G is also semiprime Goldie and $Q(R^G) = Q(R)^G$. Furthermore, it is known (see Exercises 2–4) that the minimal primes of R correspond in a one-to-one manner to the primes of $Q(R)$ and indeed that their factor rings have equal Goldie rank. In view of these remarks, we may clearly replace R by $Q(R)$ and assume that both R and R^G are semisimple. Observe that $|G|$ is now invertible in R.

Write $R = R_1 \oplus R_2 \oplus \cdots \oplus R_j$, a direct sum of simple rings, and say $T = R_2 \oplus \cdots \oplus R_j$. Since R is G-prime, these factors are permuted transitively by G and hence it is clear that the projection map onto R_1 defines an isomorphism $R^G \cong R_1^H$. Thus since $R_1 \cong R/T$, we may replace R by R_1 and G by H and then assume that $T = 0$.

Finally write $R^G = S = S_1 \oplus S_2 \oplus \cdots \oplus S_k$ as in the previous lemma. Then by that lemma we have

$$\operatorname{rank}(R/T) = \operatorname{rank} R = \sum_{i=1}^{k} (\operatorname{rank} S_i)(\operatorname{rank} f_i R).$$

Since $S_i = R^G/Q_i$ and $z_i = \operatorname{rank} f_i R_R$ is a positive integer, it remains to bound these integers. To this end, we see from Lemma 26.1(ii) that $(f_i R)^G = f_i R^G$ is an R^G-module of rank 1. Hence Theorem 26.3(i) implies that $f_i R$ is an RG-module of rank 1 and Lemma 26.5 yields the result. ∎

Let us consider two more examples, this time with $|G|^{-1} \notin R$.

First, in the example at the beginning of Section 24 we have $R = \mathrm{M}_2(K\langle x, y\rangle)$ with K a field of characteristic $p > 2$ and with G a group of automorphisms of R of order $2p^2$. Furthermore, $R^G \cong K$. Thus R^G has one prime ideal, namely 0, while R has infinitely many. This certainly precludes a reasonable correspondence between the primes of R and of R^G.

Second, suppose B is a simple domain over a field K of characteristic $p > 2$ and let $R = \mathrm{M}_2(B)$ so that R is also a simple ring. Define G to be the group of units of R generated by

$$\begin{pmatrix} 1 & 0 \\ 0 & -1 \end{pmatrix} \qquad \begin{pmatrix} 1 & 1 \\ 0 & 1 \end{pmatrix} \qquad \begin{pmatrix} 1 & x \\ 0 & 1 \end{pmatrix}$$

where x is some noncentral element of B. Then $|G| = 2p^2$ again, G acts by conjugation on R and $R^G \cong \mathbf{C}_B(x)$. Suppose in addition that $\mathbf{C}_B(x)$ is the group ring $K[\langle x \rangle] = K[x, x^{-1}]$. Then we see that R has only one prime, namely 0, but R^G has infinitely many primes and again no correspondence can exist. Note that a candidate for B is

$$B = K[x, x^{-1}, y, y^{-1} \mid xy = \lambda yx]$$

where λ is an element of infinite multiplicative order in $K^\bullet$ (see Exercise 5).

Finally we briefly consider the relationship between $\mathrm{J}(R)$ and $\mathrm{J}(R^G)$. The first result in this direction is based on the well known fact that if f is an idempotent in a ring S, then $f\mathrm{J}(S)f = \mathrm{J}(fSf)$.

Theorem 28.7. [126] Let G act on R with $|G|^{-1} \in R$. Then $\mathrm{J}(R^G) = \mathrm{J}(R) \cap R^G = \mathrm{J}(R)^G$.

Proof. Form the skew group ring RG and let $e = |G|^{-1}\dot{G}$. Then by Lemma 26.10(i)(ii) and Theorem 4.2 we have

$$e\mathrm{J}(R^G) = e\big(\mathrm{J}(RG)\big)e = e\big(\mathrm{J}(R)G\big)e = e\mathrm{J}(R)^G.$$

Thus we see that $\mathrm{J}(R^G) = \mathrm{J}(R)^G$. ∎

The above conclusion fails if $|G|^{-1} \notin R$. In fact it is easy to construct examples (see Exercise 6) with R simple but R^G not semiprime or with R^G simple but R not semiprime. Even so, there is more that can be said about this situation. Indeed, we offer a generalization of the preceding theorem from [117] using the recent clever approach of [178].

Lemma 28.8. Let $I \triangleleft R$.
    i. Suppose L_1 and L_2 are right ideals of R with $I + L_1 = I + L_2 = L_1 + L_2 = R$. Then $I + (L_1 \cap L_2) = R$.
    ii. Let $M_1, M_2, \ldots, M_k$ be maximal right ideals of R. If $I \not\subseteq M_i$ for all i, then $I + \big(\bigcap_1^k M_i\big) = R$.

Proof. (i) First $I = RI = (L_1 + L_2)I$ so $I = (I \cap L_1) + (I \cap L_2)$. Next, since $I + L_1 = R$ we see that $L_1 + (I \cap L_2) = R$ and hence

$L_2 = (L_1 \cap L_2) + (I \cap L_2)$. Finally using $I + L_2 = R$ and the above we obtain $I + (L_1 \cap L_2) = R$.

(ii) This follows easily from (i) by induction on k. ∎

Proposition 28.9. [178] *Let G act on R and let I be a G-stable right or left ideal. If $\mathrm{tr}_G(I) \subseteq \mathrm{J}(R^G)$, then $|G| \cdot I \subseteq \mathrm{J}(R)$.*

Proof. Write $n = |G|$. We proceed in a series of three steps, assuming throughout that I is G-stable.

Step 1. *If $I \lhd R$ and M is a maximal right ideal of R, then there exists $i \in I$ with $\big(n - \mathrm{tr}_G(i)\big) \cdot I \subseteq M$.*

Proof. Since $I \lhd R$, we may assume that $I \not\subseteq M$. Then $I + M = R$ and hence, for all $g \in G$, $I + M^g = R$. Part (ii) of the previous lemma now shows that $I + \bar{M} = R$ where $\bar{M}$ is the G-stable right ideal $\bar{M} = \bigcap_{g \in G} M^g$. Thus there exists $i \in I$ with $1 - i \in \bar{M}$. Furthermore, for all $g \in G$, $1 - i^g \in \bar{M}$, so summing over g yields $n - \mathrm{tr}_G(i) \in \bar{M} \subseteq M$. We conclude that $\big(n - \mathrm{tr}_G(i)\big) \cdot I \subseteq MI \subseteq M$ as required. ∎

Step 2. *If $I \lhd R$ and $\mathrm{tr}_G(I) \subseteq \mathrm{J}(R^G)$, then $nI \subseteq \mathrm{J}(R)$.*

Proof. If M is a maximal right ideal of R then, by Step 1 applied to $nI \lhd R$, there exists $ni \in nI$ with

$$\big(1 - \mathrm{tr}_G(i)\big) \cdot n^2 I = \big(n - \mathrm{tr}_G(ni)\big) \cdot nI \subseteq M.$$

Since $\mathrm{tr}_G(i) \in \mathrm{J}(R^G)$ by assumption, $1 - \mathrm{tr}_G(i)$ is a unit in the ring and hence $\big(1 - \mathrm{tr}_G(i)\big) \cdot n^2 I = n^2 I$. Thus $n^2 I \subseteq M$ for all such M so $n^2 I \subseteq \mathrm{J}(R)$. But then $(nI)^2 \subseteq \mathrm{J}(R)$ so clearly $nI \subseteq \mathrm{J}(R)$. ∎

Step 3. *If I is a right or left ideal of R and $\mathrm{tr}_G(I) \subseteq \mathrm{J}(R^G)$, then $nI \subseteq \mathrm{J}(R)$.*

Proof. Say I is a right ideal of R and let $S = I \oplus Z$ be the natural extension of I to a ring with 1. Then clearly G acts on S and I is a G-stable two-sided ideal. Furthermore, since quasi-regularity is an internal condition and $\text{tr}_G(I)$ is a right ideal of R^G contained in $\text{J}(R^G)$, we see that $\text{tr}_G(I)$ is a quasi-regular ideal of S^G. Thus $\text{tr}_G(I) \subseteq \text{J}(S^G)$ and, by Step 2 applied to $I \triangleleft S$, we have $nI \subseteq \text{J}(S)$. But again this means that nI is quasi-regular, so $nI \subseteq \text{J}(R)$ and the result follows. ∎

We can now prove

Theorem 28.10. **[117]** *Let G be a finite group acting on R. Then*

$$|G| \cdot \text{J}(R^G) \subseteq \text{J}(R) \cap R^G \subseteq \text{J}(R^G).$$

Proof. If $u \in R^G$ is invertible in R, then clearly $u^{-1} \in R^G$. Thus $\text{J}(R) \cap R^G$ is a quasi-regular ideal of R^G and $\text{J}(R) \cap R^G \subseteq \text{J}(R^G)$. For the other inclusion, let $I = \text{J}(R^G)R$. Then I is a G-stable right ideal of R with $\text{tr}_G(I) \subseteq \text{J}(R^G)$ by Lemma 26.1(ii). We conclude from the previous proposition that $|G| \cdot I \subseteq \text{J}(R)$ so $|G| \cdot \text{J}(R^G) \subseteq \text{J}(R)$. ∎

It was observed in **[178]** that Proposition 28.9 can be used to give yet another proof of Theorem 24.2. Indeed suppose I is a G-stable right ideal of R with $\text{tr}_G(I)$ nilpotent. We extend the action of G to the polynomial ring $R[\zeta]$ in the natural manner and note that $\text{tr}_G(I[\zeta])$ is still nilpotent. Thus Proposition 28.9 implies that $|G| \cdot I[\zeta] \subseteq \text{J}(R[\zeta])$ so $|G| \cdot I \subseteq \text{J}(R[\zeta]) \cap R$. But, as is well known, $\text{J}(R[\zeta]) \cap R$ is a nil ideal, so $|G| \cdot I$ is nil. Now we can use the argument of Exercise 6 of Section 26 or certain more exotic ring extensions of R (as in **[178]**) to conclude that $|G| \cdot I$ is nilpotent.

EXERCISES

1. Let e and f be idempotents in a ring R. Suppose first that $e = uv$ and $f = vu$ for some $u, v \in R$ and define $\alpha : eR \to R$ by $\alpha(er) = ver$. Prove that α determines an R-isomorphism from eR to fR. Conversely suppose that there exists an R-isomorphism $\alpha : eR \to fR$. Prove that $e = uv$ and $f = vu$ for some $u, v \in R$.

2. If R is a prime Goldie ring, prove that rank R = rank $Q(R)$.

3. Suppose R is a semiprime Goldie ring. Let T be the set of regular elements of R and let A be an essential right ideal of R. Prove that $Q(R) = RT^{-1} = A(A \cap T)^{-1}$. To this end, first observe by Lemma 26.8 that $Q(R) = R(A \cap T)^{-1}$. Then note that $A(A \cap T)^{-1}$ is an essential right ideal of $Q(R)$.

4. Let R be a semiprime Goldie ring and let $P_1, P_2, \ldots, P_n$ be its minimal primes. Prove that $Q(R) = \oplus \sum_{i=1}^{n} Q(R/P_i)$. This follows fairly easily from the previous exercise along with Exercise 2 of Section 27.

5. Let $B = K[x, x^{-1}, y, y^{-1} \mid xy = \lambda yx]$ where λ is an element of infinite multiplicative order in $K^{\bullet}$. Show first that $C_B(x) = K[x, x^{-1}]$ and hence that $\langle y \rangle$ is X-outer in its action on this ring. Next show $K[x, x^{-1}]$ is $\langle y \rangle$-simple. Conclude from Corollary 12.6 and Lemma 1.4(i) that $B = K[x, x^{-1}]\langle y \rangle$ is a simple domain.

6. Construct examples of G acting on R with R simple but R^G not semiprime or with R^G simple but R not semiprime. Suitable 2×2 matrix rings will do the trick.

7. Let $A = Z_{(2)}$ be the ring of integers localized at the ideal (2), let $R = M_2(A)$ and let G be the group of automorphisms of R generated by conjugation by the matrix $\begin{pmatrix} 0 & 1 \\ 1 & 0 \end{pmatrix}$. Show that

$$R^G = \left\{ \begin{pmatrix} a & b \\ b & a \end{pmatrix} \;\middle|\; a, b \in A \right\}$$

and that

$$I = \left\{ \begin{pmatrix} a & a \\ a & a \end{pmatrix} \;\middle|\; a \in A \right\}$$

is a quasi-regular ideal of R^G not contained in $J(R)$. This example is from [153] and is certainly relevant to Theorem 28.10.

7 Group Actions and Galois Theory

29. Traces and Truncation

The Galois theory of noncommutative rings is a natural outgrowth of the Galois theory of fields. It began with the work of [**147**] on inner automorphisms of central simple algebras and continued, for a number of years, to be concerned with rather special rings. Thus, for example the Galois theory of division rings was initiated in [**28,75,80,81**]. Furthermore, complete rings of linear transformations were investigated in [**47,146**] and paper [**187**] studied continuous transformation rings. Much of this can now be found in the book [**82**]. In addition, simple Artinian rings were considered in [**76,145**] and in a series of papers leading to the monograph [**198**]. Somewhat later, the Galois theory of separable algebras was investigated in detail. We note, in particular, the papers [**29,93,124,202**].

The best results to date are now in [**88**] and concern group actions on semiprime rings. In that generality, the proofs are long and technical and would be inappropriate for this book. Thus the goal here is merely to convey the flavor of the subject. We do this by restricting our attention to important special cases. In particular, in this section we assume that R is prime and in the next section we assume in addition that the algebra of the group is a domain. This includes, for example, the case of X-outer actions and the case where R itself is a domain. For more information the reader should consult the papers [88] and [134] or the forthcoming book [92]. The exposition here comes from [134] and [135].

Throughout this section we assume that R is prime, $Q = Q_s(R)$ and that G acts on R. As we will see below, it is frequently necessary to assume that G is infinite. So to start with, we must describe the allowable groups. As usual, we let G_{inn} be the subgroup of G consisting of those elements which act as X-inner automorphisms; thus $G_{\text{inn}} \triangleleft G$ by Lemma 12.3(ii). Furthermore recall that the algebra of the group $\mathcal{B} = \mathcal{B}(G) = \mathcal{B}_R(G)$ is defined to be the linear span of all units q of Q such that conjugation by q gives rise to an automorphism of R contained in G_{inn}. It follows easily that $\mathcal{B}$ is an algebra over C, the extended centroid of R. Note also that if $h \in G_{\text{inn}}$, $g \in G$ and $q^{-1}rq = r^h$ for all $r \in R$, then $(q^g)^{-1}r^g q^g = r^{hg}$ so conjugation by q^g gives rise to the automorphism induced by $g^{-1}hg \in G_{\text{inn}}$. We conclude therefore that $\mathcal{B}$ is stable under the natural action of G on the ring Q.

Definition. We say that G is an *M-group* if

    i. $|G : G_{\text{inn}}| < \infty$,
    ii. $\mathcal{B}$ is a semisimple finite dimensional C-algebra.

These are the finiteness conditions required for most results. In addition, note that $\mathcal{B}(G)$ is spanned by units of Q acting like elements of G on R and hence $\mathcal{B} \subseteq \mathbf{C}_Q(R^G)$. In particular, conjugation by any unit of $\mathcal{B}$ will fix R^G and it is appropriate to add these units to G. To be precise, we say that G is an *N-group* if it is an M-group satisfying the *saturation condition*

    iii. If b is any unit of $\mathcal{B}$ with $b^{-1}Rb = R$, then conjugation by b belongs to G.

This definition differs from that of [88] and [134]. The latter more restricted version of N-group, which we call an N*-group, will be discussed in the next section. Note that N-groups tend to be infinite since, if G is not outer, then the group of units of $\mathcal{B}$ will usually be infinite. By definition, any M-group can be extended to an N-group having the same fixed ring. Furthermore, if G is a finite group and R has no $|G|$-torsion, then it follows from Lemma 27.5 and Theorem 4.4 that G is an M-group.

In this section, we develop some necessary machinery. In particular, we show how to construct and truncate trace forms. These trace forms replace the trace map tr_G which was used so effectively in the case of finite groups. As we see below, appropriate trace forms exist for M-groups.

Let A be a finite dimensional algebra over any field C. If $A^\star = \mathrm{Hom}_C(A, C)$ is the *dual* of A, then $A^\star$ can be given a right A-module structure by defining the functional λa to be

$$\lambda a(z) = \lambda(az) \qquad \text{for all } z \in A.$$

Here $\lambda \in A^\star$ and $a \in A$. If $A_A \cong A_A^\star$, then A is said to be a *Frobenius algebra*. The first part of the following result asserts that $A_A^\star$ is isomorphic to the transpose of the left regular representation of A. For the second part, if V is a vector space over C, we say that a basis for V is *compatible* with the decomposition $V = V_1 \oplus V_2 \oplus \cdots \oplus V_k$ if it is a union of bases of the subspaces V_i. The third part contains necessary and sufficient conditions for A to be a Frobenius algebra.

Lemma 29.1. *Let $\{ a_1, a_2, \ldots, a_n \}$ be a C-basis for the algebra A and let $\{ a_1^\star, a_2^\star, \ldots, a_n^\star \}$ be its dual basis in $A^\star$.*

i. If $a \in A$ with $aa_i = \sum_j a_j c_{i,j}$, then $a_j^\star a = \sum_i c_{i,j} a_i^\star$. Here of course $c_{i,j} \in C$.

ii. If e is an idempotent in A, then $\{ a_i \}$ is compatible with $A = eA \oplus (1-e)A$ if and only if $\{ a_i^\star \}$ is compatible with $A^\star = A^\star e \oplus A^\star(1-e)$. Furthermore, when this occurs then $a_i \in eA$ if and only if $a_i^\star \in A^\star e$.

iii. $A_A \cong A_A^\star$ if and only if there exists $\lambda \in A^\star$ whose kernel contains no nonzero right ideal of A. Furthermore if A is semisimple, then A is Frobenius.

Proof. (i) Write $aa_i = \sum_j a_j c_{i,j}$ and $a_j^\star a = \sum_i d_{i,j} a_i^\star$ with the elements $c_{i,j}, d_{i,j} \in C$. Then

$$d_{i,j} = a_j^\star a(a_i) = a_j^\star(aa_i) = c_{i,j}.$$

(ii) Take $a = e$ in the above. Then $\{ a_1, a_2, \ldots, a_n \}$ is compatible with $A = eA \oplus (1-e)A$ if and only if the matrix $[c_{i,j}]$ is diagonal with 1 and 0 entries on the diagonal. Furthermore, by (i), this is precisely the same criterion for $\{ a_1^\star, a_2^\star, \ldots, a_n^\star \}$ to be compatible with $A^\star = A^\star e \oplus A^\star(1-e)$. Finally when this occurs, then $a_i \in eA$ if and only if $c_{i,i} = 1$ and hence if and only if $a_i^\star \in A^\star e$.

(iii) Observe that any module homomorphism $f: A \to A^\star$ is determined by $f(1) = \lambda$. Moreover $f(a) = \lambda a$ is the zero map if and only if $aA \subseteq \text{Ker}(\lambda)$. Thus f is one-to-one and hence an isomorphism if and only if the kernel of λ contains no nonzero right ideal. Now if A is semisimple, write $A = \oplus \sum_i A_i$, a ring direct sum of simple rings. Since $A^\star = \oplus \sum_i A_i^\star$, it clearly suffices, in view of (ii), to show that $A_i \cong A_i^\star$ as A_i-modules. But this is trivial since $\dim A_i = \dim A_i^\star$, both modules are completely reducible and A_i has a unique irreducible module. ∎

We remark that the condition on the existence of λ in (iii) above is actually right-left symmetric (see Exercise 1). Furthermore, if A has a 2-dimensional right ideal all of who subspaces are right ideals, then it is clear that no such λ exists and A is not Frobenius.

We will use the above lemma to construct certain trace forms. Specifically a *trace form* is a formal expression in the variable x given by

$$T(x) = \sum_{i=1}^{n} a_i x^{\sigma_i} b_i$$

where $a_i, b_i \in Q$ and $\sigma_i \in \text{Aut}(R) \subseteq \text{Aut}(Q)$. The various σ_i need not be distinct here. Of course T clearly gives rise to a linear function from Q to Q. The goal is to find suitable T, depending on G, such that $T(Q) \subseteq Q^G$. The main result in this direction is

Proposition 29.2. [88] *Let G be an M-group and let Λ be a transversal for G_{inn} in G with $1 \in \Lambda$. Then there exist trace forms*

$$\tau(x) = \sum_{\substack{1 \le i \le n \\ g \in \Lambda}} a_{i,g} x^g b_{i,g}$$

with $a_{i,g}, b_{i,g} \in B = \mathcal{B}(G)$, $n = \dim_C B$ and $\tau(Q) \subseteq Q^G$. Furthermore we have

    i. *For each $g \in \Lambda$, the sets $\{a_{i,g}\}$ and $\{b_{i,g}\}$ are C-bases of B.*
    ii. *Either basis $\{a_{i,1}\}$ or $\{b_{i,1}\}$ may be prescribed beforehand.*
    iii. *If $e \in B$ is an idempotent, then $\{a_{i,1}\}$ is compatible with $B = eB \oplus (1-e)B$ if and only if $\{b_{i,1}\}$ is compatible with $B = Be \oplus B(1-e)$. When this occurs, $a_{i,1} \in eB$ if and only if $b_{i,1} \in Be$.*

Proof. Let $\theta: B^\star \to B$ be a right B-module homomorphism and let $\{b_1, b_2, \ldots, b_n\}$ be a C-basis for B. If $\{b_1^\star, b_2^\star, \ldots, b_n^\star\}$ is the dual basis for $B^\star$, we show that

$$\tau(x) = \sum_{\substack{1 \le i \le n \\ g \in \Lambda}} \left(b_i x \theta(b_i^\star)\right)^g = \sum_{\substack{1 \le i \le n \\ g \in \Lambda}} a_{i,g} x^g b_{i,g}$$

satisfies $a_{i,g}, b_{i,g} \in B$ and $\tau(Q) \subseteq Q^G$. To this end, let us first consider $\tau_1 = \sum_i b_i x \theta(b_i^\star)$. If $b \in B$ and $bb_i = \sum_j b_j c_{i,j}$ then, by Lemma 29.1(i) and the fact that C is central in Q, we have

$$b\left(\sum_i b_i x \theta(b_i^\star)\right) = \sum_i bb_i x \theta(b_i^\star) = \sum_i \left(\sum_j b_j c_{i,j}\right) x \theta(b_i^\star)$$

$$= \sum_j b_j x \theta\left(\sum_i c_{i,j} b_i^\star\right) = \sum_j b_j x \theta(b_j^\star b)$$

for all $x \in Q$. Moreover since θ is a right B-module homomorphism, the last term above equals $\left(\sum_j b_j x \theta(b_j^\star)\right) b$. Thus for all $x \in Q$, $\tau_1(x)$ commutes with B and in particular is fixed by G_{inn}. Since Λ is a transversal for $G_{\mathrm{inn}} \lhd G$, it is now immediate that $\tau(x) = \sum_{g \in \Lambda} \tau_1(x)^g$ maps Q to Q^G.

Since G is an M-group, B is semisimple, so a right module isomorphism $\theta: B^\star \to B$ exists by Lemma 29.1(iii) and we use this θ in the above construction. With any choice of basis $\{\, b_i \,\}$, the trace form $\tau(x)$ so obtained satisfies $\tau(Q) \subseteq Q^G$ and $a_{i,g}, b_{i,g} \in B$. Now $a_{i,1} = b_i$ and $b_{i,1} = \theta(b_i^\star)$ so both $\{\, a_{i,1} \,\}$ and $\{\, b_{i,1} \,\}$ are bases for B. Moreover the basis $\{\, a_{i,1} \,\}$ may clearly be prescribed beforehand by taking $b_i = a_{i,1}$ and the basis $\{\, b_{i,1} \,\}$ may be prescribed by choosing $\{\, b_{i,1} \,\}$ to be the dual basis to $\{\, \theta^{-1}(b_{i,1}) \,\}$ in $B^{\star\star} = B$. Indeed if $b_i = \theta^{-1}(b_{i,1})^\star$, then $b_i^\star = \theta^{-1}(b_{i,1})$ so $\theta(b_i^\star) = b_{i,1}$. Thus we have (i) and (ii) since $a_{i,g} = (a_{i,1})^g$ and $b_{i,g} = (b_{i,1})^g$.

Finally let e be an idempotent in B. Since θ is an isomorphism and $b_{i,1} = \theta(b_i^\star)$, it is clear that $\{\, b_1^\star, b_2^\star, \ldots, b_n^\star \,\}$ is compatible with $B^\star = B^\star e \oplus B^\star (1-e)$ if and only if $\{\, b_{i,1}, b_{i,2}, \ldots, b_{i,n} \,\}$ is compatible with $B = Be \oplus B(1 - e)$. Therefore since $a_{i,1} = b_i$, part (iii) now follows immediately from Lemma 29.1(ii). ∎

If T is a trace form, then $T(Q) \subseteq Q$, but $T(R)$ need not be contained in R. Nevertheless we have

Lemma 29.3. *Let $T(x) = \sum_i a_i x^{\sigma_i} b_i$ be a trace form. Then there exists a nonzero ideal I of R with $T(I) \subseteq R$. Indeed, if J is any nonzero ideal of R, then there exists a nonzero ideal $L \subseteq J$ with $T(L) \subseteq J$.*

Proof. Let $0 \neq J \lhd R$ be given. Since R is prime, it suffices to assume that $T(x) = a x^\sigma b$. Now choose nonzero ideals A, B of R with $aA, Bb \subseteq R$ and set $L = J \cap (AJB)^{\sigma^{-1}}$. Then L clearly has the appropriate properties. ∎

Definition. Let S be a subring of R. We say that S has the *bimodule property* if every nonzero (R, S)- or (S, R)-subbimodule M of Q contains a nonzero ideal of R and satisfies $M \cap S \neq 0$. It follows immediately from Proposition 10.4 that $S = R$ has the bimodule property; we will use this observation freely throughout this and the next section. We remark that this definition differs from that of [88] and [**134**]. We are able to use this simplified version here because of the restricted goal of our presentation.

Now let $T(x) = \sum_{i=1}^n a_i x^{\sigma_i} b_i$ be a trace form. For any finitely many elements $r_k, s_k \in R$ we let

$$\tilde{T}(x) = \sum_k T(xr_k)s_k = \sum_{i=1}^n a_i x^{\sigma_i} \tilde{b}_i$$

where $\tilde{b}_i = \sum_k (r_k)^{\sigma_i} b_i s_k \in Q$. We call any such $\tilde{T}$ obtained in this way a *right truncation* of T. More generally, if we insist that all s_k above belong to S, then $\tilde{T}$ is a *right (R, S)-truncation* of T.

Similarly the trace form

$$T'(x) = \sum_k s_k T(r_k x) = \sum_{i=1}^n a_i' x^{\sigma_i} b_i$$

where $a_i' = \sum_k s_k a_i (r_k)^{\sigma_i}$ is called a *left (S, R)-truncation* of T.

The following result extends Proposition 12.5. The proof is similar, but somewhat more complicated because of the presence of the subring S.

Proposition 29.4. [**88**] *Let S be a subring of R satisfying the bimodule property and let $T(x) = \sum_{i=1}^n a_i x^{\sigma_i} b_i$ be a trace form with $b_1 \neq 0$ and $\sigma_1 = 1$.*

i. There exist elements $z_i \in Q$ with $z_1 = 1$, z_i either zero or a unit in Q and $z_i s = s^{\sigma_i} z_i$ for all $s \in S$. In particular, if $R = S$ and $z_i \neq 0$, then σ_i is X-inner on R.

ii. If $\bar{T}(x) = \sum_{i=1}^n a_i x^{\sigma_i} z_i$, then there exists a nonzero ideal J of R such that $\bar{T}(xj)$ is a right (R, S)-truncation of T for all $j \in J$. Furthermore, all right hand coefficients in $\bar{T}(xj)$ belong to R.

iii. There exists a right (R, S)-truncation $\tilde{T}(x) = \sum_{i=1}^n a_i x^{\sigma_i} \tilde{b}_i$ of T with $\tilde{b}_1 \in S \setminus 0$ and $\tilde{b}_i = z_i \tilde{b}_1 \in R$ for all i.

Proof. We begin with several general remarks. First, the a_i's merely play the role of place holders here. It is of no concern whether they are zero or not. Second, for any such $T = \sum_{i=1}^n a_i x^{\sigma_i} b_i$, we let the support of T be the set of subscripts i with $b_i \neq 0$. It is clear that if $\tilde{T}$ is a right (R, S)-truncation of T, then $\mathrm{Supp}\, \tilde{T} \subseteq \mathrm{Supp}\, T$. When $\mathrm{Supp}\, \tilde{T}$ is empty, we say that $\tilde{T}$ is trivial. Third, if T' is a right (R, S)-truncation of $\tilde{T}$, then T' is also a right (R, S)-truncation of T. Finally if $b_j \neq 0$ in the above, then the bimodule property implies that $Rb_j S \cap S \neq 0$. Thus there exists $\tilde{T} = \sum_{i=1}^{n} a_i x^{\sigma_i} \tilde{b}_i$ with $\tilde{b}_j \in S \setminus 0$.

The proof of the proposition proceeds by induction on the support size of T, which we may for convenience assume to be n. By the preceding remark we may further assume that $b_1 \in S \setminus 0$. If $n = 1$ then, since $\sigma_1 = 1$, the result follows with $z_1 = 1$ and J a nonzero ideal contained in $Rb_1 S$. Now suppose $n > 1$ and let $\mathcal{T}$ denote the set of all right (R, S)-truncations of T. If $\tilde{T} = \sum_i a_i x^{\sigma_i} \tilde{b}_i \in \mathcal{T}$ with $|\mathrm{Supp}\, \tilde{T}| < n$, then the result will follow by induction provided $\tilde{b}_1 \neq 0$. Thus we may assume that all such $\tilde{T} \in \mathcal{T}$ of support size less than n satisfy $\tilde{b}_1 = 0$.

One further reduction is necessary. For each i, there exists a nonzero ideal I_i of R with $I_i b_i \subseteq R$. Thus if $I = \bigcap_i (I_i)^{\sigma_i^{-1}}$, then $I \neq 0$ and $I^{\sigma_i} b_i \subseteq R$ for all i. In addition, we can choose $r \in I$ with $rb_1 \neq 0$. Then $\tilde{T} = T(xr)$ is a truncation of T with all $\tilde{b}_i \in R$ and $\tilde{b}_1 \neq 0$. We can now assume that T itself has this property. As above, we may assume in addition that $b_1 \in S \setminus 0$.

We first show that if $T' = \sum_i a_i x^{\sigma_i} b'_i \in \mathcal{T}$ with $|\mathrm{Supp}\, T'| < n$, then T' is trivial. We already know that $b'_1 = 0$ but suppose $b'_j \neq 0$ for some $j \neq 1$. By truncating T' if necessary we may assume that $b'_j \in S \setminus 0$. Since $Sb'_j R$ contains a nonzero ideal of R, we have $b_1 Sb'_j R \neq 0$ and we can choose $s \in S$ with $b_1 s b'_j \neq 0$. For this s let

$$\tilde{T}(x) = T(x)sb'_j - T'\left(x(b_j s)^{\sigma_j^{-1}}\right)$$

$$= \sum_{i=1}^{n} a_i x^{\sigma_i} \tilde{b}_i \in \mathcal{T}.$$

Here $\tilde{b}_i = b_i s b'_j - (b_j s)^{\sigma_j^{-1} \sigma_i} b'_i$. Hence since $b'_1 = 0$, we have $\tilde{b}_1 = b_1 s b'_j \neq 0$ by the choice of s. On the other hand, the above formula clearly yields $\tilde{b}_j = 0$ and this contradicts the assumptions on $\mathcal{T}$. Thus all nontrivial elements of $\mathcal{T}$ have support size n.

Now we return to T itself. Since Rb_1S is a nonzero (R, S)-bimodule, it contains a nonzero ideal J of R. Thus for each $j \in J$ there exists a right (R, S)-truncation $T_j'(x)$ of T with

$$T_j'(x) = \sum_{i=1}^{n} a_i x^{\sigma_i} b_i'(j)$$

and $b_1'(j) = j$. In fact, T_j' is unique since if T_j' and T_j'' are two such truncations with the same 1-coefficient j, then $T_j' - T_j''$ is a truncation of T with support size less than n and hence must be trivial. In other words, $T_j' = T_j''$. Thus, for each i, we see that $b_i': J \to R$ is a well-defined function. Furthermore, a similar uniqueness argument shows that each b_i' is one-to-one.

Now b_i' is surely additive, but it is not a left R-module homomorphism. Indeed by comparing $T_j'(xr)$ and $T_{rj}'(x)$ we see that $b_i'(rj) = r^{\sigma_i} b_i'(j)$. But then the composite map $(b_i')^{\sigma_i^{-1}}: J \to R$ is a left R-module homomorphism and hence represents, by Proposition 10.2(iv), an element q_i of the left Martindale ring of quotients $Q_\ell(R)$. In other words, for each $j \in J$ we have $(b_i')^{\sigma_i^{-1}}(j) = jq_i$, so $b_i'(j) = j^{\sigma_i} q_i^{\sigma_i} = j^{\sigma_i} z_i$ where we have set $z_i = q_i^{\sigma_i} \in Q_\ell(R)$. Of course $z_1 = 1$ since $b_1'(j) = j$. Note also that the image $b_i'(J)$ is a nonzero (R, S)-bimodule and hence contains a nonzero ideal of R. Thus, since b_i' is one-to-one, it follows, using the back map and the above argument, that z_i is in fact invertible in $Q_\ell(R)$.

Let $\bar{T}(x) = \sum_{i=1}^{n} a_i x^{\sigma_i} z_i$. Then, for all $j \in J$, we see that $\bar{T}(xj) = \sum_i a_i x^{\sigma_i} j^{\sigma_i} z_i = T_j'(x)$ is a truncation of $T(x)$. Thus (ii) is proved.

For (i), we must further study the elements z_i. Let $s \in S$. Then by comparing $T_j'(x)s$ and $T_{js}'(x)$, we obtain the required identity $z_i s = s^{\sigma_i} z_i$ for all $s \in S$. With this we can now show that z_i is a unit of $Q = Q_s(R)$. To start with, note that $z_i \in Q_\ell(R)$, $J^{\sigma_i} z_i \subseteq R$ and $z_i S = S^{\sigma_i} z_i$. Next, $z_i(J \cap S) = (J \cap S)^{\sigma_i} z_i \subseteq R$ and $J \cap S \neq 0$ by the bimodule property. It therefore follows that

$$z_i \cdot S(J \cap S)R = S^{\sigma_i} \cdot z_i(J \cap S) \cdot R \subseteq R.$$

But $S(J \cap S)R$ contains a nonzero ideal of R, so Proposition 10.6 implies that $z_i \in Q_s(R) = Q$. Similarly, replacing J^{σ_i} by the domain of definition of z_i^{-1}, we see that $z_i^{-1} \in Q$ and z_i is a unit of Q. Thus (i) is proved.

Finally for (iii), choose $j \in J \cap S$ with $j \neq 0$. Then $j^{\sigma_i} z_i = z_i j$ so

$$\tilde{T}(x) = \bar{T}(xj) = \sum_{i=1}^{n} a_i x^{\sigma_i} j^{\sigma_i} z_i = \sum_{i=1}^{n} a_i x^{\sigma_i} z_i j$$

and the result follows with $\tilde{b}_i = z_i j$. ∎

Since the bimodule property and $Q_s(R)$ are both right-left symmetric, the left analog of the above applies to left (S, R)-truncations. We will not bother to state the result, but we will feel free to use it when necessary.

Let $T(x) = \sum_i a_i x^{\sigma_i} b_i$ be a trace form. If any σ_i is X-inner induced by the unit q_i of Q, then for any $x \in Q$ we have $a_i x^{\sigma_i} b_i = (a_i q_i^{-1}) x (q_i b_i)$. Because of this we can usually assume that no X-inner automorphism other than $\sigma = 1$ occurs in T. Indeed we say that T is an *outer trace form* if σ_i X-inner implies $\sigma_i = 1$. Note that the traces constructed in Proposition 29.2 all have this property since $1 \in \Lambda$.

If we allow elements of C to pass across the x^{σ} factors in trace forms, then we have

Lemma 29.5. Let $T(x) = \sum_i a_i x^{\sigma_i} b_i$ be an outer trace form with $\sigma_1 = 1$ and $b_1 \neq 0$.

i. There exists a right truncation $\tilde{T}(x) = \alpha x \tilde{b}_1$ of T with $\tilde{b}_1 \in R \setminus 0$ and $\alpha = \sum' a_i c_i$. Here $c_i \in C$, $c_1 = 1$ and the sum is over $\{i \mid \sigma_i = 1\}$.

ii. Suppose that $\{a_i \mid \sigma_i = 1\}$ is C-linearly independent. If I is a nonzero ideal of R, then $T(I) \neq 0$.

Proof. (i) This follows easily by applying Proposition 29.4(i)(iii) to $T(x)$ with $S = R$. If $\tilde{b}_i \neq 0$, then σ_i is X-inner and hence $\sigma_i = 1$, by assumption. But z_i^{-1} induces σ_i so $z_i \in C$. Thus $\tilde{b}_i = c_i \tilde{b}_1$ with $c_i = z_i \in C$ and $c_1 = 1$. Letting $\sum'$ indicate the sum over $\{i \mid \sigma_i = 1\}$, we conclude that

$$\tilde{T}(x) = \sum' a_i x c_i \tilde{b}_1 = \left(\sum' a_i c_i\right) x \tilde{b}_1$$

as required.

(ii) If $T(I) = 0$, then certainly $\tilde{T}(I) = 0$ for any right truncation $\tilde{T}$ of T. In particular, if $\tilde{T}$ is as above, then $0 = \tilde{T}(I) = \alpha I \tilde{b}_1$. But $\tilde{b}_1 \neq 0$, so it follows (see Exercise 2) that $0 = \alpha = \sum' a_i c_i$ and this contradicts the linear independence of the a_i's with $\sigma_i = 1$. ∎

As an indication of the power of this machinery, we offer

Proposition 29.6. [88] *Let G be an M-group acting on the prime ring R. If I is a nonzero ideal of R, then $I \cap R^G \neq 0$.*

Proof. By Proposition 29.2 and Lemma 29.3 there is an outer trace form $\tau(x) = \sum_{i,g} a_{i,g} x^g b_{i,g}$ and a nonzero ideal $J \subseteq I$ with $\tau(J) \subseteq I \cap Q^G = I \cap R^G$. Now apply part (ii) of the previous lemma. ∎

Note that the example at the beginning of Section 24 shows that some hypothesis on G is required for the existence of fixed points, even if G is finite. Additional consequences of the machinery will be considered throughout the next section.

EXERCISES

1. Let A be a finite dimensional C-algebra and let $\lambda \in A^*$. Show that $\lambda A \neq A^*$ if and only if there exists $0 \neq b \in A$ with $\lambda A(b) = 0$. Conclude that $A_A \cong A_A^*$ if and only if there exists $\lambda \in A^*$ whose kernel contains no nonzero left ideal of A.

2. Let I be a nonzero ideal of the prime ring R and let $a, b \in Q_s(R)$. If $aIb = 0$, prove that $a = 0$ or $b = 0$. To this end, choose nonzero ideals A, B of R with $Aa, bB \subseteq R$ and observe that $(AaI)(RbB) = 0$. Now apply Lemma 10.7(ii).

3. Let G be a finite group acting on a ring R. Show that $R\dot{G}R = (RG)\dot{G}(RG)$ is equal to the skew group ring RG if and only if there exist elements $a_1, a_2, \ldots, a_n$ and $b_1, b_2, \ldots, b_n$ in R with

$$\sum_{i=1}^{n} (a_i)^g (b_i)^h = \begin{cases} 1 & \text{if } g = h \\ 0 & \text{if } g \neq h. \end{cases}$$

Here of course $g, h \in G$. When these equivalent conditions occur, the action is said to be *G-Galois*.

    4. Assume that the action of G on R is G-Galois and let $\{a_i, b_i\}$ be as above. Define the left R^G-module homomorphisms $\alpha: R \to (R^G)^n$ and $\beta: (R^G)^n \to R$ by

$$\alpha: r \mapsto \big(\mathrm{tr}_G(ra_1), \mathrm{tr}_G(ra_2), \ldots, \mathrm{tr}_G(ra_n)\big)$$

$$\beta: (s_1, s_2, \ldots, s_n) \mapsto \sum_{i=1}^{n} s_i b_i.$$

Prove that $\alpha\beta = 1$ and conclude that R is a finitely generated projective left R^G-module. By symmetry, the same is true for R as a right R^G-module.

    5. We continue with the above notation, again viewing R as a left R^G-module. Define the maps $\gamma: \mathrm{End}_{R^G}(R) \to RG$ and $\delta: RG \to \mathrm{End}_{R^G}(R)$ by

$$\gamma: f \mapsto \sum_{g \in G} g\Big(\sum_{i=1}^{n} (a_i)^g (b_i f)\Big)$$

$$\delta: \sum_{g \in G} gr_g \mapsto \sum_{g \in G} {}^g \cdot r_g.$$

Thus δ is the ring homomorphism given by Lemma 26.2(i). Show that γ and δ are inverse maps and conclude that RG is ring isomorphic to $\mathrm{End}_{R^G}(R)$.

    6. Let $R = \mathrm{M}_2(K)$ with char $K \neq 2$ and let $G = \{1, g\}$ act on R with g acting as conjugation by $\mathrm{diag}(1, -1)$. Show that this action is G-Galois even though RG is not simple.

    7. Suppose S is a G-graded ring with G finite and let $R = S\#G^*$. Show that the action of G on R, as described in Theorem 2.5, is G-Galois. Then use the conclusion of Exercise 5 to prove that $(S\#G^*)G = RG \cong \mathrm{M}_G(S)$. This is the alternate proof of duality contained in [152] and [200].

30. The Galois Correspondence

We continue with the notation of the previous section. Thus R is a prime ring, $Q = Q_s(R)$ is its symmetric Martindale ring of quotients and $C = \mathbf{Z}(Q)$ is the extended centroid of R. Furthermore, G acts on R with algebra of the group $B = \mathcal{B}(G)$, a C-subalgebra of Q. The goal here is to obtain the usual sort of Galois correspondence between certain subgroups of G and appropriate intermediate rings containing R^G. As we indicated earlier, these results will not be given in their full generality; we will make certain simplifying assumptions. These will still maintain the flavor of the proofs, but will avoid some of the unpleasant technicalities.

We begin with

Proposition 30.1. [88] *Let G be an M-group acting on a prime ring R. Then $\mathbf{C}_Q(R^G) = B$.*

Proof. Certainly $\mathbf{C}_Q(R^G) \supseteq B$, since B is spanned by elements which induce the X-inner automorphisms of G. We consider the reverse inclusion. Let $\beta \in \mathbf{C}_Q(R^G)$.

Let e be a primitive idempotent of the algebra B and let $\tau(x) = \sum_{i,g} a_{i,g} x^g b_{i,g}$ and I be as in Proposition 29.2 and Lemma 29.3. Furthermore, we can assume that the C-basis $\{ a_{i,1} \}$ is chosen compatibly with the decomposition $B = eB \oplus (1-e)B$.

If $T(x)$ is defined by $T(x) = \beta e\tau(x) - \tau(x)\beta e$ then, since $\beta e \in \mathbf{C}_Q(R^G)$ and $\tau(I) \subseteq R^G$, we see that T vanishes on I. Furthermore, in the expression $\beta e\tau(x)$, we can delete all those $a_{i,1}$ in $(1-e)B$ and we use $\sum'$ to denote such a deleted sum. Since T is an outer trace form, Lemma 29.5(ii) implies that the left-hand coefficients of

$$T(x) = {\sum}' \beta e a_{i,g} x^g b_{i,g} - \sum_{i,g} a_{i,g} x^g b_{i,g} \beta e$$

corresponding to $g = 1$ are C-linearly dependent. Thus there exist $c_i, d_i \in C$, not all zero, with

$$\beta e {\sum}' c_i a_{i,1} = \sum_i d_i a_{i,1}.$$

In particular, some c_i must be nonzero. Furthermore, the $a_{i,1}$ in the left-hand sum all belong to eB and thus we have $\beta \alpha = \sum_i d_i a_{i,1}$

where $\alpha = \sum' c_i a_{i,1}$ is necessarily a nonzero element of eB. Since e is primitive and B is semisimple, $e \in \alpha B$ and we conclude immediately that $\beta e \in B$.

Finally if $1 = e_1 + e_2 + \cdots + e_n$ is a decomposition of 1 into orthogonal primitive idempotents in B, then each $\beta e_i \in B$ and hence $\beta \in B$. ∎

If S is a subring of R, let

$$\mathrm{Gal}(R/S) = \{\, \sigma \in \mathrm{Aut}(R) \mid \sigma \text{ centralizes } S \,\}.$$

Thus $\mathrm{Gal}(R/S)$ is certainly a subgroup of $\mathrm{Aut}(R)$. We can now quickly prove the first main result on Galois theory.

Theorem 30.2. (Galois Group) [88] *Let G be an N-group acting on the prime ring R. Then $\mathrm{Gal}(R/R^G) = G$.*

Proof. Certainly $\mathrm{Gal}(R/R^G) \supseteq G$. In the other direction, suppose $\sigma \in \mathrm{Gal}(R/R^G)$. Let $\tau(x) = \sum a_{i,g} x^g b_{i,g}$ and I be as in Proposition 29.2 and Lemma 29.3. If $T(x)$ is defined by

$$T(x) = \tau(x) - \tau(x)^\sigma = \sum_{i,g} a_{i,g} x^g b_{i,g} - \sum_{i,g} (a_{i,g})^\sigma x^{g\sigma} (b_{i,g})^\sigma$$

then T vanishes on I since $\tau(I) \subseteq R^G$.

If $g\sigma$ is not an X-inner automorphism of R for any g above, then the only X-inner automorphisms in $T(x)$ occur when $g = 1$ and in the first sum. However, this contradicts Lemma 29.5(ii). Thus $g\sigma$ is X-inner for some $g \in G$ and say this automorphism is induced by the unit $q \in Q$. Since both g and σ fix R^G, it follows that $q \in C_Q(R^G) = B$ by the previous proposition. But then q is a unit of B normalizing R and hence q gives rise to an automorphism in G, since G is an N-group. In other words, $g\sigma \in G$ and we conclude that $\sigma \in G$. ∎

It turns out that the N-group condition is not quite strong enough to prove the best possible results. We actually require N*-groups.

Definition. Let G act on the prime ring R. We say that G is an $N^\star$-*group* if G is an M-group satisfying

iv. If b is any unit of $\mathcal{B}(G)$, then $b^{-1}Rb = R$ and conjugation by b belongs to G.

Notice that an $N^\star$-group is necessarily an N-group, but the condition is definitely more restrictive. We will consider some examples after Theorem 30.4.

Now we have already seen, in Proposition 30.1, how idempotents of B come into play in Galois theory arguments. Later proofs, in the presence of idempotents, become even more technical. To avoid this, we will make the simplifying assumption throughout the remainder of this section that $B = \mathcal{B}(G)$ is a domain and hence a division ring. Notice that this is always the case if G is X-outer, since then $B = C$, or if R is a domain, since then Q is a domain by Lemma 10.7(i). As we mentioned previously, the reader can consult [88], [134] or [92] for information on the general case.

Proposition 30.3. [88] *Let G be an M-group acting on the prime ring R and assume that B is a domain. Let S be a subring of R containing R^G.*
    i. S is prime.
    ii. S has the bimodule property.
    iii. Suppose G is an $N^\star$-group and $H = \mathrm{Gal}(R/S)$. Then S contains a nonzero ideal of R^H.

Proof. (i) Suppose $s, t \in S$ with $sSt = 0$ and $t \neq 0$. Let $\tau(x) = \sum_{i,g} a_{i,g} x^g b_{i,g}$ with $b_{1,1} = 1$ and let $0 \neq I \lhd R$ be as in Proposition 29.2 and Lemma 29.3. Then $\tau(I) \subseteq R^G \subseteq S$ so $s\tau(I)t = 0$. In particular, if $T(x)$ is the trace form $T(x) = s\tau(x)t$, then T vanishes on I. Notice that T is outer and has as a right-hand coefficient $b_{1,1}t = t \neq 0$. It therefore follows from Lemma 29.5(ii) that there exist suitable $c_i \in C$, not all zero, with $\sum_i sa_{i,1}c_i = 0$. But $\{a_{i,1}\}$ is C-linearly independent, so $\sum_i a_{i,1}c_i$ is a nonzero element of B and hence a unit of B. Since this element annihilates s, we conclude that $s = 0$.

(ii) Let $M \neq 0$ be an (S, R)-subbimodule of Q and choose $m \in M \setminus 0$. Let $\tau(x) = \sum_{i,g} a_{i,g} x^g b_{i,g}$ and I be as in Proposition 29.2 and Lemma 29.3. Since $\tau(I) \subseteq R^G \subseteq S$, it follows that $\tau(I)m \subseteq SM \subseteq M$. Thus if $T(x) = \tau(x)m$, then $T(I) \subseteq M$. Furthermore, since $MR \subseteq M$, it follows that if $\tilde{T}$ is any right (R, R)-truncation of T, then $\tilde{T}(I) \subseteq M$. Now let $\tilde{T}(x) = \alpha x \tilde{b}_1$ be given as in Lemma 29.5(i). Then $\tilde{b}_1 \in R \setminus 0$ and α is a nonzero element of B since $\{ a_{i,1} \}$ is C-linearly independent. Thus $\alpha I \tilde{b}_1 \subseteq M$. Note that α is invertible in B. Since $\alpha^{-1} J \subseteq R$ for some nonzero ideal J of R, we have $\alpha^{-1} JI \subseteq I$ and hence

$$JI\tilde{b}_1 R = \alpha(\alpha^{-1}JI)\tilde{b}_1 R \subseteq (\alpha I \tilde{b}_1)R \subseteq MR = M.$$

Thus M contains the nonzero ideal $JI\tilde{b}_1 R$. A similar argument, using the left analog of Lemma 29.5(i) works for (R, S)-bimodules. Finally, by Proposition 29.6, any nonzero ideal of R meets R^G and hence S nontrivially. We conclude therefore that S has the bimodule property.

(iii) Since H fixes $S \supseteq R^G$, Theorem 30.2 implies that $H \subseteq G$. Thus $G_{\mathrm{inn}} H = HG_{\mathrm{inn}}$ is a subgroup of G and we can choose a transversal Λ for G_{inn} in G with $1 \in \Lambda$ and $\Lambda \cap (G_{\mathrm{inn}}H) \subseteq H$. Using this Λ, we apply Proposition 29.2 and Lemma 29.3 as usual to obtain a trace form $\tau(x) = \sum_{i,g} a_{i,g} x^g b_{i,g}$ and a nonzero ideal I of R with $\tau(I) \subseteq R^G$. Since $R^G \subseteq S$ it follows that, for any right (R, S)-truncation $\tilde{T}$ of τ, we have $\tilde{T}(I) \subseteq S$. By (ii) above, S has the bimodule property and thus Proposition 29.4(i)(ii) applies; we use its notation, allowing for double subscripts.

Suppose $z_{i,g} \neq 0$. Then $z_{i,g}$ is a unit of Q and $z_{i,g}s = s^g z_{i,g}$ for all $s \in S$. Since $S \supseteq R^G$ and g fixes R^G, we see that $z_{i,g} \in \mathbf{C}_Q(R^G) = B$, by Proposition 30.1. But then, since G is an N*-group, conjugation by $z_{i,g}$ induces an element w of G_{inn} so $s^{gw} = (z_{i,g})^{-1} s^g z_{i,g} = s$ and $gw \in H$. It follows that $g \in \Lambda \cap (HG_{\mathrm{inn}}) \subseteq H$, by assumption, and we conclude that $w \in H$ so $z_{i,g} \in \mathbf{C}_Q(R^H)$.

Now consider $\bar{T}(x) = \sum_{i,g} a_{i,g} x^g z_{i,g}$. Then, by the above, we see that $\bar{T}: Q \to Q$ is a right R^H-module homomorphism. Furthermore, by Proposition 29.4(ii), there exists a nonzero ideal J of R with $\bar{T}(xj)$ a right (R, S)-truncation of τ for all $j \in J$. Thus $\bar{T}(Ij) \subseteq S$ for each

j so $\bar{T}(IJ) \subseteq S$. Since $\bar{T}$ is a right R^H-module homomorphism, $IJ \neq 0$ and $S \subseteq R^H$, it follows from Lemma 29.5(ii) that $V = \bar{T}(IJ)$ is a nonzero right ideal of R^H contained in S. Similarly, using the left analogs of these results, we find a nonzero left ideal U of R^H contained in S. But S is prime, so USV is a nonzero two-sided ideal of R^H contained in S and the result follows. ∎

We say that S is an *ideal-cancellable subring* of R if, for all nonzero ideals I of S and elements $r \in R$, $Ir \subseteq S$ implies that $r \in S$. We can now obtain the second main result on Galois theory.

Theorem 30.4. (Correspondence) [88] *Let G be an N*-group of automorphisms of the prime ring R and assume that $\mathcal{B}(G)$ is a domain. Then the maps $H \mapsto R^H$ and $S \mapsto \mathrm{Gal}(R/S)$ yield a one-to-one correspondence between the N*-subgroups H of G and the ideal-cancellable subrings S with $R \supseteq S \supseteq R^G$.*

Proof. Let H be a subgroup of G. We first show that $S = R^H$ is ideal-cancellable. Let I be a nonzero ideal of S. Since $H \subseteq G$ we have $S \supseteq R^G$ so S satisfies the bimodule property by the previous proposition. In particular, RI contains a nonzero ideal of R and hence $\mathrm{r}_R(I) = 0$. Now suppose $Ir \subseteq S$. If $h \in H$ and $s \in I$, then $sr \in S$ so $sr = (sr)^h = s^h r^h = sr^h$ and $I(r - r^h) = 0$. Thus $r = r^h$ for all $h \in H$ and $r \in R^H = S$. In addition, if we assume that H is an N*-group, then it follows that $H = \mathrm{Gal}(R/S)$, by Theorem 30.2.

Conversely suppose $S \supseteq R^G$ is ideal-cancellable and let $H = \mathrm{Gal}(R/S)$. Then H fixes R^G so $H \subseteq G$ by Theorem 30.2 again. Moreover, it is clear that H is an N*-group, since B is a division ring and G satisfies (iv). By Proposition 30.3(iii), S contains a nonzero ideal I of R^H. Thus I is also an ideal of S and if $r \in R^H$, then $Ir \subseteq I \subseteq S$. We conclude from the ideal-cancellable property that $r \in S$ and thus $S = R^H$ as required. ∎

We remark that the ideal-cancellable property can be restated in terms of the left Martindale ring of quotients. Indeed, if $S \supseteq R^G$ then it can be shown that $Q_\ell(S)$ is contained naturally in $Q_\ell(R)$. With this embedding, S is ideal-cancellable if and only if $S = Q_\ell(S) \cap R$

(see Exercise 3). Note that, if $\mathcal{B}(G)$ is not a domain, then the ideal-cancellable condition is not sufficient for S to be a fixed ring. The precise properties required can be found in [88], [134] or [92].

We now consider several examples of interest. Let D be a division algebra with center K and let $D\langle x, y \rangle$ be the free D-algebra on two generators x and y. This ring is of course a domain with group of units $D^{\bullet} = D \setminus 0$. Furthermore, it is symmetrically closed by Theorem 13.4, since domains are necessarily cohesive. In particular, $K = \mathbf{Z}(D\langle x, y \rangle)$ is the extended centroid. For convenience, we denote the augmentation ideal $xD\langle x, y \rangle + yD\langle x, y \rangle$ by $D\langle x, y \rangle'$.

Now let F be any subdivision ring of D and set $R = F + D\langle x, y \rangle'$. Then R is a domain containing an ideal of $D\langle x, y \rangle$ so, by Lemma 10.8(iii), we have

$$Q_s(R) = Q_s(D\langle x, y \rangle) = D\langle x, y \rangle.$$

Thus we know the unit group of $Q_s(R)$ and it follows that the group G of all X-inner automorphisms of R is induced by conjugation by

$$\left\{ d \in D^{\bullet} \mid d^{-1}Fd = F \right\}.$$

In other words, $G \cong \mathbf{N}_{D^{\bullet}}(F^{\bullet})/K^{\bullet}$ since only elements of K will centralize $D\langle x, y \rangle'$.

For our first example take $F = K$ and assume $1 < \dim_K D < \infty$. Since $d^{-1}Kd = K$ for all $d \in D^{\bullet}$, it follows that all such d induce elements of G. Thus $\mathcal{B}(G) = D$ and G is an $\mathrm{N}^{\star}$-group. Notice that G is not inner on R since the only units of R are in $K^{\bullet}$.

For a more interesting example take $D = K + Ki + Kj + Kk$ to be the quaternion division algebra over the real field K and set $F = K + Ki = K[i]$. If $d \in D^{\bullet}$, then $d^{-1}Fd = F$ if and only if $d^{-1}id = \pm i$ and hence if and only if $d \in F \cup Fj$. Thus the elements of G are induced by all nonzero $d \in F \cup Fj$ so G is an N-group on R but not an $\mathrm{N}^{\star}$-group. Note that $\mathcal{B}(G) = F + Fj = D$. Since R^G is centralized by $\mathcal{B}(G) = D$, it follows easily that $R^G = K\langle x, y \rangle$.

Now let $E = K + K(i + j) = K[i + j] \subseteq D$ and let S be the intermediate ring $S = K + E\langle x, y \rangle'$. Note that $d \in D^{\bullet}$ centralizes S if and only if $d \in \mathbf{C}_D(E) = E$. But $E \cap (F \cup Fj) = K$, so

$H = \text{Gal}(R/S) = \langle 1 \rangle$ and $R^H = R^{\langle 1 \rangle} = R$. Clearly S contains no nonzero ideal of $R^H = R$ so we see that Proposition 30.3(iii) does indeed require that G be an N*-group. Furthermore, it is easy to see that $S \supseteq R^G$ is ideal-cancellable, but S is not a fixed ring. Thus Theorem 30.4 also requires that G be an N*-group.

The latter example indicates a failure of the theory in its present state. Certain automorphisms which should be present are missing. We expect all of $D^\bullet$ to act on R but, by accident, some of the elements do not normalize R. This problem can be eliminated by suitably redefining what we mean by an allowable automorphism. In the case of prime rings R, an appropriate candidate is the set of all automorphisms σ of $Q_s(R)$ with the property that σ and σ^{-1} both send some ideal of R into R. In other words, one weakens the condition $R^\sigma = R$ to $I^\sigma \subseteq R$ and $I^{\sigma^{-1}} \subseteq R$. This is presumably the approach being taken in [92].

We close this section by considering some applications of this work to the relationship between R and the fixed ring R^G. First we have

Proposition 30.5. [88] *Let G be an M-group acting on the prime ring R. If $\mathcal{B}(G)$ is a domain, then $Q_\ell(R)^G = Q_\ell(R^G)$ and $Q_s(R)^G = Q_s(R^G)$.*

Proof. Note that G acts on $Q_\ell(R)$ and $Q_s(R)$ and $R^G \subseteq Q_s(R)^G \subseteq Q_\ell(R)^G$. Moreover R^G is prime by Proposition 30.3(i). We show first that $Q_\ell(R)^G \supseteq R^G$ satisfies (i)–(iv) of Proposition 10.2. To this end, let $q \in Q_\ell(R)^G$. If $0 \neq I \triangleleft R$ with $Iq \subseteq R$, then $I \cap R^G$ is a nonzero ideal of R^G, by Proposition 29.6, and clearly $(I \cap R^G)q \subseteq R^G$. This of course yields (ii). For (iii), suppose Y is a nonzero ideal of R^G with $Yq = 0$. Then $RYq = 0$ and RY is a nonzero (R, R^G)-bimodule. It follows from Proposition 30.3(ii) that RY contains a nonzero ideal of R, so $q = 0$ and (iii) is proved.

For (iv), let $f : Y \to R^G$ be a left R^G-module homomorphism with $0 \neq Y \triangleleft R^G$. We extend f to a map $f^* : RY \to R$, in the natural manner, by $f^* : \sum_k r_k y_k \mapsto \sum_k r_k(y_k f)$ where $r_k \in R$ and $y_k \in Y$. To see that this is well defined, suppose $\sum_k r_k y_k = 0$ and choose $\tau = \sum_{i,g} a_{i,g} x^g b_{i,g}$ and I as in Proposition 29.2 and Lemma 29.3

with $b_{1,1} = 1$. Note that $b_{i,g} \in B$ centralizes R^G and hence Y. Furthermore, for all $g \in \Lambda$ we have $0 = \left(\sum_k r_k y_k\right)^g = \sum_k (r_k)^g y_k$. It therefore follows that the right truncation $T(x) = \sum_k \tau(x r_k) y_k$ has all right-hand coefficients equal to zero and hence $T(I) = 0$. In addition, since $\tau(I) \subseteq R^G$ and f is a left R^G-module homomorphism, we see that

$$T'(x) = \sum_k \tau(x r_k)(y_k f) = \sum_{i,g} a_{i,g} x^g b'_{i,g}$$

also vanishes on I. But T' is an outer trace form, so Lemma 29.5(ii) yields $0 = b'_{1,1} = \sum_k r_k(y_k f)$. We have therefore shown that f^* is well defined.

Again, by the bimodule property, RY contains a nonzero ideal J of R and $f^*: J \to R$ is a left R-module homomorphism. It follows that there exists $q \in Q_\ell(R)$ with $jf^* = jq$ for all $j \in J$. In particular, if $y \in Y$ and $j \in J$, then $jyq = (jy)f^* = j(yf)$ so $J(yq - yf) = 0$ and $yf = yq$. Finally, since $yq \in R^G$ for all $y \in Y$, it follows easily that $q \in Q_\ell(R)^G$ and we conclude from Proposition 10.2 that $Q_\ell(R)^G = Q_\ell(R^G)$. The result for $Q_s(R^G)$ is now immediate from the above, Proposition 29.6 and Proposition 10.6. ∎

Finally we specialize to the case of finite X-outer groups. Recall that any such group G is necessarily an N^*-group. Furthermore, if R is a simple ring, then $Q_s(R) = R$ so outer and X-outer are equivalent properties.

Lemma 30.6. *Let G be a finite group of outer automorphisms of the simple ring R. Then R^G is simple if and only if R contains an element of trace 1, that is if and only if $1 \in \mathrm{tr}_G(R)$.*

Proof. By Lemmas 26.1 and 29.5(ii), $\mathrm{tr}_G(R)$ is a nonzero ideal of R^G. In particular, if R^G is simple, then $\mathrm{tr}_G(R) = R^G$ and $1 \in \mathrm{tr}_G(R)$.

Conversely, suppose R is simple and $1 \in \mathrm{tr}_G(R)$. If I is a nonzero ideal of R^G, then IR is a nonzero (R^G, R)-submodule of R and hence contains, by Proposition 30.3(ii), a nonzero ideal of R. Thus $IR = R$ and $1 \in \mathrm{tr}_G(R) = \mathrm{tr}_G(IR) \subseteq I$. We conclude therefore that $I = R^G$ and that R^G is simple. ∎

Theorem 30.7. *Let G be a finite group of outer automorphisms of the simple ring R and suppose that $1 \in \mathrm{tr}_G(R)$. Then the maps $H \mapsto R^H$ and $S \mapsto \mathrm{Gal}(R/S)$ yield a one-to-one correspondence between the subgroups H of G and the intermediate rings $S \supseteq R^G$. In particular, there are only finitely many intermediate rings and they are all simple.*

Proof. If H is a subgroup of G and Λ is a left transversal for H in G, then it follows easily that $\mathrm{tr}_G(r) = \mathrm{tr}_H(\mathrm{tr}_\Lambda(r))$ and hence that $\mathrm{tr}_G(R) \subseteq \mathrm{tr}_H(R)$. In particular, we now know that $1 \in \mathrm{tr}_H(R)$ and hence that R^H is simple by the previous lemma.

Conversely let $S \supseteq R^G$ be any intermediate ring and let $H = \mathrm{Gal}(R/S) \subseteq G$. By Proposition 30.3(iii), S contains a nonzero ideal of its overring R^H. But R^H is simple, so we conclude that $S = R^H$. The result follows from Theorem 30.4. ∎

The hypothesis that R contains an element of trace 1 is trivially satisfied if $|G|^{-1} \in R$ or if R is a division ring. Indeed if R is a division ring, then so is R^G and Lemma 29.5(ii) yields this fact. Thus the Galois correspondence for fields is a consequence of the above, as is the analogous result for divisions rings in [**80**]. More generally, if R is simple Artinian and G is outer, then there always exists an element of trace 1, as the following lemma shows. Therefore the Galois correspondence of [**76**] and [**144**] can also be recovered from Theorem 30.7.

Lemma 30.8. *Let R be a simple Artinian ring and let G be a finite group of outer automorphisms of R. Then R^G is simple Artinian and $1 \in \mathrm{tr}_G(R)$.*

Proof. Set $I = \mathrm{tr}_G(R)$ so that I is a nonzero ideal of R^G by Lemmas 26.1 and 29.5(ii). According to Proposition 30.3(i)(ii), R^G is prime and satisfies the bimodule property. In particular, IR contains a nonzero ideal of R so $IR = R$ and $\ell_R(I) = 0$.

We show first that if J is any nonzero right ideal of R^G, then J contains a nonzero minimal right ideal. Since $R = \mathrm{M}_n(D)$, the ring of $n \times n$ matrices over the division ring D, we can choose $a \in J \setminus 0$ to have minimal rank as a matrix in R. Then aI is a nonzero minimal right ideal of R^G contained in J. To see this, choose $b \in aI$ with $b \neq 0$. Then $bR \subseteq aIR = aR$, so $aR = bR$ by the minimality of the rank of a. Applying the trace map yields

$$aI = a \operatorname{tr}_G(R) = \operatorname{tr}_G(aR) = \operatorname{tr}_G(bR) = b \operatorname{tr}_G(R) \subseteq bR^G$$

and aI is indeed minimal.

Now in any semiprime ring, a minimal one-sided ideal is generated by an idempotent. Thus if J_1 is a minimal right ideal of R^G, then $J_1 = e_1 R^G$ for some idempotent e_1 and $R^G = e_1 R^G \oplus (1 - e_1) R^G = J_1 \oplus J_1'$. Assuming $J_1' \neq 0$, we can find a minimal right ideal $e_2 R^G = J_2$ contained in J_1'. Since J_2 is a direct summand of R^G, it is a direct summand of J_1' so $R^G = J_1 \oplus J_2 \oplus J_2'$. We continue in this manner and observe that the procedure must stop after at most n steps, since $R = \mathrm{M}_n(D)$ cannot contain more that n mutually orthogonal idempotents. Thus R^G is a finite sum of minimal right ideals and hence is Artinian. Finally, since R^G is also prime, the ring is simple Artinian and $1 \in \operatorname{tr}_G(R)$ by Lemma 30.6. ∎

The remaining aspects of the Galois theory will be considered in the next section.

EXERCISES

1. Let G be an M-group on R and let $e \neq 0, 1$ be an idempotent of B. If $0 \neq I$ is an ideal of R with $eI \subseteq R$, show that eI is an nonzero (R^G, R)-subbimodule of R which contains no nonzero ideal of R. Thus Proposition 30.3(ii) fails, according to the definition of the bimodule property given here, if $\mathcal{B}(G)$ is not a domain.

2. Show by example that Proposition 30.3(i) also fails if $\mathcal{B}(G) = B$ is not a domain. It is shown in [**134**] that if G is an M-group, then R^G is semiprime with minimal primes in one-to-one correspondence with the G-centrally primitive idempotents of B.

3. Let G be an N*-group with $\mathcal{B}(G)$ a domain and let $S \supseteq R^G$ with $H = \operatorname{Gal}(R/S)$. Use Proposition 30.5 and Lemma 10.8(iii)

to show that $Q_\ell(S) = Q_\ell(R^H) \subseteq Q_\ell(R)$ and $Q_s(S) = Q_s(R^H) \subseteq Q_s(R)$. Conclude that $S = R^H$ if and only if $S = Q_\ell(S) \cap R$ and that this is equivalent to S being ideal-cancellable.

4. Let R be a prime ring and let G be the set of all $\sigma \in$ $\mathrm{Aut}(Q_s(R))$ such that $I^\sigma \subseteq R$ and $I^{\sigma^{-1}} \subseteq R$ for some nonzero ideal I of R depending on σ. Show first that I^σ contains a nonzero ideal of R and then that G is a subgroup of $\mathrm{Aut}(Q_s(R))$. Furthermore, prove that G contains all inner automorphisms of $Q_s(R)$.

5. Let $R = K[x, x^{-1}, y, y^{-1} \mid xy = \lambda yx]$ be the simple ring described in Exercise 5 of Section 28 and let σ be the K-algebra automorphism of R determined by $x^\sigma = x^{-1}$ and $y^\sigma = y^{-1}$. Show that σ is well defined, has order 2 and is X-outer. If $G = \langle \sigma \rangle$ and $\mathrm{char}\, K = 2$, prove that $1 \notin \mathrm{tr}_G(R)$ and hence that R^G is not simple by Lemma 30.6. This is an example of [150]. Are there any nontrivial intermediate rings?

6. Suppose σ is a field automorphism of K of finite order ≥ 3 and let F be the fixed field $F = K^{\langle \sigma \rangle}$. Set $R = \mathrm{M}_2(K)$ so that $G = \langle \mathrm{GL}_2(K), \sigma \rangle$ acts on R. Show that G is an N*-group with $B = \mathrm{M}_2(K)$ and $R^G = F$, embedded as scalar matrices. Now define $S = \{ \mathrm{diag}(a, a^\sigma) \mid a \in K \}$ so that $S \supseteq R^G$ and $S \cong K$. Show that S is ideal-cancellable, but not a fixed ring.

31. Almost Normal Subgroups

In this section we complete our study of the formal aspects of the Galois theory of rings. Here we are concerned with normal subgroups H of the Galois group G and with the action of G/H on the fixed ring R^H. The notation will be as before. We start with

Theorem 31.1. (Extension) [88] [134] *Let G be an N*-group acting on the prime ring R with $B = \mathcal{B}(G)$ a domain and let $S \supseteq R^G$ be an intermediate ring. If $\varphi\colon S \to R$ is a ring isomorphism into, with φ the identity on R^G, then φ is the restriction of some $g \in G$.*

Proof. Let us recall some notation and observations from the beginning of the proof of Proposition 30.3(iii). First, $H = \mathrm{Gal}(R/S) \subseteq$

G and Λ is a transversal for G_{inn} in G chosen with $1 \in \Lambda$ and $\Lambda \cap (G_{\text{inn}}H) \subseteq H$. Next, $\tau(x) = \sum_{i,g} a_{i,g} x^g b_{i,g}$ and $0 \neq I \lhd R$ are given as usual by Proposition 29.2 and Lemma 29.3. Third, there exist elements $z_{i,g} \in Q$ such that if $z_{i,g} \neq 0$, then $g \in H$ and $z_{i,g}$ is a unit of $\mathbf{C}_Q(R^H)$. Finally if $\bar{T}(x) = \sum_{i,g} a_{i,g} x^g z_{i,g}$, then there exists a nonzero ideal J of R with $\bar{T}(xj)$ a right (R, S)-truncation of τ for all $j \in J$.

Now for each i, g we construct a map $\theta_{i,g} : J \to Q$ as follows. Let $j \in J$ and write $\bar{T}(xj) = \sum_k \tau(xr_k) s_k$ with $r_k \in R$ and $s_k \in S$. Then we set

$$\theta_{i,g}(j) = \sum_k (r_k)^g b_{i,g} (s_k)^\varphi.$$

We must first show that $\theta_{i,g}$ is well defined. To this end, suppose $\bar{T}(xj)$ is also equal to $\sum_k \tau(x\bar{r}_k) \bar{s}_k$ with $\bar{r}_k \in R$ and $\bar{s}_k \in S$. Then for any $y \in I$ we have

$$\sum_k \tau(yr_k) s_k = \sum_k \tau(y\bar{r}_k) \bar{s}_k$$

and, since $\tau(I) \subseteq R^G$, we obtain by applying φ

$$\sum_k \tau(yr_k)(s_k)^\varphi = \sum_k \tau(y\bar{r}_k)(\bar{s}_k)^\varphi.$$

In other words, the trace form

$$\tilde{T}(x) = \sum_k \tau(xr_k)(s_k)^\varphi - \sum_k \tau(x\bar{r}_k)(\bar{s}_k)^\varphi = \sum_{i,g} a_{i,g} x^g \tilde{b}_{i,g}$$

vanishes on I. Note that $\tilde{T}(x^{g^{-1}})$ is an outer trace form for any $g \in \Lambda$ and that the left-hand coefficients corresponding to the automorphism 1 are C-linearly independent. Thus Lemma 29.5(ii) implies that

$$0 = \tilde{b}_{i,g} = \sum_k (r_k)^g b_{i,g} (s_k)^\varphi - \sum_k (\bar{r}_k)^g b_{i,g} (\bar{s}_k)^\varphi$$

and $\theta_{i,g}$ is indeed well defined.

Now it is clear that $\theta_{i,g} : J \to Q$ is additive. Furthermore, since $\bar{T}(xrj)$ can be obtained from $\bar{T}(xj)$ by replacing x by xr, we have easily $\theta_{i,g}(rj) = r^g\theta_{i,g}(j)$. Observe that $\theta_{i,g}(J) \subseteq Rb_{i,g}R$ and that for some nonzero ideal L of R we have $L^g b_{i,g} \subseteq R$. Hence $\theta_{i,g}(LJ) = L^g\theta_{i,g}(J) \subseteq R$ and, by replacing J with LJ if necessary, we can assume that $\theta_{i,g}: J \to R$. Since the composite map $(\theta_{i,g})^{g^{-1}}: J \to R$ is a left R-module homomorphism, there exists $\bar{q}_{i,g} \in Q_\ell(R)$ with $\theta_{i,g}(j)^{g^{-1}} = j\bar{q}_{i,g}$. Equivalently, $\theta_{i,g}(j) = j^g q_{i,g}$ where we set $q_{i,g} = (\bar{q}_{i,g})^g \in Q_\ell(R)$.

As we observed earlier, if $z_{i,g} \neq 0$ then $g \in H$ and $z_{i,g}$ is a unit of Q centralizing R^H. It follows that $\bar{T}(xjs) = \bar{T}(xj)s$ for $s \in S$ and hence that $\theta_{i,g}(js) = \theta_{i,g}(j)s^\varphi$. Thus

$$j^g s^g q_{i,g} = \theta_{i,g}(js) = \theta_{i,g}(j)s^\varphi = j^g q_{i,g}s^\varphi$$

and, since this holds for all $j \in J$, we have $s^g q_{i,g} = q_{i,g}s^\varphi$. With this, we can now show that $q_{i,g} \in Q$. Indeed, it follows from $J^g q_{i,g} \subseteq R$ and $S^g q_{i,g} = q_{i,g}S^\varphi$ that

$$q_{i,g}S^\varphi(J \cap S)^\varphi R = S^g \cdot (J \cap S)^g q_{i,g} \cdot R \subseteq R.$$

But S and S^φ satisfy the bimodule property, by Proposition 30.3(ii), so $J \cap S \neq 0$ and $S^\varphi(J \cap S)^\varphi R$ contains a nonzero ideal of R. Thus since $q_{i,g} \in Q_\ell(R)$, Proposition 10.6 implies that $q_{i,g} \in Q_s(R) = Q$. Note further that both φ and g fix R^G and hence $q_{i,g} \in \mathbf{C}_Q(R^G) = B$, by Proposition 30.1.

It remains to find some i, g with $q_{i,g} \neq 0$. To this end, observe that $\bar{T}$ does not vanish on IJ, by Lemma 29.5(ii), and say $\bar{T}(yj) \neq 0$ with $y \in I$ and $j \in J$. If $\bar{T}(xj) = \sum_k \tau(xr_k)s_k$ with $r_k \in R$ and $s_k \in S$, then $0 \neq \bar{T}(yj) = \sum_k \tau(yr_k)s_k$. But $\tau(yr_k) \in R^G \subseteq S$ and φ is one-to-one and fixes R^G; thus

$$0 \neq \sum_k \tau(yr_k)^\varphi(s_k)^\varphi = \sum_k \tau(yr_k)(s_k)^\varphi.$$

In particular, for some i, g we must have $0 \neq \sum_k (r_k)^g b_{i,g}(s_k)^\varphi = \theta_{i,g}(j)$. It follows for this i, g that $\theta_{i,g} \neq 0$ and hence that $q_{i,g} \neq 0$. But then, since G is an N*-group, $q_{i,g}$ is a unit of the division ring B which induces an automorphism $g_0 \in G_{\mathrm{inn}}$. We conclude, therefore, from $s^g q_{i,g} = q_{i,g}s^\varphi$ that φ is the restriction of gg_0 to S and the result follows. ∎

Now suppose H is an N-subgroup of G. Then R^H is a prime subring of R and it is natural to consider $\mathrm{Gal}(R^H/R^G)$. We show below that this group is isomorphic to N/H where $N = \mathbf{N}_G(H)$ and that N/H is an N*-group in its action. The proof of the latter is routine, but somewhat tedious.

Lemma 31.2. *Let G be an N*-group in its action on the prime ring R and assume that $\mathcal{B}(G) = B$ is a domain. Let H be an N-subgroup of G and set $N = \mathbf{N}_G(H)$.*

i. The restriction map yields an isomorphism between the groups N/H and $\mathrm{Gal}(R^H/R^G)$.

ii. $\mathcal{B}_{R^H}(N/H) = B^H \subseteq B$ and hence it is a domain.

iii. N/H is an N-group in its action on R^H.*

Proof. (i) It is trivial to see, from Theorem 30.2, that $g \in G$ stabilizes R^H if and only if $g \in N = \mathbf{N}_G(H)$. In particular, by way of restriction, we obtain a homomorphism from N to $\mathrm{Gal}(R^H/R^G)$. Since H is an N-group, the kernel of this map is H, by Theorem 30.2 again. Moreover, the map is onto by the previous theorem. Thus $N/H \cong \mathrm{Gal}(R^H/R^G)$.

(ii) According to Proposition 30.5, $Q_s(R^H) = Q_s(R)^H = Q^H$. Now suppose that q is a unit of Q^H which acts, by conjugation, like an element of N on R^H. Then q centralizes R^G and hence $q \in B$, by Proposition 30.1. It follows that $\mathcal{B}_{R^H}(N/H) \subseteq B \cap Q^H = B^H$. Conversely, suppose q is a unit of B^H. Since G is an N*-group, conjugation by q on R gives rise to an automorphism $g \in G_{\mathrm{inn}}$. Furthermore, $q^{-1}R^H q \subseteq R \cap Q^H = R^H$. Thus $g \in N$ and conjugation by q on R^H gives rise to an automorphism in $(N/H)_{\mathrm{inn}}$. We conclude in particular that $\mathcal{B}_{R^H}(N/H) = B^H$.

Note also that G acts on the field $C = \mathbf{Z}(Q)$ and that G_{inn} acts trivially. Thus, since G/G_{inn} is finite, it follows from the Galois theory of fields that $F = C^G$ is a subfield of C with $(C : F) < \infty$. Hence $\dim_F B^H < \infty$. But $F \subseteq \mathbf{Z}(Q^H)$ and thus B^H is finite dimensional over the extended centroid $\mathbf{Z}(Q^H)$ of R^H.

(iii) It is clear from (ii) above that $\mathcal{B}_{R^H}(N/H)$ is a finite dimensional division algebra over the extended centroid of R^H. Furthermore, every nonzero element of this division ring gives rise, via conjugation, to an X-inner automorphism in N/H. Thus we need only prove that the subgroup of N/H consisting of X-inner automorphisms has finite index. For this, let $Z = \mathbf{C}_B(R^H)$ and write $\mathcal{Z}$ for the multiplicative group $Z^{\bullet}$. Next, let $\mathcal{N}$ be the multiplicative group of all units of B which give rise to elements of N_{inn} in their action on R. Our goal is to show that $\mathcal{Z} \triangleleft \mathcal{N}$ and that $\mathcal{Z}(\mathcal{N} \cap B^H)$ is a subgroup of $\mathcal{N}$ of finite index. Once this is proved, the result will follow quickly. Note that Z and B^H commute elementwise, since $Z = \mathcal{B}(H)$ is spanned by units which act by conjugation on Q like elements of the group $H_{\mathrm{inn}} \subseteq H$.

Suppose $h \in H$ and $y \in \mathcal{N}$. If y gives rise to the automorphism $a \in N$, then $y^{-1}y^h$ is a unit of B which gives rise to the automorphism $a^{-1}a^h = a^{-1}h^{-1}ah \in H$, since N normalizes H. Thus $y^{-1}y^h \in \mathbf{C}_B(R^H) = Z$ and we conclude that $y^h = yz$ for some $z \in \mathcal{Z}$. This observation is used twice in the proof.

Since H is an N-subgroup of G and G is an N*-group, all units of Z give rise to automorphisms in H and hence in N. Thus $\mathcal{Z}$ is a subgroup of $\mathcal{N}$ and in fact $\mathcal{Z} \triangleleft \mathcal{N}$ since $H \triangleleft N$. This implies that $\mathcal{Z}(\mathcal{N} \cap B^H)$ is a subgroup of $\mathcal{N}$. Now $\mathcal{N}$ acts on Z by conjugation and hence also on T, the center of Z. Since T is an extension field of C of finite degree and $\mathcal{N}$ centralizes C, we conclude that $\mathcal{N}_1 = \mathcal{N} \cap \mathbf{C}_B(T)$ has finite index in $\mathcal{N}$. Thus it suffices to show that $\mathcal{Z}(\mathcal{N} \cap B^H)$, which is clearly a subgroup of $\mathcal{N}_1$, has finite index in $\mathcal{N}_1$.

Let $y \in \mathcal{N}_1$. Then conjugation by y yields an automorphism of the finite dimensional division algebra Z which fixes its center T. By the Skolem-Noether theorem, this automorphism must be inner on Z. Thus there exists $z \in \mathcal{Z}$ such that $z^{-1}y$ centralizes Z. In other words, $\mathcal{N}_1 = \mathcal{Z}\mathcal{N}_2$ where $\mathcal{N}_2 = \mathcal{N}_1 \cap \mathbf{C}_B(Z)$. Thus, we have

$$\mathcal{N}_2 \cap \mathcal{Z}(\mathcal{N} \cap B^H) = (\mathcal{Z} \cap \mathcal{N}_2)(\mathcal{N} \cap B^H) = T^{\bullet}(\mathcal{N} \cap B^H)$$

and it suffices to show that $|\mathcal{N}_2 : T^{\bullet}(\mathcal{N} \cap B^H)| < \infty$. Note that H acts on B, Z and T and that H_{inn} acts trivially on T. Thus if $L = \mathbf{C}_H(T)$, then $L \supseteq H_{\mathrm{inn}}$ so $|H/L| < \infty$. Furthermore, since $H \subseteq N$, it is clear that H acts on $\mathcal{N}$ and then on $\mathcal{N}_2$. We show below that $|\mathcal{N}_2 : \mathcal{N}_2 \cap B^L| < \infty$ and then that $\mathcal{N}_2 \cap B^L = T^{\bullet}(\mathcal{N} \cap B^H)$. This will surely yield our goal.

Observe that H_{inn} centralizes $\mathcal{N}_2$, so the finite group L/H_{inn} acts on $\mathcal{N}_2$. Fix $h \in L$ and, for each $y \in \mathcal{N}_2$, write $y^h = y\lambda(y)$ where $\lambda(y) \in \mathcal{Z}$ by the observation of the second paragraph. Since $y^h, y \in \mathcal{N}_2$, we see that $\lambda(y) \in \mathcal{Z} \cap \mathcal{N}_2 = T^\bullet$ and thus λ maps $\mathcal{N}_2 \to T^\bullet$. Indeed, if $y_1, y_2 \in \mathcal{N}_2$ then, since $\mathcal{N}_2$ centralizes T, we have

$$y_1 y_2 \lambda(y_1 y_2) = (y_1 y_2)^h = (y_1)^h (y_2)^h$$
$$= y_1 \lambda(y_1) y_2 \lambda(y_2) = y_1 y_2 \lambda(y_1) \lambda(y_2)$$

and $\lambda \colon \mathcal{N}_2 \to T^\bullet$ is actually a linear character. Furthermore, since $h \in L$ acts trivially on $T^\bullet$, we have easily $y^{h^m} = y\lambda(y)^m$ for any integer m. But $|L/H_{\text{inn}}| < \infty$, so $h^n \in H_{\text{inn}}$ for some $n \geq 1$ and thus λ is a homomorphism from $\mathcal{N}_2$ into the finite group of nth roots of unity in T. We conclude therefore that h centralizes a subgroup of finite index in $\mathcal{N}_2$, namely the kernel of λ. Since this is true for each element $h \in L$ and since L/H_{inn} is finite, we deduce that $|\mathcal{N}_2 : \mathcal{N}_2 \cap B^L| < \infty$.

Note that $\mathcal{N}_2 \cap B^L \supseteq T^\bullet(\mathcal{N} \cap B^H)$ and fix an element $y \in \mathcal{N}_2 \cap B^L$. Then for each $h \in H$ we have, by the observation of the second paragraph again, $y^h = y\mu(h)$ where $\mu(h) \in \mathcal{Z}$. Again since $y^h, y \in \mathcal{N}_2$, we see that $\mu(h) \in \mathcal{Z} \cap \mathcal{N}_2 = T^\bullet$ and thus μ maps H to $T^\bullet$. Indeed, since L acts trivially on $y \in \mathcal{N}_2 \cap B^L$, μ is actually a map from the finite group H/L to $T^\bullet$. Next suppose $h_1, h_2 \in H$. Then

$$y\mu(h_1 h_2) = y^{h_1 h_2} = \big(y\mu(h_1)\big)^{h_2} = y\mu(h_2)\mu(h_1)^{h_2}$$

so μ satisfies Noether's equation $\mu(h_1 h_2) = \mu(h_1)^{h_2}\mu(h_2)$. Therefore, by the above remarks and the fact that H/L acts faithfully on the field T, we conclude from [83, page 75] that μ is a trivial crossed homomorphism. In other words, there exists $t \in T^\bullet$ with $\mu(h) = t/t^h$ for all $h \in H$. But then $y^h = y\mu(h) = yt(t^h)^{-1}$ implies that $yt \in \mathcal{N} \cap B^H$. Hence, since $yt = ty$, we have $y = t^{-1} \cdot ty \in T^\bullet(\mathcal{N} \cap B^H)$. Thus $\mathcal{N}_2 \cap B^L = T^\bullet(\mathcal{N} \cap B^H)$ and, as indicated above, this yields our goal.

Finally since $|N : N_{\text{inn}}| < \infty$, it follows that the restriction of $\mathcal{N}$ to R^H is a subgroup of finite index in N/H. Hence the restriction of $\mathcal{Z}(\mathcal{N} \cap B^H)$ also has finite index. But $\mathcal{Z} \subseteq \mathbf{C}_B(R^H)$ restricts to 1 and $\mathcal{N} \cap B^H$ restricts to the subgroup of X-inner automorphisms in N/H. We conclude therefore that $(N/H)_{\text{inn}}$ has finite index in N/H and the result follows. ∎

Since $\mathcal{B}(G)$ is a domain, it follows that any subgroup of G is an M-group in its action on R. In particular, this applies to N above. On the other hand, N need not be an N*-subgroup; we will consider some examples after the next theorem.

Definition. Let G be an N*-group in its action on the prime ring R. If N is an M-subgroup of G, then N can be completed to an N*-subgroup $\tilde{N}$ of G by adjoining to N the action of all units of $\mathcal{B}(N)$. Thus clearly $\mathcal{B}(N) = \mathcal{B}(\tilde{N})$ and $R^N = R^{\tilde{N}}$ since any element of R fixed by N is fixed by all units of $\mathcal{B}(N)$. We say that H is *almost normal* in G if for $N = \mathbf{N}_G(H)$ we have $\tilde{N} = G$. In addition, we say that the extension R^H/R^G is N*-*Galois* if $\text{Gal}(R^H/R^G)$ is an N*-group in its action on R^H with fixed ring equal to R^G.

We can now quickly prove

Theorem 31.3. [134] *Let G be an N*-group of automorphisms of the prime ring R and assume that $B = \mathcal{B}(G)$ is a domain. If H is an N-subgroup of G, then the following are equivalent.*

    i. H is almost normal in G.

    ii. R^H is N-Galois over R^G with algebra of the group a domain.*

    iii. $\text{Gal}(R^H/R^G)$ acts on R^H with fixed ring R^G.

Proof. (i) $\Rightarrow$ (ii) If $N = \mathbf{N}_G(H)$, then by assumption $\tilde{N} = G$. According to Lemma 31.2, $N/H = \text{Gal}(R^H/R^G)$ is an N*-group in its action on R^H with algebra of the group a domain. Furthermore, we have
$$(R^H)^{(N/H)} = R^N = R^{\tilde{N}} = R^G.$$
By definition, R^H is N*-Galois over R^G.

    (ii) $\Rightarrow$ (iii) This is obvious.

    (iii) $\Rightarrow$ (i) By Lemma 31.2 again, $\text{Gal}(R^H/R^G) = N/H$ where $N = \mathbf{N}_G(H)$ and note that $\tilde{N}$ is an N*-subgroup of G. By assumption, we have
$$R^G = (R^H)^{\text{Gal}(R^H/R^G)} = (R^H)^{N/H} = R^N = R^{\tilde{N}}.$$

Thus by Theorem 30.4, $\tilde{N} = G$ and H is almost normal in G. ∎

We remark that most of the Galois theory results we have considered extend to the case of arbitrary N*-action without any additional assumption on $\mathcal{B}(G)$. Of course, other properties do come into play. For example, in the characterization of fixed rings $S \supseteq R^G$, S must satisfy more than just ideal cancellation. But the above theorem is different; it is more restrictive in nature and certain assumptions on both $\mathcal{B}(G)$ and $\mathcal{B}(H)$ are needed even in the general case.

Now let us consider some examples to show that, in the previous theorem, N need not be an N*-group in its action on R. To start with, let $D = K + Ki + Kj + Kk$ be the quaternion division algebra over the real field K and let G be the group of all inner automorphisms of D. Then G is an N*-group with $\mathcal{B}(G) = D$ and $D^G = K$. Suppose H is the subgroup of G consisting of all automorphisms induced by nonzero elements of $F = K + Ki = K[i]$. Then H is an N*-group and $\mathcal{B}(H) = F = D^H$. Note that F/K is Galois, but only admitting outer automorphisms. Furthermore, $N = \mathbf{N}_G(H) = H \cup Hj$ so $N \neq G$. Of course $\mathcal{B}(N) = D$ and $\tilde{N} = G$.

This is actually part of a more general phenomenon. Namely, let D be any finite dimensional division algebra with center K and let G be the group of all inner automorphisms. Then, as above, $\mathcal{B}(G) = D$ and $D^G = K$. Furthermore, if H is an N-subgroup of G, then $\mathcal{B}(H) = F$ is a division K-subalgebra. Now observe that if $H \triangleleft G$, then $F^\bullet \triangleleft D^\bullet$ and the Cartan-Brauer-Hua theorem (see [82, page 186]) implies that either $F = K$ or $F = D$. In other words, $H \triangleleft G$ if and only if $H = \langle 1 \rangle$ or $H = G$.

The following is an immediate consequence of Lemma 31.2(ii) and Theorem 31.3.

Corollary 31.4. *Let G be a finite group of X-outer automorphisms of the prime ring R. If H is a subgroup of G, then $H \triangleleft G$ if and only if the extension R^H/R^G is Galois with $\mathrm{Gal}(R^H/R^G)$ a finite group of X-outer automorphisms.*

For our final application of traces and truncation we allow $\mathcal{B}(G)$ to be a general semisimple algebra. The additional work required for this is fairly minimal.

Theorem 31.5. [88] *Let G act as an M-group on the prime ring R. Then there exists a nonzero ideal A of R with the property that, for any $a \in A$, aR is contained in a finitely generated right R^G-submodule of R and Ra is contained in a finitely generated left R^G-submodule of R. In particular, if R is simple, then R is finitely generated as a right and as a left R^G-module.*

Proof. For each $\alpha \in B$, define

$$A_\alpha = \left\{ r \in R \ \middle| \ \alpha R r \subseteq \sum_{i=1}^{n} R^G r_i \text{ for some } n < \infty \text{ and } r_i \in R \right\}.$$

It follows easily that A_α is a two-sided ideal of R. The goal is to show that $A_1 \neq 0$. To this end, define $W = \{ \alpha \in B \mid A_\alpha \neq 0 \}$. Then $0 \in W$ and W is closed under addition since clearly $A_\alpha \cap A_\beta \subseteq A_{\alpha+\beta}$ for $\alpha, \beta \in B$. Moreover, if $0 \neq J \triangleleft R$ with $\beta J \subseteq R$, then the inclusion $\alpha\beta Jr \subseteq \alpha Rr$ implies that $A_{\alpha\beta} \subseteq JA_\alpha$. Thus W is a right ideal of B and hence a C-subalgebra of $B = \mathcal{B}(G)$.

Suppose by way of contradiction that $W \neq B$. As usual, let $\tau = \sum_{i,g} a_{i,g} x^g b_{i,g}$ and $0 \neq I \triangleleft R$ be given as in Proposition 29.2 and Lemma 29.3. Furthermore, assume that the basis $\{ a_{i,1} \}$ is chosen compatibly with $B = W' \oplus W$ where W' is a complementary C-subspace and with $a_{1,1} \in W'$. By Lemma 29.5(i), there exists a right (R, R)-truncation $\tilde{T}$ of τ with

$$\alpha x \tilde{b}_{1,1} = \tilde{T}(x) = \sum_{k} \tau(x r_k) s_k.$$

Here $\alpha = \sum_i c_i a_{i,1} \in B$ with $c_i \in C$ and $c_1 = 1$. Moreover, $\tilde{b}_{1,1} \in R \setminus 0$. Since $\tau(I) \subseteq R^G$, we see that $\alpha I \tilde{b}_{1,1} \subseteq \sum_k R^G s_k$ so $I \tilde{b}_{1,1} \subseteq A_\alpha$ and hence $A_\alpha \neq 0$. On the other hand, by the choice of the basis $\{ a_{i,1} \}$ and the fact that $c_1 = 1$, it follows that $\alpha \notin W$, a contradiction.

We conclude therefore that $W = B$ so $1 \in W$ and $A_1 \neq 0$. Similarly A_1', the right analog of A_1, is also nonzero and the result follows with $A = A_1 \cap A_1'$. ∎

As we mentioned above, if R is simple, then R is finitely generated as both a right and a left R^G-module. However, if R is not simple, this need not be the case. For example, let $R = K\langle x, y \rangle$ be the free algebra on two generators over the field K of characteristic not 2 and let σ be the X-outer automorphism of order 2 determined by $x^\sigma = -x$ and $y^\sigma = y$. If $G = \{1, \sigma\}$, then $|G| = 2$ and R^G is spanned by all monomials containing x an even number of times. Suppose R is finitely generated as a right R^G-module. Then for some n we have $R = \sum_\mu \mu R^G$ where μ runs over all monomials of degree $\leq n$. But $y^n x$ cannot belong to the right-hand side, so we have a contradiction.

In case R is a simple Artinian ring and G is a finite group of outer automorphisms, we can sharpen the above result using skew group ring techniques. Indeed we have

Theorem 31.6. [146] *Let G be a finite group of automorphisms of the simple Artinian ring R and assume that the skew group ring RG is simple. For example, this occurs if G is outer on R. Then there exists a division ring D and an integer k with $R^G \cong \mathrm{M}_k(D)$, $RG \cong \mathrm{M}_{k|G|}(D)$ and $1 \leq k \leq \mathrm{rank}\, R_R$. Furthermore, R is a free right or left R^G-module on $|G|$ generators.*

Proof. If G is outer on R, then RG is simple by Corollary 12.6.

Now assume that RG is simple. Since R is Artinian and G is finite, RG is Artinian and hence $RG \cong \mathrm{M}_n(D)$ for some division ring D. In the following, all dimensions will be computed with D acting on the left. In particular, we have $\dim_D RG = n^2$. Furthermore, if V is an irreducible right RG-module, then $\mathrm{End}_{RG}(V) = D$, acting on the left, and $\dim_D V = n$. By Lemma 26.2(i), R is a cyclic RG-module and thus R_{RG} is isomorphic to $V^{\oplus k}$, the direct sum of k copies of V. Clearly $1 \leq k \leq \mathrm{rank}\, R_R$. By computing dimensions we get $kn = \dim_D V^{\oplus k} = \dim_D R$ and hence $kn|G| = \dim_D RG = n^2$. Thus $k|G| = n$.

Again, since $R_{RG} \cong V^{\oplus k}$, it follows that

$$\mathrm{End}_{RG}(R) = \mathrm{End}_{RG}(V^{\oplus k}) \cong \mathrm{M}_k(D),$$

acting on the left. Furthermore, if W is an irreducible left $M_k(D)$-module, then $\dim_D W = k$. Thus R, as a left $M_k(D)$-module, is a direct sum of

$$\dim_D R/\dim_D W = kn/k = n = k|G|$$

copies of W. But $W^{\oplus k}$ is a free left $M_k(D)$-module, so we see that $R = (W^{\oplus k})^{\oplus |G|}$ is a free left $M_k(D)$-module on $|G|$ generators. Finally, $\mathrm{End}_{RG}(R) = R^G$, by Lemma 26.2(i) again, so the result follows. ∎

Note that the above yields an alternate proof of Lemma 30.8.

EXERCISES

1. Suppose D is a noncommutative division ring, finite dimensional over its center K, and let $S = D\langle X\rangle$ be the free D-ring on the variables in the set X. As usual, let $D\langle X\rangle'$ be the augmentation ideal of S and set $R = K + D\langle X\rangle' \subseteq S$. Then, as in the example following Theorem 30.4, $G = D^\bullet/K^\bullet$ is the full group of X-inner automorphisms of R and it is an N^*-group in its action on R. For which sets X is R a finitely generated R^G-module?

2. Let G be a finite group of X-outer automorphisms of the prime ring R and, in the skew group ring RG, define $I = R\dot{G}R \cap R$. Prove that $0 \neq I \triangleleft R$ and that I is contained in the ideal A of Theorem 31.5. For the latter, let $a \in I$ so $a = \sum_{i=1}^n r_i\dot{G}s_i$. Now compute $\dot{G}Ra$ or $aR\dot{G}$.

3. Let $R = M_2(C)$ where C is the field of complex numbers and let σ act on R by

$$\sigma\colon \begin{pmatrix} a & b \\ c & d \end{pmatrix} \mapsto \begin{pmatrix} \bar{d} & -\bar{c} \\ -\bar{b} & \bar{a} \end{pmatrix}$$

where $^-$ denotes complex conjugation. Prove that σ is an outer automorphism of order 2 and determine $R^{\{1,\sigma\}}$. In particular, observe that k can be strictly less than rank R_R in Theorem 31.6.

The following examples, with K a field, indicate some of the limitations necessary in extending Theorem 31.1 to general N*-groups. Verify the details.

4. Let $R = M_4(K)$ and $G = GL_4(K)$ so that $R^G = K$. Set $S_1 = \{\,\mathrm{diag}(a, a, b, b) \mid a, b \in K\,\}$ and let $S_2 = \{\,\mathrm{diag}(a, b, b, b) \mid a, b \in K\,\}$. Then the natural isomorphism $\varphi\colon S_1 \to S_2$ cannot be extended to an element of G.

5. Let $\sigma \neq 1$ be an automorphism of K of finite order and set $R = M_2(K)$ and $G = \langle GL_2(K), \sigma \rangle$ so that $R^G = F$, the fixed field of σ. If S is the subring of diagonal matrices, then the isomorphism $\varphi\colon S \to S$ given by $\varphi\colon \mathrm{diag}(a, b) \mapsto \mathrm{diag}(a^\sigma, b)$ cannot be extended to an element of G.

6. Let $T = K\langle x, y, z \rangle$ be the free algebra over $K \neq GF(2)$ and let Sym_3 act on T by permuting the generators. Set $R = M_2(T)$ and let $G = GL_2(K) \times \mathrm{Sym}_3$ act on R. Then $R^G = T^{\mathrm{Sym}_3}$, $C = K$ and $B = M_2(K)$. Now let H be the subgroup of G generated by $GL_1(K) \times GL_1(K)$ and $\begin{pmatrix} 0 & 1 \\ 1 & 0 \end{pmatrix}\sigma$ where σ is the transposition $(x\,y)$. It follows that $S = R^H = \{\,\mathrm{diag}(a, a^\sigma) \mid a \in T\,\}$. If $\varphi\colon S \to S$ is defined by $\varphi\colon \mathrm{diag}(a, a^\sigma) \to \mathrm{diag}(a^\tau, a^{\tau\sigma})$, where τ is the transposition $(y\,z)$, then φ cannot be extended to an element of G.

32. Free Rings and Subrings

We close this chapter by applying some of the Galois theory results of the previous sections to the special case of free rings. We begin, by working in the larger class of rings satisfying appropriate weak algorithms. These were considered at the end of Section 13 in the context of filtered rings. Here we are mainly concerned with graded rings and, in this case, the relevant definitions simplify.

To start with, let $R = \oplus \sum_{i=0}^{\infty} R_i$ be an arbitrary ring graded by Z^+, the semigroup of positive integers. In particular, $1 \in R_0$ and $R_i R_j \subseteq R_{i+j}$. Furthermore, R has a natural degree function and, by convention, $\deg 0 = -\infty$. We restrict our attention to $\mathcal{H} = \bigcup_{i=0}^{\infty} R_i$, the set of homogeneous elements of R. A set $\{\,a_1, a_2, \ldots, a_m\,\} \subseteq \mathcal{H}$

is said to be *right dependent* if there exist elements $b_i \in \mathcal{H}$, not all zero, with $\sum_{i=1}^m a_i b_i = 0$. Moreover, the element $a \in \mathcal{H}$ is said to be *right dependent on* $\{a_1, a_2, \ldots, a_m\}$ if there exist $b_i \in \mathcal{H}$ with $a = \sum_{i=1}^m a_i b_i$. Finally, R satisfies the *n-term weak algorithm* if given any right dependent set $\{a_1, a_2, \ldots, a_m\} \subseteq \mathcal{H}$ with $m \leq n$ and $\deg a_1 \leq \deg a_2 \leq \cdots \leq \deg a_m$, then some a_i is right dependent on $\{a_1, a_2, \ldots, a_{i-1}\}$. If R satisfies the n-term weak algorithm then, by definition, it satisfies the n'-term weak algorithm for all $n' \leq n$. Furthermore, it is easy to see that R satisfies the 1-term weak algorithm if and only if it is a domain.

Of course, any Z^+-graded ring R is filtered by the partial sums $\sum_{i=0}^j R_i$ and it is an easy exercise to verify that the above definition of the n-term weak algorithm agrees with that of Section 13. Indeed the degree inequalities of the original formulation merely indicate that certain leading terms vanish.

Now let G be a group of automorphisms of the Z^+-graded ring $R = \oplus \sum_{i=0}^\infty R_i$. We say that G is *homogeneous* if it stabilizes all the homogeneous components R_i. Note that this implies that $R^G = \oplus \sum_{i=0}^\infty (R_i)^G$ is a graded subring of R. With all this notation behind us, we can now prove

Theorem 32.1. [89] [97] *Let G be a group of homogeneous automorphisms of the Z^+-graded ring R. If R satisfies the n-term weak algorithm with respect to this grading, then so does R^G.*

Proof. We know that R^G is a graded subring of R and we show that it satisfies the n-term weak algorithm with respect to this grading. To this end, suppose $m \leq n$ and $\{a_1, a_2, \ldots, a_m\}$ is a right dependent set consisting of homogeneous elements of R^G with $\deg a_1 \leq \deg a_2 \leq \cdots \leq \deg a_m$. Since R^G is a graded subring of R, $\{a_i\}$ is also right dependent in R and we choose $k \leq m$ minimal with $\{a_1, a_2, \ldots, a_k\}$ right dependent. By assumption, R satisfies the n-term weak algorithm so there exist homogeneous $b_i \in R$ with $a_k = \sum_{i=1}^{k-1} a_i b_i$.

Now let $g \in G$. Since $\{a_i\} \subseteq R^G$, we have

$$a_k = \sum_{i=1}^{k-1} (a_i b_i)^g = \sum_{i=1}^{k-1} a_i (b_i)^g$$

and thus $0 = \sum_{i=1}^{k-1} a_i \big(b_i - (b_i)^g\big)$. But $\{a_1, a_2, \ldots, a_{k-1}\}$ is not right dependent, by the choice of k, and $b_i - (b_i)^g$ is homogeneous, since g is a homogeneous automorphism. Thus $b_i = (b_i)^g$ for all i and all $g \in G$. In other words, $b_i \in R^G$, so a_k is dependent on $\{a_1, a_2, \ldots, a_{k-1}\}$ in R^G and the result follows. ∎

If R is a Z^+-graded ring, we say that R satisfies the *weak algorithm* if it satisfies the n-term weak algorithm for all n. The graded analog of Theorem 13.12 is given below. Notice that, if $R = K\langle X \rangle$ is the free K-algebra on the set X of variables, then R can be Z^+-graded by making each $x \in X$ homogeneous of some arbitrarily assigned positive degree. For example, the usual grading is given by $\deg x = 1$ for all $x \in X$. Furthermore R_0, the 0-component of R, is equal to the field K by our assumption that all generators have positive degree.

Theorem 32.2. [41] *Let $R = \oplus \sum_{i=0}^{\infty} R_i$ be a Z^+-graded ring with $R_0 = K$ a central subfield. Then R is the free associative K-algebra on a right independent homogeneous generating set if and only if R satisfies the weak algorithm.*

Proof. Suppose first that $R = K\langle X \rangle$ is graded as above and let $\{a_1, a_2, \ldots, a_k\} \subseteq \mathcal{H}$ be right dependent. Say $\deg a_1 \leq \deg a_2 \leq \cdots \leq \deg a_k$ and that $\sum_{i=1}^{k} a_i b_i = 0$ with $b_i \in \mathcal{H}$ not all zero. We can of course assume that all terms in the latter sum have the same degree. Thus after deleting zero terms if necessary, we have $\deg b_k \leq \deg b_{k-1} \leq \cdots \leq \deg b_1$. Now let μ be an X-monomial which occurs in b_k and, for each i, write $b_i = b_i' \mu + b_i''$ where μ is not a trailing term of any monomial in the support of b_i''. Then the dependence relation yields

$$\left(\sum_{i=1}^{k} a_i b_i' \right) \mu + \sum_{i=1}^{k} a_i b_i'' = 0.$$

Furthermore, $\deg b_i \geq \deg b_k = \deg \mu$ implies that μ is not a right factor of any X-monomial in $\sum_{i=1}^{k} a_i b_i''$. Thus $\sum_{i=1}^{k} a_i b_i' = 0$. But b_k' is a nonzero element of the field $K = R_0$, so we can solve for a_k and this direction is proved.

Conversely, suppose R satisfies the weak algorithm. We first find X. To this end, note that $\sum_{0<j<n} R_j R_{n-j}$ is a K-subspace of R_n and we can choose $X_n = \{x_{n,1}, x_{n,2}, \ldots\}$ to be a complementary basis. Then $X = \bigcup_{n=1}^{\infty} X_n = \{x_{n,i}\}$ is a set of homogeneous elements of R which we claim freely generates R over K. It is clear that K and X at least generate R. Indeed, if the components R_j for $j < n$ are all in the subring generated by K and X, then the definition of X_n implies that the same is true for R_n.

We must now show that all *formally different* X-monomials are K-linearly independent. We proceed by induction on the number m of monomials μ in a dependence relation $\sum_{j=1}^{m} k_j \mu_j = 0$ with $k_j \in K \backslash 0$. The case $m = 1$ is clear since R is a domain. Furthermore, if all μ_j start with the same $x_{n,i} \in X$, then we can cancel this common factor, again since R is a domain. Thus we may assume that the μ_j's start with at least two different members of X. In addition, we can assume that all μ_j's have the same degree in R.

Finally, by grouping the μ_j's according to their leading factor, the dependence relation becomes $\sum_{n,i} x_{n,i} \alpha_{n,i} = 0$ where each $\alpha_{n,i}$ is a K-linear sum of formally distinct monomials. Furthermore, there are less than m such monomials in each $\alpha_{n,i}$ so, by induction, we have $\alpha_{n,i} \neq 0$. It follows that the set X is right dependent and, since R satisfies the weak algorithm, we obtain

$$x_{n,i} = \sum_{\substack{j \leq n \\ k}} x_{j,k} \beta_{j,k} \qquad (*)$$

for some n, i with each $\beta_{j,k}$ homogeneous. In addition, we can assume that all summands satisfy $\deg x_{j,k} \beta_{j,k} = n$. In particular, if $j = n$ then $\beta_{j,k} \in K$ while if $j < n$ then $x_{j,k} \beta_{j,k} \in R_j R_{n-j}$. But then $(*)$ yields a K-linear dependence of the set X_n modulo $\sum_{0<j<n} R_j R_{n-j}$ and this contradicts the definition of X_n. ∎

As an immediate consequence of the previous two results we have

Corollary 32.3. [89] [97] *Let $R = K\langle X \rangle$ be a free K-algebra graded with $R_0 = K$. If G is a group of homogeneous automorphisms of R, then R^G is a free K^G-algebra on homogeneous generators.*

It is not known whether the fixed ring of a free algebra is free without the assumption that the group acting is homogeneous.

Again let R be a Z^+-graded ring. The interesting cases of the n-term weak algorithm begin with $n = 2$. For example, we have the following correspondence theorem.

Theorem 32.4. Let $R = \oplus \sum_{i=0}^{\infty} R_i$ be a Z^+-graded ring satisfying the 2-term weak algorithm and suppose G is a homogeneous N^*-group of automorphisms of R. Then the maps $H \mapsto R^H$ and $S \mapsto \mathrm{Gal}(R/S)$ yield a one-to-one correspondence between the N^*-subgroups H of G and the intermediate rings $S \supseteq R^G$ which satisfy the 2-term weak algorithm with respect to any filtration.

Proof. Since R is a domain, Theorem 30.4 applies. In particular, if H is an N^*-subgroup of G then, by Theorem 32.1, $S = R^H$ is a graded subring of R satisfying the 2-term weak algorithm and $H = \mathrm{Gal}(R/S)$. In the other direction, if $S \supseteq R^G$ is given, then certainly $H = \mathrm{Gal}(R/S)$ is an N^*-subgroup of G and the goal is to show that $S = R^H$. By Proposition 30.3(iii), we know at least that $R^H \supseteq S \supseteq I$ where I is a nonzero ideal of R^H. If S is symmetrically closed then, by Lemma 10.8(iii), $Q_s(R^H) = Q_s(S) = S$ so $S = R^H$ as required. On the other hand, if S is not symmetrically closed, then Theorem 13.11 implies that $S = D[x; \sigma, \delta]$ is a generalized polynomial ring in the variable x over the division ring D. Note that this ring clearly has a division algorithm and hence every right or left ideal is principal. In particular, $I = \alpha S$ for some nonzero $\alpha \in S$. Finally, if $r \in R^H$ then $\alpha r \in I = \alpha S$ so $\alpha r = \alpha s$ for some $s \in S$. But R is a domain and $\alpha \neq 0$, so $r = s \in S$ and hence $R^H = S$ in this case also. ∎

Suppose G acts on $R = K\langle X \rangle$. According to Corollary 13.6, R admits no nonidentity X-inner automorphism. In particular, G is an M-group in its action if and only if G is finite.

Theorem 32.5. [89] Let $R = K\langle X \rangle$ be a free K-algebra graded so that its generators are homogeneous and $R_0 = K$. Moreover, let G be a finite group of homogeneous automorphisms. Then the maps

$H \mapsto R^H$ and $S \mapsto \mathrm{Gal}(R/S)$ *yield a one-to-one correspondence between the subgroups H of G and the free algebras $S \supseteq R^G$.*

Proof. As we observed above, G is a finite X-outer group and hence an N^*-group in its action on R. By Corollary 32.3, each R^H is a free algebra and, by Theorem 32.2, each free algebra S satisfies the 2-term weak algorithm. Thus Theorem 32.4 now yields the result. ∎

Notice that if $S \supseteq R^G$ and S satisfies the 2-term weak algorithm, then $S = R^H$ so S is in fact a free algebra.

Again, let G be a homogeneous group of automorphisms of $K\langle X \rangle$. For example, if we take the usual grading in $K\langle X \rangle$ and assume that G consists of K-algebra automorphisms, then G homogeneous means that each $g \in G$ acts *linearly*. That is, for all $x \in X$, we have $x^g = \sum_{y \in X} c_{x,y} y$ with $c_{x,y} \in K$. On the other hand, an example of a homogeneous nonlinear action is as follows. Take $X = \{x, y\}$ and define $g \colon K\langle X \rangle \to K\langle X \rangle$ by $x^g = -x + y^2$ and $y^g = y$. Notice that g has order 2 and that g is homogeneous if we define $\deg x = 2$ and $\deg y = 1$. In any case, since R^G is free, it is natural to consider the size of its generating set and, in particular, whether it is finite or infinite. This has been settled at least for linear actions by finite groups. For the proof of this result, we first need some definitions.

To start with, let $R = K\langle X \rangle$ have the usual grading. Then the component R_t is spanned by all monomials $y_1 y_2 \cdots y_t$ with $y_i \in X$. Moreover, if $\sigma \in \mathrm{Sym}_t$, then σ acts linearly on R_t by defining

$$(y_1 y_2 \cdots y_t)^\sigma = y_{1\sigma} y_{2\sigma} \cdots y_{t\sigma}.$$

In this way, Sym_t acts as *place permutations* on R_t. Note that Sym_t is not a group of algebra automorphisms; it only acts on the single component R_t.

Next, if $X = \{x_1, x_2, \ldots, x_d\}$, then an element of particular interest in R is the standard polynomial

$$\delta = S_d(x_1, x_2, \ldots, x_d) = \sum_{\sigma \in \mathrm{Sym}_d} (-)^\sigma x_{1\sigma} x_{2\sigma} \cdots x_{d\sigma}.$$

Note also that if G acts linearly on R, then each element of G is represented by a matrix in its action on $R_1 = KX$ and, in this way, G is a subgroup of $\mathrm{GL}_d(K)$. In particular, we can speak about the determinant of each element of G. We record a few basic facts. In the following, part (i) is clear and part (ii) follows immediately by considering the effect of elementary row operations on δ.

Lemma 32.6. *Let G act linearly on $R = K\langle x_1, x_2, \ldots, x_d\rangle$.*

i. G commutes with all place permutations under the usual grading.

ii. If $g \in G$, then $\delta^g = (\det g)\delta$, where δ is the standard polynomial in $x_1, x_2, \ldots, x_d$.

The proof we offer of the next result is from [**48**]. It significantly simplifies the original arguments.

Theorem 32.7. [**49**] [**91**] *Let G be a finite group of linear automorphisms of the free algebra $K\langle X\rangle$. Then $K\langle X\rangle^G$ is finitely generated if and only if X is finite and G acts as scalar matrices.*

Proof. Write $R = K\langle X\rangle$ and consider the usual grading on R. Suppose first that X is finite and G acts as scalar matrices. Then $G = \langle g\rangle$ is cyclic with $x^g = \epsilon x$ for all $x \in X$ and some $\epsilon \in K^\bullet$ of order e. It follows that R^G is spanned by all monomials of length divisible by e and therefore it is generated by the finitely many monomials of length precisely e.

Now suppose that R^G is finitely generated. If X is infinite, then $R^G \subseteq K\langle W\rangle$ for some finite subset W of X and there are infinitely many free intermediate rings $S \supseteq R^G$. But this contradicts Theorem 32.5 and thus X must be finite; say $|X| = d$. If $\tilde{K}$ is the algebraic closure of K, then certainly G acts on $\tilde{K}\langle X\rangle = \tilde{K}\otimes_K K\langle X\rangle$, $G \subseteq \mathrm{GL}_d(\tilde{K})$ and $\tilde{K}\langle X\rangle^G = \tilde{K}\otimes_K K\langle X\rangle^G$ is finitely generated. Replacing K by $\tilde{K}$ if necessary, we may therefore assume that K is algebraically closed.

Fix g in G. The goal is to show that g is represented by a scalar matrix. Since K is algebraically closed, there is a basis Y of $R_1 = KX$ with respect to which the matrix of g is in Jordan form.

Note that Y also freely generates R as a K-algebra and with the same grading. Thus we may assume that $Y = X$. Let A denote the set of nonidentity X-monomials which occur in the supports of the finitely many generators of R^G. Then R^G is certainly contained in the K-subalgebra of R generated by A. Moreover, since A is finite, we can choose the integer n so that $|G|$ divides n and n is larger than the degrees of the elements in A.

Let $\delta = S_d(x_1, x_2, \ldots, x_d) \in R_d$ be the standard polynomial. If $h \in G$ then, by Lemma 32.6(ii), $\delta^h = (\det h)\delta$ and hence $(\delta^n)^h = (\det h)^n \delta^n = \delta^n$ since $|G|$ divides n. Thus $\delta^n \in (R_{nd})^G$. Furthermore, since the action of G commutes with the place permutations on R_{nd}, it follows that every place permutation of δ^n is also contained in $(R_{nd})^G$. But R^G is contained in the K-algebra generated by the monomials in A, so we conclude that every monomial of length nd occurring in the support of a place permutation of δ^n is a product of elements of A. In particular, this applies to every place permutation of $(x_1 x_2 \cdots x_d)^n$.

Suppose first that g is not a diagonal matrix. Then by considering the top of a nontrivial Jordan block and by appropriately labeling the variables, we can assume that $x_1^g = \lambda x_1 + x_2$, $x_2^g = \lambda x_2 + x_3$ and that the remaining x_i^g do not involve x_1 or x_2. Here of course $\lambda \in K^\bullet$ and the x_3 term above may be missing if the block has size 2. Now $x_1^n x_2^n \cdots x_d^n$ is a place permutation of $(x_1 x_2 \cdots x_d)^n$ and hence is a product of monomials in A, each of which is different from 1 and has degree at most n. It follows that $x_1^m \in A$ for some m with $1 \le m \le n$. In other words, x_1^m occurs in a generator r for R^G, which we may assume to be homogeneous of degree m; furthermore, we may take the coefficient of x_1^m in r to be 1. Thus, for some $k \in K$,

$$r = x_1^m + k x_1^{m-1} x_2 + \alpha,$$

where α involves the other monomials of degree m. Now it follows easily from the nature of the action of g that

$$r^g = (\lambda x_1 + x_2)^m + k(\lambda x_1 + x_2)^{m-1}(\lambda x_2 + x_3) + \alpha^g$$
$$= \lambda^m x_1^m + \lambda^{m-1} x_1^{m-1} x_2 + k\lambda^m x_1^{m-1} x_2 + \beta,$$

where β is a sum of terms not involving x_1^m or $x_1^{m-1}x_2$. But $r^g = r$, so comparing coefficients yields $\lambda^m = 1$ and $k = \lambda^{m-1} + k\lambda^m$, a contradiction.

Thus g must be diagonal and say $x_i^g = \lambda_i x_i$. Suppose that for every monomial $v \in R$ there exists an appropriate $x \in X$ with $vx \notin A$. Then starting with $v = 1$, we can construct a monomial u of degree n all of whose initial segments are not in A. But observe that there is a place permutation of $(x_1 x_2 \cdots x_d)^n$ which starts with u and this monomial is a product of elements of A, a contradiction. In other words, there exists a monomial w with $wx_i \in A$ for all i. Note that g sends any monomial to a scalar multiple of itself. Thus $w^g = \mu w$ for some $\mu \in K^\bullet$. Moreover, since g fixes all generators of R^G, it fixes all monomials in their supports and hence g fixes all members of A. We conclude that $wx_i = (wx_i)^g = \mu\lambda_i wx_i$ for all i, so $\lambda_i = \mu^{-1}$ and g is indeed represented by a scalar matrix. This completes the proof. ∎

If G is allowed to be infinite, then the above is definitely false. For example, if just one $g \in G$ acts as a scalar matrix on X with eigenvalue of infinite order, then $K\langle X\rangle^G = K$.

Suppose $R = K\langle X\rangle$ is any free algebra graded so that each component R_n is finite dimensional over K. Then one can describe the degrees of the generators X by means of a *Hilbert series*. In the following, we just briefly sketch some of the basic ideas and we compute some examples. Define the two power series in ζ by

$$H(R) = \sum_{n=0}^{\infty}(\dim_K R_n)\zeta^n$$

and

$$H(X) = \sum_{n=0}^{\infty}|X \cap R_n|\zeta^n.$$

Note that $H(R)$ has constant term 1 since $R_0 = K$ and, for the same reason, $H(X)$ has constant term 0.

Lemma 32.8. *The Hilbert series are related by*

$$H(R) = \frac{1}{1 - H(X)}.$$

Proof. For $n > 0$, R_n has a basis consisting of the monomials w of degree n and every such monomial is uniquely writable as $w = w'x$ with $x \in X$. If $\deg x = i$, then w' is a monomial of degree $n - i$. Furthermore, for fixed i there are $|X \cap R_i|$ choices for x and $\dim_K R_{n-i}$ choices for the monomial w'. Thus

$$\dim_K R_n = \sum_{i=1}^{n} \dim_K R_{n-i} \cdot |X \cap R_i|.$$

Since $\dim_K R_0 = 1$, we conclude that $H(R) = H(R) \cdot H(X) + 1$ and the lemma is proved. ∎

Suppose for example that $|X| = d$ and that R is graded in the usual manner. Then $H(X) = d\zeta$ so

$$H(R) = \frac{1}{1 - H(X)} = \frac{1}{1 - d\zeta} = \sum_{n=0}^{\infty} d^n \zeta^n$$

and we obtain the not surprising result that $\dim_K R_n = d^n$.

More interesting examples arise when we consider the fixed ring R^G. Let $X = \{x, y\}$ and let $G = \{1, g\}$ act on $R = K\langle X \rangle$ with $x^g = x$ and $y^g = -y$. Here of course $\operatorname{char} K \neq 2$. We grade R as usual and consider the graded free ring R^G. It is clear that R^G is spanned by all monomials containing an even number of y-factors and thus R^G is freely generated by the monomials

$$Y = \{x\} \cup \{yx^iy \mid i \geq 0\}.$$

In other words, $|Y \cap (R^G)_n| = 1$ for all $n \geq 1$ so $H(Y) = \sum_{n=1}^{\infty} \zeta^n = \zeta/(1 - \zeta)$ and hence

$$H(R^G) = \frac{1}{1 - H(Y)} = 1 + \frac{\zeta}{1 - 2\zeta} = 1 + \sum_{n=1}^{\infty} 2^{n-1} \zeta^n.$$

It follows that $\dim_K(R^G)_n = 2^{n-1}$ for all $n \geq 1$.

In case G is finite and $\operatorname{char} K = 0$, there is another way to compute $H(R^G)$. Note that, since $G \subseteq \operatorname{GL}_d(K)$, we can speak about the matrix trace of the elements of G. We have

Theorem 32.9. [49] *Let* $R = K\langle X \rangle$ *be a finitely generated free algebra over the field* K *of characteristic 0 and let* G *be a finite group acting linearly on* R. *Then*

$$\mathrm{H}(R^G) = |G|^{-1} \sum_{g \in G} \frac{1}{1 - \mathrm{tr}(g)\zeta}$$

where $\mathrm{tr}(g)$ *is the matrix trace of* $g \in G$.

Proof. We begin with two observations. Suppose first that V is any finite dimensional module for the group algebra $K[G]$ and let e be the idempotent $e = |G|^{-1}\dot{G}$. Then $V^G = Ve$, so

$$\dim_K V^G = \mathrm{tr}_V(e) = |G|^{-1} \sum_{g \in G} \mathrm{tr}_V(g).$$

Here, of course, tr_V denotes the matrix trace of the representation of G on V. Next it is clear that, as K-vector spaces, $R_n = (R_1)^{\otimes n}$, the n-fold tensor product of R_1 over K. It then follows from this, by choosing appropriate bases if necessary, that $\mathrm{tr}_{R_n}(g) = \mathrm{tr}_{R_1}(g)^n = \mathrm{tr}(g)^n$ for all $g \in G$.

With these facts in hand, we can now quickly prove the result. Indeed, by the above,

$$\dim_K (R_n)^G = |G|^{-1} \sum_{g \in G} \mathrm{tr}_{R_n}(g) = |G|^{-1} \sum_{g \in G} \mathrm{tr}(g)^n$$

and thus

$$\mathrm{H}(R^G) = |G|^{-1} \sum_{g \in G} \sum_{n=0}^{\infty} \mathrm{tr}(g)^n \zeta^n = |G|^{-1} \sum_{g \in G} \frac{1}{1 - \mathrm{tr}(g)\zeta}$$

as required. ∎

In particular, if char $K = 0$ in the example discussed immediately before this theorem, then $\mathrm{tr}(1) = 2$ and $\mathrm{tr}(g) = 0$ so

$$\mathrm{H}(R^G) = \frac{1}{2}\left(1 + \frac{1}{1 - 2\zeta}\right) = 1 + \frac{\zeta}{1 - 2\zeta}$$

as before.

Another example of interest occurs in characteristic $p > 0$. Let $X = \{x_1, x_2, \ldots, x_p\}$ and let g act on $R_1 = KX$ as a Jordan block of size p with eigenvalues 1. Then $g^p = 1$ and one can show that, for all $n \geq 1$, the Jordan form of g on R_n is a sum of these blocks of full size p. It follows that if $G = \langle g \rangle$, then $\dim_K (R_n)^G = (\dim_K R_n)/p = p^{n-1}$ for all $n \geq 1$ and hence

$$H(R^G) = 1 + \sum_{n=1}^{\infty} p^{n-1} \zeta^n = 1 + \frac{\zeta}{1 - p\zeta}.$$

If R^G is freely generated by the homogeneous set Y, then Lemma 32.8 yields

$$H(Y) = 1 - \frac{1}{H(R^G)} = \frac{\zeta}{1 - (p-1)\zeta}$$

and thus $|Y \cap (R^G)_n| = (p-1)^{n-1}$ for all $n \geq 1$.

Finally, we remark that the dichotomy of Theorem 32.7 also manifests itself in the prime correspondence of Theorem 28.3. This is apparent in the following two results which we offer without proof. Note that the assumption $|G|^{-1} \in K$ is required for the prime correspondence to exist.

Theorem 32.10. [137] *Let G act as scalars on the free K-algebra $R = K\langle x_1, x_2, \ldots, x_d \rangle$ with $|G|^{-1} \in K$. Let Q be a prime ideal of R^G and let P be a prime of R lying over Q. If either R^G/Q is finite dimensional over K or satisfies a polynomial identity, then the same is true of R/P.*

Theorem 32.11. [137] *Let G act linearly, but not as scalars, on the free K-algebra $R = K\langle x_1, x_2, \ldots, x_d \rangle$ with $|G|^{-1} \in K$. Then there exist uncountably many primes Q of R^G with $\dim_K R^G/Q = 1$ such that if P is any prime of R lying over Q, then R/P does not even satisfy a polynomial identity.*

Some aspects of the proof are considered in 4–7 below.

EXERCISES

1. Show that the two definitions of the n-term weak algorithm agree in the case of Z^+-graded rings.

2. Let $R = D[x; \sigma, \delta]$ be a generalized polynomial ring over the division ring D and assume that R is Z^+-graded with $R_0 = D$. Prove that R is in fact a skew polynomial ring over D in the variable $y = x - d$ for some $d \in D$. Now suppose that G is a group of homogeneous automorphisms of R. Show that G stabilizes each set Dy^n and deduce that R^G is either D^G or a skew polynomial ring over D^G. In either case, R^G is Noetherian.

3. Let R be a Z^+-graded ring satisfying the 2-term weak algorithm and let G act as an M-group on R. If G is homogeneous, prove that the division ring R_0 is a finite module over R_0^G. As a first step, show that G is an M-group in its action on R_0 if R is symmetrically closed.

In the remaining problems, G is a finite group acting on R with $|G|^{-1} \in R$. We consider the prime correspondence of Theorem 28.3.

4. Let J be a G-stable ideal of R and set $\bar{R} = R/J$. Show that the correspondence between the primes of $\bar{R}$ and $\bar{R}^G$ is consistent with the correspondence in R.

5. Let $I \triangleleft R^G$ and define $\mathrm{T}(I)$ to be the unique largest ideal of R with $\mathrm{tr}_G(\mathrm{T}(I)) \subseteq I$. Prove that $\mathrm{T}(I)$ is G-stable. Furthermore, if Q is a prime ideal of R^G, show that $\mathrm{T}(Q) = \bigcap_{g \in G} P^g$ where P lies over Q. For the latter, go back to the original proof of the correspondence or apply the conclusion of Exercise 4 to $\bar{R} = R/T$.

6. Suppose $R = K\langle x_1, x_2, \ldots, x_d \rangle$ and G acts as scalars on R so that we have $G = \langle g \rangle$ and $x_i^g = \epsilon x_i$ where ϵ has order e. Let $I \triangleleft R^G$. If $\alpha_1, \alpha_2, \ldots, \alpha_e \in I$ and $\tau_1, \tau_2, \ldots, \tau_e$ have degree 1, show that $\alpha_1 \tau_1 \alpha_2 \tau_2 \cdots \alpha_e \tau_e \in \mathrm{T}(I)$.

7. Let $R = K\langle x_1, x_2, \ldots, x_d \rangle$ with K a countable field and suppose G acts linearly, but not as scalars, on R. Show that there exist uncountably many primes Q of R^G with $\dim_K R^G/Q = 1$ and only countably many primes P of R with $\dim_K R/P < \infty$. For the latter, observe that $\dim_K R/P < \infty$ implies that R/P is contained isomorphically in $\mathrm{M}_n(K)$ for some n.

8 Grothendieck Groups and Induced Modules

33. Grothendieck Groups

The goal of this chapter is essentially to prove one marvelous theorem. This result of [**140, 142**] asserts that the Grothendieck group of a Noetherian crossed product $R*G$ is determined in a well defined manner by the sub-crossed products $R*H$ for the various finite subgroups $H \subseteq G$. While this is very pretty in its own right, it also has far reaching and surprising applications. In particular, it solves two outstanding group ring problems concerning zero divisors and Goldie rank. Moreover, these results cannot be proved just for group rings; the proofs are inductive and make strong use of the elementary fact that $R*G = (R*N)*(G/N)$ for $N \triangleleft G$. In other words, these are truly crossed product theorems.

Our exposition, for the most part, is based on the papers [**33**] and [**54**] which considerably simplify the original argument. As we will see, the proof requires some familiarity with homological concepts, in particular with the projective dimension of a module and the derived functor Tor. To aid the reader in this, we will discuss the necessary background material, indicating without proof or with brief sketches the basic properties required. The latter are actually all quite elementary and can be found for example in [**190**]. Finally we remark that the study of Grothendieck groups becomes somewhat unnatural in the case of non-Noetherian rings. Thus we will usually assume that the rings being considered are right Noetherian.

Definition. Let R be a right Noetherian ring and let $\mathcal{F}$ be the free additive abelian group whose free generators are the isomorphism classes of the finitely generated right R-modules. Notice that every finitely generated R-module is isomorphic to $R^{\oplus n}/N$ for some integer n and some submodule N of $R^{\oplus n}$. Because of this, the collection of isomorphism classes of such modules is a set and we can choose representatives of the individual classes from among the modules $R^{\oplus n}/N$. In other words, the construction of $\mathcal{F}$ involves no set theoretic difficulties.

If A is a finitely generated R-module, that is if $A \in \bmod R$, we let $\tilde{A}$ denote the generator of $\mathcal{F}$ which is its equivalence class. Then we define the subgroup $\mathcal{R}$ of $\mathcal{F}$ to be generated by all expressions of the form $\tilde{B} - \tilde{A} - \tilde{C}$ if $0 \to A \to B \to C \to 0$ is a short exact sequence with $A, B, C \in \bmod R$. The group $\mathcal{F}/\mathcal{R}$ is called the *Grothendieck group* of R and is denoted by $\mathbf{G_0}(R)$. Furthermore, we denote the image of $\tilde{A}$ in $\mathbf{G_0}(R)$ by $[A]$. Thus $\mathbf{G_0}(R)$ is generated by all such $[A]$ subject to the relations $[B] = [A] + [C]$ for all $0 \to A \to B \to C \to 0$.

Let us compute some examples.

Lemma 33.1. *i. Let $\mathcal{A}$ be an additive abelian group and consider any map $\theta\colon \bmod R \to \mathcal{A}$. Suppose further that $\theta(B) = \theta(A) + \theta(C)$ for all $0 \to A \to B \to C \to 0$. Then θ determines a group homomorphism $\theta\colon \mathbf{G_0}(R) \to \mathcal{A}$ by defining $\theta([A]) = \theta(A)$.*

ii. Suppose R is right Artinian and that $V_1, V_2, \ldots, V_k$ are representatives of the isomorphism classes of the finitely many irreducible R-modules. Then $\mathbf{G_0}(R)$ is the free abelian group on the generators $[V_1], [V_2], \ldots, [V_k]$. Furthermore, composition length determines a homomorphism $\rho\colon \mathbf{G_0}(R) \to Z$.

Proof. (i) From the exact sequence $0 \to 0 \to 0 \to 0 \to 0$ we see that $\theta(0) = 0$ and then from $0 \to 0 \to B \to C \to 0$ we see that $B \cong C$ implies that $\theta(B) = \theta(C)$. Since $\mathcal{F}$ is free abelian, we can now defined $\theta \colon \mathcal{F} \to \mathcal{A}$ by $\theta(\tilde{A}) = \theta(A)$ and, since $\theta(\mathcal{R}) = 0$, this map factors through $\mathbf{G_0}(R) = \mathcal{F}/\mathcal{R}$.

(ii) Since every finitely generated R-module has a finite composition series, it follows that $\mathbf{G_0}(R)$ is spanned by $[V_1], [V_2], \ldots, [V_k]$. The goal is to show that these are independent. To this end, let $\mathcal{A}$ be the free abelian group on the generators $v_1, v_2, \ldots, v_k$ and for each $A \in \mathbf{mod}\, R$ define $\theta(A) = \sum_j m_j v_j \in \mathcal{A}$, where m_j is the multiplicity of V_j as a composition factor of A. Then $\theta(A)$ is well defined and respects short exact sequences. Thus by (i), θ determines a homomorphism $\theta \colon \mathbf{G_0}(R) \to \mathcal{A}$. But $\theta([V_i]) = v_i$ and the v_i are independent, so the same must be true of the elements $[V_i]$. Finally for ρ, we need only observe that the composition length is a well defined integer which respects short exact sequences. ∎

Next we consider the behavior of $\mathbf{G_0}$ under a change of rings. To start with, we recall that the tensor product is a *right exact functor*. That is, if $0 \to A \to B \to C \to 0$ is any exact sequence of right R-modules and if $_R M$ is a left R-module, then at least the abbreviated sequence

$$A \otimes_R M \to B \otimes_R M \to C \otimes_R M \to 0$$

is exact in general. We then say that $_R M$ is *flat* if for all $0 \to A \to B \to C \to 0$ we have

$$0 \to A \otimes_R M \to B \otimes_R M \to C \otimes_R M \to 0$$

exact. Note that $\otimes_R$ commutes with arbitrary direct sums and that $A \otimes R \cong A$ via the map $a \otimes r \to ar$. It then follows that every projective left R-module is flat. Furthermore, flatness is local so that every *directed limit* of flat modules is flat. Specifically, if M has a family $\{\, M_i \,\}$ of flat submodules with the property that each finite subset of M is in some M_i, then M is flat (see [**190**, Corollary 3.29 and Theorem 3.30]). Of course, there are analogous definitions and properties for flat right R-modules.

Now suppose $\theta \colon R \to S$ is a ring homomorphism, not necessarily onto. Then S can be viewed as a left R-module via $r \cdot s = \theta(r)s$

for $r \in R$ and $s \in S$. Furthermore, this yields a map from right R-modules to right S-modules by $A \mapsto A \otimes_R S$. Observe that if $A = \sum_{i=1}^n a_i R$ is finitely generated, then $A \otimes_R S$ is generated by the elements $a_i \otimes 1$ and hence is a finitely generated S-module. We have

Lemma 33.2. *Let $\theta \colon R \to S$ be a ring homomorphism and assume that, in this way, $_R S$ is a flat left R-module. Then $A \mapsto A \otimes_R S$ determines a group homomorphism $\theta' \colon \mathbf{G_0}(R) \to \mathbf{G_0}(S)$ given by $[A] \mapsto [A \otimes_R S]$. Furthermore, if θ is onto, then so is θ' and we have $A \otimes_R S \cong A/AI$ where $I = \mathrm{Ker}(\theta)$.*

Proof. Note that if A_R is finitely generated, then so is $A \otimes_R S$. Thus we obtain a map θ' from **mod** R to $\mathbf{G_0}(S)$ by $\theta' \colon A \mapsto [A \otimes_R S]$. The flatness of $_R S$ and the relations in $\mathbf{G_0}(S)$ imply that θ' respects short exact sequences. We conclude from Lemma 33.1(i) that $\theta' \colon \mathbf{G_0}(R) \to \mathbf{G_0}(S)$ exists. Finally suppose θ is onto and that B is a finitely generated S-module. Then, by way of the homomorphism $R \to S \to \mathrm{End}(B)$, we can view B as an R-module and it will also be finitely generated. Since $B \otimes_R S \cong B$ via the map $b \otimes s \mapsto bs$, it follows that θ' is onto. On the other hand, if A is an R-module, then $A \otimes_R S = A \otimes_R (R/I) \cong A/AI$ via the map $a \otimes (r + I) \mapsto ar + AI$. This completes the proof. ∎

If $R \subseteq S$ then, as in Section 3, we denote $A \otimes_R S$ by $A^{|S}$ and call it the *induced* S-module. We will use the above result in some of the cases listed below.

Lemma 33.3. *The following ring extensions are flat, that is the overring is a flat left module for the subring.*

i. *$R \subseteq R*G$, where G is any semigroup.*

ii. *$R*H \subseteq R*G$, where H is a subgroup of the group G.*

iii. *$R \subseteq RT^{-1}$, where T is a right divisor set of regular elements of R. Furthermore, the map $\mathbf{G_0}(R) \to \mathbf{G_0}(RT^{-1})$ is onto.*

Proof. Cases (i) and (ii) follow since the larger ring is a free module over the smaller one. For (iii), we observe that $RT^{-1} = \bigcup_{t \in T} Rt^{-1}$ and that each Rt^{-1} is a free, and hence flat, left R-module. Since common denominators exist in this right ring of fractions, it follows that $\{\, Rt^{-1} \mid t \in T \,\}$ is a directed family of flat R-submodules and hence RT^{-1} is flat. Finally if $B = \sum_1^n b_i RT^{-1}$ is a finitely generated right RT^{-1}-module, then $A = \sum_1^n b_i R$ is a finitely generated R-submodule with $A \otimes_R RT^{-1} \cong B$. We conclude that the induced map $\mathbf{G_0}(R) \to \mathbf{G_0}(RT^{-1})$ is onto. ∎

In particular, (iii) above applies to the classical right quotient ring $Q(R)$ if it exists. On the other hand, the extension $R \subseteq Q_s(R)$ is not flat, in general, as can be seen from the following example of [**100**]. Let $F = K\langle x, y \rangle$ be a free K-algebra on two generators and let I be the ideal of F generated by x and y^2. If $R = K + I \subseteq F$ then, by Lemma 10.8(iii) and Theorem 13.4, $Q_s(R) = Q_s(F) = F$. On the other hand, $_R F$ is not flat (see Exercise 1).

Another change of ring result occurs in the context of Z^+-graded rings. To start with, let $R = \sum_{i=0}^{\infty} R_i$ be such a ring. Then we can form the *graded Grothendieck group* $\mathbf{gr\, G_0}(R)$ by restricting our attention to the finitely generated graded R-modules $\mathbf{gr\, mod}\, R$. Specifically, $\mathbf{gr\, G_0}(R)$ is the additive abelian group generated by the symbols $[A]$ for each $A \in \mathbf{gr\, mod}\, R$ subject to the relations $[B] = [A] + [C]$ for all short exact sequences $0 \to A \to B \to C \to 0$ with graded homomorphisms. The formal construction of $\mathbf{gr\, G_0}(R)$ is of course analogous to the construction of $\mathbf{G_0}(R)$ and it enjoys similar properties.

Suppose again that $R = \sum_{i=0}^{\infty} R_i$ is Z^+-graded and form the polynomial ring $R[t]$. Then $R[t]$ can be Z^+-graded by *total degree*, that is $R_i t^j$ has grade $i + j$. In particular, $R[t]_0 = R_0$. The following is a key result. Notice that the ring homomorphism considered cuts across the grade.

Proposition 33.4. *Let* $R = \sum_{i=0}^{\infty} R_i$ *be a right Noetherian Z^+-graded ring and let $R[t]$ be Z^+-graded by total degree. Then the ring homomorphism $R[t] \to R$ given by $t \mapsto 1$ yields a group epimorphism* $\mathbf{gr\, G_0}(R[t]) \to \mathbf{G_0}(R)$.

Proof. Note that $R[t]$ is right Noetherian and that the kernel of the homomorphism $\theta \colon R[t] \to R$ is the principal ideal $I = (t - 1)R[t]$

generated by the central element $t-1$. As above, θ determines a map from the finitely generated $R[t]$-modules to the finitely generated R-modules given by

$$A \mapsto A \otimes_{R[t]} R \cong A/AI = A/A(t-1).$$

In particular, we get a map from $\mathbf{gr\,mod}\,R[t]$ to $\mathbf{G_0}(R)$ by defining $\theta'(A) = [A/A(t-1)] \in \mathbf{G_0}(R)$. This will in turn yield a group homomorphism from $\mathbf{gr\,G_0}(R[t])$ to $\mathbf{G_0}(R)$ provided we show that θ' respects graded short exact sequences.

To this end, let $0 \to A \to B \xrightarrow{\beta} C \to 0$ be a short exact sequence of graded modules and graded homomorphisms. We may suppose that A is in fact a submodule of B. The assumption that the homomorphisms are graded then implies that A is a graded submodule of B. Now the combined map $B \to C \to C/C(t-1)$ is certainly an epimorphism with $B(t-1)$ in its kernel. Thus we have an epimorphism $\bar{\beta}: B/B(t-1) \to C/C(t-1)$ with kernel $U/B(t-1)$ and clearly $U \supseteq A + B(t-1)$. We now show that equality occurs. For this, let $u \in U$. Then $\beta(u) \in C(t-1)$ so, since β is onto, $\beta(u) = \beta(b)(t-1)$ for some $b \in B$. But then $u - b(t-1) \in \mathrm{Ker}(\beta) = A$ and $U \subseteq A + B(t-1)$. It follows that

$$\mathrm{Ker}(\bar{\beta}) = U/B(t-1) = \big(A + B(t-1)\big)/B(t-1)$$
$$\cong A/\big(A \cap B(t-1)\big)$$

and it remains to determine the intersection $A \cap B(t-1)$.

Certainly $A \cap B(t-1) \supseteq A(t-1)$. Conversely suppose $\sum_i a_i \in A$ and $\sum_i b_i \in B$ are elements written in terms of their homogeneous components with

$$\sum_i a_i = \Big(\sum_i b_i\Big)(t-1) \in A \cap B(t-1).$$

Then by comparing homogenous components we have, setting $b_{-1} = 0$, $a_n = b_{n-1}t - b_n$ for all integers $n \geq 0$. Thus $b_0 = -a_0 \in A$ and, by induction, $b_n = b_{n-1}t - a_n \in A$ for all n. Therefore $\big(\sum_i b_i\big)(t-1) \in A(t-1)$ and we conclude that $A \cap B(t-1) = A(t-1)$. Hence

$$\mathrm{Ker}(\bar{\beta}) \cong A/\big(A \cap B(t-1)\big) = A/A(t-1)$$

and we obtain a short exact sequence

$$0 \to A/A(t-1) \to B/B(t-1) \to C/C(t-1) \to 0.$$

In other words, in $\mathbf{G_0}(R)$ we have $\theta'(B) = \theta'(A) + \theta'(C)$ and thus, by the graded analog of Lemma 33.1(i), θ' determines a group homomorphism $\mathbf{gr}\,\mathbf{G_0}(R[t]) \to \mathbf{G_0}(R)$ as required.

Now we show that this map is an epimorphism. Let V' be a finitely generated R-module. Then, for some integer n, we can map the free R-module $F' = \oplus \sum_1^n f_i' R$ onto V' with kernel U'. Let $F = \oplus \sum_1^n f_i R[t]$ be a free $R[t]$-module on the generators $f_1, f_2, \ldots, f_n$, graded so that each f_i is homogeneous of degree 0. Now each $u' \in U'$ can be written uniquely as $u' = \sum_i f_i' r_i = \sum_{i,j} f_i' r_{i,j}$ where $r_i = \sum_j r_{i,j} \in \sum_j R_j$ is the decomposition of r_i into its homogeneous components with $r_{i,j} \in R_j$. If m is the largest such j which occurs with $r_{i,j} \neq 0$, we define $\eta(u') \in F$ by

$$\eta(u') = \sum_{i,j} f_i r_{i,j} t^{m-j}.$$

Then $\eta(u')$ is a homogeneous element of F of degree m and we let U be the $R[t]$-submodule of F generated by all these $\eta(u')$ for $u' \in U'$. Clearly U is a graded submodule of F and it is finitely generated since $R[t]$ is right Noetherian. Furthermore, if $M = F/U$ then M is a finitely generated graded $R[t]$-module with $0 \to U \to F \to M \to 0$ exact. It follows by the result of the preceding paragraph that

$$0 \to U/U(t-1) \to F/F(t-1) \to M/M(t-1) \to 0$$

is an exact sequence of R-modules.

Consider the additive group epimorphism $F \to F'$ given by $f_i r t^j \mapsto f_i' r$ for all $r \in R$. The kernel of this map is clearly $F(t-1)$ and hence we obtain an R-isomorphism $F/F(t-1) \to F'$. Under this map, the image of $\eta(u') + F(t-1)$ is precisely u' and therefore $U/U(t-1) = U/(U \cap F(t-1))$ maps onto U'. Thus $V' \cong F'/U' \cong M/M(t-1)$ so $[V']$ is in the image of $\theta': \mathbf{gr}\,\mathbf{G_0}(R[t]) \to \mathbf{G_0}(R)$ and the result follows. ∎

It is apparent from the last part of the argument that the above result is an analog of the classical trick of adding a new variable to introduce homogeneous coordinates.

Now let us return to arbitrary right Noetherian rings, but this time we restrict our attention to finitely generated projective modules. In this case, we get a further group of interest. Specifically, let R be a right Noetherian ring and let **pr mod** R denote the family of finitely generated projective R-modules. Then $\mathbf{K_0}(R)$ is the additive abelian group generated by the symbols $[P]$ for each $P \in \mathbf{pr\,mod}\,R$ subject to the relations $[P] = [Q] + [Q']$ for all short exact sequences $0 \to Q \to P \to Q' \to 0$. Again, the formal construction of this group is analogous to the construction of $\mathbf{G_0}(R)$ and it enjoys similar properties. Furthermore, since Q' is projective in the above, the sequence splits and is equivalent to $P \cong Q \oplus Q'$. Under certain circumstances, $\mathbf{K_0}(R)$ is naturally isomorphic to $\mathbf{G_0}(R)$. To prove this fact, we require some definitions.

If A_R is an R-module, then a *projective resolution* for A is a long exact sequence

$$\cdots \to P_2 \to P_1 \to P_0 \to A \to 0$$

with the P_i projective. For convenience, we abbreviate this by $\mathbf{P} \to A \to 0$ and we write $\mathbf{P}_n$ for the nth module P_n. Now it is clear that every module has such a resolution. Indeed, start with A and map a free module P_0 onto it, obtaining $P_0 \to A \to 0$. Then map a free module P_1 onto $\text{Ker}(P_0 \to A)$ to get $P_1 \to P_0 \to A \to 0$. Continuing in this manner, we clearly construct a free and hence projective resolution for A. Furthermore, if R is Noetherian and A is finitely generated, then this process yields a projective resolution for A consisting of finitely generated projective (or free) modules. A projective resolution is said to be *finite* if the P_n's are eventually all zero. We record some basic properties.

Lemma 33.5. *Let*

$$0 \to X \to P_n \to \cdots \to P_1 \to P_0 \to A \to 0$$

and

$$0 \to Y \to Q_n \to \cdots \to Q_1 \to Q_0 \to A \to 0$$

be exact sequences with P_i and Q_i projective. Then

$$X \oplus (P_{n-1} \oplus P_{n-3} \oplus \cdots) \oplus (Q_n \oplus Q_{n-2} \oplus \cdots)$$
$$\cong Y \oplus (Q_{n-1} \oplus Q_{n-3} \oplus \cdots) \oplus (P_n \oplus P_{n-2} \oplus \cdots).$$

In particular, X is projective if and only if Y is.

Proof. We proceed by induction on $n \geq 0$. Let B be the kernel of the map $P_0 \to A$ and let $C = \text{Ker}(Q_0 \to A)$. Then we have

$$0 \to B \to P_0 \to A \to 0$$

$$0 \to C \to Q_0 \to A \to 0$$

and therefore $B \oplus Q_0 \cong C \oplus P_0$ by Schanuel's lemma (see [**190**, Theorem 3.41] or Exercise 2). Furthermore we have

$$0 \to X \to P_n \to \cdots \to P_2 \to P_1 \to B \to 0$$

$$0 \to Y \to Q_n \to \cdots \to Q_2 \to Q_1 \to C \to 0$$

and hence

$$0 \to X \to P_n \to \cdots \to P_2 \to P_1 \oplus Q_0 \to B \oplus Q_0 \to 0$$

$$0 \to Y \to Q_n \to \cdots \to Q_2 \to Q_1 \oplus P_0 \to C \oplus P_0 \to 0.$$

But $B \oplus Q_0 \cong C \oplus P_0$ so the result now follows by induction. ∎

Notice that the above says that if $\mathbf{P} \to A \to 0$ and $\mathbf{Q} \to A \to 0$ are two projective resolutions for A, then $\text{Ker}(\mathbf{P}_n \to \mathbf{P}_{n-1})$ is projective if and only if $\text{Ker}(\mathbf{Q}_n \to \mathbf{Q}_{n-1})$ is. Furthermore, when this occurs, then the truncated series

$$0 \to \text{Ker}(\mathbf{P}_n \to \mathbf{P}_{n-1}) \to \mathbf{P}_n \to \cdots \to \mathbf{P}_1 \to \mathbf{P}_0 \to A \to 0$$

is a finite projective resolution for A. In particular, if A has some finite resolution, then any projective resolution can be suitably truncated to a finite one.

As an immediate consequence we have

Lemma 33.6. *Let A be a finitely generated R-module with R a right Noetherian ring. If A has a finite projective resolution, then it has one consisting of finitely generated projective modules.*

The following is [**190**, Lemma 6.11].

Lemma 33.7. *Let $0 \to A \to B \to C \to 0$ be a short exact sequence of R-modules and let $\mathbf{P} \to A \to 0$ and $\mathbf{Q} \to C \to 0$ be projective resolutions. Then there exists a projective resolution $\mathbf{P} \oplus \mathbf{Q} \to B \to 0$ such that the diagram*

$$
\begin{array}{ccccccccc}
0 & \to & \mathbf{P} & \to & \mathbf{P} \oplus \mathbf{Q} & \to & \mathbf{Q} & \to & 0 \\
  &   & \downarrow &   & \downarrow & & \downarrow & & \\
0 & \to & A & \to & B & \to & C & \to & 0 \\
  &   & \downarrow &   & \downarrow & & \downarrow & & \\
  &   & 0 &   & 0 & & 0 & &
\end{array}
$$

commutes. Here $(\mathbf{P} \oplus \mathbf{Q})_n = \mathbf{P}_n \oplus \mathbf{Q}_n$ and the maps in $0 \to \mathbf{P} \to \mathbf{P} \oplus \mathbf{Q} \to \mathbf{Q} \to 0$ are the natural ones determined by these direct sums.

A module A is said to have *finite projective dimension* if it has a finite projective resolution. Indeed the *projective dimension* of A, $\mathrm{pd}_R A$, is the minimal n for which

$$0 \to P_n \to \cdots \to P_1 \to P_0 \to A \to 0$$

exists. In particular, $\mathrm{pd}_R A = 0$ if and only if A is projective. Furthermore, if $\mathbf{Q} \to A \to 0$ is any projective resolution then, by our previous comments, $\mathrm{pd}_R A$ is the minimal n with $\mathrm{Ker}(\mathbf{Q}_{n-1} \to \mathbf{Q}_{n-2})$ projective. We can now prove

Proposition 33.8. *Let R be a right Noetherian ring. Then there is a natural map $c \colon \mathbf{K_0}(R) \to \mathbf{G_0}(R)$ given by $[P] \mapsto [P]$. If, in addition, every finitely generated R-module has finite projective dimension, then c is an isomorphism.*

Proof. The existence of c is obvious from the projective analog of Lemma 33.1(i). Assume now that each finitely generated R-module has finite projective dimension. We show that c is an isomorphism by constructing a back map. To this end, let A be a finitely generated R-module and, by Lemma 33.6, let $0 \to P_n \to \cdots \to P_1 \to P_0 \to A \to 0$ be a finite projective resolution consisting of finitely generated modules. We define

$$\theta(A) = \sum_i (-)^i [P_i] \in \mathbf{K_0}(R).$$

In view of Lemma 33.5 with $X = Y = 0$, $\theta(A)$ is well defined, independent of the particular resolution chosen. Furthermore, by Lemma 33.7 and the relations in $\mathbf{K_0}(R)$, it follows that θ respects short exact sequences. Thus by Lemma 33.1(i), θ determines a group homomorphism $\theta \colon \mathbf{G_0}(R) \to \mathbf{K_0}(R)$. Now it is trivial to see that θc is the identity on $\mathbf{K_0}(R)$. Moreover, since $[A] = \sum_i (-)^i [P_i]$ in $\mathbf{G_0}(R)$, we see that $c\theta$ is the identity on $\mathbf{G_0}(R)$. Thus $\theta = c^{-1}$ and c is an isomorphism. ∎

The homomorphism c above is called the *Cartan map*. Now suppose R is Artinian with $P_1, P_2, \ldots, P_k$ representatives of the isomorphism classes of principal indecomposable R-modules. Then, as in Lemma 33.1(ii), we see that $\mathbf{K_0}(R)$ is the free abelian group on the elements $[P_1], [P_2], \ldots, [P_k]$. Thus since the P_i are in one-to-one correspondence with the irreducible R-modules $V_1, V_2, \ldots, V_k$, we see that $\mathbf{K_0}(R)$ and $\mathbf{G_0}(R)$ are abstractly isomorphic. However, c itself need not be an isomorphism; indeed it is not even necessarily one-to-one (see Exercise 4).

Lemma 33.9. *i. Let $0 \to A \to B \to C \to 0$ be given. If any two of these modules have finite projective dimension, then so does the third.*

ii. Let $B = \oplus \sum_i A_i$ and let k be an integer. Then $\mathrm{pd}_R B \leq k$ if and only if $\mathrm{pd}_R A_i \leq k$ for all i.

Proof. (i) If $\mathrm{pd}_R A$ and $\mathrm{pd}_R C$ are finite, then certainly so is $\mathrm{pd}_R B$ by Lemma 33.7. Now suppose $\mathrm{pd}_R A$ and $\mathrm{pd}_R B$ are finite. Choose a finite projective resolution $\mathbf{P} \to A \to 0$ for A and any resolution $\mathbf{Q} \to C \to 0$ for C. Then by Lemma 33.7 again, $\mathbf{P} \oplus \mathbf{Q} \to B \to 0$

is a resolution for B. But $\mathrm{pd}_R B < \infty$ so this resolution eventually has projective kernels. In addition, since $\mathbf{P}_n = 0$ for all sufficiently large n, it follows that $\mathrm{Ker}(\mathbf{Q}_n \to \mathbf{Q}_{n-1})$ is eventually projective. Therefore C has finite projective dimension. The argument with $\mathrm{pd}_R B$, $\mathrm{pd}_R C < \infty$ is similar.

(ii) Now let $\mathbf{P_i} \to A_i \to 0$ be projective resolutions. Then, since $B = \oplus \sum_i A_i$, the obvious direct sum yields a projective resolution $\oplus \sum_i \mathbf{P_i} \to B \to 0$. Moreover, the kernel of the nth map in $\oplus \sum_i \mathbf{P_i}$ is the direct sum of the corresponding kernels in the various $\mathbf{P_i}$. Since a direct sum is projective if and only if each summand is, the result clearly follows. ∎

We consider one last ring extension. Let $R*G$ be a crossed product with $G = \langle x_1, x_2, \ldots, x_t \rangle$ the free abelian semigroup on the variables x_i. In other words, $R*G$ is an *iterated* skew polynomial ring and we have a homomorphism $R*G \to R$ given by $\bar{x}_i \mapsto 0$ for all i. In particular, if M is an R-module, then via the combined map $R*G \to R \to \mathrm{End}(M)$, we see that M is also an $R*G$-module.

Lemma 33.10. *Let $R*G$ be as above with $G = \langle x_1, x_2, \ldots, x_t \rangle$ a free abelian semigroup. If M is any R-module of finite projective dimension, then M also has finite projective dimension when viewed as an $R*G$-module.*

Proof. It suffices to assume, by induction on t, that $G = \langle x \rangle$ has one generator. Write $S = R*G$ and consider the short exact sequence $0 \to \bar{x}S \to S \to R \to 0$ determined by $\bar{x} \mapsto 0$. Since S and $\bar{x}S \cong S$ are free as right S-modules, we have $\mathrm{pd}_S R \le 1$ and indeed, by Lemma 33.9(ii), $\mathrm{pd}_S P \le 1$ for any projective R-module P. Finally let

$$0 \to P_n \to \cdots \to P_1 \to P_0 \to M \to 0$$

be a finite projective resolution for M as an R-module. Then as S-modules, each P_i has finite projective dimension and it follows from Lemma 33.9(i) and induction that the same is true for M. ∎

We remark that $\mathbf{K_0}(R)$ is the first of a series of $\mathbf{K}$-groups associated with the ring R (see [**179**] for details).

EXERCISES

1. Let F be any K-algebra, let $I \lhd F$ and set $R = K + I \subseteq F$. Assume that $2 \leq \dim_K F/I < \infty$ and $1 \leq \dim_K I/I^2 < \infty$. Show that $_R F$ is not flat. To this end, consider $0 \to I \to R \to R/I \to 0$ and observe that $(R/I) \otimes F \cong F/I$ and $R \otimes F \cong F$. Thus if $0 \to I \otimes F \to R \otimes F \to (R/I) \otimes F \to 0$ is exact, then $V = I \otimes F \cong I$. But V has as a homomorphic image

$$(I/I^2) \otimes F \cong (\dim_K I/I^2) \cdot (R/I) \otimes F \cong (\dim_K I/I^2) \cdot F/I$$

and this has larger dimension than I/I^2, a contradiction.

2. Let $0 \to X \to P \overset{\alpha}{\to} A \to 0$ and $0 \to Y \to Q \overset{\beta}{\to} A \to 0$ be given with P and Q projective. Let W be the R-submodule of $P \oplus Q$ consisting of all (p, q) with $\alpha(p) = \beta(q)$. Observe that the projection map $W \to Q$ is an epimorphism with kernel $(X, 0)$. Deduce that $X \oplus Q \cong W \cong Y \oplus P$.

3. Let R be a right Artinian ring with principal indecomposable modules $P_1, P_2, \ldots, P_k$ and corresponding irreducible modules $V_1, V_2, \ldots, V_k$. The matrix of $c: \mathbf{K_0}(R) \to \mathbf{G_0}(R)$ with respect to the bases $\{[P_i]\}$ and $\{[V_i]\}$ is called the *Cartan matrix* of R. Describe this matrix in case R is semisimple or $R = K[x \mid x^m = 0]$. In particular, note that c need not be surjective.

4. Let K be a field and write $A(a, b)$ and $B(a, b)$ for the sets of 4×4 upper triangular matrices

$$A(a, b) = \begin{pmatrix} a & * & 0 & 0 \\ & b & 0 & 0 \\ & & b & * \\ & & & a \end{pmatrix} \qquad B(a, b) = \begin{pmatrix} a & * & * & * \\ & b & * & * \\ & & b & * \\ & & & a \end{pmatrix}$$

with the $*$ entries arbitrary. Now let R and S be the 4- and 8-dimensional subalgebras of $M_4(K)$ given by

$$R = \{ A(a, b) \mid a, b \in K \} \qquad S = \{ B(a, b) \mid a, b \in K \}.$$

Observe that $R/J(R) \cong K \oplus K \cong S/J(S)$ and describe the Cartan matrices of R and S. Conclude that c need not be injective in general.

5. If $0 \to A \to B \to C \to 0$ is given, then the projective dimensions of any two of these modules bound the dimension of the third. Obtain this information by sharpening the proof of Lemma 33.9(i).

6. Let $R*G$ be a crossed product with $G = \langle x_1, x_2, \ldots, x_t \rangle$ the free abelian semigroup on the variables x_i. If M is an R-module, show that $\mathrm{pd}_{R*G} M \le \mathrm{pd}_R M + t$.

34. Graded Rings

In this section, we are concerned for the most part with Z^+-graded rings. Specifically if $R = \oplus \sum_{i=0}^{\infty} R_i$ is Z^+-graded and right Noetherian, our goal is to show that the induced map $A \mapsto A \otimes_{R_0} R$ yields an isomorphism from $\mathbf{G_0}(R_0)$ to $\mathbf{G_0}(R)$. The proof of this key result relies on the *derived functor* Tor.

Let R be any ring and let A_R and $_R M$ be right and left R-modules respectively. Then A and M determine a sequence of additive abelian groups denoted by $\mathrm{Tor}_n^R(A, M) = \mathrm{Tor}_n(A, M)$ for $n = 0, 1, 2, \ldots$. These Tor groups enjoy many of the properties of the tensor product and indeed $\mathrm{Tor}_0(A, M) = A \otimes_R M$. However, their construction is somewhat more complicated.

To start with suppose

$$\cdots \to X_{n+1} \overset{d_{n+1}}{\to} X_n \overset{d_n}{\to} X_{n-1} \to \cdots \to X_1 \overset{d_1}{\to} X_0 \to 0$$

is a *zero sequence* or *complex* of additive abelian groups. By definition, this means that the maps d_n are homomorphisms and that $d_n d_{n+1} = 0$ for all n. In particular, $\mathrm{Im}(d_{n+1}) \subseteq \mathrm{Ker}(d_n)$ and the *homology groups* $\mathrm{H}_n(\mathbf{X}) = \mathrm{Ker}(d_n)/\mathrm{Im}(d_{n+1})$ are a measure of the failure of exactness of the complex $\mathbf{X}$ at the nth module X_n.

Now let

$$\cdots \to Q_n \to \cdots \to Q_1 \to Q_0 \to M \to 0$$

be any (left) projective resolution for $_R M$. By tensoring this resolution with A_R and ignoring the first term $A \otimes M$, we obtain

$$\cdots \to A \otimes Q_n \to \cdots \to A \otimes Q_1 \to A \otimes Q_0 \to 0$$

and the groups $\mathrm{Tor}_n(A, M)$ are precisely the homology groups of this complex. A few basic properties are as follows (see [**190**, Section 8] for details).

First, $\mathrm{Tor}_n(A, M)$ is well defined, independent of the particular projective resolution for M. In fact, Tor can also be computed in a similar manner by using projective resolutions for A. Specifically, if

$$\cdots \to P_n \to \cdots \to P_1 \to P_0 \to A \to 0$$

is such a resolution, then the groups $\mathrm{Tor}_n(A, M)$ are also the homology groups of the complex

$$\cdots \to P_n \otimes M \to \cdots \to P_1 \otimes M \to P_0 \otimes M \to 0.$$

In view of this, properties of Tor are clearly right-left symmetric. Morover, Tor_n commutes with arbitrary direct sums and is *functorial* in the sense that any R-module homomorphism $A \to B$ gives rise to a natural group homomorphism $\mathrm{Tor}_n(A, M) \to \mathrm{Tor}_n(B, M)$. Finally, we know that the tensor product is a right exact functor, but it need not be exact. The key property of Tor is that it takes up the slack. Indeed, we have the following *long exact Tor sequence* (see [**190**, Theorem 8.3]).

Lemma 34.1. *Let $_RM$ be a left R-module and let $0 \to A \to B \to C \to 0$ be a short exact sequence of right R-modules. Then there exists a long exact sequence*

$$\cdots \to \mathrm{Tor}_{n+1}(C, M) \xrightarrow{\delta} \mathrm{Tor}_n(A, M) \to \mathrm{Tor}_n(B, M)$$
$$\to \mathrm{Tor}_n(C, M) \xrightarrow{\delta} \mathrm{Tor}_{n-1}(A, M) \to \mathrm{Tor}_{n-1}(B, M) \to$$
$$\cdots \to \mathrm{Tor}_1(C, M) \xrightarrow{\delta} A \otimes M \to B \otimes M \to C \otimes M \to 0.$$

Here the maps $\mathrm{Tor}_n(A, M) \to \mathrm{Tor}_n(B, M) \to \mathrm{Tor}_n(C, M)$ are given by the functorial property of Tor and the various δ's are called *connecting homomorphisms*. It follows easily from the construction of Tor that if $_RM$ is flat, then $\mathrm{Tor}_n(A, M) = 0$ for all $n \geq 1$ and all right R-modules A. Conversely, if $\mathrm{Tor}_1(A, M) = 0$ for all A, then the preceding lemma implies that M is flat. Moreover, we have

Lemma 34.2. *Let $_R M$ be a left R-module.*

i. Suppose $0 \to A \to B \to C \to 0$ is a short exact sequence of right R-modules and that $\mathrm{Tor}_{n+1}(C, M) = 0 = \mathrm{Tor}_n(B, M)$ for some integer $n \geq 0$. Then $\mathrm{Tor}_n(A, M) = 0$.

ii. If $\mathrm{pd}_R M = n$, then $\mathrm{Tor}_k(A, M) = 0$ for all $k \geq n + 1$.

Proof. Part (i) follows from Lemma 34.1 since $\mathrm{Tor}_0(A, M) = A \otimes M$. Part (ii) follows by constructing the groups $\mathrm{Tor}_k(A, M)$ from the finite projective resolution

$$\cdots \to 0 \to 0 \to Q_n \to \cdots \to Q_1 \to Q_0 \to M \to 0.$$

This completes the proof. ∎

Now suppose that $R = R_0 \oplus I$ where $I \triangleleft R$ and R_0 is a subring with the same 1. Then $\bar{R} = R/I$ is a right and left R-module and an R_0-module with $_{R_0}\bar{R} \cong {}_{R_0}R_0$. As usual, if A is any right R_0-module, then we are concerned with the induced right R-module $A \otimes_{R_0} R$.

Lemma 34.3. *Let $R = R_0 \oplus I$ with $I \triangleleft R$, set $\bar{R} = R/I$ and let A be any right R_0-module.*

i. $(A \otimes_{R_0} R) \otimes_R \bar{R} \cong A$, where this is a functorial isomorphism as right R_0-modules.

ii. If $_{R_0}R$ is flat, then $\mathrm{Tor}_n^R(A \otimes_{R_0} R, \bar{R}) = 0$ for all $n \geq 1$.

Proof. (i) By the associativity of tensor products we have

$$(A \otimes_{R_0} R) \otimes_R \bar{R} \cong A \otimes_{R_0} (R \otimes_R \bar{R})$$
$$\cong A \otimes_{R_0} \bar{R} \cong A \otimes_{R_0} R_0 \cong A$$

as required.

(ii) Here we compute the various Tor groups directly from the definition. Let

$$\cdots \to P_n \to \cdots \to P_1 \to P_0 \to A \to 0$$

be a projective resolution for the R_0-module A. Since $_{R_0}R$ is flat, by assumption, the sequence

$$\cdots \to P_n \otimes R \to \cdots \to P_1 \otimes R \to P_0 \otimes R \to A \otimes R \to 0$$

is also exact. Note further that $R_0 \otimes_{R_0} R \cong R$ and that $\otimes$ commutes with arbitrary direct sums. It therefore follows that each $P_n \otimes R$ is a projective R-module and hence that the above sequence is in fact a projective resolution for $A \otimes R$.

To compute $\mathrm{Tor}_n^R(A, \bar{R})$, we tensor this resolution with $\bar{R}$ and find the homology groups. But by (i) above, we have the functorial isomorphism $(P_n \otimes_{R_0} R) \otimes_R \bar{R} \cong P_n$ and hence the new series is just

$$\cdots \to P_n \to \cdots \to P_1 \to P_0 \to A \to 0,$$

the original exact sequence. Since the abbreviated series

$$\cdots \to P_n \to \cdots \to P_1 \to P_0 \to 0$$

is clearly exact at all P_n with $n \geq 1$, the homology groups are trivial at those modules and therefore $\mathrm{Tor}_n(A \otimes R, \bar{R}) = 0$ for all $n \geq 1$. ∎

Now let $R = \oplus \sum_{i=0}^{\infty} R_i$ be a Z^+-graded ring and define $I = R_+ = \oplus \sum_{i=1}^{\infty} R_i$. Then $I \triangleleft R$ and $R = R_0 \oplus I$ so the above result applies. We fix this notation throughout and also write $\bar{R} = R/I = R/R_+$. Recall that $A = \oplus \sum_{i=0}^{\infty} A_i$ is a graded R-module if $A_i R_j \subseteq A_{i+j}$. The following is the graded version of Nakayama's lemma. It is surprisingly powerful.

Lemma 34.4. *Let R be a Z^+-graded ring and let A be a graded R-module. If $A = AR_+$, then $A = 0$.*

Proof. Since $A = A(R_+)^n$ for all $n \geq 0$, it follows that A has no elements of grade $< n$. ∎

Observe that if A is an R_0-module, then

$$B = A \otimes_{R_0} R = \oplus \sum_{i=0}^{\infty} A \otimes_{R_0} R_i$$

is a graded R-module with components $B_i = A \otimes_{R_0} R_i$. Moreover, in this way, $R_0 \otimes_{R_0} R$ is graded isomorphic to R.

Lemma 34.5. Let $R = \oplus \sum_{i=0}^{\infty} R_i$ be a Z^+-graded ring and let A be a graded R-module generated by its kth component A_k for some integer k. If $\mathrm{Tor}_1(A, \bar{R}) = 0$, then $A = A_k \otimes_{R_0} R$. Furthermore, if $_{R_0}R$ is flat, then $\mathrm{Tor}_n(A, \bar{R}) = 0$ for all $n \geq 1$.

Proof. Since $A = A_k R$, it is clear that A has no elements of degree less than k. Let $\xi: A_k \otimes_{R_0} R \to A_k R = A$ be the natural graded epimorphism. The goal is to show that ξ is one-to-one.

To start with, let $x \in \mathrm{Ker}(\xi)$. Since $R = R_0 \oplus I$, we can write

$$x = \sum_{r_t \in R_0} a_t \otimes r_t + \sum_{r'_t \in I} a'_t \otimes r'_t$$

with $a_t, a'_t \in A_k$. Then we have

$$0 = \xi(x) = \sum_{r_t \in R_0} a_t r_t + \sum_{r'_t \in I} a'_t r'_t \in A_k + \sum_{i>k} A_i.$$

It follows that $\sum_t a_t r_t = 0$ so $\sum_t a_t \otimes r_t = \left(\sum_t a_t r_t\right) \otimes 1 = 0$ and thus

$$x = \sum_{r'_t \in I} a'_t \otimes r'_t \in A_k \otimes I = (A_k \otimes R)I.$$

In other words, $\mathrm{Ker}(\xi) \subseteq (A_k \otimes R)I$.

Now tensor the exact sequence $0 \to \mathrm{Ker}(\xi) \to A_k \otimes R \to A \to 0$ with $\bar{R} = R/I$. Since $B \otimes_R (R/I) \cong B/BI$ for any R-module B, Lemma 34.1 yields the exact sequence

$$\mathrm{Tor}_1(A, \bar{R}) \to \mathrm{Ker}(\xi)/\mathrm{Ker}(\xi)I \overset{i}{\to} (A_k \otimes R)/(A_k \otimes R)I.$$

But $\mathrm{Tor}_1(A, \bar{R}) = 0$ by assumption, so the map i is in fact an injection. On the other hand, we have shown that $\mathrm{Ker}(\xi) \subseteq (A_k \otimes R)I$ so i must also be 0. It follows that $\mathrm{Ker}(\xi)/\mathrm{Ker}(\xi)I = 0$ and the graded version of Nakayama's lemma yields $\mathrm{Ker}(\xi) = 0$ as required. In particular, $A \cong A_k \otimes_{R_0} R$ and Lemma 34.3(ii) yields the result. ∎

We also need

Lemma 34.6. *Let A be a graded R-module and let $k \geq 0$ be an integer. If $B = \sum_{i<k} A_i R$, then $B \cap AI = BI$.*

Proof. Clearly $B \cap AI \supseteq BI$. For the converse, first note that

$$B = \sum_{i<k} A_i(R_0 + I) = \left(\sum_{i<k} A_i\right) + BI.$$

Thus since $BI \subseteq AI$, it clearly suffices to show that $\left(\sum_{i<k} A_i\right) \cap AI \subseteq BI$. To this end, let x belong to the latter intersection. Then $x \in AI$, so we can write $x = \sum_j a_j r_j$ with $a_j \in A$ and $r_j \in I$ all homogeneous. Furthermore, since $\deg x < k$, we can assume that each summand satisfies $\deg a_j + \deg r_j < k$. But then $a_j r_j \in BI$, and the lemma is proved. ∎

We now come to the first main result of this section and we follow the proof in [**33**].

Theorem 34.7. [179] *Let $R = \oplus \sum_{i=0}^{\infty} R_i$ be a right Noetherian Z^+-graded ring. Assume that $_{R_0}R$ is flat and that $\bar{R} = R/R_+$ has finite projective dimension as a left R-module. Then the map $A \mapsto A \otimes_{R_0} R$ induces an epimorphism $\mathbf{G_0}(R_0) \to \mathbf{gr}\,\mathbf{G_0}(R)$.*

Proof. As we observed earlier, if $E \in \mathbf{mod}\,R_0$, then $E \otimes_{R_0} R \in \mathbf{gr}\,\mathbf{mod}\,R$. Thus since $_{R_0}R$ is flat, Lemma 33.1(i) implies that $E \mapsto E \otimes_{R_0} R$ determines a group homomorphism $\theta\colon \mathbf{G_0}(R_0) \to \mathbf{gr}\,\mathbf{G_0}(R)$. The goal is to show that θ is onto. Note also that $\mathrm{Tor}_n(E \otimes_{R_0} R, \bar{R}) = 0$ for all $n \geq 1$ by Lemma 34.3(ii).

Let $\mathcal{A}$ denote the family of all finitely generated graded R-modules A such that $\mathrm{Tor}_n(A, \bar{R}) = 0$ for all $n \geq 1$. We begin by showing that $[A] \in \mathbf{gr}\,\mathbf{G_0}(R)$ is contained in the image of θ for all such $A \in \mathcal{A}$. To this end, let A be given. Since A is finitely generated, we have $A = \sum_{i=0}^{k} A_i R$ for some integer k and we proceed by induction on k. Set $B = \sum_{i<k} A_i R = \sum_{i<k} B_i R$. If $C = A/B$, then $0 \to B \to A \to C \to 0$ is an exact sequence sequence with graded homomorphisms and by the preceding lemma

we have $0 \to B/BI \to A/AI \to C/CI \to 0$ exact. On the other hand, if we tensor the original sequence with $\bar{R} = R/I$, then the long exact Tor sequence yields

$$\cdots \to \mathrm{Tor}_1(A, \bar{R}) \to \mathrm{Tor}_1(C, \bar{R}) \to B/BI \to A/AI \to C/CI \to 0.$$

But $\mathrm{Tor}_1(A, \bar{R}) = 0$ and the map $B/BI \to A/AI$ is injective, so we conclude that $\mathrm{Tor}_1(C, \bar{R}) = 0$. Now it is clear that $C = C_k R$ and thus, by Lemma 34.5, we have $C = C_k \otimes_{R_0} R$ and $\mathrm{Tor}_n(C, \bar{R}) = 0$ for all $n \geq 1$. Finally, by Lemma 34.2(i) applied to $0 \to B \to A \to C \to 0$, we see that $\mathrm{Tor}_n(B, \bar{R}) = 0$ for all $n \geq 1$ so $B \in \mathcal{A}$. It therefore follows by induction on k that $[B]$ is in the image of θ. But then $[A] = [B] + [C] \in \mathrm{Im}(\theta)$ and this part is proved.

Next we show that the image of θ contains $[A]$ for certain additional modules A. Specifically, for each integer $m \geq 1$, set $\mathcal{A}_m$ equal to the family of all $A \in \mathbf{gr\,mod}\,R$ with $\mathrm{Tor}_n(A, \bar{R}) = 0$ for all $n \geq m$. Thus for example $\mathcal{A}_1 = \mathcal{A}$ and we show by induction on m that if $A \in \mathcal{A}_m$, then $[A] \in \mathrm{Im}(\theta)$; the case $m = 1$ has of course already been proved. Now suppose $A \in \mathcal{A}_m$ with $m > 1$ and let $0 \to B \to F \to A \to 0$ be a graded short exact sequence with F a finitely generated free R-module. Since $R = R_0 \otimes_{R_0} R$ and Tor commutes with direct sums, it follows that $F \in \mathcal{A}$ and hence, by Lemma 34.2(i), that $B \in \mathcal{A}_{m-1}$. By induction, $[B] \in \mathrm{Im}(\theta)$ and, since clearly $[F] \in \mathrm{Im}(\theta)$, the same is true of $[A] = [F] - [B]$.

Finally we use the hypothesis that $_R\bar{R}$ has finite projective dimension; say $\mathrm{pd}_R\bar{R} = t$. Then, by Lemma 34.2(ii), $\mathrm{Tor}_n(A, \bar{R}) = 0$ for all right R-modules A and all $n \geq t + 1$. In other words, $\mathbf{gr\,mod}\,R = \mathcal{A}_{t+1}$ and, with this, the result follows. ∎

We remark that the map $\mathbf{G_0}(R_0) \to \mathbf{gr\,G_0}(R)$ is in fact an isomorphism. We will prove the one-to-one aspect after the next key result.

Corollary 34.8. [179] *Let $R = \oplus \sum_{i=0}^{\infty} R_i$ be a right Noetherian Z^+-graded ring. Assume that $_{R_0}R$ is flat and that $\bar{R} = R/R_+$ has finite projective dimension as a left R-module. Then the map $A \mapsto A \otimes_{R_0} R$ induces an epimorphism $\mathbf{G_0}(R_0) \to \mathbf{G_0}(R)$.*

Proof. Form the polynomial ring $S = R[t]$ and let S be Z^+-graded by total degree. Then S is right Noetherian, since R is, and $S_0 = R_0$. Furthermore, $_{R_0}R$ is flat so it is clear that $_{S_0}S$ is also flat and $\mathrm{pd}_R \bar{R} < \infty$ so it follows from Lemma 33.10 that $\mathrm{pd}_S \bar{S}$ is also finite. We can now apply Theorem 34.7 to conclude that the map $A \mapsto A \otimes_{S_0} S$ yields an epimorphism $\mathbf{G_0}(R_0) = \mathbf{G_0}(S_0) \to \mathbf{gr}\,\mathbf{G_0}(S)$.

Furthermore Proposition 33.4 implies that the ring homomorphism $R[t] \to R$ given by $t \mapsto 1$ yields an epimorphism $\mathbf{gr}\,\mathbf{G_0}(S) = \mathbf{gr}\,\mathbf{G_0}(R[t]) \to \mathbf{G_0}(R)$. Thus we obtain the epimorphism $\mathbf{G_0}(R_0) \to \mathbf{gr}\,\mathbf{G_0}(R[t]) \to \mathbf{G_0}(R)$ and all that remains is to identify this combined map. To this end, let $A \in \mathbf{mod}\,R_0$. Then we see that

$$[A] \mapsto [A \otimes_{R_0} R[t]] \mapsto [(A \otimes_{R_0} R[t]) \otimes_{R[t]} R]$$

where R is viewed as a left $R[t]$-module via the above ring homomorphism. Since the associativity of tensor product yields

$$(A \otimes_{R_0} R[t]) \otimes_{R[t]} R \cong A \otimes_{R_0} R,$$

the composite map is determined by $[A] \mapsto [A \otimes_{R_0} R]$ and the result follows. ∎

It turns out that the epimorphisms of the preceding two results are in fact both isomorphisms. For this we require one more elementary property of Tor. Suppose A_R and $_RM$ are R-modules and assume that M is an (R, S)-bimodule. Then $\mathrm{Tor}_n^R(A, M)$ is a right S-module and the maps of Lemma 34.1 are all S-homomorphisms. Indeed, for the most part, this is just a consequence of the functorial nature of Tor. Furthermore, if R is right Noetherian, $A_R \in \mathbf{mod}\,R$ and $M_S \in \mathbf{mod}\,S$, then $\mathrm{Tor}_n^R(A, M)$ is a finitely generated right S-module (see Exercise 5).

Proposition 34.9. [179] *The epimorphisms in Theorem 34.7 and Corollary 34.8 are both isomorphisms.*

Proof. The arguments are identical in the two cases, so we just consider Corollary 34.8. Here we have an epimorphism $\theta \colon \mathbf{G_0}(R_0) \to$

$\mathbf{G_0}(R)$ and we show that θ is an isomorphism by constructing a back map.

Since the module ${}_R \bar{R}$ has finite projective dimension t, by assumption, Lemma 34.2(ii) yields $\operatorname{Tor}_k^R(A, \bar{R}) = 0$ for all $k \geq t + 1$. Furthermore, $\bar{R}$ is an (R, R_0)-bimodule and thus each $\operatorname{Tor}_n^R(A, \bar{R})$ is a right R_0-module. Since $\bar{R}_{R_0} \cong (R_0)_{R_0}$, it follows from the above comments that $A \in \mathbf{mod}\, R$ implies that $\operatorname{Tor}_n^R(A, \bar{R}) \in \mathbf{mod}\, R_0$ and we define the map $\mathbf{mod}\, R \to \mathbf{G_0}(R_0)$ by

$$A \mapsto \sum_{k=0}^{t} (-)^k [\operatorname{Tor}_k^R(A, \bar{R})] .$$

Now it is an easy consequence of the long exact Tor sequence that this map respects short exact sequences and hence, by Lemma 33.1(i), we obtain a group homomorphism $\varphi \colon \mathbf{G_0}(R) \to \mathbf{G_0}(R_0)$ given by

$$\varphi \colon [A] \mapsto \sum_{k=0}^{t} (-)^k [\operatorname{Tor}_k^R(A, \bar{R})] .$$

Finally if $B \in \mathbf{mod}\, R_0$ then Lemma 34.3(ii) implies that we have $\operatorname{Tor}_n^R(B \otimes_{R_0} R, \bar{R}) = 0$ for all $n \geq 1$. Hence since Tor_0 is the tensor product, we see that

$$\varphi\theta \colon [B] \mapsto [(B \otimes_{R_0} R) \otimes_R \bar{R}] = [B] ,$$

by Lemma 34.3(i). Thus $\varphi\theta$ is the identity on $\mathbf{G_0}(R_0)$ and θ is one-to-one as required. $\blacksquare$

We remark that an appropriate $\mathbf{K_0}$ analog of Theorem 34.7 holds and is a simple consequence of the graded version of Nakayama's lemma (see Exercises 6–8). On the other hand, there is no $\mathbf{K_0}$ analog of Corollary 34.8 without additional assumptions on the ring. A counterexample can be found at the end of Section 36. We close with a few comments on $\mathbf{K_0}(R)$ which are relevant to that example.

Let R be a ring. If $A, B \in \mathbf{mod}\, R$, then A and B are said to be *stably isomorphic* if $A \oplus R^n \cong B \oplus R^n$ for some finitely generated free R-module R^n. Furthermore, A is *stably free* if $A \oplus R^n \cong R^m$ for some $m, n \geq 0$.

Lemma 34.10. *Let P and Q be finitely generated projective R-modules.*

i. $[P] = [Q]$ in $\mathbf{K_0}(R)$ if and only if P and Q are stably isomorphic.

ii. $[P]$ is in the cyclic subgroup of $\mathbf{K_0}(R)$ generated by $[R]$ if and only if P is stably free.

Proof. (i) If P and Q are stably isomorphic, then certainly $[P] = [Q]$ in $\mathbf{K_0}(R)$. For the converse, let $\mathcal{F}$ be the free additive abelian group on the isomorphism classes $\tilde{A}$ of finitely generated projective R-modules. Furthermore, let $\mathcal{R}$ be the subgroup of $\mathcal{F}$ generated by the relations $\tilde{B} - \tilde{A} - \tilde{C}$ with $B \cong A \oplus C$. Then $\mathbf{K_0}(R) \cong \mathcal{F}/\mathcal{R}$ and $[A] = \tilde{A} + \mathcal{R}$. In particular, if $[P] = [Q]$ then $\tilde{P} - \tilde{Q}$ can be written as a finite sum and difference of relations, say

$$\tilde{P} - \tilde{Q} = \sum_i (\tilde{B}_i - \tilde{A}_i - \tilde{C}_i) + \sum_j (\tilde{D}_j + \tilde{F}_j - \tilde{E}_j)$$

with $B_i \cong A_i \oplus C_i$ and $E_j \cong D_j \oplus F_j$. Thus

$$\tilde{P} + \sum_i \tilde{A}_i + \sum_i \tilde{C}_i + \sum_j \tilde{E}_j = \tilde{Q} + \sum_i \tilde{B}_i + \sum_j \tilde{D}_j + \sum_j \tilde{F}_j.$$

and this means that the summands on the left must match, term for term, with the summands on the right. Hence $P \oplus V \cong Q \oplus V$ where V is the direct sum

$$V \cong \oplus \sum_i A_i \oplus \sum_i C_i \oplus \sum_j E_j$$

$$\cong \oplus \sum_i B_i \oplus \sum_j D_j \oplus \sum_j F_j.$$

But V is projective, so $V \oplus U \cong R^n$ for some n and therefore P is stably isomorphic to Q.

(ii) If P is stably free with say $P \oplus R^n \cong R^m$, then $[P] = (m - n)[R]$ in $\mathbf{K_0}(R)$. Conversely suppose $[P]$ is contained in the cyclic subgroup of $\mathbf{K_0}(R)$ generated by $[R]$ and say $[P] = n[R]$. If $n \geq 0$, then P is stably isomorphic to R^n and hence P is stably free.

On the other hand, if $n < 0$, then $P \oplus R^{(-n)}$ is stably isomorphic to 0 and again P is stably free. ∎

EXERCISES

1. Let M be a left R-module. Show that M is flat if and only if $\mathrm{Tor}_1(A, M) = 0$ for all A and, when this occurs, that $\mathrm{Tor}_n(A, M) = 0$ for all A and all $n \geq 1$.

2. Suppose that R is a commutative domain with field of fractions Q. If A is a right R-module, prove that $\mathrm{Tor}_1(A, Q/R)$ is isomorphic to the *torsion submodule* of A, namely

$$\{\, a \in A \mid ar = 0 \text{ for some } r \in R \setminus 0 \,\}.$$

3. Let $0 \to A_n \to \cdots \to A_1 \to A_0 \to 0$ be an exact sequence of finitely generated R-modules. Show that $\sum_{i=0}^{n}(-)^i[A_i] = 0$ in $\mathbf{G_0}(R)$.

4. Let X be a partially ordered multiplicative semigroup with the property that $xy, yx > y$ for all $x \in X^{\#} = X \setminus \{1\}$ and $y \in X$. If $R*X$ is a right Noetherian crossed product, show that $I = R*X^{\#} \lhd R*X$ with $R*X/I \cong R$. Furthermore, prove that the obvious analog of Theorem 34.7 holds in this context.

5. Assume that R is right Noetherian, $A_R \in \mathbf{mod}\,R$ and that $_RM_S$ is finitely generated S-module. Use an appropriate projective resolution for A_R to show that $\mathrm{Tor}_n^R(A, M)$ is a finitely generated S-module.

6. Let R be a Z^+-graded ring and let A and P be graded R-modules with P projective. Suppose $\theta: A \to P$ is a graded homomorphism which induces an isomorphism of the modules $A/AI \to P/PI$. Use the graded analog of Nakayama's lemma to show first that θ is onto and then that θ is an isomorphism. For the latter, note that $\mathrm{Ker}(\theta)$ is a graded submodule of A contained in AI and that $A = P' \oplus \mathrm{Ker}(\theta)$ for some submodule (not necessarily graded) $P' \cong P$.

7. Again let R be a Z^+-graded ring and let P be a graded R-module which is projective. Note that PI is a graded submodule of P

and set $Q_i = P_i/(P_i \cap PI)$ so that $P/PI = \oplus \sum_i Q_i$. Show first that each Q_i is a projective R_0-module and hence that $Q = \oplus \sum_i Q_i \otimes_{R_0} R$ is a projective R-module. Furthermore, if Q is graded so that $Q_i \otimes R_j$ has degree $i + j$, prove that Q/QI is graded isomorphic to P/PI.

8. Continuing with the above notation, show that there exists a *graded* homomorphism $\theta: Q \to P$ which induces the isomorphism $Q/QI \to P/PI$. Conclude that θ itself is an isomorphism.

35. Group Extensions

In this and the next section, we prove the key result of [**142**] on the Grothendieck group of Noetherian crossed products. As will be apparent, almost all the work involves the study of abelian-by-finite groups. Thus suppose $\Lambda \triangleleft \Gamma$ with Λ abelian and Γ/Λ finite. Then the proof divides naturally into three cases according to whether the action of Γ/Λ on Λ is (1) free, (2) rationally free or (3) arbitrary. Case (1) follows fairly easily from Corollary 34.8 once we observe that such groups split, and case (3) uses a Tor argument similar to that of the proof of Theorem 34.7. Case (2) is by far the most interesting; it is here that the finite subgroups miraculously appear as the stabilizers in a certain permutation action. In this section, we consider the first two cases.

To start with, we need some general information on group extensions and we sketch the necessary background material in the special case of abelian normal subgroups. Let Γ be an arbitrary group, Λ a normal abelian subgroup and set $G = \Gamma/\Lambda$. Then Γ acts on Λ by conjugation and Λ acts trivially, so we obtain a group homomorphism $G \to \mathrm{Aut}(\Lambda)$. In other words, Λ is a module for the integral group ring $Z[G]$ (even though we continue to view Λ multiplicatively). Now suppose that we know G, Λ and the structure of Λ as a $Z[G]$-module. The goal is to understand the possibilities for Γ. One such, of course, is the *split extension* $\Gamma = \Lambda \rtimes G$, but there are certainly other groups which we now proceed to describe.

For each $x \in G$, choose $\bar{x} \in \Gamma$ to be a representative of the Λ-coset corresponding to x. Then $\Gamma = \bigcup_{x \in G} \bar{x}\Lambda$ and $\bar{x}\bar{y} \in \overline{xy}\Lambda$ for all $x, y \in G$. Specifically, let us write $\bar{x}\bar{y} = \overline{xy}\alpha(x, y)$ where $\alpha(x, y) \in$

Λ. Notice that the map $\alpha\colon G \times G \to \Lambda$ completely determines the structure of Γ. Indeed if $x, y \in G$ and $a, b \in \Lambda$, then

$$\bar{x}a \cdot \bar{y}b = \bar{x}\bar{y}a^y b = \overline{xy}\alpha(x, y)a^y b$$

where of course a^y denotes the image of a under the action of y. Furthermore, the associativity of this multiplication is easily seen (Exercise 1) to be equivalent to the equation

(2-cocycle) $$\alpha(xy, z)\alpha(x, y)^z = \alpha(x, yz)\alpha(y, z)$$

for all $x, y, z \in G$. Functions which satisfy this condition are called 2-*cocycles* and we denote the set of all such α by $\mathrm{C}^2(G, \Lambda)$. Note that, since Λ is abelian, $\mathrm{C}^2(G, \Lambda)$ is a subgroup of the group of all functions from $G \times G$ to Λ with the operation being pointwise multiplication. We remark that the similarity between the 2-cocycle equation and the twisting relation for crossed products is certainly not surprising in view of the fact that $Z[\Gamma] = Z[\Lambda]*(\Gamma/\Lambda) = Z[\Lambda]*G$.

Now suppose that we choose different coset representatives, say $\tilde{x} \in \Gamma$. Then for each $x \in G$ we have $\tilde{x} = \bar{x}\delta(x)$ for some $\delta(x) \in \Lambda$ and the 2-cocycle β associated with $\{\tilde{x}\}$ is given by

$$\beta(x, y) = \alpha(x, y) \cdot \delta(xy)^{-1}\delta(x)^y\delta(y).$$

Notice that the 2-cocycle α corresponding to a splitting of Γ has $\alpha(x, y) = 1$ for all $x, y \in G$ and thus if β is defined by

(2-coboundary) $$\beta(x, y) = \delta(xy)^{-1}\delta(x)^y\delta(y)$$

for any function $\delta\colon G \to \Lambda$, then β is also a 2-cocycle, but a special one called a 2-*coboundary*. Now we let $\mathrm{B}^2(G, \Lambda)$ denote the subgroup of $\mathrm{C}^2(G, \Lambda)$ consisting of these 2-coboundaries and we define $\mathrm{H}^2(G, \Lambda)$ to be the factor group

$$\mathrm{H}^2(G, \Lambda) = \mathrm{C}^2(G, \Lambda)/\mathrm{B}^2(G, \Lambda).$$

From the above discussion, it is clear that there is a one-to-one correspondence between extensions of Λ by G and the elements of this 2nd-*cohomology group* $\mathrm{H}^2(G, \Lambda)$. Thus we have

Lemma 35.1. *Let Λ be a normal abelian subgroup of Γ so that Λ is a $Z[G]$-module for $G = \Gamma/\Lambda$. If Λ' is a $Z[G]$-module containing Λ, then there exists a group $\Gamma' \supseteq \Gamma$ with $\Lambda' \lhd \Gamma'$, $\Gamma' = \Gamma\Lambda'$, $\Lambda' \cap \Gamma = \Lambda$ and $\Gamma'/\Lambda' = \Gamma/\Lambda = G$ acting appropriately on Λ'.*

Proof. The structure of Γ gives rise to a 2-cocycle $\alpha\colon G \times G \to \Lambda$. By viewing α as a map to Λ', it is clear that $\alpha \in C^2(G, \Lambda')$ and thus α determines a group Γ' with $\Lambda' \lhd \Gamma'$ and $\Gamma'/\Lambda' \cong G$. In fact, if $\{\bar{x} \mid x \in G\}$ is the set of coset representatives for Λ' in Γ' obtained in this manner, then $\Gamma' = \bigcup_{x \in G} \bar{x}\Lambda'$ and $\Gamma \cong \bigcup_{x \in G} \bar{x}\Lambda$ so the result follows. ∎

Furthermore, we have

Lemma 35.2. *Let Λ be a $Z[G]$-module with G finite.*
   i. $H^2(G, \Lambda)$ has exponent dividing $|G|$.
   ii. If Λ is finitely generated, then $H^2(G, \Lambda)$ is finite.
   iii. If Λ is a free $Z[G]$-module, then $H^2(G, \Lambda) = \langle 1 \rangle$.

Proof. (i) Let $\alpha \in C^2(G, \Lambda)$ be given and multiply the 2-cocycle equation over all $x \in G$. Since Λ is commutative, we obtain $\delta(z)\delta(y)^z = \delta(yz)\alpha(y, z)^n$, where $\delta(y) = \prod_{x \in G} \alpha(x, y)$ and $n = |G|$. Thus α^n is a 2-coboundary.

(ii) This follows from (i) and the fact that $C^2(G, \Lambda)$ is a finitely generated abelian group, being a subgroup of the direct product of $|G|^2$ copies of Λ.

(iii) Since Λ is a free $Z[G]$-module, we can write Λ as the direct product $\Lambda = \prod_{z \in G} \Lambda_z$ where the Λ_z are subgroups of Λ permuted regularly by G. In particular, if a_z denotes the z-component of $a \in \Lambda$, then $a = \prod_{z \in G} a_z$ and $(a^g)_z = (a_{zg^{-1}})^g$ for any $z, g \in G$. Now let $\alpha \in C^2(G, \Lambda)$, replace z by z^{-1} in the 2-cocycle equation and conjugate by z to obtain

$$\alpha(xy, z^{-1})^z \cdot \alpha(x, y) = \alpha(x, yz^{-1})^z \cdot \alpha(y, z^{-1})^z.$$

Reading off the z-components then yields

$$\alpha_1(xy, z^{-1})^z \cdot \alpha(x, y)_z = \alpha_1(x, yz^{-1})^z \cdot \alpha_1(y, z^{-1})^z$$

where, by definition, $\alpha_1(x,y) = \alpha(x,y)_1$. Finally, since $\alpha(x,y) = \prod_{z\in G}\alpha(x,y)_z$ it follows, by multiplying the above over all $z \in G$, that $\delta(xy)\alpha(x,y) = \delta(x)^y\delta(y)$ where $\delta(x) = \prod_{z\in G}\alpha_1(x,z^{-1})^z$. Thus α is a 2-coboundary as required. ∎

We can now handle case (1).

Proposition 35.3. [142] *Let $R*\Gamma$ be a crossed product with R right Noetherian and Γ a finitely generated abelian-by-finite group. Specifically, let Λ be a normal abelian subgroup of Γ with Γ/Λ finite and assume that Λ is a free $Z[\Gamma/\Lambda]$-module. Then $\Gamma = \Lambda \rtimes G$ for some finite subgroup $G \subseteq \Gamma$ and the map $A \mapsto A \otimes_{R*G} R*\Gamma$ yields an epimorphism $\mathbf{G_0}(R*G) \to \mathbf{G_0}(R*\Gamma)$.*

Proof. The assumption that Λ is a free $Z[\Gamma/\Lambda]$-module is used in two different ways. First, by Lemma 35.2(iii), $\mathrm{H}^2(\Gamma/\Lambda,\Lambda) = \langle 1 \rangle$ and thus the extension splits; say $\Gamma = \Lambda \rtimes G$ for some finite subgroup $G \subseteq \Gamma$. Second, since Λ is now a free $Z[G]$-module, we can choose a free generating set $\{x_1, x_2, \ldots, x_n\}$ for Λ which is permuted by G. It follows that if X is the subsemigroup of Λ generated by the x_k's, then X is the free abelian semigroup of rank n and $R*X$ is a skew polynomial ring in n variables. Furthermore, if X_i denotes the set of monomials in X of degree i, then $R*X$ is Z^+-graded with i^{th} component $R*X_i$. But observe that G normalizes X and in fact each X_i. Thus $S = R*(XG) = R*(GX)$ is a subring of $R*\Gamma$ and indeed S is Z^+-graded with i^{th} component $S_i = R*(X_iG) = R*(GX_i)$. In particular, since $X_0 = \{1\}$, we have $S_0 = R*G$.

We show now that $S = \oplus\sum_{i=0}^{\infty} S_i$ satisfies the hypotheses of Corollary 34.8. First, S is right Noetherian since $R*X$ is, according to Lemma 1.7, and since $S = \bar{G} \cdot R*X$ is a finitely generated right $R*X$-module. Next we note that $_{S_0}S$ is flat since, in fact, $_{S_0}S$ is free with $\bar{X}$ as a free basis. Finally, since $_R R$ is free, it follows from Lemma 33.10 that $R*X/(R*X)_+ = R$ has finite projective dimension as a left $R*X$-module and we can choose

$$0 \to Q_m \to \cdots \to Q_1 \to Q_0 \to R \to 0$$

to be an appropriate finite (left) projective resolution. Now S is a free right $R*X$-module with basis $\bar{G}$, so tensoring this resolution with S_{R*X} yields the exact sequence

$$0 \to S \otimes Q_m \to \cdots \to S \otimes Q_1 \to S \otimes Q_0 \to S \otimes_{R*X} R \to 0.$$

Furthermore, since $S \otimes_{R*X} (R*X) \cong {}_S S$ and $\otimes$ commutes with arbitrary direct sums, we see that each $S \otimes_{R*X} Q_i$ is projective. In other words, the above is a finite (left) projective resolution for the module

$$S \otimes_{R*X} R = S \otimes_{R*X} \left(R*X/(R*X)_+ \right) \cong S/S_+$$

and $\mathrm{pd}_S(S/S_+) < \infty$ as required.

We can now apply Corollary 34.8 to conclude that the map $A \mapsto A \otimes_{R*G} S$ yields an epimorphism $\mathbf{G_0}(R*G) = \mathbf{G_0}(S_0) \to \mathbf{G_0}(S)$. Furthermore, since $R*\Gamma = S\bar{X}^{-1}$, it follows from Lemma 33.3(iii) that the map $B \mapsto B \otimes_S R*\Gamma$ yields an epimorphism $\mathbf{G_0}(S) \to \mathbf{G_0}(R*\Gamma)$. Thus the combined map

$$A \mapsto A \otimes_{R*G} S \mapsto (A \otimes_{R*G} S) \otimes_S R*\Gamma \cong A \otimes_{R*G} R*\Gamma$$

determines an epimorphism $\mathbf{G_0}(R*G) \to \mathbf{G_0}(R*\Gamma)$ and the proposition is proved. ∎

We now move on to case (2). Here the goal is to show that $\mathbf{G_0}(R*\Gamma)$ is spanned by the images, under induction, of $\mathbf{G_0}(R*G_i)$ for finitely many finite subgroups G_i of Γ. The basic idea of the proof is to embed $R*\Gamma$ in a larger, better behaved ring where the result is already known and then to translate this information back to $R*\Gamma$. The proof is in fact conceptually easy but it can get technically complicated. We simplify matters by doing most of the preliminary work in the context of arbitrary rings. We begin with a trivial observation.

Suppose that $T \subseteq S$ are rings and that u is a unit of S. Then $T^u = u^{-1}Tu$ is also a subring of S and, since $T \cong T^u$, there is a natural correspondence between the modules of the two subrings. Specifically, if A is a right T-module, then $A^u = \{ a^u \mid a \in A \}$ is an isomorphic copy of the additive abelian group A with the module structure given by $a^u \cdot t^u = (at)^u$.

Lemma 35.4. *Let $T \subseteq S$ be rings and let u be a unit of S. If A is a right T-module, then $A^u \otimes_{T^u} S \cong A \otimes_T S$. Furthermore, $A \in \operatorname{mod} T$ if and only if $A^u \in \operatorname{mod} T^u$.*

Proof. The appropriate maps between the two tensor products are given by $a^u \otimes s \mapsto a \otimes us$ and $a \otimes s \mapsto a^u \otimes u^{-1}s$. The last part is clear. ∎

Now we consider the classical version of the *Morita correspondence*. Let S be any ring and let $e \neq 0$ be an idempotent in S. If A is a right eSe-module, then $\alpha(A) = A \otimes_{eSe} eS$ is a right S-module, and if B is a right S-module, then $\beta(B) = Be$ is a right eSe-module. For convenience, we call α and β the *Morita maps* determined by e. Note that if S is right Noetherian, then so is eSe since $I \mapsto IS$ is a one-to-one inclusion preserving map from the set of right ideals I of eSe to the right ideals of S. Thus, in this context, we can consider both $\mathbf{G_0}(S)$ and $\mathbf{G_0}(eSe)$. Part (iii) of the following lemma is not really required; we include it for the sake of completeness.

Lemma 35.5. *Let $e \neq 0$ be an idempotent in the ring S and assume that $SeS = fS$ where f is a central idempotent of S.*

    i. The Morita maps yield a one-to-one correspondence between the right eSe-modules A and the right S-modules B with $B = Bf$.

    ii. $A \in \operatorname{mod} eSe$ if and only if $\alpha(A) \in \operatorname{mod} S$.

    iii. A is projective if and only if $\alpha(A)$ is projective.

    iv. If S is right Noetherian and $f = 1$, then $\mathbf{G_0}(S)$ and $\mathbf{G_0}(eSe)$ are isomorphic with isomorphisms determined by the Morita maps.

Proof. Since $e \in fS$, $e(fS) = eS$ and $e(fS)e = eSe$, it clearly suffices to assume that $f = 1$. Thus $S = SeS$ and we have $1 = \sum_1^n s_i e s_i'$ for suitable $s_i, s_i' \in S$. Note also that if N is a right S-module with $Ne = 0$, then $0 = N(SeS) = NS$ and $N = 0$.

    (i) We show that the composite maps $\beta\alpha$ and $\alpha\beta$ are both the identity and the first is clear since $(A \otimes_{eSe} eS)e = A \otimes_{eSe} eSe \cong A$.

For the second, we need to prove that $Be \otimes_{eSe} eS \cong B$ and at least we have a map $\theta \colon Be \otimes eS \to B$ given by $\theta \colon be \otimes es \mapsto (be)(es) = bes$. Furthermore, $\text{Im}(\theta) = BeS = B(SeS) = B$ so θ is onto and we have $\text{Ker}(\theta)e = 0$ since $(Be \otimes eS)e = Be \otimes eSe \cong Be$. Thus $\text{Ker}(\theta) = 0$ and θ is indeed an isomorphism.

(ii) If A is a finitely generated eSe-module, then $\alpha(A) = A \otimes eS$ is finitely generated since eS is a cyclic right S-module. On the other hand, if $B = \sum_1^m b_j S$ is a finitely generated S-module, then $\beta(B) = Be = \sum_1^m b_j Se$ and it suffices to show Se is a finitely generated right eSe-module. But $1 = \sum_1^n s_i es_i'$ implies that $Se = \sum_1^n s_i(eSe)$, so the result follows.

(iii) Since α and β commute with direct sums, it suffices to show that $\alpha(eSe)$ and $\beta(S)$ are projective and the first is clear since $\alpha(eSe) = eS$. For the second, $\beta(S) = Se$ and we define the right eSe-module homomorphisms $\sigma \colon (eSe)^n \to Se$ and $\tau \colon Se \to (eSe)^n$ by

$$\sigma \colon (t_1, t_2, \ldots, t_n) \mapsto \sum_1^n s_i t_i$$

and

$$\tau \colon se \mapsto (es_1' se, es_2' se, \ldots, es_n' se).$$

Since $\sum_1^n s_i es_i' = 1$, it follows that $\sigma\tau(se) = se$ and therefore that Se is projective.

(iv) As we observed, if S is right Noetherian, then so is eSe. In view of (i) and (ii), we need only show that the maps $A \mapsto [\alpha(A)] \in \mathbf{G_0}(S)$ and $B \mapsto [\beta(B)] \in \mathbf{G_0}(eSe)$ respect short exact sequences. The latter is clear since $B_1 \subseteq B_2$ implies that $B_2 e \cap B_1 = B_1 e$. For the former, suppose $A_1 \subseteq A_2$ and let N denote the kernel of the map $A_1 \otimes eS \to A_2 \otimes eS$. Since $A \otimes eSe \cong A$, it is clear that $Ne = 0$ and thus that $N = 0$. This completes the proof. ∎

The next lemma is a combination of the previous two. It is slightly tedious to state, but it is precisely what is needed. Note that if $T \subseteq S$ is an extension of rings and if $e \in T$, then we also have the ring extension $eTe \subseteq eSe$.

Lemma 35.6. *Suppose we are given*

   i. a right Noetherian ring S and an idempotent $e \in S$ with
$S = SeS$,
   ii. subrings $T_i \subseteq S$ and idempotents $e_i \in T_i$ such that $T_i e_i T_i = f_i T_i$ with f_i a central idempotent of T_i,
   iii. units u_i of S with $(e_i)^{u_i} = u_i^{-1} e_i u_i = e$.
If $\mathbf{G_0}(S)$ is generated by the elements $[B_i f_i \otimes_{T_i} S]$ for all i and $B_i \in \mathbf{mod}\, T_i$, then $\mathbf{G_0}(eSe)$ is generated by the elements $[A_i \otimes_{e(T_i)^{u_i} e}\, eSe]$ for all i and all $A_i \in \mathbf{mod}\, e(T_i)^{u_i} e$.

Proof. By Lemma 35.4 we have

$$(B_i)^{u_i}(f_i)^{u_i} \otimes_{(T_i)^{u_i}} S \cong B_i f_i \otimes_{T_i} S$$

so these modules determine the same element of $\mathbf{G_0}(S)$. Thus without loss of generality, we can now assume that each $u_i = 1$ and $e_i = e$.

   By Lemma 35.5(iv), β determines an epimorphism from $\mathbf{G_0}(S)$ to $\mathbf{G_0}(eSe)$. Thus since $\beta(B_i f_i \otimes_{T_i} S) = B_i f_i \otimes_{T_i} Se$, we see that $\mathbf{G_0}(eSe)$ is generated by the elements $[B_i f_i \otimes_{T_i} Se]$ for all i and $B_i \in \mathbf{mod}\, T_i$. In addition, by Lemma 35.5(i)(ii), each $B_i f_i$ is isomorphic to $A_i \otimes_{eT_i e} eT_i$ for some appropriate $A_i \in \mathbf{mod}\, eT_i e$. In other words, $\mathbf{G_0}(eSe)$ is generated by the elements $[M]$ with modules M of the form

$$B_i f_i \otimes_{T_i} Se \cong (A_i \otimes_{eT_i e} eT_i) \otimes_{T_i} Se$$
$$\cong A_i \otimes_{eT_i e} (eT_i \otimes_{T_i} Se).$$

But $eT_i \otimes_{T_i} Se \cong eSe$ since $\otimes$ commutes with direct sums and $T_i = eT_i \oplus (1-e)T_i$. Thus

$$B_i f_i \otimes_{T_i} Se \cong A_i \otimes_{eT_i e} eSe$$

and the result follows. $\blacksquare$

   Now we return to crossed products. Suppose $R*\Gamma$ is given and Γ' is a group containing Γ as a subgroup of finite index. Then we can not, in general, hope to extend $R*\Gamma$ to some $R*\Gamma'$ since, for example, the action of Γ on R might not extend. Nevertheless, we have the following observation reminiscent of duality. As it turns out, we do not really care that $R^n * \Gamma'$ is a matrix ring, but it does offer a way to construct this ring without having to check associativity.

Lemma 35.7. *Let $\Gamma' \supseteq \Gamma$ be groups with $|\Gamma' : \Gamma| = n < \infty$ and let $R * \Gamma$ be given. Then $\mathrm{M}_n(R * \Gamma)$ is a crossed product $R^n * \Gamma'$. Moreover*

i. $R^n = \mathrm{diag}(R, R, \ldots, R)$.

ii. The group of trivial units of $R^n * \Gamma'$ transitively permutes the matrix idempotents $e_{1,1}, e_{2,2}, \ldots, e_{n,n}$ by conjugation.

iii. If Ω' is any subgroup of Γ', then

$$e_{1,1}(R^n * \Omega')e_{1,1} = R * (\Omega' \cap \Gamma) \cdot e_{1,1} \cong R * (\Omega' \cap \Gamma)$$

with $R * \Gamma$ embedded as scalar matrices.

Proof. Let $\mathcal{T}$ be a right transversal for Γ in Γ' and assume that $1 \in \mathcal{T}$. Since $|\mathcal{T}| = n$, we can index the rows and columns of $\mathrm{M}_n(R * \Gamma)$ by the elements of $\mathcal{T}$ with the first row and column corresponding to 1. In particular, for each $t, t' \in \mathcal{T}$ we have the usual matrix units $e_{t,t'} \in \mathrm{M}_n(R * \Gamma)$.

We can now grade $\mathrm{M}_n(R * \Gamma)$ by assigning to each expression $r \bar{g} e_{t,t'}$ the grade $t^{-1} g t' \in \Gamma'$. Here of course $r \in R$, $g \in \Gamma$ and $t, t' \in \mathcal{T}$. It follows immediately that, in this way, $\mathrm{M}_n(R * \Gamma)$ becomes at least a Γ'-graded ring. Furthermore, $t^{-1} g t' = 1$ if and only if $t = t'$ and $g = 1$. Thus the identity component in this grading is $R^n = \mathrm{diag}(R, R, \ldots, R)$.

Now let $x \in \Gamma'$ and observe that for each $t \in \mathcal{T}$ there exists $t' \in \mathcal{T}$ with $tx \in \Gamma t'$; say $tx = x_t t'$ with $x_t \in \Gamma$. Indeed, the map $t \mapsto t'$ corresponds to the permutation action of x on the right cosets of Γ and thus $\tilde{x} = \sum_{t \in \mathcal{T}} \bar{x}_t e_{t,t'}$ is clearly an invertible matrix which is homogeneous of grade x. It follows from this that $\mathrm{M}_n(R * \Gamma)$ is not only Γ'-graded, but it is in fact a crossed product $R^n * \Gamma'$.

It remains to prove (ii) and (iii) and for the former we need only observe that the idempotents $e_{t,t}$ are central in R^n and that $e_{t,t}\tilde{x} = \bar{x}_t e_{t,t'} = \tilde{x} e_{t',t'}$. Finally if $s = e_{1,1} \cdot r \bar{g} e_{t,t'} \cdot e_{1,1} \neq 0$, then $t = t' = 1$, $s = r \bar{g} e_{1,1} \in R * \Gamma \cdot e_{1,1}$ and $r \bar{g} e_{t,t'}$ has grade $t^{-1} g t' = g \in \Gamma$. With this, (iii) follows and the proof is complete. ∎

We can now quickly handle case (2). Note that if Λ is a finitely generated free abelian group and a $Z[G]$-module, then $\Lambda \otimes_Z Q$ is naturally a module for the rational group algebra $Q[G]$. We say that Λ is a *rationally free* $Z[G]$-module, if $\Lambda \otimes_Z Q$ is free as a $Q[G]$-module.

Proposition 35.8. [142] *Let $R*\Gamma$ be a crossed product with R right Noetherian and Γ a finitely generated abelian-by-finite group. Specifically, let Λ be a torsion free normal abelian subgroup of Γ with Γ/Λ finite and assume that Λ is a rationally free $Z[\Gamma/\Lambda]$-module. Then there exist finitely many finite subgroups G_i of Γ such that $\mathbf{G_0}(R*\Gamma)$ is generated by the images of the various $\mathbf{G_0}(R*G_i)$ under the induced module map.*

Proof. In this paragraph we temporarily view Λ additively. By assumption, $\Lambda \otimes_Z Q$ is a free $Q[\Gamma/\Lambda]$-module and thus it has a Q-basis $\{\mu_1, \mu_2, \ldots, \mu_k\}$ which is permuted in regular orbits by the group Γ/Λ. Note that each of the finitely many generators of $\Lambda \subseteq \Lambda \otimes Q$ is a Q-linear combination of the μ_i's. Thus if d is an appropriate common denominator, then $\Lambda \subseteq \Lambda' = \sum_1^k Z\mu_i/d$. In other words, Λ is contained in the finitely generated free $Z[\Gamma/\Lambda]$-module Λ'. In addition, since $\Lambda' \subseteq \Lambda \otimes Q$, we see that Λ'/Λ is torsion and hence finite.

We can now apply Lemma 35.1 to conclude that there exists a group $\Gamma' \supseteq \Gamma$ with $\Lambda' \triangleleft \Gamma'$ and $\Gamma'/\Lambda' = \Gamma/\Lambda$ acting appropriately on Λ'. In addition, $\Gamma' = \Gamma\Lambda'$ and $\Lambda' \cap \Gamma = \Lambda$ so $|\Gamma' : \Gamma| = |\Lambda' : \Lambda| = n < \infty$. Set $S = M_n(R*\Gamma)$. Then, by Lemma 35.7, $S = R^n*\Gamma'$ is a crossed product of Γ' over the right Noetherian ring R^n. Thus since Λ' is a free $Z[\Gamma'/\Lambda']$-module, we conclude from Proposition 35.3 that $\Gamma' = \Lambda' \rtimes G$ for some finite subgroup $G \subseteq \Gamma'$. Moreover, if $T = R^n*G \subseteq S$, then the map $B \mapsto B \otimes_T S$ yields an epimorphism $\mathbf{G_0}(T) \to \mathbf{G_0}(S)$. Set $e = e_{1,1} \in S$ and note that $S = SeS$.

By Lemma 35.7, the trivial units of R^n*G permute the set $\{e_{j,j}\}$ of diagonal idempotents by conjugation. This action is not necessarily transitive so, for each orbit $\mathcal{O}_i$, we choose $e_i \in \mathcal{O}_i$ and let f_i denote the sum of the members of the orbit. Then $1 = f_1 + f_2 + \cdots$ is a decomposition of $1 \in T = R^n*G$ into orthogonal central idempotents. Hence, since $\mathbf{G_0}(T) \to \mathbf{G_0}(S)$ is surjective, it follows that

$\mathbf{G_0}(S)$ is generated by elements of the form $[B_i f_i \otimes_T S]$ for all i and $B_i \in \mathbf{mod}\, T$. Moreover $e_i T \subseteq f_i T$ and in fact $T e_i T = f_i T$ since f_i is a sum of T-conjugates of e_i. Finally, by Lemma 35.7(ii) again, for each i there exists a trivial unit $\tilde{x}_i \in R^n * \Gamma'$ with $x_i \in \Gamma'$ and $\tilde{x}_i^{-1} e_i \tilde{x}_i = e_{1,1} = e$.

In other words, the hypothesis of Lemma 35.6 is satisfied with all $T_i = T$. We conclude therefore that $\mathbf{G_0}(eSe)$ is spanned by elements of the form $[A_i \otimes_{eT^{\tilde{x}_i}e} eSe]$ for all i and $A_i \in \mathbf{mod}\, eT^{\tilde{x}_i}e$. But observe that $eSe = R * \Gamma \cdot e$, by Lemma 35.7(iii), and that $T^{\tilde{x}_i} = R^n * G^{x_i}$ so $eT^{\tilde{x}_i}e = R * G_i \cdot e$ where $G_i = G^{x_i} \cap \Gamma$. Thus $\mathbf{G_0}(R * \Gamma \cdot e)$ is generated by the images of the various $\mathbf{G_0}(R * G_i \cdot e)$ under the induced module map. Since $R * \Gamma \cdot e \cong R * \Gamma$ and each G_i is a finite subgroup of Γ, the result follows. ∎

We will deal with case (3) and prove the main result in the next section.

EXERCISES

1. Show that each element of $\mathrm{H}^2(G, \Lambda)$ gives rise to a group Γ which is an extension of Λ by G. For this, one must prove not only the associativity of Γ but also the existence of 1 and of inverses.

2. Let R be right Noetherian and let $U, V \in \mathbf{mod}\, R$. Show that $[U] = [V]$ in $\mathbf{G_0}(R)$ if and only if there exist two short exact sequences $0 \to A \to B \to C \to 0$ and $0 \to A' \to B' \to C' \to 0$ with $U \oplus A' \oplus B \oplus C' \cong V \oplus A \oplus B' \oplus C$. This requires a slight modification of the proof of Lemma 34.10(i).

3. Suppose $R[\Gamma]$ is a group ring with R right Noetherian and $\Gamma = H \rtimes G$ polycyclic-by-finite. Assume that R has finite projective dimension as a left $R[H]$-module. Use the argument of Proposition 34.9 to show that the induced module map $\mathbf{G_0}(R[G]) \to \mathbf{G_0}(R[\Gamma])$ is one-to-one.

4. Show that the proof of Proposition 35.8 from Proposition 35.3 works equally well for $\mathbf{K_0}$. In other words, if there is a $\mathbf{K_0}$ analog of the case (1) result, then there will also be an analog in case (2).

5. Let G be a finite group of X-outer automorphisms of the simple ring R and assume that $|G|^{-1} \in R$. What does Lemma 35.5 say about the relationship between R-modules and those of the fixed ring R^G?

6. Let $1 = e_1 + e_2 + \cdots + e_n$ be a decomposition of $1 \in R$ into orthogonal idempotents and let G be a group of units of R which transitively permutes the set $\{\, e_i \,\}$ by conjugation. For each i choose $g_i \in G$ with $e_i = g_i^{-1} e_1 g_i$ and then define $e_{i,j} = g_i^{-1} e_1 g_j$. Show that the latter elements multiply like matrix units and then prove that $R \cong \mathrm{M}_n(S)$ with $S \cong e_1 R e_1$.

36. The Induction Theorem

The goal of this section is to prove the induction theorem. Specifically we show that if $R*\Gamma$ is a crossed product with R right Noetherian and Γ polycyclic-by-finite, then $\mathbf{G_0}(R*\Gamma)$ is generated by the images under the induced module map of the Grothendieck groups $\mathbf{G_0}(R*G_i)$ for finitely many finite subgroups $G_i \subseteq \Gamma$. As indicated earlier, most of the work here involves abelian-by-finite groups. Furthermore, such groups split into three cases and the first two have already been dealt with. Thus it remains to consider case (3) and for this we follow the argument of [**33**]. Again the idea is to introduce a larger, better behaved ring, but this time $R*\Gamma$ will not be a subring but rather a homomorphic image.

We start with an observation which has already proved useful, namely *transitivity of induction*. Stated below in its full generality, it is still an immediate consequence of the (functorial) associativity of tensor product.

Lemma 36.1. *Let* $\theta\colon R \to S$ *and* $\varphi\colon S \to T$ *be ring homomorphisms and let* A *be a right* R-module. *Then* $(A \otimes_R S) \otimes_S T \cong A \otimes_R T$ *where* ${}_R T$ *is obtained from the ring homomorphism* $\varphi\theta\colon R \to T$. *Furthermore, if* ${}_R S$ *and* ${}_S T$ *are flat, then so is* ${}_R T$.

Next we require a slight extension of Lemma 34.3(ii). Since the proof is essentially the same, we just briefly sketch it here.

Lemma 36.2. *Let T be a subring of S with $_TS$ flat. If A is a right T-module and M is a left S-module, then*

$$\operatorname{Tor}_i^S(A \otimes_T S, M) \cong \operatorname{Tor}_i^T(A, M)$$

for all $i \geq 0$. In particular if $_TM$ is flat, then $\operatorname{Tor}_i^S(A \otimes_T S, M) = 0$ for all $i \geq 1$.

Proof. Let $\mathbf{P} \to A \to 0$ be a projective resolution for the T-module A. Since $_TS$ is flat, it follows that $\mathbf{P} \otimes_T S \to A \otimes_T S \to 0$ is a projective resolution for $A \otimes S$. Now, by definition, the groups $\operatorname{Tor}_i^T(A, M)$ come from the homology of the complex $\mathbf{P} \otimes_T M \to 0$ while the groups $\operatorname{Tor}_i^S(A \otimes S, M)$ come from the homology of the complex $(\mathbf{P} \otimes_T S) \otimes_S M \to 0$. But

$$(\mathbf{P}_n \otimes_T S) \otimes_S M \cong \mathbf{P}_n \otimes_T M$$

for all n, so the two complexes are naturally isomorphic and hence have the same homology. ∎

The following result is the analog of Lemma 35.6 in the context of homomorphisms.

Lemma 36.3. *Suppose we are given*

    i. a right Noetherian ring S and an epimorphism $\varphi \colon S \to S^\varphi$ such that $_SS^\varphi$ has finite projective dimension,

    ii. a subring S_0 of S with $S = S_0 \oplus \operatorname{Ker}(\varphi)$ and $_{S_0}S$ flat,

    iii. subrings $T_i \subseteq S$ with $_{T_i}S$ and $_{T_i}S^\varphi$ both flat.

If $\mathbf{G}_0(S)$ is generated by the elements $[B_i \otimes_{T_i} S]$ for all i and $B_i \in \operatorname{\mathbf{mod}} T_i$, then $\mathbf{G}_0(S^\varphi)$ is generated by the elements $[A_i \otimes_{T_i^\varphi} S^\varphi]$ for all i and $A_i \in \operatorname{\mathbf{mod}} T_i^\varphi$.

Proof. Notice that both S and S^φ are right Noetherian and that if $A \in \operatorname{\mathbf{mod}} S$, then $\operatorname{Tor}_i^S(A, S^\varphi) \in \operatorname{\mathbf{mod}} S^\varphi$. Furthermore, since $_SS^\varphi$ has finite projective dimension, Lemma 34.2(ii) implies that, for each A, only finitely many of these Tor groups can be nonzero. In particular, it makes sense to define $\eta \colon \operatorname{\mathbf{mod}} S \to \mathbf{G}_0(S^\varphi)$ by

$$\eta(A) = \sum_{i=0}^{\infty} (-)^i [\operatorname{Tor}_i^S(A, S^\varphi)]$$

and it then follows from Lemma 34.1 that η respects short exact sequences. Thus η determines a group homomorphism $\eta \colon \mathbf{G_0}(S) \to \mathbf{G_0}(S^\varphi)$.

Now suppose T is any subring of S with $_T S$ and $_T S^\varphi$ both flat. If $B \in \mathbf{mod}\, T$, then $B \otimes_T S \in \mathbf{mod}\, S$ and we compute $\eta([B \otimes_T S])$. Using both flatness assumptions, it follows from Lemma 36.2 that $\mathrm{Tor}_i^S(B \otimes_T S, S^\varphi) = 0$ for all $i \geq 1$. Thus since Tor_0 is the tensor product, we have

$$\eta([B \otimes_T S]) = [(B \otimes_T S) \otimes_S S^\varphi] = [B \otimes_T S^\varphi].$$

Note that $_{S_0}S$ is flat, by assumption, and that $_{S_0}S^\varphi \cong {}_{S_0}S_0$ since $S = S_0 \oplus \mathrm{Ker}(\varphi)$. Thus, $_{S_0}S^\varphi$ is also flat and the above implies that $\eta([B \otimes_{S_0} S]) = [B \otimes_{S_0} S^\varphi]$ for all $B \in \mathbf{mod}\, S_0$. But φ restricts to a ring isomorphism $\varphi \colon S_0 \to S^\varphi$ so it follows that every finitely generated S^φ-module is isomorphic to a suitable $B \otimes_{S_0} S^\varphi$. In other words, $\mathbf{G_0}(S^\varphi)$ is spanned by the images under η of the various $[B \otimes_{S_0} S]$ and thus η is certainly an epimorphism.

Hence, since $\mathbf{G_0}(S)$ is generated by the elements $[B_i \otimes_{T_i} S]$ for all i and $B_i \in \mathbf{mod}\, T_i$, it follows that $\mathbf{G_0}(S^\varphi)$ is generated by the images of these elements under η. But, by assumption, $_{T_i}S$ and $_{T_i}S^\varphi$ are flat so $\eta([B_i \otimes_{T_i} S]) = [B_i \otimes_{T_i} S^\varphi]$ by the observation of the second paragraph. Thus since

$$B_i \otimes_{T_i} S^\varphi \cong (B_i \otimes_{T_i} T_i^\varphi) \otimes_{T_i^\varphi} S^\varphi$$

and $B_i \otimes_{T_i} T_i^\varphi \in \mathbf{mod}\, T_i^\varphi$, the result follows. ∎

Now we return to crossed products where we begin by constructing the necessary larger ring. Suppose $\Gamma' = \Lambda' \rtimes \Gamma$ is a split extension of Λ' by Γ. Then Γ can be viewed as both a subgroup of Γ' and as a homomorphic image. To avoid this confusion, let us write Γ_0 for the subgroup so that $\Gamma' = \Lambda'\Gamma_0$ and reserve Γ for the homomorphic image.

Lemma 36.4. *Let $\Gamma' = \Lambda' \rtimes \Gamma$ be a split extension of Λ' by Γ and let $R*\Gamma$ be given. Write $\Gamma' = \Lambda'\Gamma_0$ where Γ_0 is a complement for*

Λ' and let $R*\Gamma_0$ be the crossed product obtained naturally from the isomorphism $\Gamma_0 \cong \Gamma$.

 i. $R*\Gamma_0$ extends to a crossed product $R*\Gamma'$ with $R*\Lambda' = R[\Lambda']$ an ordinary group ring.

 ii. The map $\varphi: \Gamma' \to \Gamma$ extends to an epimorphism $\varphi: R*\Gamma' \to R*\Gamma$ with $R*\Gamma' = R*\Gamma_0 \oplus \mathrm{Ker}(\varphi)$.

 iii. If R has finite projective dimension as a left $R[\Lambda']$-module, then $R*\Gamma$ has finite projective dimension as a left $R*\Gamma'$-module.

Proof. Since $R*\Gamma$ is given, it is endowed with a twisting function $\tau: \Gamma \times \Gamma \to \mathrm{U}(R)$ and an action $\sigma: \Gamma \to \mathrm{Aut}(R)$ satisfying conditions (i) and (ii) of Lemma 1.1. In particular, if $\varphi: \Gamma' \to \Gamma$ is any map then, by way of composition, we obtain $\tau': \Gamma' \times \Gamma' \to \mathrm{U}(R)$ and $\sigma': \Gamma' \to \mathrm{Aut}(R)$. Furthermore, if φ is a group homomorphism, then it is trivial to verify that (i) and (ii) of Lemma 1.1 are also satisfied by these functions. It follows that τ' and σ' give rise to a crossed product $R*\Gamma'$ and that φ extends to a ring epimorphism $R*\Gamma' \to R*\Gamma$.

Observe that the restriction of φ to Γ_0 yields the given isomorphism $\Gamma_0 \cong \Gamma$ and hence that $\varphi: R*\Gamma_0 \to R*\Gamma$ is the natural isomorphism between these two crossed products. In addition, since $\bar{1} = 1 \in R*\Gamma$ and $\Lambda' = \mathrm{Ker}(\varphi)$, it follows that the twisting and action in $R*\Lambda' \subseteq R*\Gamma'$ are trivial and hence that $R*\Lambda' = R[\Lambda']$. Finally if $\mathbf{P} \to R \to 0$ is a finite projective resolution for R as a left $R[\Lambda']$-module, then

$$(R*\Gamma') \otimes_{R[\Lambda']} \mathbf{P} \to (R*\Gamma') \otimes_{R[\Lambda']} R \to 0$$

is a finite projective resolution for $(R*\Gamma') \otimes_{R[\Lambda']} R \cong R*\Gamma$. This completes the proof. ∎

We also need an analog of Lemma 33.10 for group rings. Let $R[\Lambda]$ be a group ring with $\Lambda = \langle x_1, x_2, \ldots, x_t \rangle$ free abelian on the generators x_i. Then we have a homomorphism $R[\Lambda] \to R$ given by $x_i \mapsto 1$ for all i. In particular, if M is an R-module then, via the combined map $R[\Lambda] \to R \to \mathrm{End}(M)$, we see that M is also an $R[\Lambda]$-module.

Lemma 36.5. *Let $R[\Lambda]$ be as above with $\Lambda = \langle x_1, x_2, \ldots, x_t \rangle$ a free abelian group. If M is any R-module of finite projective dimension, then M also has finite projective dimension when viewed as an $R[\Lambda]$-module.*

Proof. It suffices to assume, by induction on t, that $\Lambda = \langle x \rangle$ is infinite cyclic. Write $S = R[\Lambda]$ and consider the short exact sequence $0 \to (x-1)S \to S \to R \to 0$ determined by $x \mapsto 1$. Since both S and $(x-1)S \cong S$ are free right S-modules, we have $\mathrm{pd}_S R \le 1$ and indeed, by Lemma 33.9(ii), $\mathrm{pd}_S P \le 1$ for any projective R-module P. Finally let

$$0 \to P_n \to \cdots \to P_1 \to P_0 \to M \to 0$$

be a finite projective resolution for M as an R-module. Then, as S-modules, each P_i has finite projective dimension so it follows from Lemma 33.9(i) and induction that the same is true for M. An analogous proof works for left modules. ∎

We can now handle case (3).

Proposition 36.6. [142] *Let $R*\Gamma$ be a crossed product with R right Noetherian and Γ a finitely generated abelian-by-finite group. Then there exist finitely many finite subgroups G_i of Γ such that $\mathbf{G_0}(R*\Gamma)$ is generated by the images of the various $\mathbf{G_0}(R*G_i)$ under the induced module map.*

Proof. By assumption, Γ has a finitely generated normal abelian subgroup Λ of finite index. Moreover, if the torsion subgroup of Λ has order n, then $\Lambda^n = \{ x^n \mid x \in \Lambda \}$ is a characteristic torsion free subgroup of Λ of finite index. Thus we may replace Λ by Λ^n if necessary and assume that Λ is free abelian.

In this paragraph, we will temporarily view Λ additively. Then $\Lambda \otimes_Z Q$ is a finite dimensional module for the rational group ring $Q[\Gamma/\Lambda]$. Thus since the latter ring is semisimple Artinian, there exists a finite dimensional $Q[\Gamma/\Lambda]$-module V with $V \oplus (\Lambda \otimes_Z Q)$ free. Let $\mathcal{B}$ be a Q-basis for V and let Λ' be the Z-submodule of V

generated by the finite set $\mathcal{B} \cdot (\Gamma/\Lambda)$. It follows that Λ' is a finitely generated $Z[\Gamma/\Lambda]$-module with $\Lambda' \otimes_Z Q = V$ and thus $\Lambda' \oplus \Lambda$ is a rationally free $Z[\Gamma/\Lambda]$-module.

Now revert to multiplicative notation and observe that Γ acts on Λ' with Λ acting trivially. Thus if $\Gamma' = \Lambda' \rtimes \Gamma$, then Γ' has a normal abelian subgroup $\Omega' \cong \Lambda' \times \Lambda$ of finite index and Ω' is a rationally free $Z[\Gamma'/\Omega']$-module since $\Gamma'/\Omega' \cong \Gamma/\Lambda$. By Lemma 36.4, there exists a crossed product $S = R*\Gamma'$ such that the map $\varphi \colon \Gamma' \to \Gamma$ extends to a ring homomorphism $\varphi \colon R*\Gamma' \to R*\Gamma = S^\varphi$. Furthermore, $S = S_0 \oplus \mathrm{Ker}(\varphi)$ where $S_0 = R*\Gamma_0$. Since Λ' is free abelian and ${}_R R$ is free, it follows from Lemma 36.5 that R has finite projective dimension as a left $R[\Lambda']$-module. Thus, by Lemma 36.4(iii), S^φ has finite projective dimension as a left S-module.

The structure of Γ' and Proposition 35.8 imply that there exist finitely many finite subgroups G_i of Γ' such that $\mathbf{G_0}(R*\Gamma')$ is generated by the elements $[B_i \otimes_{R*G_i} S]$ for all i and $B_i \in \mathbf{mod}\, R*G_i$. In particular, setting $T_i = R*G_i \subseteq R*\Gamma'$, we see immediately that the hypotheses of Lemma 36.3 are all satisfied. Indeed S is right Noetherian, by Proposition 1.6, and certainly ${}_{S_0} S$ and ${}_{T_i} S$ are free and hence flat. Furthermore, since G_i is finite, we have $G_i \cap \Lambda' = \langle 1 \rangle$ so $\varphi \colon R*G_i \to (R*G_i)^\varphi = R*G_i^\varphi$ is an isomorphism. It follows that $S^\varphi = R*\Gamma$ is a free and hence flat left T_i-module.

Finally, we conclude from Lemma 36.3 that $\mathbf{G_0}(S^\varphi)$ is generated by the elements $[A_i \otimes_{T_i^\varphi} S^\varphi]$ for all i and $A_i \in \mathbf{mod}\, T_i^\varphi$. Thus, since $T_i^\varphi = R*G_i^\varphi$ and the map $\mathbf{G_0}(T_i^\varphi) \to \mathbf{G_0}(S^\varphi)$ is well defined, it follows that $\mathbf{G_0}(R*\Gamma)$ is generated by the images of the various $\mathbf{G_0}(R*G_i^\varphi)$. This completes the proof. ∎

At this point, it is a simple matter to prove the induction theorem. However, in order to give that result a more precise formulation, we require a few additional observations on group extensions. These will yield information on the conjugacy classes of finite subgroups of a polycyclic-by-finite group.

Let Γ be an arbitrary group which is the split extension $\Gamma = \Lambda G$ of a normal abelian subgroup Λ by some complement G. As usual, assume that we know the structure of Λ as a module over $Z[G] \cong Z[\Gamma/\Lambda]$. The question now is to determine all possible complements for Λ in Γ. Suppose that H is another such complement. Since each coset of Λ has a unique representative in H and in G, there exists a function $\alpha\colon G \to \Lambda$ with $H = \{\, x\alpha(x) \mid x \in G \,\}$. Furthermore, since H is a subgroup we have $x\alpha(x) \cdot y\alpha(y) = xy\alpha(xy)$ and hence

(1-cocycle) $$\alpha(x)^y \alpha(y) = \alpha(xy)$$

for all $x, y \in G$. Functions which satisfy this condition are called 1-*cocycles* and we denote the set of all such α by $C^1(G,\Lambda)$. Since Λ is abelian, $C^1(G,\Lambda)$ is a group under pointwise multiplication.

Now suppose that we replace H by a Γ-conjugate to obtain a new complement for Λ. Since $\Gamma = \Lambda H = H\Lambda$, this conjugation can be achieved by an element $c^{-1} \in \Lambda$ and the 1-cocycle associated with cHc^{-1} is given by $\beta(x) = \alpha(x)c^{-1}c^x$. In particular, 1-cocycles satisfying

(1-coboundary) $$\beta(x) = c^{-1}c^x$$

for some $c \in \Lambda$ are called 1-*coboundaries* and we let $B^1(G,\Lambda)$ denote the subgroup of $C^1(G,\Lambda)$ consisting of these functions. It follows from the above discussion, that the elements of the 1st-*cohomology group*

$$H^1(G,\Lambda) = C^1(G,\Lambda)/B^1(G,\Lambda)$$

are in one-to-one correspondence with the conjugacy classes of complements for Λ in Γ.

The next result is analogous to Lemma 35.2 with a similar proof which we just sketch.

Lemma 36.7. *Let Λ be a $Z[G]$-module with $|G| = n$ finite.*
  i. $H^1(G,\Lambda)$ *has exponent dividing* $|G|$.
  ii. *If Λ is finitely generated, then $H^1(G,\Lambda)$ is finite.*
  iii. *If Λ is a free $Z[G]$-module, then $H^1(G,\Lambda) = \langle 1 \rangle$.*
  iv. *If $G = \langle g \rangle$ is cyclic, then*

$$H^1(G,\Lambda) \cong \{\, a \in \Lambda \mid a^{1+g+\cdots+g^{n-1}} = 1 \,\}/\Lambda^{g-1}.$$

Proof. (i) Let $\alpha \in C^1(G,\Lambda)$ be given and multiply the 1-cocycle equation over all $x \in G$. Thus if $c = \prod_{x \in G}\alpha(x) \in \Lambda$, then $c^y\alpha(y)^n = c$ and α^n is a 1-coboundary. Part (ii) is clear.

(iii) As in Lemma 35.2, we replace y by y^{-1} and conjugate the 1-cocycle equation by y. Reading off the y-components then yields $\alpha(x)_y \cdot \alpha_1(y^{-1})^y = \alpha_1(xy^{-1})^y$ where, by definition, $\alpha_1(x) = \alpha(x)_1$. Now multiply over all $y \in G$ to obtain $\alpha(x)c = c^x$ where $c = \prod_y \alpha_1(y^{-1})^y \in \Lambda$.

(iv) Let $\alpha \in C^1(G, \Lambda)$ and write $\alpha(g) = a \in \Lambda$. Then it follows from the cocycle equation that, for all $i \geq 0$, $\alpha(g^i) = a^{1+x+\cdots+x^{i-1}}$. But $g^n = g^0 = 1$, so we must have $a^{1+x+\cdots+x^{n-1}} = 1$. Since α is a coboundary if and only if $a = c^{-1}c^g \in \Lambda^{g-1}$, the result follows. ∎

We remark that $H^1(G, \Lambda)$ and $H^2(G, \Lambda)$ are part of an infinite series of cohomology groups starting with $H^0(G, \Lambda)$. However, only these first three have true group theoretic interpretations. We also note that part (iii) above explains why we can choose any complement G for Λ in Proposition 35.3. Indeed, all such complements are conjugate in that case. As a consequence of part (ii) we have

Lemma 36.8. *If* Γ *is a polycyclic-by-finite group, then* Γ *has only finitely many conjugacy classes of finite subgroups.*

Proof. Let us first assume that Γ has a torsion free normal abelian subgroup Λ of finite index. If G is a finite subgroup of Γ, then $\Lambda G/\Lambda$ is a subgroup of the finite group Γ/Λ and hence there are only finitely many possibilities for the group ΛG. Because of this, it suffices to assume that $\Gamma = \Lambda G$ and to show that there are only finitely many conjugacy classes of finite subgroups H with $\Lambda H = \Gamma$. But observe that Λ is torsion free and H is finite, so $\Lambda \cap H = \langle 1 \rangle$ and H is a complement for Λ in Γ. Thus since $H^1(G, \Lambda)$ is finite, by Lemma 36.7(ii), the result follows in this case.

For the general polycyclic-by-finite group Γ, we proceed by induction on the Hirsch number and we may clearly assume that Γ is infinite. Then Γ has a torsion free normal abelian subgroup $\Lambda \neq \langle 1 \rangle$ and, by induction, Γ/Λ has only finitely many conjugacy classes of finite subgroups, say with representatives Γ_i/Λ for $i = 1, 2, \ldots, t$. If G is any finite subgroup of Γ, then $\Lambda G/\Lambda$ is conjugate to some Γ_i/Λ, so G is conjugate to a finite subgroup of Γ_i. But, by the above, Γ_i

has only finitely many conjugacy classes of finite subgroups and thus the result follows. ∎

It is time to prove the *induction theorem*. Notice, in the following, how the simple crossed product property $R*\Gamma = (R*N)*(\Gamma/N)$ with $N \triangleleft \Gamma$ allows for an almost immediate reduction to the abelian-by-finite case.

Theorem 36.9. [140] [142] *Let $R*\Gamma$ be a crossed product with R right Noetherian and Γ a polycyclic-by-finite group. Suppose $G_1, G_2, \ldots, G_t$ are representatives of the conjugacy classes of the maximal finite subgroups of Γ. Then $\mathbf{G_0}(R*\Gamma)$ is generated by the images of the various $\mathbf{G_0}(R*G_i)$ under the induced module map.*

Proof. We first show that $\mathbf{G_0}(R*\Gamma)$ is generated by the images of the Grothendieck groups $\mathbf{G_0}(R*G_i)$ for finitely many finite subgroups $G_i \subseteq \Gamma$. The proof of this proceeds by induction on the Hirsch number of Γ and of course we may assume that Γ is infinite. Then Γ has a normal torsion free abelian subgroup $\Lambda \neq \langle 1 \rangle$ and observe that

$$R*\Gamma = (R*\Lambda)*(\Gamma/\Lambda) = S*\tilde{\Gamma}$$

with $\tilde{\Gamma} = \Gamma/\Lambda$ and with $S = R*\Lambda$ right Noetherian by Proposition 1.6. Thus by induction, $\tilde{\Gamma}$ has finitely many finite subgroups $\tilde{\Gamma}_i$ such that $\mathbf{G_0}(S*\tilde{\Gamma})$ is generated by the images of the various $\mathbf{G_0}(S*\tilde{\Gamma}_i)$ under the induced module map.

Note that $\tilde{\Gamma}_i = \Gamma_i/\Lambda$ for some subgroup $\Gamma_i \subseteq \Gamma$ and that $S*\tilde{\Gamma}_i = R*\Gamma_i$. Thus, since $S*\tilde{\Gamma} = R*\Gamma$, the above says that $\mathbf{G_0}(R*\Gamma)$ is generated by the images of the various $\mathbf{G_0}(R*\Gamma_i)$ under the induced module map. But each Γ_i is abelian-by-finite, so Proposition 36.6 applies. We conclude that there exist finitely many finite subgroups $G_{i,j} \subseteq \Gamma_i$ such that the induced module map $\oplus \sum_j \mathbf{G_0}(R*G_{i,j}) \to \mathbf{G_0}(R*\Gamma_i)$ is surjective. Thus the composite map

$$\oplus \sum_{i,j} \mathbf{G_0}(R*G_{i,j}) \to \oplus \sum_i \mathbf{G_0}(R*\Gamma_i) \to \mathbf{G_0}(R*\Gamma)$$

is also surjective and, by transitivity of induction, this first key observation is proved.

We have therefore shown that there exist finitely many finite subgroups $G_i \subseteq \Gamma$ with the induced module map $\oplus \sum_i \mathbf{G_0}(R*G_i) \to \mathbf{G_0}(R*\Gamma)$ surjective. Now observe that if $G \subseteq H \subseteq \Gamma$ then, by transitivity of induction again, the image of $\mathbf{G_0}(R*G)$ is contained in the image of $\mathbf{G_0}(R*H)$. In other words, we can replace each G_i by some $H_i \supseteq G_i$ with H_i a maximal finite subgroup of Γ. Furthermore, by Lemma 35.4, conjugate subgroups give rise to the same image in $\mathbf{G_0}(R*\Gamma)$. Thus we require only one representative from each conjugacy class of maximal finite subgroups and, with this, the result follows. ∎

It is natural to ask if an analogous result holds for $\mathbf{K_0}(R*\Gamma)$. The answer is "no". A close look at the proof shows that all reductions from case (1) to (2) to (3) to Theorem 36.9 do in fact hold for $\mathbf{K_0}$. The problem is that case (1) and Corollary 34.8 fail in this context. We close by constructing an appropriate commutative counterexample. To start with, we have

Lemma 36.10. *Let R be a commutative domain with quotient field K and let P and Q be R-submodules of K.*

i. If $PQ = R$, then P and Q are both finitely generated projective R-modules.

ii. If P is stably free, then $P = cR$ for some $c \in P$.

Proof. (i) We use the fact that $PQ \subseteq R$ and that $1 = \sum_1^n p_i q_i$ with $p_i \in P$ and $q_i \in Q$. Define the R-module maps $\sigma \colon R^n \to P$ and $\tau \colon P \to R^n$ by

$$\sigma \colon (r_1, r_2, \ldots, r_n) \mapsto \sum_{i=1}^n p_i r_i$$

$$\tau \colon p \mapsto (q_1 p, q_2 p, \ldots, q_n p).$$

Then $\sigma\tau(p) = p$ for all $p \in P$, so P is finitely generated and projective.

(ii) Suppose $P \neq 0$ is stably free and that $P \oplus R^n = V = R^m$. Since $PK = RK = K$, it follows, by computing the K-dimension of VK, that $m = n + 1$. Now let $\{ v_0, v_1, \ldots, v_n \}$ be an R-basis for $V = R^{n+1}$ and let $\{ w_1, \ldots, w_n \}$ be a basis for $W = R^n \subseteq V$. Then, for each $1 \leq j \leq n$, we have $w_j = \sum_{i=0}^{n} v_i r_{i,j}$ with $r_{i,j} \in R$ and we study the $(n+1) \times n$ matrix $M = [r_{i,j}]$. Notice that if I is any maximal ideal of R, then

$$V/VI = (R/I)^n \oplus (P/PI)$$

so the elements $w_1, \ldots, w_n$ are (R/I)-linearly independent modulo VI. It follows that if $\bar{M} = [\bar{r}_{i,j}]$, where $\bar{r}_{i,j} = r_{i,j} + I \in R/I$, then $\bar{M}$ must have rank n as a matrix over the field R/I. Hence at least one of the maximal minors of $\bar{M}$ is not zero. Since I is an arbitrary maximal ideal, this implies that the set $\{ m_0, m_1, \ldots, m_n \}$ of maximal minors of M is *unimodular*, that is $\sum_{i=0}^{n} m_i R = R$. In other words, we can find elements $r_{i,0} \in R$ so that the augmented $(n+1) \times (n+1)$ matrix $M^* = [r_{i,j}]$ has determinant 1 and hence is invertible. Finally, if we define $w_0 = \sum_{i=0}^{n} v_i r_{i,0}$, then $\{ w_0, w_1, \ldots, w_n \}$ must also be an R-basis for V. Thus $V = w_0 R \oplus W$ and $P \cong V/W \cong w_0 R$ is free of rank 1. ∎

With this we can prove

Proposition 36.11. [192] *Let R be a commutative domain with quotient field K and suppose there exists $a \in K \setminus R$ with $a^2, a^3 \in R$. If S is the polynomial ring $S = R[t]$ or the group ring $S = R[T]$ with $T = \langle t \rangle$ infinite cyclic, then S has a finitely generated projective module which is not stably free.*

Proof. We will just consider the group ring $R[T]$. The polynomial ring proof is identical except at one point where it is in fact slightly simpler.

Let $f = at \in K[T]$ and let $P = (1 + f, 1 + f + f^2)$ be the $R[T]$-submodule of $K[T]$ generated by $1 + f$ and $1 + f + f^2$. Similarly let $Q = (1 - f, 1 - f + f^2)$ and observe that

$$PQ = (1 - f^2, 1 + f^3, 1 - f^3, 1 + f^2 + f^4) \subseteq R[T]$$

since $a^2, a^3, a^4 \in R$. Furthermore, we have $2 = (1 + f^3) + (1 - f^3) \in PQ$ and $3 = (1 - f^2)(2 + f^2) + (1 + f^2 + f^4) \in PQ$. Thus $1 \in PQ$ and P is a finitely generated projective $R[T]$-module by Lemma 36.10(i).

Now suppose, by way of contradiction, that P is stably free. Then Lemma 36.10(ii) implies that $P = cR[T]$ for some $c \in K[T]$. Moreover, we have $P \cdot K[T] = K[T]$, since $(1 + f + f^2) - (1 + f)f = 1$, and thus $P = cR[T]$ implies that $K[T] = cK[T]$. In other words, c is a unit of $K[T]$ so, since K is a field and $T = \langle t \rangle$ is infinite cyclic, $c = kt^n$ for some $k \in K \setminus 0$. But t^n is a unit of $R[T]$ so we have $P = kR[T]$ and furthermore $Q = k^{-1}R[T]$ since $PQ = R[T]$. Finally $k(1 - at) \in kQ = R[T]$ so $k \in R$ and then $1 + at \in P = kR[T] \subseteq R[T]$ yields $a \in R$, a contradiction. We conclude therefore that P is not stably free. ∎

Now for the example. Let F be a field and let R be the sub-ring of the power series ring $F[[x]]$ consisting of all elements with x-coefficient 0. Then R is Noetherian since $R = 1 \cdot F[[x^2]] + x^3 \cdot F[[x^2]]$ and R is certainly local. Thus all finitely generated projective R-modules are free and $\mathbf{K_0}(R)$ is the cyclic group generated by $[R]$. On the other hand, if $S = R[t]$ or $R[T]$ as above, then S has a finitely generated projective module which is not stably free. This follows from Proposition 36.11 since $x^2, x^3 \in R$ but $x \notin R$. Thus, by Lemma 34.10(ii), $\mathbf{K_0}(S)$ is properly larger than its cyclic subgroup generated by $[S]$ and hence the induced module map $\mathbf{K_0}(R) \to \mathbf{K_0}(S)$ is not surjective. Finally notice that if $S = R[t]$, then S is Z^+-graded with $S_0 = R$. On the other hand, if $S = R[T]$ then, since T is torsion free, the only finite subgroup of T is $G = \langle 1 \rangle$ and $R[G] = R$. Thus we see that neither Corollary 34.8 nor Theorem 36.9 has a $\mathbf{K_0}$ analog without additional assumptions on the ring.

EXERCISES

1. Show that the reductions from case (2) to case (3) and from case (3) to Theorem 36.9 work equally well for $\mathbf{K_0}$.

2. Let $R * G$ be a crossed product with the property that every finitely generated right module has finite projective dimension. Show that the same is true for $R * H$ with H any subgroup of G. To this end, use Lemma 3.10 to show that any $R * H$-module is a direct summand of the restriction of an $R * G$-module.

3. Let $R*\Gamma$ be a crossed product with R right Noetherian and Γ polycyclic-by-finite. If all finitely generated $R*\Gamma$-modules have finite projective dimension, prove that a precise $\mathbf{K_0}$ analog of Theorem 36.9 holds. For this, use the previous exercise along with Proposition 33.8.

4. Let Λ be a $Z[G]$-module with $G = \langle g \rangle$ cyclic. If $|G| = n$, prove that $H^2(G, \Lambda) = \Lambda^G / \Lambda^{1+g+\cdots+g^{n-1}}$ where $\Lambda^G = \mathbf{C}_\Lambda(G)$. If G is infinite, show that $H^1(G, \Lambda) = \Lambda / \Lambda^{g-1}$.

5. Let Λ be a $Z[G]$-module and assume that G is a free group. Show that $H^2(G, \Lambda) = \langle 1 \rangle$.

6. Let R be a commutative ring and let P and Q be ideals of R with $P + Q = R$. Prove that $P \oplus Q \cong R \oplus (PQ)$ as R-modules. In particular, if R is a domain and PQ is principal, conclude that both P and Q are projective.

9 Zero Divisors and Idempotents

37. Zero Divisors and Goldie Rank

We start this chapter by discussing the famous *zero divisor problem*. Let $K[G]$ be a group algebra and suppose G has a nonidentity element x of finite order n. Then the equation

$$(1 - x)(1 + x + \cdots + x^{n-1}) = 1 - x^n = 0$$

shows that $K[G]$ has nontrivial zero divisors. On the other hand, if G has no such element x, that is if G is torsion free, then $K[G]$ has at least no obvious divisors of zero. Because of this, and with frankly very little supporting evidence, it was conjectured that if G is torsion free then $K[G]$ must be a domain. Amazingly, the conjecture has held up for over forty years.

Progress on this problem divides naturally and historically into four stages. The first is the purely group theoretic approach concerned with ordered groups, unique product groups and their analogs. Specifically, one asks if the torsion free assumption on G implies other properties of G which are more relevant to the zero divisor question.

The answer here has been a hodge-podge of definitions and unsatis-factory results. In some sense, this era was finally brought to an end by the recent paper [183] where small cancellation theory was used to construct a torsion free group which does not have the unique product property. Indeed even polycyclic examples are now known to exist. The structure of the following group is well known. What is new here is the fact that the unique product property fails.

Proposition 37.1. [177] *Let G be the group*

$$G = \langle x, y \mid x^{-1}y^2x = y^{-2}, \ y^{-1}x^2y = x^{-2} \rangle.$$

Then G is a torsion free abelian-by-finite group which does not have the unique product property.

Proof. Set $z = xy$, $a = x^2$, $b = y^2$ and $c = z^2$. Then clearly $a^z = a^{-1}$ and $b^z = b^{-1}$. Moreover, we have $(xy)^{-1} = y^{-1}x^{-1} = b^{-1}ayx$ so $c^{-1} = (xy)^{-2} = yxyx$ and hence $c^x = c^{y^{-1}} = c^{-1}$. It follows that $D = \langle a, b, c \rangle$ is a normal abelian subgroup of G. Furthermore, G/D is generated by $\bar{x}$ and $\bar{y}$ with $\bar{x}^2 = \bar{y}^2 = (\bar{x}\bar{y})^2 = 1$ and hence G/D is at most a fours group.

We next observe that D is free abelian of rank 3 and that $|G/D| = 4$. To this end, consider the two rational matrices $X, Y \in$ GL$_4(Q)$ given by

$$X = \begin{pmatrix} 0 & 1 & 0 & 0 \\ 2 & 0 & 0 & 0 \\ 0 & 0 & 0 & 1/2 \\ 0 & 0 & 1 & 0 \end{pmatrix} \qquad Y = \begin{pmatrix} 0 & 0 & 1 & 0 \\ 0 & 0 & 0 & 1 \\ 2 & 0 & 0 & 0 \\ 0 & 1/2 & 0 & 0 \end{pmatrix}.$$

Then it is easy to see that $X^{-1}Y^2X = Y^{-2}$ and $Y^{-1}X^2Y = X^{-2}$ so there is a well defined epimorphism $\sigma \colon G \to \langle X, Y \rangle$ given by $\sigma(x) = X$ and $\sigma(y) = Y$. If $Z = XY$, we have

$$\sigma(a) = X^2 = \text{diag}(2, 2, 1/2, 1/2)$$
$$\sigma(b) = Y^2 = \text{diag}(2, 1/2, 2, 1/2)$$
$$\sigma(c) = Z^2 = \text{diag}(2, 1/2, 1/2, 2)$$

and these matrices generate a free abelian group of rank 3. Furthermore, since $\sigma(D)$ is diagonal and $1, X, Y, Z$ are not diagonal multiples of each other, we have $|G/D| \geq 4$ and hence G/D is a fours group.

Now we note that G is torsion free. To start with, if $g \in G$ has finite order then, since $g^2 \in D$ and D is free abelian, we must have $g^2 = 1$. Next, we consider the D coset containing g. Suppose for example that $g \in Dx$ and write $g = dx$ with $d \in D$. Then we have $1 = g^2 = dxdx = dd^x a$ and this is a contradiction since a occurs with even exponent in dd^x. Similarly $g \notin Dy$ or Dz. Thus $g \in D$ and therefore $g = 1$.

Finally for the unique product property, let S be the subset of G given by $S = Ax \cup By \cup C$ with

$$A = \left\{ 1, a^{-1}, a^{-1}b, b, a^{-1}c^{-1}, c \right\} \qquad C = \left\{ c, c^{-1} \right\}$$

$$B = \left\{ 1, a, b^{-1}, b^{-1}c, c, ab^{-1}c \right\}.$$

Then $|S| = 14$ and $S^2 = S \cdot S$ has no unique product. For the latter, observe that $A, B, C \subseteq D$ and that D is commutative. Thus the intersections of S^2 with the four cosets of D are given by

$$S^2 \cap Dx = AxC \cup CAx = \left(AC^x \cup AC \right)x$$

$$S^2 \cap Dy = ByC \cup CBy = \left(BC^y \cup BC \right)y$$

$$S^2 \cap Dz = AxBy \cup ByAx = \left(AB^x \cup A^y Ba^{-1}bc^{-1} \right)z$$

$$S^2 \cap D = AxAx \cup ByBy \cup C^2 = AA^x a \cup BB^y b \cup C^2.$$

In particular, since $C^x = C^y = C$, the elements in $S^2 \cap Dx$ and $S^2 \cap Dy$ all occur with even multiplicities. Unfortunately, there is no such simple explanation for the other two cosets. The 72 and 76 products, respectively, must be checked by hand or computer to prove that all multiplicities are larger than 1. Since these computations all occur in the free abelian group D, there are no technical difficulties involved. ∎

In view of this result, the zero divisor problem for the group G above cannot be settled by group theoretic means. It was, however, dealt with during the second stage of progress by the introduction of certain ring theoretic machinery. Specifically, this stage is based on the theorem of [42] which asserts that if R_1 and R_2 are both n-firs containing a common division ring D, then the coproduct $R_1 \coprod_D R_2$ over D is also an n-fir. This turns out to be relevant since a ring is a 1-fir if and only if it is a domain (see Exercise 8 of Section 13) and indeed we have

Theorem 37.2. [42] [99] *Let K be a field, G a group and suppose that G has subgroups A, B and $N \triangleleft G$ with $G/N = (A/N) * (B/N)$. If $K[A]$ and $K[B]$ are domains and if $K[N]$ is an Ore domain, then $K[G]$ has no zero divisors.*

Here of course $(A/N) * (B/N)$ denotes the free product of the two groups. Now, in general, free products are quite complicated and are certainly not Noetherian. However, it is easy to see that $C_2 * C_2$, the free product of two groups of order 2, is isomorphic to the infinite dihedral group and it turns out that this is a basic building block of torsion free supersolvable groups. Recall that a group G is said to be *supersolvable* if it has a finite normal series

$$\langle 1 \rangle = G_0 \subseteq G_1 \subseteq \cdots \subseteq G_n = G$$

with each quotient G_{i+1}/G_i cyclic. Thus any such group is necessarily polycyclic. As a consequence we have

Corollary 37.3. [61] *If G is a torsion free supersolvable group, then the group algebra $K[G]$ is a domain.*

In particular, this applies to the example of Proposition 37.1. The third stage of progress handled all polycyclic-by-finite groups. It began with the following general observation.

Theorem 37.4. [203] *Let R be a ring satisfying*
    i. R is right Noetherian and semiprime,
    ii. all finitely generated R-modules have finite projective dimension,

iii. all finitely generated projective R-modules are stably free.
Then R is a domain.

In particular, if $R = K[G]$ with G a torsion free polycyclic-by-finite group, then R is certainly right Noetherian and it is also semiprime, by Theorems 5.3 and 5.4. Furthermore, a key result of [193] (see [161, Theorem 10.3.12]) asserts that $K[G]$ has *finite global dimension*, that is all R-modules have finite projective dimension. Thus (i) and (ii) above are satisfied. While we now know that (iii) also holds (see Exercise 4), this fact was not available when the zero divisor applications were originally obtained. Indeed, (iii) was circumvented by either a suitable localization or by somehow dropping down to subgroup of finite index. The work began with [26] where these ideas were applied, using localization, to handle all torsion free abelian-by-finite groups in characteristic 0 and abelian-by-(finite p) groups in characteristic p. Then [55] used the fact that (iii) held for poly-Z groups to handle all torsion free polycyclic-by-finite groups in characteristic 0. That paper also considered some groups in characteristic $p > 0$, but it remained for [32] to complete the result. Combining all this, we have

Theorem 37.5. [26] [55] [32] *Let K be a field and let G be a torsion free polycyclic-by-finite group. Then the group algebra $K[G]$ is a domain.*

Some additional results on more general solvable groups were obtained by [195], but the above theorem essentially ended the third stage. The techniques used in its proof were powerful, but not powerful enough to handle Noetherian group rings $D[G]$ with D a division ring. We are now in the fourth stage of progress and this is all based on the upper bound for $\mathbf{G_0}(R*G)$ given in Theorem 36.9. Indeed, the applications can be obtained quite quickly and elegantly by techniques which are not really new. What is new here and crucial is Theorem 36.9 itself. We begin with the necessary variant of Theorem 37.4.

Theorem 37.6. [203] *Let R be a semiprime right Noetherian ring. If $\mathbf{G_0}(R)$ is the cyclic group generated by $[R]$, then R is a domain.*

Proof. By Goldie's theorem, R has a classical right ring of quotients $Q = Q(R)$ which is semisimple Artinian and, by Lemma 33.3(iii), the induced module map $A \mapsto A \otimes_R Q$ gives rise to an epimorphism $\theta \colon \mathbf{G_0}(R) \to \mathbf{G_0}(Q)$. In particular, the assumption on $\mathbf{G_0}(R)$ translates to the assertion that $\mathbf{G_0}(Q)$ is generated by $[Q]$. Lemma 33.1(ii) now implies first that Q has only one irreducible module and then that Q_Q is itself irreducible. Thus Q is a divison ring and R is a domain. ∎

We remark that it suffices to assume in the above that $\mathbf{G_0}(R)$ is generated by $[R]$ modulo its torsion subgroup. This follows since the torsion disappears in $\mathbf{G_0}(Q(R))$. We will need the following result on Ore localization. It actually follows as in Proposition 12.4(i), but we offer an alternate approach here.

Lemma 37.7. *Let $R*G$ be a crossed product and let T be a right divisor set of regular elements of R. If T is G-stable, then T is a right divisor set of regular elements of $R*G$ and $(R*G)T^{-1} = (RT^{-1})*G$.*

Proof. It suffices to show that T is a right divisor set in $R*G$; the remainder of the argument is then trivial. Thus suppose $t \in T$ and $\alpha = \sum \bar{x} a_x \in R*G$. Since T is G-stable, we have $t^{\bar{x}} \in T$ for all $x \in \mathrm{Supp}\,\alpha$ and we can write the finitely many elements $(t^{\bar{x}})^{-1} a_x \in RT^{-1}$ all with a common right denominator $s \in T$. In particular, if $(t^{\bar{x}})^{-1} a_x = b_x s^{-1}$, then $t^{\bar{x}} b_x = a_x s$ and hence $t\left(\sum \bar{x} b_x\right) = \alpha s$ as required. ∎

It is now a simple matter to prove

Lemma 37.8. *Let $R*G$ be a crossed product with R an Ore domain and with G a locally polycyclic-by-finite group. Suppose that $R*H$ is a domain for all finite subgroups $H \subseteq G$. Then $R*G$ is an Ore domain.*

Proof. Since the property of being an Ore domain can be checked two elements at a time, it suffices to prove this result for the finitely generated subgroups of G. In other words, we may suppose that G

itself is polycyclic-by-finite. Furthermore, in view of the previous lemma, we can localize and assume that R is a divison ring. Thus $R*G$ is right Noetherian and it is semiprime by Proposition 8.3(ii).

Now let H be any finite subgroup of G. Then $R*H$ is an Artinian domain and hence a divison ring, so Lemma 33.1(ii) implies that $\mathbf{G_0}(R*H)$ is the cyclic group generated by $[R*H]$. Hence since $R*H \otimes_{R*H} R*G = R*G$, the induced module maps sends $\mathbf{G_0}(R*H)$ to the cyclic subgroup of $\mathbf{G_0}(R*G)$ generated by $[R*G]$.

Finally, by Theorem 36.9, $\mathbf{G_0}(R*G)$ is generated by the images of finitely many such $\mathbf{G_0}(R*H)$ and thus $\mathbf{G_0}(R*G)$ is the cyclic group generated by $[R*G]$. Theorem 37.6 now implies that $R*G$ is a domain and then it is an Ore domain by Goldie's theorem. ∎

If G is torsion free, then the only possibility for H above is the identity subgroup. Thus $R*H = R$ is a domain, and Lemma 37.7 already extends Theorem 37.5. But more can be done; we can consider arbitrary solvable groups. Indeed we study groups G having a finite subnormal series

$$\langle 1 \rangle = G_0 \vartriangleleft G_1 \vartriangleleft \cdots \vartriangleleft G_n = G$$

with each quotient G_{i+1}/G_i locally polycyclic-by-finite. For convenience, we let $\mu(G)$ denote the minimal such n for the group G. Thus $\mu(G) = 0$ if and only if $G = \langle 1 \rangle$ and $\mu(G) = 1$ if and only if G is a nonidentity locally polycyclic-by-finite group. We list a few basic properties of μ.

Lemma 37.9. *Let G be as above and let H be a subgroup of G. Then $\mu(H) \le \mu(G)$ and $\mu(G/H) \le \mu(G)$ if $H \vartriangleleft G$. Moreover if $\langle 1 \rangle \ne H \vartriangleleft G$ and $|G/H| < \infty$, then $\mu(G) = \mu(H)$.*

Proof. The first observations are clear. Now suppose $\langle 1 \rangle \ne H \vartriangleleft G$ with $|G/H| < \infty$. Since $n = \mu(H) \le \mu(G)$, we need only obtain the reverse inequality. By assumption, $n \ge 1$ so there exists $N \vartriangleleft H$ with H/N locally polycyclic-by-finite and with $\mu(N) \le n - 1$. If T is a finite transversal for H in G, then $M = \bigcap_{t \in T} N^t \vartriangleleft G$ and H/M is locally polycyclic-by-finite since $H/M \hookrightarrow \prod_{t \in T} H/N^t$. Moreover,

$\mu(M) \leq \mu(N) \leq n-1$, so it clearly suffices to show that $\bar{G} = G/M$ is locally polycyclic-by-finite. To this end, let $\bar{X}$ be a finitely generated subgroup of $\bar{G}$. Then $|\bar{X}/(\bar{X} \cap \bar{H})| < \infty$, so $\bar{X} \cap \bar{H}$ is also finitely generated (see Exercise 5) and hence $\bar{X} \cap \bar{H}$ is polycyclic-by-finite. Thus $\bar{X}$ is also polycyclic-by-finite and the lemma is proved. $\blacksquare$

We now come to the main results of this section. Notice how the simple crossed product property $R*G = (R*N)*(G/N)$ with $N \triangleleft G$ allows for a fairly easy inductive argument.

Theorem 37.10. [94] [141] *Let $R*G$ be a crossed product with R an Ore domain and assume that G has a finite subnormal series*

$$\langle 1 \rangle = G_0 \triangleleft G_1 \triangleleft \cdots \triangleleft G_n = G$$

with each quotient G_{i+1}/G_i locally polycyclic-by-finite. If $R*H$ is a domain for every finite subgroup $H \subseteq G$, then $R*G$ is an Ore domain.

Corollary 37.11. [94] [141] *Let $R*G$ be a crossed product with R an Ore domain and assume that G has a finite subnormal series*

$$\langle 1 \rangle = G_0 \triangleleft G_1 \triangleleft \cdots \triangleleft G_n = G$$

with each quotient G_{i+1}/G_i locally polycyclic-by-finite. If G is torsion free, then $R*G$ is an Ore domain.

Proof. We prove the theorem by induction on $\mu(G)$, starting with the trivial case $\mu(G) = 0$. Now if $\mu(G) = n \geq 1$, then there exists $N \triangleleft G$ with $\mu(N) \leq n - 1$ and G/N locally polycyclic-by-finite. Thus, by Lemma 1.3, we have

$$R*G = (R*N)*(G/N) = S*(G/N)$$

where, by induction, $S = R*N$ is an Ore domain. Furthermore, if H/N is a finite subgroup of G/N, then $S*(H/N) = R*H$ and there are two cases to consider. If $N \neq \langle 1 \rangle$ then, by the previous lemma, $\mu(H) = \mu(N) \leq n-1$ so $S*(H/N) = R*H$ is a domain by induction.

On the other hand, if $N = \langle 1 \rangle$ then H is a finite subgroup of G so $S*(H/N) = R*H$ is a domain by assumption. We can now apply Lemma 37.8 to conclude that $S*(G/N) = R*G$ is an Ore domain and the theorem is proved. The corollary is of course an immediate consequence. ∎

Paper [94] obtains a transfinite version of the above by allowing for more general types of subnormal series (see Exercises 7 and 8 for details). Now we move on to consider the *Goldie rank problem*. Specifically, let $R*G$ be given with R right Noetherian and G polycyclic-by-finite and let us assume that $R*G$ is prime. By Goldie's theorem, the classical right ring of quotients $Q(R*G)$ is isomorphic to $M_n(D)$, a full matrix ring over a division ring D. The integer n here is called the *Goldie rank* of $R*G$ and the problem is to determine this rank. In the case of ordinary group algebras $K[G]$, it was conjectured in [52] and [188] that n is the least common multiple of the orders of the finite subgroups of G. Special cases of this were verified in [103, 170, 189] before the conjecture was finally settled in the affirmative in [140] using, as usual, Theorem 36.9. More generally, the Goldie rank of $R*G$ is the least common multiple of certain parameters determined by the finite subgroups of G and, for this result, we follow [105].

Let R be a semiprime Noetherian ring so that $Q(R)$ exists and is semisimple Artinian. If A is a finitely generated R-module, then $A \otimes_R Q(R)$ is a finitely generated $Q(R)$-module and hence has finite composition length. The *reduced rank* of A is then given by

$$\rho_R(A) = \text{composition length of } A \otimes_R Q(R).$$

We remark that the reduced rank can be defined for all Noetherian rings, but we only require the semiprime case here. Since $_RQ(R)$ is flat, by Lemma 33.3(iii), it follows that ρ_R respects short exact sequences and hence gives rise to a group homomorphism $\rho_R: \mathbf{G_0}(R) \to Z$. More generally, suppose S is an overring of R with S_R finitely generated. Then any finitely generated S-module restricts to a finitely generated R-module and again this yields a group homomorphism $\rho_R: \mathbf{G_0}(S) \to Z$. Finally the *normalized reduced rank* χ_R is defined by

$$\chi_R(A) = \rho_R(A)/\rho_R(R)$$

so that $\chi_R(A)$ is a rational number and $\chi_R(R) = 1$. Again χ_R gives rise to a group homomorphism $\chi_R\colon \mathbf{G_0}(S) \to Q$.

Now suppose $R*G$ is a prime crossed product with R right Noetherian and G polycyclic-by-finite. If $N \lhd G$, then we have $R*G = (R*N)*(G/N)$ and certainly $R*N$ is also right Noetherian. Thus Lemmas 14.1(i) and 14.2(i) imply that $R*N$ is G-prime and hence semiprime. In particular, it makes sense to consider χ_{R*N}. Note further that if $|G/N| < \infty$, then $R*G$ is a finitely generated right $R*N$-module. In the following lemma, we determine the normalized reduced rank of certain induced $R*G$-modules.

Lemma 37.12. *Let $R*G$ be a prime crossed product with R right Noetherian and G polycyclic-by-finite. Suppose N is a normal subgroup of finite index in G and that $H \subseteq G$ is finite.*

    i. *If $A \in \mathbf{mod}\, R*G$, then $|G : N|\, \chi_{R*G}(A) = \chi_{R*N}(A)$.*

    ii. *If $B \in \mathbf{mod}\, R$, then $\chi_{R*N}(B \otimes_R R*N) = \chi_R(B)$.*

    iii. *If $C \in \mathbf{mod}\, R*H$, then $\chi_{R*G}(C \otimes_{R*H} R*G) = |H|^{-1}\chi_R(C)$.*

Proof. (i) Since $R*G = (R*N)*(G/N)$, it clearly suffices, in this part only, to assume that $N = \langle 1 \rangle$ and that G is finite. Moreover, since $\chi_{R*G}(A)$ and $\chi_R(A)$ are obtained by localizing the rings involved, there is no harm in first doing a partial localization by a regular Ore set T provided $T \subseteq R$. In other words, by Lemma 37.7, we may now assume that $R = Q(R)$ is Artinian. But then, since G is finite, $R*G$ is also Artinian; thus $R*G = Q(R*G)$ and no further localization is required.

Now $R*G$ is prime, so it is a simple Artinian ring with a unique irreducible module M. Indeed, if $A \in \mathbf{mod}\, R*G$, then $A \cong n \cdot M$, the direct sum of $n = \rho_{R*G}(A)$ copies of M. Furthermore, $A_{|R} \cong n \cdot M_{|R}$ so

$$\rho_R(A) = n \cdot \rho_R(M) = \rho_{R*G}(A) \cdot \rho_R(M).$$

In particular, this holds for $A = R*G$ which, as an R-module, is free on $|G|$ generators. Thus we have

$$|G|\, \rho_R(R) = \rho_R(R*G) = \rho_{R*G}(R*G) \cdot \rho_R(M)$$

and dividing the first equation by the second yields

$$|G|^{-1}\chi_R(A) = \chi_{R*G}(A),$$

as required.

   (ii) Again, there is no harm in localizing R since the equation

$$\left(B \otimes_R \mathrm{Q}(R)\right) \otimes_{\mathrm{Q}(R)} \left(\mathrm{Q}(R)*G\right) \cong \left(B \otimes_R R*G\right) \otimes_{R*G} \left(\mathrm{Q}(R)*G\right)$$

shows that this localization commutes with the induced module map. Thus we can assume that R is semisimple Artinian. Moreover, we note that the group $\mathcal{G}$ of trivial units of $R*G$ acts as automorphisms on R, $R*N$ and $\mathrm{Q}(R*N)$ and hence it permutes the modules of each ring. Indeed, if $B \in \mathbf{mod}\, R$ and $g \in \mathcal{G}$, then we have

$$B^g \otimes_R \mathrm{Q}(R*N) \cong \left(B \otimes_R \mathrm{Q}(R*N)\right)^g$$

via the map $b^g \otimes s^g \mapsto (b \otimes s)^g$ for all $b \in B$ and $s \in \mathrm{Q}(R*N)$. Since conjugate modules have the same composition length, we conclude that

$$\rho_{R*N}(B^g \otimes_R R*N) = \rho_{R*N}(B \otimes_R R*N).$$

   Now $R*G$ is prime by assumption, so R is G-prime and hence G-simple. This implies that all irreducible R-modules are $\mathcal{G}$-conjugate and hence, when induced to $R*N$, yield the same reduced rank. Suppose M is a fixed irreducible R-module. If $B \in \mathbf{mod}\, R$, then B is a direct sum of $\rho_R(B)$ irreducible R-modules and hence, by the above observation,

$$\rho_{R*N}(B \otimes_R R*N) = \rho_R(B) \cdot \rho_{R*N}(M \otimes_R R*N).$$

Furthermore, if $B = R$, then $B \otimes_R R*N = R*N$ so

$$\rho_{R*N}(R*N) = \rho_R(R) \cdot \rho_{R*N}(M \otimes_R R*N).$$

Finally, dividing the first of these equations by the second yields

$$\chi_{R*N}(B \otimes_R R*N) = \chi_R(B)$$

and part (ii) is proved.

(iii) Here we use the induced and restricted module notation of Section 3. Let N be a poly-Z normal subgroup of G of finite index and let $C \in \mathbf{mod}\, R*H$. Then by (i) above

$$\chi_{R*G}(C \otimes_{R*H} R*G) = |G : N|^{-1}\chi_{R*N}((C^{|R*G})_{|R*N})$$

and we apply the Mackey decomposition to the latter module. Notice that N is torsion free so $N \cap H = \langle 1 \rangle$ and, since $N \lhd G$, the (H, N)-double cosets in G are merely the right cosets of the subgroup HN. In particular, if $\mathcal{D}$ is a right transversal for HN in G, then Lemma 3.10 yields

$$(C^{|R*G})_{|R*N} \cong \oplus \sum_{d \in \mathcal{D}} \left[(C \otimes \bar{d})_{|R*(H^d \cap N)}\right]^{|R*N}$$

$$\cong \oplus \sum_{d \in \mathcal{D}} (C^{\bar{d}}_{|R})^{|R*N}$$

since $N \cap H^d = \langle 1 \rangle$ and $C \otimes \bar{d} \cong C^{\bar{d}}$. Thus since all $C^{\bar{d}}_{|R}$ have the same reduced rank, we conclude from (ii) above that

$$\chi_{R*N}((C^{|R*G})_{|R*N}) = |\mathcal{D}|\,\chi_{R*N}((C_{|R})^{|R*N})$$

$$= |\mathcal{D}|\,\chi_R(C).$$

But $|\mathcal{D}| = |G : HN| = |G : N|/|H|$, so the result follows. ∎

We need one last bit of notation. Let $R*H$ be a crossed product with R semiprime Noetherian and H finite. Since $\rho_R \colon \mathbf{G_0}(R*H) \to Z$ is a group homomorphism, we have $\mathrm{Im}(\rho_R) = mZ$ for some positive integer $m = m(R*H)$. Thus, m divides $\rho_R(R*H) = \rho_R(R) \cdot |H|$ and we define the *index* of $R*H$ to be the integer

$$\mathrm{ind}(R*H) = \rho_R(R*H)/m(R*H).$$

Lemma 37.13. *Let $R*H$ be as above.*

i. $\mathrm{ind}(R*H) = \mathrm{ind}(Q(R)*H)$.

ii. *If R is Artinian, then $m(R*H)$ is the greatest common divisor of the ranks $\rho_R(A)$ of all irreducible $R*H$-modules A.*

iii. $R*H$ is a domain if and only if $\mathrm{ind}(R*H) = 1$.

iv. If $R*H = R[H]$, then $\mathrm{ind}(R*H) = \rho_R(R) \cdot |H|$.

Proof. Part (i) follows since ρ_R factors through the epimorphism $\mathbf{G_0}(R*H) \to \mathbf{G_0}(Q(R)*H)$. In (ii), since R and $R*H$ are both Artinian, $\mathbf{G_0}(R*H)$ is generated by the elements $[A]$ for all irreducible $R*H$-modules A. Thus

$$mZ = \mathrm{Im}(\rho_R) = \sum_A \rho_R(A)\,Z$$

and it follows that m is the greatest common divisor of the various $\rho_R(A)$. Part (iii) is now immediate from the above and (iv) follows since $R[H]$ has a module A with $\rho_R(A) = 1$. ∎

We remark that there are many opportunities for $\mathrm{ind}(R*H)$ to equal $\rho_R(R) \cdot |H|$. For example, by Lemma 26.2(i), we can take $R*H$ to be a skew group ring with R a domain. Alternately, $R*H$ can be a twisted group algebra $K^t[H]$ with the degrees of its irreducible modules relatively prime. We now prove

Theorem 37.14. [105] *Let $R*G$ be a prime crossed product with R right Noetherian and G polycyclic-by-finite. If $G_1, G_2, \ldots, G_k$ are representatives of the conjugacy classes of the maximal finite subgroups of G, then the Goldie rank of $R*G$ is the least common multiple of the indices $\mathrm{ind}(R*G_i)$.*

Proof. We compute the image of $\mathbf{G_0}(R*G)$ under the map χ_{R*G}. To start with, if $R*G$ has Goldie rank n, then $Q(R*G) = \mathrm{M}_n(D)$ and it follows easily that $\mathrm{Im}(\chi_{R*G}) = n^{-1}Z$. On the other hand, by Theorem 36.9, $\mathbf{G_0}(R*G) = \sum_i W_i$ where W_i is the image of $\mathbf{G_0}(R*G_i)$ under the induced module map. In particular, if $\mathrm{ind}(R*G_i) = w_i$, then it follows from Lemma 37.12(iii) that

$$\chi_{R*G}(W_i) = |G_i|^{-1} \chi_R(\mathbf{G_0}(R*G_i))$$
$$= |G_i|^{-1} \cdot m(R*G_i)/\rho_R(R) \cdot Z = w_i^{-1} Z.$$

In other words, $n^{-1}Z = \sum_i w_i^{-1}Z$. Now $w_i^{-1} \in n^{-1}Z$ implies that $w_i | n$ and thus we have $w | n$ where w is the least common multiple of the w_i. On the other hand, since $w/n \in \sum_i (w/w_i)Z \subseteq Z$, we see that $n|w$ and the result follows. ∎

As an immediate consequence of this and Lemma 37.13(iv) we have

Corollary 37.15. **[140]** *Let $R[G]$ be a prime group ring with R right Noetherian and G polycyclic-by-finite. If $G_1, G_2, \ldots, G_k$ are representatives of the conjugacy classes of the maximal finite subgroups of G, then the Goldie rank of $R[G]$ is equal to $\rho_R(R)$ times the least common multiple of the orders of the subgroups G_i.*

Finally **[94]** extends aspects of this to more general groups. For example we quote the following result without proof.

Theorem 37.16. **[94]** *Let G be a group having a finite subnormal series*

$$\langle 1 \rangle = G_0 \lhd G_1 \lhd \cdots \lhd G_n = G$$

with each quotient G_{i+1}/G_i locally polycyclic-by-finite. Then the group ring $R[G]$ has a right Artinian quotient ring if and only if R has a right Artinian quotient ring and the finite subgroups of G have bounded order.

EXERCISES

1. Verify the group theoretic relations in the proof of Proposition 37.1. Write a computer program to show that all multiplicities in S^2 are larger than 1.

2. Show that $C_2 * C_2$ is isomorphic to the infinite dihedral group D. Prove that the group G of Proposition 37.1 can be embedded in the direct product $D \times D \times D$. Conclude that G is supersolvable.

3. Let R be a right Noetherian ring with finite global dimension. If $\mathbf{G_0}(R)$ is the cyclic group generated by $[R]$, prove that all finitely generated projective R-modules are stably free. For this, use the isomorphism of $\mathbf{K_0}(R)$ with $\mathbf{G_0}(R)$ given by Proposition 33.8.

4. If G is a torsion free polycyclic-by-finite group, prove that all finitely generated projective modules for the group algebra $K[G]$ are stably free. This is a result of [140].

5. Let G be a group generated by $x_1, x_2, \ldots, x_k$ and let H be a subgroup of finite index with right transversal $\{y_1, y_2, \ldots, y_t\}$. For each i, j there exists $h_{i,j} \in H$ with $y_i x_j = h_{i,j} y_{i'}$ and we let H_0 be the subgroup of H generated by all $h_{i,j}$. If $W = \bigcup_{i=1}^{t} H_0 y_i$, prove that $WG = W$ and hence that $W = G$. Since $H_0 \subseteq H$, conclude that $H = H_0$ is finitely generated.

6. Suppose $(a/b)Z = \sum_1^t (a_i/b_i)Z$ with g.c.d.$(a, b) = 1$ and g.c.d.$(a_i, b_i) = 1$. Show that $a = $ g.c.d.$\{a_i\}$ and $b = $ l.c.m.$\{b_i\}$.

Let Λ be a well ordered set. A subnormal series for G of type Λ is a collection $\{G_\lambda \mid \lambda \in \Lambda\}$ of subgroups such that

   i. if λ is not a limit ordinal, then $G_{\lambda-1} \triangleleft G_\lambda$,
   ii. if λ is a limit ordinal, then $G_\lambda = \bigcup_{\sigma < \lambda} G_\sigma$,
   iii. $G_0 = \langle 1 \rangle$ and $G = \bigcup_{\lambda \in \Lambda} G_\lambda$.

The quotients in the series are the groups $G_\lambda / G_{\lambda-1}$.

7. Let G be a group having a subnormal series as above with all quotients locally polycyclic-by-finite. Define a transfinite subnormal length $\mu(G)$ and note that $\mu(G)$ cannot be a limit ordinal if G is finitely generated. Show that the analog of Lemma 37.9 holds for finitely generated groups.

8. State an appropriate analog of Theorem 37.10 and prove it by transfinite induction. This is a result of [94].

38. The Zalesskii-Neroslavskii Example

A question posed in [50] has given rise to some extremely interesting mathematics. Specifically, it was asked whether there exists a simple Noetherian ring having zero divisors but no idempotents. The affirmative answer given in [208] is a particular twisted group algebra $K^t[G]$ of an abelian-by-finite group G over a field K of characteristic 2. This ring is now known as the *Zalesskii-Neroslavskii example*. Most of its basic properties are quite easy to obtain; the main difficulty is the absence of idempotents and the original proof of this fact was very computational. Somewhat later, it was shown in [**196**] that this same ring is not Morita equivalent to a domain. The recent paper [**106**] now contains a slick noncomputational proof of both of these results by studying the Grothendieck group of the ring. Moreover, [**104**] offers analogous examples in all characteristics $p > 0$. We follow the approach of [**106**].

To start with, we study skew group rings $S = RG$ which are algebras over a field K. By this we mean precisely that $K \subseteq \mathbf{Z}(R)$ and that G acts trivially on K. In particular, $S \supseteq KG = K[G]$, the ordinary group algebra of G over K. This structure allows us to copy a basic ingredient from the representation theory of finite groups, namely the tensor product module with diagonal action. Since S is a skew group ring, there is as usual no need to use overbars on the group elements.

Suppose A is a right S-module and B is a right $K[G]$-module. Then $A \otimes_K B$ can be made into an S-module by defining

$$(a \otimes b)(xr) = axr \otimes bx$$

for all $a \in A, b \in B, r \in R$ and $x \in G$. Indeed, it is clear that the above formula defines, for each $xr \in S$, a K-endomorphism $\theta(xr) \colon A \otimes B \to A \otimes B$ and hence we obtain an additive map $\theta \colon S \to \mathrm{End}_K(A \otimes B)$. To see that θ is multiplicative, we need only observe that $\theta(xr)\theta(yt) = \theta(xr \cdot yt) = \theta(xyr^y t)$ and this is trivial to check. Thus we have confirmed that $A \otimes_K B$ is an S-module. To avoid confusion with the usual tensor product, we write this module as $A \odot B = A \odot_K B$. In other words, $A \odot B$ has the K-vector space structure of $A \otimes_K B$, but its S-module structure is given by the *diagonal action*

$$(a \odot b)(xr) = axr \odot bx.$$

Furthermore, it is easy to see that if $\alpha \colon A \to A'$ is an S-module homomorphism and $\beta \colon B \to B'$ is a $K[G]$-module homomorphism, then

$$\alpha \odot \beta \colon A \odot B \to A' \odot B'$$

is an S-module homomorphism. This is the content of Lemma 38.1(i) below.

If G is a finite group then, by Lemma 26.2(i), R is a right S-module. Indeed this is true even if G is infinite; we merely define $u \cdot xr = u^x r$ for all $u, r \in R$ and $x \in G$. Then as above, this gives rise to an additive map $\phi \colon S \to \mathrm{End}(R)$ and one checks easily that $\phi(xr)\phi(yt) = \phi(xr \cdot yt) = \phi(xyr^y t)$. It is this S-module structure for R which is used in part (iii) below. Part (ii) is an aspect of *Frobenius reciprocity*.

Lemma 38.1. *Let $S = RG$ be a skew group ring which is an algebra over the field K.*

    i. *If A is an S-module and B is a $K[G]$-module, then $A \odot_K B$ is an S-module which is functorial in each factor.*

    ii. *If A is an S-module, then $A_{|R} \otimes_R S \cong A \odot_K K[G]$.*

    iii. *$R \odot_K K[G] \cong S$.*

Proof. Part (i) has already been discussed and for (ii) we need only observe that the map $A_{|R} \otimes_R S \to A \odot_K K[G]$ given by $a \otimes rx \mapsto arx \odot x$ is an S-isomorphism with back map $a \odot x \mapsto ax^{-1} \otimes x$. Finally for (iii), let $A = R$ and observe that $A_{|R} = R_R$ is the regular R-module. Thus by (ii) we have $R \odot_K K[G] \cong R_{|R} \otimes_R S \cong S$ and the lemma is proved. ∎

If S is any ring and $\mathcal{A}$ is an abelian group, then an additive map $\eta \colon S \to \mathcal{A}$ is called a *trace* if $\eta(st) = \eta(ts)$ for all $s, t \in S$. It is clear that, for any such η, we have $\mathrm{Ker}(\eta) \supseteq [S, S]$, the linear span of all Lie commutators $[s, t] = st - ts$. Thus η factors through the additive homomorphism $\tau \colon S \to S/[S, S]$ which is itself a trace. The following key result uses a few simple properties of group algebras. First, $K[G]$ has a one-dimensional irreducible module V_1 called the *principal module*. Specifically, each $g \in G$ acts like the identity on V_1, so $V_1 \cong K[G]/I$ where I is the augmentation ideal of the algebra. Moreover, if G is a finite p-group and $\mathrm{char}\,K = p > 0$, then I is nilpotent by [**161**, Lemma 3.1.6] and hence V_1 is the unique irreducible $K[G]$-module. Thus, in this case, $K[G]_{K[G]}$ has a composition series of length $|G|$ with all composition factors isomorphic to V_1.

Theorem 38.2. [**104**] [**106**] *Let $S = RG$ be a skew group ring which is an algebra over the field K. Assume that*

i. char $K = p > 0$ and G is a nonidentity finite p-group,

ii. R is right Noetherian and all finitely generated projective R-modules are stably free,

iii. $1 \notin [S, S]$.

If $\lambda \colon \mathbf{G_0}(S) \to Z$ is any group homomorphism, then $\lambda([P]) \in pZ$ for all finitely generated projective S-modules P.

Proof. Notice that S is a finitely generated free right R-module and that R is right Noetherian. Thus, since G is finite, S is also right Noetherian and it makes sense to consider $\mathbf{G_0}(S)$. Also note that if A is any S-module then, by (ii) and the discussion above, the S-module $A \odot K[G]$ has a series of length $|G|$ with factors isomorphic to $A \odot V_1 \cong A$. Hence if A is finitely generated, then in $\mathbf{G_0}(S)$ we have $[A \odot K[G]] = |G| \cdot [A]$. On the other hand, if A is projective then the series splits so $A \odot K[G] \cong A^{|G|}$.

Now let P be a finitely generated projective S-module. Then $P_{|R}$ is a finitely generated projective R-module and hence it is stably free by assumption; say $P_{|R} \oplus R^m \cong R^n$ with $m, n \geq 0$. If R also denotes the S-module of Lemma 38.1(iii) then, by parts (ii) and (iii) of that lemma, we have

$$
\begin{aligned}
(P \odot K[G]) \oplus S^m &\cong (P \odot K[G]) \oplus (R \odot K[G])^m \\
&\cong \left(P_{|R} \oplus (R_{|R})^m \right) \otimes_R S \\
&\cong R^n \otimes_R S \cong S^n.
\end{aligned}
$$

Moreover, since P is projective, we have $P \odot K[G] \cong P^{|G|}$. Thus $P^{|G|} \oplus S^m \cong S^n$ and, by adding additional S summands if necessary, we can assume that $p \mid m$. The first goal is to show that $p \mid n$.

Since G is a nonidentity p-group, $p \mid |G|$ and $S^n \cong P^{|G|} \oplus S^m \cong X^p$ for some S-module X. In particular,

$$
S' = \mathrm{M}_n(S) \cong \mathrm{End}_S(S^n) \cong \mathrm{End}_S(X^p) \cong \mathrm{M}_p(T) = T'
$$

where $T = \mathrm{End}_S(X)$. If $\{ e_{i,j} \}$ denotes the set of matrix units of $\mathrm{M}_p(T)$, then we have $1 = \sum_{i=1}^{p}[e_{i,1}, e_{1,i}]$ since char $K = p$. In other words, $1 \in [T', T']$ and hence, via the above isomorphism, we have $1 \in [S', S']$. On the other hand, consider the combined map

$$
\sigma \colon S' \xrightarrow{\ tr\ } S \xrightarrow{\ \tau\ } S/[S, S]
$$

where tr denotes the usual matrix trace (that is, the sum of the diagonal entries) and τ is the trace map of S discussed above. Then it is easy to see that σ is also a trace map so $\sigma(1) = 0$ since $1 \in [S', S']$. But $\sigma(1) = \tau(n) = n \cdot \tau(1)$ and $\tau(1) \neq 0$ by assumption (iii). Thus, since char $K = p > 0$, we conclude that $p \mid n$ as required.

Finally, since $S \cong R \odot K[G]$ we have $[S] = |G| \cdot [R]$ in $\mathbf{G_0}(S)$ and then $P^{|G|} \oplus S^m \cong S^n$ yields $|G| \cdot [P] = (n-m)|G| \cdot [R]$. Hence if $\lambda \colon \mathbf{G_0}(S) \to Z$ is any group homomorphism then, since Z is torsion free and $p \mid (n-m)$, we have $\lambda([P]) = (n-m)\lambda([R]) \in pZ$ and the result follows. ∎

Two rings S and T are said to be *Morita equivalent* if their categories of modules are equivalent. The key theorem of [**143**] gives necessary and sufficient conditions for this to occur and in fact describes the equivalence. We will not need that result; we only require a few simple observations which follow directly from the existence of the category equivalence. These are all contained in the following lemma which is an immediate consequence of [**6**, Propositions 21.4, 21.6(2), 21.8(2) and Corollary 21.9]. Here, we let $\mathbf{K_0^c}(S) \subseteq \mathbf{G_0}(S)$ denote the image of $\mathbf{K_0}(S)$ under the Cartan map c.

Lemma 38.3. *Let S and T be Morita equivalent rings with S right Noetherian. Then T is right Noetherian and there exists an isomorphism $\theta \colon \mathbf{G_0}(S) \to \mathbf{G_0}(T)$ which sends $\mathbf{K_0^c}(S)$ to $\mathbf{K_0^c}(T)$.*

With this, we can quickly prove

Corollary 38.4. [**104**] [**106**] *Let $S = RG$ satisfy the hypothesis of Theorem 38.2. Then S is not Morita equivalent to a domain. Furthermore, if the reduced rank of S satisfies $\rho_S(S) \leq p$, then $\rho_S(S) = p$ and S contains no nontrivial idempotents.*

Proof. Suppose T is a domain Morita equivalent to S. Then T is right Noetherian and the regular module T_T has reduced rank $\rho_T(T) = 1$. In other words, there is a homomorphism $\rho_T \colon \mathbf{G_0}(T) \to Z$ and a finitely generated projective T-module P with $\rho_T([P]) = 1$. On the other hand, by Theorem 38.2, if $\lambda \colon \mathbf{G_0}(S) \to Z$ is any map, then

$\lambda \colon \mathbf{K}_0^c(S) \to pZ$. Thus we have a contradiction, by Lemma 38.3, and the first fact is proved.

Now suppose $\rho_S(S) \leq p$. By Theorem 38.2 again, we know that $p \mid \rho_S(P)$ for all finitely generated projective S-modules P. In particular, this implies that $\rho_S(S) = p$ and that if $P \neq 0$ then $\rho_S(P) \geq p = \rho_S(S)$. Thus S cannot have a nontrivial idempotent e since otherwise $P = eS$ would be a projective module having smaller rank than S. $\blacksquare$

To use this result, we obviously need some sufficient conditions for (ii) and (iii) of Theorem 38.2 to hold. Part (iii) is easy.

Lemma 38.5. *If $S = K^t[G]$ is a twisted group algebra, then $1 \notin [S, S]$.*

Proof. Since $[S, S]$ is the K-linear span of the Lie commutators $[\bar{x}, \bar{y}]$ with $x, y \in G$, it suffices to show that $1 \notin \operatorname{Supp} [\bar{x}, \bar{y}]$. In fact, since $\operatorname{Supp} [\bar{x}, \bar{y}] \subseteq \{xy, yx\}$, we need only consider the case $y = x^{-1}$. But then $\bar{x}\bar{y} = k \in K^\bullet$ so $\bar{x}^{-1} = k^{-1}\bar{y}$ and hence $\bar{x}$ and $\bar{y}$ commute. Thus $[\bar{x}, \bar{y}] = 0$ here and the lemma is proved. $\blacksquare$

The next result predates Theorem 36.9 and was originally proved using the techniques of Section 33.

Theorem 38.6. [67] [193] *Let $R*G$ be a crossed product with G a poly-Z group and with R a ring of finite global dimension.*

    i. $R*G$ has finite global dimension.

    ii. If R is right Noetherian, then the induced module map yields an epimorphism $\mathbf{K}_0(R) \to \mathbf{K}_0(R*G)$.

Proof. (i) By induction on the Hirsch number, it suffices to assume that $G = \langle x \rangle$ is infinite cyclic. Of course, this implies that $S = R*\langle x \rangle = R\langle x \rangle$ is a skew group ring.

Let B be an R-module and let $\mathbf{P} \to B \to 0$ be a finite projective resolution for B which exists since R has finite global dimension. Then, since $_RS$ is free and hence flat, $\mathbf{P} \otimes_R S \to B \otimes_R S \to 0$ is a finite projective resolution for $B \otimes_R S$. In other words, we now know that any induced S-module has finite projective dimension.

Finally, let A be an arbitrary S-module and observe that there is an epimorphism $\alpha: A_{|R} \otimes_R S \to A$ given by $a \otimes s \mapsto as$ for all $a \in A$ and $s \in S$. Since $_R S$ is free with basis $\{\, x^i \mid i \in Z \,\}$, we have $A_{|R} \otimes S = \oplus \sum_i A \otimes x^i$ and

$$L = \mathrm{Ker}(\alpha) = \left\{ \sum_i a_i \otimes x^i \ \Big| \ \sum_i a_i x^i = 0 \right\}.$$

Furthermore, if $(A_{|R})^x$ is the conjugate R-module determined by the automorphism x, then the map $\beta: (A_{|R})^x \otimes_R S \to L$ given by $a^x \otimes x^i \mapsto a \otimes x^{i+1} - ax \otimes x^i$ is easily seen (Exercise 4) to be an S-module isomorphism. Thus

$$0 \to (A_{|R})^x \otimes S \xrightarrow{\ \beta\ } A_{|R} \otimes S \xrightarrow{\ \alpha\ } A \to 0$$

is a short exact sequence of S-modules. But both $A_{|R} \otimes S$ and $(A_{|R})^x \otimes S$ have finite projective dimension and hence so does A by Lemma 33.9(i).

(ii) Here we note that $S = R*G$ is Noetherian and that, by Proposition 33.8 and the above, the maps $c: \mathbf{K_0}(R) \to \mathbf{G_0}(R)$ and $c^{-1}: \mathbf{G_0}(S) \to \mathbf{K_0}(S)$ are isomorphisms. Furthermore, by Theorem 36.9, the induced module map yields an epimorphism $\mathbf{G_0}(R) \to \mathbf{G_0}(S)$. This follows since $\langle 1 \rangle$ is the unique finite subgroup of G. Hence the combined map $\mathbf{K_0}(R) \to \mathbf{K_0}(S)$ is also an epimorphism and this is clearly the induced module map restricted to $\mathbf{K_0}$. ∎

Now we begin to construct the examples. Let K be a field containing an element λ of infinite multiplicative order and let

$$A_n = \langle x_1, x_2, \ldots, x_n, y_1, y_2, \ldots, y_n \rangle$$

be a free abelian group on the $2n$ generators x_i and y_i. Then we let $R_n = K^t[A_n]$ be the twisted group algebra with $\bar{x}_i \bar{y}_i = \lambda \bar{y}_i \bar{x}_i$ for all i and such that all other pairs of generators commute. Note that $A_n \cong A_1 \times A_1 \times \cdots \times A_1$ and that $R_n \cong R_1 \otimes R_1 \otimes \cdots \otimes R_1$. Note also that since A_n is an ordered group, all units of R_n are trivial. In particular, if G is a group of automorphisms of R_n, then G acts on the group $\mathcal{A}_n$ of trivial units and then on $A_n \cong \mathcal{A}_n / K^\bullet$.

Lemma 38.7. *Let A_n and R_n be as above and let G be a finite group of K-algebra automorphisms of R_n such that the inherited action of G on A_n is faithful. Then the skew group ring $R_n G$ is a simple Noetherian ring and it is a twisted group algebra $K^t[A_n \rtimes G]$.*

Proof. We first study $R = R_n$. Set $X = \langle x_1, x_2, \ldots, x_n \rangle$ and $Y = \langle y_1, y_2, \ldots, y_n \rangle$. Then $A = A_n = X \times Y$ and R is the iterated skew group ring $R = (KX)Y$. Note that $KX = K[X]$ is an ordinary group algebra, so it is a commutative domain, and we consider the action of Y on this ring. To start with, KX is Y-simple. Indeed, suppose I is a nonzero Y-stable ideal of R and let $\alpha = \sum k_x \bar{x} \in I$ be a nonzero element of minimal support size. We may assume that $1 \in \text{Supp } \alpha$ and then, for all $y \in Y$, we have $\text{Supp } (\alpha^{\bar{y}} - \alpha) \subset \text{Supp } \alpha$. The minimality of $\text{Supp } \alpha$ now implies that α is centralized by Y and hence, since λ has infinite multiplicative order, that $\alpha \in K^{\bullet}$. Thus $I = KX$ as required. Next, we note that Y acts faithfully on KX. Thus since KX is a commutative domain, Lemma 10.3(i) implies that Y is X-outer in its action. It now follows easily, from Corollary 12.6 and the comments preceding Lemma 14.1, that R is a simple ring; it is certainly a Noetherian domain by Proposition 1.6.

Since R is a simple ring, we have $Q_s(R) = R$ and hence all X-inner automorphisms are inner. But A is an ordered group, so all units of R are trivial. It follows from this, since A is abelian, that the X-inner automorphisms of R centralize A in the inherited action. In particular, if G is a finite group of K-algebra automorphisms of R acting faithfully on A, then G must be X-outer and hence, by Corollary 12.6 again, the skew group ring RG is simple. Since this ring is also clearly Noetherian and isomorphic to $K^t[A \rtimes G]$, the result follows. ∎

We can now quickly construct examples in characteristic $p > 0$; no examples are known in characteristic 0.

Theorem 38.8. [208] [196] *Let $R = R_1 = K^t[A_1]$ be as above with K a field of characteristic 2. Suppose $G = \{1, \sigma\}$ is a group of order 2 which acts on R as K-algebra automorphisms by $(\bar{x}_1)^{\sigma} = \bar{x}_1^{-1}, (\bar{y}_1)^{\sigma} = \bar{y}_1^{-1}$. Then the skew group ring $S = RG$ is a simple*

Noetherian ring with nonzero nilpotent elements. Furthermore, S has no nontrivial idempotents and it is not Morita equivalent to a domain.

Theorem 38.9. **[104]** *Let K be a field of characteristic $p > 0$ and let $R = R_p = K^t[A_p]$ be as above. Suppose G is a group of order p which acts on $R \cong R_1 \otimes R_1 \otimes \cdots \otimes R_1$ by cyclically permuting the p factors. Then the skew group ring $S = RG$ is a simple Noetherian ring with nonzero nilpotent elements. Furthermore, S has no nontrivial idempotents and it is not Morita equivalent to a domain.*

Proof. It is clear that G acts as K-algebra automorphisms in the second set of examples. For the first, since $\bar{x}_1 \bar{y}_1 = \lambda \bar{y}_1 \bar{x}_1$, we have

$$(\bar{x}_1)^\sigma (\bar{y}_1)^\sigma = \bar{x}_1^{-1} \bar{y}_1^{-1} = (\bar{y}_1 \bar{x}_1)^{-1} = (\lambda^{-1} \bar{x}_1 \bar{y}_1)^{-1}$$
$$= \lambda \bar{y}_1^{-1} \bar{x}_1^{-1} = \lambda (\bar{y}_1)^\sigma (\bar{x}_1)^\sigma$$

and again $G \subseteq \text{Aut}_K(R)$. In either case, since G acts faithfully on $A = A_1$ or $A = A_p$, Lemma 38.7 implies that $S = RG$ is a simple Noetherian ring. Furthermore, since R is a domain and $|G| = p$, the reduced ranks satisfy $\rho_S(S) \leq \rho_R(S) = p$.

Now note that $S \cong K^t[A \rtimes G]$ so $1 \notin [S, S]$ by Lemma 38.5. Furthermore, since $R = K^t[A]$, Theorem 38.6 implies that the induced module map yields an epimorphism $\mathbf{K_0}(K) \to \mathbf{K_0}(R)$. But $\mathbf{K_0}(K)$ is the cyclic group generated by $[K]$, so $\mathbf{K_0}(R)$ is the cyclic group generated by $[K \otimes_R R] = [R]$ and hence Lemma 34.10(ii) implies that all finitely generated projective R-modules are stably free. The results now follow from Corollary 38.4 and the fact that the augmentation ideal of $K[G] \subseteq S$ is nilpotent. ∎

The first ring $S = RG$ above is the Zalesskii-Neroslavskii example. It of course varies somewhat depending on the choice of the field K and the element $\lambda \in K^\bullet$. In the next sections, we will explicitly compute $\mathbf{G_0}(S)$ along with other examples. To do this, we will need to know that the fixed ring R^G is Noetherian. We prove a more general result below.

Suppose $A = Z^k$ is the free additive abelian group on k generators. Then A can be ordered *lexicographically* by defining

$$(a_1, a_2, \ldots, a_k) <_\ell (b_1, b_2, \ldots, b_k)$$

if and only if $a_1 = b_1$, $a_2 = b_2$, $\ldots$, $a_{i-1} = b_{i-1}$ and $a_i < b_i$ for some subscript $1 \le i \le k$. In this way, A becomes an ordered group. Alternately, we have the *product ordering*, where

$$(a_1, a_2, \ldots, a_k) \le_\pi (b_1, b_2, \ldots, b_k)$$

if and only if $a_i \le b_i$ for all i. Note that this is only a partial order, but it again respects the group addition. Let $N = \{0, 1, 2, \ldots\}$ denote the set of nonnegative integers. Then $N^k \subseteq Z^k = A$ and we have

Lemma 38.10. *Any sequence $\{x_1, x_2, \ldots\}$ of elements in N^k contains a subsequence $\{y_1, y_2, \ldots\}$ with $y_i \le_\pi y_{i+1}$ for all i. In particular, if M is a nonempty subset of N^k, then M has only finitely many minimal elements and each element of M contains at least one of these.*

Proof. The natural ordering on N has the property that every sequence has a nondecreasing subsequence. Therefore, the same holds for N^k, by projecting the given sequence into each of the k summands and, at each step, taking an appropriate subsequence. It follows that N^k cannot contain (i) infinitely many incomparable elements under $\le_\pi$ or (ii) a strictly decreasing sequence. Thus the facts on $M \subseteq N^k$ follow immediately. ∎

We close with

Proposition 38.11. [77] *Suppose $R = D*A$ is a crossed product with A a torsion free abelian group of rank $k < \infty$ and with D a division ring. Let the group $G = \{1, \sigma\}$ act on R in such a way that D is centralized and the inherited action of σ on A is inversion, that is $a^\sigma = a^{-1}$ for all $a \in A$. Then R is right and left Noetherian as a module over the fixed ring R^G.*

Proof. We need some definitions. First, let us think of $A = Z^k$ as being additive and define a linear ordering on A as follows. For each $a = (a_1, a_2, \ldots, a_k) \in A$, let $\|a\| = \sum_i |a_i| \geq 0$. Then we say that $a \preceq b$ if and only if either (i) $\|a\| < \|b\|$ or (ii) $\|a\| = \|b\|$ and $a \leq_\ell b$ where $\leq_\ell$ denotes the lexicographical ordering of A. Note that $\preceq$ does not respect addition, but it is a well ordering.

It will be necessary to keep track of the signs of the entries in $a = (a_1, a_2, \ldots, a_k)$. For this, we let $\epsilon = (\epsilon_1, \epsilon_2, \ldots, \epsilon_k)$ denote a *multi-sign*, so that each $\epsilon_i = \pm$, and we set

$$A_\epsilon = \left\{ (a_1, a_2, \ldots, a_k) \in A \ \middle| \ \begin{array}{l} a_i \geq 0 \text{ if } \epsilon_i = + \\ a_i < 0 \text{ if } \epsilon_i = - \end{array} \right\}.$$

It is clear that the 2^k sets A_ϵ, obtained in this manner, partition A.

For convenience, we let $\{\, e(1), e(2), \ldots, e(k)\,\}$ denote the canonical Z-basis of A. In other words, $e(i)$ has a 1 in its ith entry and 0's elsewhere.

Claim 1. *Let M be a finite nonempty subset of A and let $m = \max M$ be the maximal element of M under the ordering $\preceq$. If $m \in A_\epsilon$ with $\epsilon = (\epsilon_1, \epsilon_2, \ldots, \epsilon_k)$, then*

$$m + \epsilon_i e(i) = \max \big(M \pm e(i) \big).$$

Furthermore, if $m + \epsilon_i e(i) = a + \delta e(i)$ for some $a \in M$ and $\delta = \pm$, then we have $a = m$ and $\delta = \epsilon_i$.

Proof. Suppose that $a \in M$, $\delta = \pm$ and $m + \epsilon_i e(i) \preceq a + \delta e(i)$. The goal is to show that $a = m$ and $\delta = \epsilon_i$. To start with, we have

$$\|a + \delta e(i)\| \leq \|a\| + 1 \leq \|m\| + 1 = \|m + \epsilon_i e(i)\|$$

since $a \preceq m$ and $m \in A_\epsilon$. But, by assumption, $a + \delta e(i) \succeq m + \epsilon_i e(i)$ so $\|a + \delta e(i)\| \geq \|m + \epsilon_i e(i)\|$ and we must have equality throughout. In particular, $\|a + \delta e(i)\| = \|a\| + 1 = \|m + \epsilon_i e(i)\|$ and therefore $a + \delta e(i) \geq_\ell m + \epsilon_i e(i)$. Furthermore, $\|a\| = \|m\|$ and hence $a \leq_\ell m$ since $a \preceq m$. Now, if either $\epsilon_i = +$ or $\epsilon_i = -$ and $\delta = -$, then $a \leq_\ell m$ clearly yields $a + \delta e(i) \leq_\ell m + \epsilon_i e(i)$. Thus these two elements are equal and it follows easily that $a = m$ and $\delta = \epsilon_i$. All that remains now is the case $\epsilon = -$, $\delta = +$ and we show that this cannot occur by comparing components. If $a_j < m_j$ for some $j < i$, then certainly $a + \delta e(i) <_\ell m + \epsilon_i e(i)$, a contradiction. Thus since $a \le_\ell m$, we must have $a_j = m_j$ for all $j < i$ and hence $a_i \le m_i$. But then $a_i \le m_i < 0$, since $\epsilon_i = -$, and this contradicts the fact that $\delta = +$ and $\|a + \delta e(i)\| = \|a\| + 1$. $\blacksquare$

We continue with the proof of the proposition and introduce additional notation. If $0 \ne r \in R$, we let $\deg r = \max (\mathrm{Supp}\, r) \in A$ using the ordering $\preceq$. In addition, if I is a nonzero R^G-submodule of R, we let $\deg I$ equal the set of degrees of the nonzero elements of I. In the following, $N = \{\, 0, 1, 2, \ldots \,\}$ will again denote the set of nonnegative integers.

Claim 2. *Let I, J be nonzero R^G-submodules of R.*
  i. *If $I \subset J$, then $\deg I \subset \deg J$.*
  ii. *If $0 \ne r \in I$ and $\deg r \in A_\epsilon$ with $\epsilon = (\epsilon_1, \epsilon_2, \ldots, \epsilon_k)$, then*

$$\deg I \supseteq \deg r + (\epsilon_1 N, \epsilon_2 N, \ldots, \epsilon_k N).$$

Proof. (i) Certainly $I \subset J$ implies $\deg I \subseteq \deg J$. Now suppose, by way of contradiction, that $\deg I = \deg J$ and choose $r \in J \setminus I$ of minimal degree. Then $\deg r \in \deg J = \deg I$ implies that there exists $s \in I$ with $\deg s = \deg r$. Furthermore, note that $D \subseteq R^G$ by assumption, so I is a D-vector space. Thus since D is a division ring, we can multiply s by an appropriate element of D so that the leading coefficients of r and s match. But then $r - s \in J \setminus I$ has smaller degree than that of r and this is a contradiction.

(ii) Write $\mathrm{Supp}\, r = M$ and $m = \max M \in A_\epsilon$. If $\overline{e(i)}$ denotes the element of R corresponding to $e(i)$, then we have

$$t_i = \overline{e(i)} + \overline{e(i)}^{\,\sigma} \in R^G$$

and $\mathrm{Supp}\, t_i = \{\, e(i), -e(i) \,\}$ since, by assumption, the inherited action of σ on A is inversion. Furthermore, $r t_i \in I \cdot R^G = I$ and

$$\mathrm{Supp}\, r t_i \subseteq (\mathrm{Supp}\, r) \pm e(i) = M \pm e(i).$$

Now, by Claim 1, $m + \epsilon_i e(i) = \max\left(M \pm e(i)\right)$ and indeed, for all $a \in M$ and $\delta = \pm$, we have $a + \delta e(i) = m + \epsilon_i e(i)$ if and only if $a = m$ and $\delta = \epsilon_i$. Thus since D is a division ring, it follows that $m + \epsilon_i e(i) \in \operatorname{Supp} rt_i$ and then that $m + \epsilon_i e(i) = \deg rt_i$. In other words, $\deg r + \epsilon_i e(i) = \deg rt_i \in \deg I$. Furthermore, since $m + \epsilon_i e(i) \in A_\epsilon$ and $rt_i \in I$, we can clearly continue this process and, in so doing, we obtain

$$\deg r + \sum_i n_i \epsilon_i e(i) = \deg r(t_1)^{n_1}(t_2)^{n_2} \cdots (t_k)^{n_k} \in \deg I$$

as required. ∎

It is now a simple matter to complete the proof of the proposition. Suppose by way of contradiction that $0 \subset I_1 \subset I_2 \subset \cdots$ is an infinite strictly increasing chain of R^G-submodules of R. Then, by Claim 2(i), $\deg I_1 \subset \deg I_2 \subset \cdots$ is strictly increasing, so we can choose elements $a(p) \in \deg I_p \setminus \deg I_{p-1}$. By considering a suitable subchain if necessary, we can assume that all $a(p)$ belong to the same A_ϵ with $\epsilon = (\epsilon_1, \epsilon_2, \ldots, \epsilon_k)$ and then, for any $a = (a_1, a_2, \ldots, a_k) \in A_\epsilon$, we write

$$\epsilon a = (\epsilon_1 a_1, \epsilon_2 a_2, \ldots, \epsilon_k a_k) \in (N, N, \ldots, N).$$

Finally, suppose $p < q$. Then $a(p) + (\epsilon_1 N, \epsilon_2 N, \ldots, \epsilon_k N) \subseteq \deg I_p$, by Claim 2(ii), so $a(q) \notin a(p) + (\epsilon_1 N, \epsilon_2 N, \ldots, \epsilon_k N)$ and hence $\epsilon a(q) \notin \epsilon a(p) + (N, N, \ldots, N)$. In other words, in the product ordering on N^k we have $\epsilon a(p) \not\leq_\pi \epsilon a(q)$. But then the infinite sequence $\{\epsilon a(1), \epsilon a(2), \ldots\} \subseteq N^k$ cannot have a nondecreasing subsequence in the product order and this contradicts the previous lemma. ∎

EXERCISES

1. Let $I = \omega(K[G])$ be the augmentation ideal of the group algebra $K[G]$ of the nonidentity group G. Show that I is nilpotent if and only if $\operatorname{char} K = p > 0$ and G is a finite p-group. For this, first observe that $r_{K[G]}(I) \neq 0$ if and only if G is finite.

2. Let $S = RG$ be as in Lemma 38.1. Suppose $\lambda: G \to K^\bullet$ is a linear character of G and let V_λ be its corresponding 1-dimensional $K[G]$-module. Prove that $R \odot_K V_\lambda \cong R$ if and only if there exists a unit u of R with $u^g = \lambda(g)u$ for all $g \in G$. To this end, note that any such S-isomorphism is an R-isomorphism and hence is determined by a suitable left multiplication.

3. Suppose S is a semiprime right Noetherian ring and that P is a nonzero projective (or flat) S-module. Use the inclusion $0 \to S \to Q(S)$ to show that $\rho_S(P) \neq 0$.

4. Verify the details of Theorem 38.6. Specifically show that the map $\beta: (A_{|R})^x \otimes_R S \to L$ is an S-module isomorphism. To see that β is onto, proceed by induction on the breadth of the elements of L.

5. Let G act on the crossed product $D*H$, normalizing the group $\mathcal{H}$ of trivial units. If D is a division ring, prove that G normalizes D and hence acts on $H \cong \mathcal{H}/D^\bullet$.

6. Let $S = RG$ be the Zalesskii-Neroslavskii example and observe that the action of G on R is G-Galois (see Exercises 3, 4 and 5 of Section 29). Deduce that R_{R^G} is a finitely generated indecomposable projective module over the Noetherian domain R^G. Note that indecomposability follows from the fact that $\text{End}_{R^G}(R) \cong RG = S$ has no nontrivial idempotents.

7. Continue with the above notation and observe, since $\text{char} K = 2$, that $0 \to R^G \to R \xrightarrow{\text{tr}_G} T \to 0$ is an exact sequence of R^G-modules with $T = \text{tr}_G(R) \subseteq R^G$. Deduce that $\rho_{R^G}(R) = 2$. This and the results of the preceding exercise are from [**77**].

39. Almost Injective Modules

In this section we develop some techniques for explicitly computing the Grothendieck group of the Zalesskii-Neroslavskii example and certain similar twisted group algebras. Since Theorem 36.9 already yields an upper bound for $\mathbf{G_0}$, the goal here is to obtain an appropriate lower bound. We do this by studying the restrictions of the modules of the larger ring to group algebras of finite subgroups. Since these restricted modules are most likely infinitely generated, it is then necessary to isolate within them a certain finite parameter. A number of the preliminary lemmas below, and in particular Proposition 39.3, hold in the more general context of *quasi-Frobenius rings*, that is rings whose projective and injective modules coincide. However, we have opted to consider only group algebras here, since this allows us to proceed in a fairly self-contained manner.

Let $K[G]$ denote the group algebra of a finite group G. If A and B are $K[G]$-modules then they determine, as in the previous section, a tensor module $A \odot_K B$ with diagonal action. Specifically $(a \odot b)g = ag \odot bg$ for all $a \in A$, $b \in B$ and $g \in G$. Recall that a ring R is said to be *self-injective* if the regular module R_R is injective.

Lemma 39.1. *Let $K[G]$ be a group algebra with G finite and let A be a right $K[G]$-module.*

i. *$K[G]$ is self-injective.*

ii. *$A \odot_K K[G]$ is a free $K[G]$-module on $\dim_K A$ generators.*

iii. *The map $\alpha: A \to A \odot_K K[G]$ given by $\alpha(a) = \sum_{g \in G} a \odot g$, for all $a \in A$, is a $K[G]$-isomorphism into. In particular, if A is irreducible, then A is isomorphic to a submodule of $K[G]$.*

Proof. (i) Let $f: K[G] \to K$ denote the map which reads off the identity coefficient. Then f is certainly a linear functional and, for all $\gamma \in K[G]$, we have $\gamma = \sum_{g \in G} f(\gamma g^{-1})g$. Now let $A \subseteq B$ be $K[G]$-modules and let $\sigma: A \to K[G]$ be a $K[G]$-homomorphism. Then $f\sigma: A \to K$ is a linear functional which can certainly be extended to $(f\sigma)^\star: B \to K$. We now define $\sigma^\star: B \to K[G]$ by $\sigma^\star(b) = \sum_{g \in G}(f\sigma)^\star(bg^{-1})g$. It is easy to verify that $\sigma^\star$ is a $K[G]$-module homomorphism which extends σ.

(ii) If $\{ a_i \mid i \in I \}$ is a K-basis for A, then so is $\{ a_i g \mid i \in I \}$ for any fixed $g \in G$. Furthermore, since $A \odot_K K[G] = \oplus \sum_{g \in G} A \odot g$, it then follows that $\{ a_i g \odot g \mid g \in G, i \in I \}$ is a K-basis for $A \odot K[G]$ and hence that $\{ a_i \odot 1 \mid i \in I \}$ is a $K[G]$-basis for the module.

(iii) It is clear that α is a one-to-one $K[G]$-homomorphism. Thus $A \cong \alpha(A) \subseteq A \odot_K K[G]$ and the latter is isomorphic to a direct sum of copies of $K[G]$. In particular, if A is irreducible, then the projection of $\alpha(A)$ into one of the direct summands must be nontrivial and hence an embedding. ∎

In the following lemma we let $E(A) = E_R(A)$ denote the injective hull of the R-module A and we use the well known fact (see Exercise 2) that, in the case of Noetherian rings, a direct sum of injectives is injective. For convenience, we say that a module is *injective-free* if it contains no nonzero injective submodule.

Lemma 39.2. *Let $K[G]$ be the group algebra of the finite group G and let V be a $K[G]$-module.*

i. If A is an irreducible $K[G]$-module, then $E(A)$ is isomorphic to a direct summand of $K[G]$.

ii. If the socle of V is written as $\operatorname{soc} V \cong \oplus \sum_{i \in I} A_i$ *with each A_i irreducible, then* $E(V) \cong \oplus \sum_{i \in I} E(A_i)$.

iii. V is projective if and only if it is injective.

iv. $V \cong V_1 \oplus V_2$ where V_2 is injective and V_1 is injective-free.

Proof. (i) By the previous lemma, A is isomorphic to a submodule of $K[G]$ and $K[G]$ is injective. Thus $E(A)$ is isomorphic to a submodule of $K[G]$ and hence to a direct summand.

(ii) Since $K[G]$ is Artinian, $\operatorname{soc} V$ is essential in V and hence $E(V) = E(\operatorname{soc} V) \cong \oplus \sum_i E(A_i)$.

(iii) Since $K[G]_{K[G]}$ is injective, it follows that any projective $K[G]$-module is also injective. Conversely, if V is injective and is described as in (ii) above, then $V = E(V) \cong \oplus \sum_i E(A_i)$. But each $E(A_i)$ is projective by (i), so we conclude that V itself is projective.

(iv) By Zorn's lemma, we can define V_2 to be a maximal direct sum of injective submodules of V. Then V_2 is injective, so $V = V_1 \oplus V_2$ for some submodule V_1 and it is clear that V_1 is injective-free. ∎

We say that a $K[G]$-module V is *almost injective* if $E(V)/V$ is finitely generated. For such *a.i.* modules V, $[E(V)/V]$ is a well defined element of $\mathbf{G_0}(K[G])$ and we let $\theta(V)$ denote the image of $-[E(V)/V]$ in $\bar{\mathbf{G}}_0(K[G]) = \mathbf{G_0}(K[G])/\mathbf{K_0^c}(K[G])$. Note that, if V is finitely generated, then so is $E(V)$ by Lemma 39.2(i)(ii) and hence $\theta(V)$ is the image of $[V] - [E(V)]$ in $\bar{\mathbf{G}}_0(K[G])$. But $E(V)$ is projective, so $[E(V)] \in \mathbf{K_0^c}(K[G])$ and $\theta(V)$ is in fact the image of $[V]$. We will not need all of the following proposition; it is included however for its potential usefulness.

Proposition 39.3. [115] *Let $K[G]$ be a group algebra of the finite group G.*

    i. Let V be a $K[G]$-module and write $V = V_1 \oplus V_2$ where V_2 is injective and V_1 is injective-free. Then V is a.i. if and only if V_1 is finitely generated and, in this case, $\theta(V) = [V_1] + \mathbf{K_0^c}(K[G])$.

    ii. Let $0 \to A \to B \to C \to 0$ be a short exact sequence of $K[G]$-modules. If any two of these modules are a.i., then so is the third and we have $\theta(B) = \theta(A) + \theta(C)$.

Proof. (i) Since V_2 is projective and $\mathrm{E}(V) = \mathrm{E}(V_1) \oplus V_2$, it clearly suffices to assume that $V_2 = 0$. In other words, we can suppose that V is injective-free. If V is finitely generated then, as we observed above, $\mathrm{E}(V)$ is finitely generated and V is certainly a.i. Conversely, suppose V is a.i. In the notation of Lemma 39.2(ii), we have $V \subseteq \mathrm{E}(V) = \oplus \sum_{i \in I} \mathrm{E}(A_i)$ and we let π_i denote the projection into the ith summand. Then $\pi_i(V) \neq \mathrm{E}(A_i)$ since $\mathrm{E}(A_i)$ is projective and V, being injective-free, does not contain an isomorphic copy of $\mathrm{E}(A_i)$. Since the finitely generated module $\mathrm{E}(V)/V$ has as a homomorphic image the direct sum $\oplus \sum_{i \in I} \mathrm{E}(A_i)/\pi_i(V)$, we conclude that I is finite. It follows that $\mathrm{E}(V)$ is finitely generated and hence so is V.

    (ii) We may suppose that $A \subseteq B$ and $C = B/A$. Let A_2 be an injective submodule of A. Then $A_2 \subseteq A \subseteq B$ yields $B = B' \oplus A_2$ and $A = A' \oplus A_2$ where $A' = B' \cap A$. Since $C \cong B'/A'$ and the module A_2 is both injective and projective, it follows easily that we need only consider $0 \to A' \to B' \to C \to 0$. In other words, we may assume that $A_2 = 0$ and hence, by Lemma 39.2(iv), that A is injective-free.

    Next write $C = C_1 \oplus C_2$ as in Lemma 39.2(iv) and let B_1 be the complete inverse image of C_1 in B. Since C_2 is projective, there exists a submodule B_2 of B which maps isomorphically to C_2. It follows that $B = B_1 \oplus B_2$ and that $A \subseteq B_1$. Since $B_2 \cong C_2$ are both injective and projective, it follows easily that we need only consider $0 \to A \to B_1 \to C_1 \to 0$. In other words, we may assume that $C_2 = 0$ and hence that C is injective-free.

    Now if A and C are a.i. then, by (i) above, they are both finitely generated. Hence B is also finitely generated and a.i. Furthermore, $[B] = [A] + [C]$ in $\mathbf{G_0}(K[G])$ and therefore $\theta(B) = \theta(A) + \theta(C)$ in

$\bar{\mathbf{G}}_0(K[G])$.

Finally suppose B is a.i. so that $\mathrm{E}(B)/B$ is finitely generated. Now $\mathrm{E}(B) \supseteq \mathrm{E}(A)$ yields $\mathrm{E}(B) = \mathrm{E}(A) \oplus D$ for some injective sub-module D and hence

$$C = B/A \subseteq \mathrm{E}(B)/A = \bar{A} \oplus D$$

where $\bar{A} = \mathrm{E}(A)/A$. Furthermore, $(\bar{A} \oplus D)/C \cong \mathrm{E}(B)/B$ is finitely generated. In particular, if C is a.i., then C is finitely generated and hence so is $\bar{A}$. On the other hand, if A is a.i., then $\bar{A}$ is finitely generated and hence so is $\mathrm{E}(\bar{A})$. But then $(\mathrm{E}(\bar{A}) \oplus D)/C$ is finitely generated and, since $\mathrm{E}(C) \subseteq \mathrm{E}(\bar{A}) \oplus D$, we conclude that $\mathrm{E}(C)/C$ is finitely generated. This completes the proof. ∎

If A and B are $K[G]$-modules, then $\mathrm{Hom}_K(A, B)$ is also a $K[G]$-module with a *conjugate action*. Specifically, if $f \in \mathrm{Hom}_K(A, B)$ and $g \in G$, then $f^g \colon A \to B$ is defined by $f^g(a) = f(ag^{-1})g$ for all $a \in A$. It is clear that $f^g \in \mathrm{Hom}_K(A, B)$ and that $(f^g)^h = f^{gh}$ for all $g, h \in G$. Thus $C = \mathrm{Hom}_K(A, B)$ is indeed a $K[G]$-module. Note that $f^g(a)g^{-1} = f(ag^{-1})$ and thus $f^g = f$ for all $g \in G$ if and only if $f \in \mathrm{Hom}_{K[G]}(A, B)$. In other words,

$$\mathrm{Hom}_{K[G]}(A, B) = C^G = \{\, f \in C \mid f^g = f \text{ for all } g \in G \,\}.$$

Of course, if G is finite, then we have the trace map $\mathrm{tr}_G \colon C \to C^G$ given by $\mathrm{tr}_G(f) = \sum_{g \in G} f^g$.

Lemma 39.4. *Let G be a finite group and let V be a $K[G]$-module.*

i. Suppose X is an irreducible $K[G]$-module and $\pi \colon X \to V$ is a $K[G]$-homomorphism. Then $\pi \in \mathrm{tr}_G\big(\mathrm{Hom}_K(X, V)\big)$ if and only if π extends to a homomorphism $\pi^\star \colon \mathrm{E}(X) \to V$.

ii. V is a.i. if and only if, for all irreducible $K[G]$-modules X,

$$\mathrm{Hom}_{K[G]}(X, V)/\mathrm{tr}_G\big(\mathrm{Hom}_K(X, V)\big)$$

is finite dimensional over K.

Proof. (i) Suppose first that π is a trace and say $\pi = \mathrm{tr}_G(f) = \sum_{g \in G} f^g$. Then we can define $\beta \colon X \odot_K K[G] \to V$ by $\beta(x \odot g) =$

$f(xg^{-1})g$, for all $x \in X$, $g \in G$, and it follows easily that β is a $K[G]$-homomorphism. Furthermore, if $\alpha: X \to X \odot_K K[G]$ is the map of Lemma 39.1(iii), then

$$\beta\alpha(x) = \beta\left(\sum_g x \odot g\right) = \sum_g f(xg^{-1})g$$

so $\beta\alpha = \mathrm{tr}_G(f) = \pi$. Now $X \odot K[G]$ is free and hence injective, so α extends to a map $\alpha^\star: \mathrm{E}(X) \to X \odot K[G]$. Thus π extends to a map $\pi^\star: \mathrm{E}(X) \to V$ as required.

Conversely if π extends to a map $\mathrm{E}(X) \to V$ then, since $\mathrm{E}(X)$ is a direct summand of $K[G]$, π extends to a map $\pi^\star: K[G] \to V$. Let $f: K[G] \to K \subseteq K[G]$ be the functional which reads off the identity coefficient. Then $\pi^\star f$ maps $K[G]$ to V and, for all $\gamma \in K[G]$, we have

$$\pi^\star(\gamma) = \pi^\star\left(\sum_{g \in G} f(\gamma g^{-1})g\right)$$
$$= \sum_{g \in G} \pi^\star f(\gamma g^{-1}) \cdot g = \big(\mathrm{tr}_G(\pi^\star f)\big)(\gamma).$$

Thus $\pi^\star = \mathrm{tr}_G(\pi^\star f)$ and restricting these functions to X yields the result.

(ii) Write $V = V_1 \oplus V_2$ with V_2 injective and V_1 injective-free. Then $\mathrm{Hom}_K(X, V) = \mathrm{Hom}_K(X, V_1) \oplus \mathrm{Hom}_K(X, V_2)$ is a $K[G]$-module direct sum and it follows from (i) that $\mathrm{Hom}_{K[G]}(X, V_2) = \mathrm{tr}_G\big(\mathrm{Hom}_K(X, V_2)\big)$. Thus

$$\mathrm{Hom}_{K[G]}(X, V)/\mathrm{tr}_G\big(\mathrm{Hom}_K(X, V)\big)$$
$$\cong \mathrm{Hom}_{K[G]}(X, V_1)/\mathrm{tr}_G\big(\mathrm{Hom}_K(X, V_1)\big)$$

and, with this, it clearly suffices to assume that $V_2 = 0$. In other words, we now suppose that $V = V_1$ is injective-free. Furthermore, by (i) again, this implies that $\mathrm{tr}_G\big(\mathrm{Hom}_K(X, V)\big) = 0$. Indeed, if $\pi: X \to V$ is a nonzero trace, then π extends to a map $\pi^\star: \mathrm{E}(X) \to V$. But X is irreducible, so π is an embedding, and X ess $\mathrm{E}(X)$ implies that $\pi^\star$ is also an embedding, a contradiction. Thus, by Proposition 39.3(i), we must show that V is finitely generated if and only if $\operatorname{Hom}_{K[G]}(X,V)$ is finite dimensional over K for all irreducible $K[G]$-modules X.

If V is finitely generated, then $\dim_K V < \infty$ and $\operatorname{Hom}_{K[G]}(X,V)$ is certainly finite dimensional. For the converse note that, since X is irreducible, we have $\operatorname{Hom}_{K[G]}(X,V) = \operatorname{Hom}_{K[G]}(X,\operatorname{soc} V)$. In particular, $\operatorname{Hom}_{K[G]}(X,V)$ is finite dimensional over K if and only if $\operatorname{soc} V$ does not contain an infinite direct sum of copies of X. Thus if $\operatorname{Hom}_{K[G]}(X,V)$ is finite dimensional for all the finitely many irreducible $K[G]$-modules X, then $\operatorname{soc} V$ is a finite direct sum of irreducible $K[G]$-modules. In other words, $\operatorname{soc} V$ is finitely generated and, since $V \subseteq \operatorname{E}(V) = \operatorname{E}(\operatorname{soc} V)$, we conclude that V is finitely generated. ∎

To proceed further, it is necessary to reformulate part (ii) above in terms of tensor products. If $K = V_1$ is the principal $K[G]$-module, then $X^* = \operatorname{Hom}_K(X,K)$ is called the *contragredient* of X. Some properties are as follows.

Lemma 39.5. *Let X be a finite dimensional $K[G]$-module.*

  i. $X^{**} \cong X$.

  ii. X *is irreducible if and only if* X^* *is.*

  iii. *If V is any $K[G]$-module, then* $\operatorname{Hom}_K(X,V) \cong V \odot_K X^*$.

Proof. (i) We know that, as vector spaces, X and X^{**} are canonically isomorphic via $x \mapsto \sigma(x)$ where $\sigma(x)(\lambda) = \lambda(x)$ for all $\lambda \in X^*$. Furthermore, it is easy to see that σ is a $K[G]$-homomorphism.

(ii) In view of (i), it suffices to show that if X is not irreducible, then neither is X^*. For this, note that if Y is a $K[G]$-homomorphic image of X, then Y^* is naturally embedded in X^*.

(iii) Define $\sigma: V \odot_K X^* \to \operatorname{Hom}_K(X,V)$ so that $\sigma(v \odot \lambda)(x) = v \cdot \lambda(x)$ for all $v \in V$, $\lambda \in X^*$ and $x \in X$. It then follows easily that σ is a $K[G]$-module homomorphism. Furthermore, suppose $\{x_1, x_2, \ldots, x_n\}$ is a K-basis for X and let $\{x_1^*, x_2^*, \ldots, x_n^*\}$ be its dual basis in X^*. Then we define $\tau: \operatorname{Hom}_K(X,V) \to V \odot_K X^*$ by $\tau(f) = \sum_{i=1}^n f(x_i) \odot x_i^*$ for all $f \in \operatorname{Hom}_K(X,V)$. Since $\tau = \sigma^{-1}$, it follows that σ is a $K[G]$-isomorphism. ∎

As above, if V is a $K[G]$-module then we let

$$V^G = \{\, v \in V \mid vg = v \text{ for all } g \in G \,\}.$$

In addition, the map $\mathrm{tr}_G \colon V \to V^G$ is given by $\mathrm{tr}_G(v) = \sum_{g \in G} vg$.

The following is an immediate consequence of Lemmas 39.4(ii) and 39.5(ii)(iii).

Proposition 39.6. [115] *Let V be a $K[G]$-module with G a finite group. Then V is almost injective if and only if, for all irreducible $K[G]$-modules X,*

$$(V \odot X)^G / \mathrm{tr}_G(V \odot X)$$

is finite dimensional over K.

Recall that RG is said to be a K-algebra skew group ring if $K \subseteq \mathbf{Z}(R)$ and G acts as K-algebra automorphism on R. It then follows that $RG \supseteq K[G]$. Furthermore, $\mathrm{tr}_G(R)$ is an ideal of the fixed ring R^G and R^G is a K-subalgebra of R. In the following we assume that R_{R^G} is a Noetherian module. Since $R^G \subseteq R$, this clearly means that R^G is a right Noetherian ring and that R is finitely generated as a right R^G-module.

If S is any right Noetherian ring then, as in the case of $K[G]$, we let $\mathbf{K_0^c}(S) \subseteq \mathbf{G_0}(S)$ denote the image of the Cartan map c and we set $\bar{\mathbf{G}}_0(S) = \mathbf{G_0}(S)/\mathbf{K_0^c}(S)$.

Theorem 39.7. [115] *Let RG be a K-algebra skew group ring with G finite. Assume that R is a Noetherian right R^G-module and that $R^G/\mathrm{tr}_G(R)$ is finite dimensional over K. Furthermore, let $S \supseteq RG$ be an extension of rings with S right Noetherian and finitely generated as a right RG-module.*

i. If V is a finitely generated S-module, then $V_{|K[G]}$ is an almost injective $K[G]$-module.

ii. There is a homomorphism $\theta \colon \mathbf{G_0}(S) \to \bar{\mathbf{G}}_0(K[G])$ given by

$$\theta \colon [V] \mapsto -[\mathrm{E}(V')/V'] + \mathbf{K_0^c}(K[G])$$

where $V' = V_{|K[G]}$.

iii. If in addition $S_{K[G]}$ is projective, then θ factors through $\bar{\mathbf{G}}_0(S)$ so there exists a homomorphism $\bar{\theta}: \bar{\mathbf{G}}_0(S) \to \bar{\mathbf{G}}_0(K[G])$ given by

$$\bar{\theta}: [V] + \mathbf{K}_0^c(S) \mapsto -[E(V')/V'] + \mathbf{K}_0^c(K[G])$$

where $V' = V_{|K[G]}$.

Proof. (i) We start with a few simple observations. Let V be any RG-module. Then it is easy to see that V^G and $\mathrm{tr}_G(V)$ are both R^G-submodules of V. Furthermore, if $v \in V^G$ and $r \in R$, we have

$$\begin{aligned} v \, \mathrm{tr}_G(r) &= \sum_{g \in G} v r^g = \sum_{g \in G} v g^{-1} \cdot rg \\ &= \sum_{g \in G} vr \cdot g = \mathrm{tr}_G(vr). \end{aligned}$$

Thus $V^G \cdot \mathrm{tr}_G(R) \subseteq \mathrm{tr}_G(V)$ and we conclude that $V^G/\mathrm{tr}_G(V)$ is a module for the ring $R/\mathrm{tr}_G(R)$.

Now assume in addition that V is finitely generated. Since G is finite and R_{RG} is a Noetherian module, it follows in turn that $V_{|R}$ and $V_{|R^G}$ are also finitely generated. Moreover, since R^G is a Noetherian ring, we conclude that $V^G \subseteq V_{|R^G}$ is finitely generated and hence, by the above, so is the $R^G/\mathrm{tr}_G(R)$-module $V^G/\mathrm{tr}_G(V)$. But, by assumption, the latter ring is a finite dimensional K-algebra and therefore $V^G/\mathrm{tr}_G(V)$ is finite dimensional over K.

Furthermore, if X is any irreducible $K[G]$-module, then $V \odot_K X$ is also finitely generated since, as an R-module, it is a direct sum of $\dim_K X < \infty$ copies of V. Thus, by the above, $(V \odot X)^G/\mathrm{tr}_G(V \odot X)$ is finite dimensional over K for all such X, and we conclude from the previous proposition that $V_{|K[G]}$ is a.i.

Finally note that S_{RG} is finitely generated. Thus if W is any finitely generated S-module, then $V = W_{|RG}$ is also finitely generated and the above implies that $W_{|K[G]} = V_{|K[G]}$ is almost injective.

(ii) If $0 \to A \to B \to C \to 0$ is a short exact sequence of finitely generated S-modules, then $0 \to A' \to B' \to C' \to 0$ is also exact, where $'$ denotes the restriction to $K[G]$. But these modules are almost injective, so Proposition 39.3(ii) implies that

$\theta(B') = \theta(A') + \theta(C')$. It therefore follows from Lemma 33.1(i) that $[A] \mapsto \theta(A')$ determines a homomorphism from $\mathbf{G_0}(S)$ to $\bar{\mathbf{G}}_0(K[G])$.

(iii) Now suppose in addition that $S_{K[G]}$ is projective. If V is any projective S-module, then it follows that $V' = V_{|K[G]}$ is projective and hence injective. Thus $\theta(V') = 0$, so $[V] \mapsto 0$ and θ factors through $\bar{\mathbf{G}}_0(S)$. ∎

To use the above, we need some information about $\mathbf{G_0}(K[G])$. Suppose first that $\operatorname{char} K = 0$. Then $K[G]$ is a semisimple Artinian ring, so every $K[G]$-module is projective and $\bar{\mathbf{G}}_0(K[G]) = 0$. On the other hand, we have

Lemma 39.8. *Let* $\operatorname{char} K = p > 0$ *and let* G *be a* p-group.

  i. $\mathbf{G_0}(K[G])$ *is the infinite cyclic group generated by* $[V_1]$, *where* V_1 *is the principal* $K[G]$-module.

  ii. $\bar{\mathbf{G}}_0(K[G]) \cong Z/(|G| \cdot Z)$.

Proof. As we observed earlier, V_1 is the unique irreducible $K[G]$-module and thus (i) follows from Lemma 33.1(ii). Furthermore, since the augmentation ideal of $K[G]$ is nilpotent, the regular representation of $K[G]$ is indecomposable. Thus $\mathbf{K}_0^c(K[G])$ is the subgroup of $\mathbf{G_0}(K[G])$ generated by $[K[G]] = |G| \cdot [V_1]$ and (ii) is proved. ∎

As a first application, we consider a simple but interesting example. Suppose that $\Gamma = \langle x, y \mid y^{-1}xy = x^{-1}, y^2 = 1 \rangle$ is the infinite dihedral group and that $S = K[\Gamma]$ is a characteristic 2 group algebra. Now Γ has two conjugacy classes of maximal finite subgroups, and we can take $G_1 = \langle y \rangle$ and $G_2 = \langle xy \rangle$ as representatives of these classes. If W_i denotes the principal $K[G_i]$-module then, by the previous lemma, $\mathbf{G_0}(K[G_i])$ is generated by $[W_i]$ and $2[W_i] = [K[G_i]]$. Furthermore, by Theorem 36.9, $\mathbf{G_0}(S)$ is generated by the images of $\mathbf{G_0}(K[G_i])$ for $i = 1, 2$ under the induced module map. Thus setting $\alpha_i = [W_i^{|S}] \in \mathbf{G_0}(S)$, we see that $\mathbf{G_0}(S)$ is generated by α_1 and α_2. But these are not independent. Indeed, since $2[W_i] = [K[G_i]]$ in $\mathbf{G_0}(K[G_i])$, we have

$$2\alpha_i = 2[W_i^{|S}] = [K[G_i]^{|S}] = [S]$$

so putting $\beta = \alpha_1 - \alpha_2$ yields $\mathbf{G_0}(S) = \langle \alpha_1, \beta \rangle$ with $2\alpha_1 = [S]$ and $2\beta = 0$. The goal is to show that α_1 has infinite order and that $\beta \neq 0$ so $\mathbf{G_0}(S) \cong Z \oplus (Z/2Z)$.

To start with, let $R = K[\langle x \rangle] \subseteq S$ so that S is the skew group ring $S = RG_i$ for $i = 1, 2$. If ρ_R denotes the reduced rank as an R-module, then $\rho_R \colon \mathbf{G_0}(S) \to Z$ and $\rho_R([S]) = 2$. Thus $\rho_R(\alpha_1) = 1$ and α_1 certainly has infinite order. Next we consider β. Since $W_1 \cong (1 + y)K[G_1]$ and $W_2 \cong (1 + xy)K[G_2]$, it follows that

$$W_1|^S \oplus W_2|^S \cong (1 + y)K[\Gamma] \oplus (1 + xy)K[\Gamma].$$

On the other hand, since y and xy generate Γ and char $K = 2$, the augmentation ideal I of $K[\Gamma]$ is given by

$$I = (1 + y)K[\Gamma] + (1 + xy)K[\Gamma]$$

and this is easily seen to be a direct sum (Exercise 4). Thus $\alpha_1 + \alpha_2 = [I]$ and, if V denotes the principal $K[\Gamma]$-module, then

$$\alpha_1 + \alpha_2 + [V] = [I] + [V] = [S] = 2\alpha_1.$$

In other words, $\beta = \alpha_1 - \alpha_2 = [V]$.

Now R^{G_1} has as a K-basis the set $\{1\} \cup \{ x^i + x^{-i} \mid i > 0 \}$. It then follows easily that $R^{G_1} = K[x + x^{-1}]$ is Noetherian and R is a finitely generated R^{G_1}-module with generating set $\{1, x\}$. In addition, $\dim_K R^{G_1}/\mathrm{tr}_{G_1}(R) = 1$. Thus Theorem 39.7(ii) applies and we have a homomorphism $\theta \colon \mathbf{G_0}(S) \to \bar{\mathbf{G}}_0(K[G_1]) \cong Z/2Z$ as described in that result. Since $\beta = [V]$ and $V_{|K[G_1]} = W_1$ is a finitely generated $K[G_1]$-module, we have

$$\theta(\beta) = \begin{cases} [W_1] + \mathbf{K}_0^c(K[G_1]), & \text{in } \bar{\mathbf{G}}_0(K[G_1]); \\ 1 + 2Z, & \text{in } Z/2Z. \end{cases}$$

Thus $\theta(\beta) \neq 0$, so $\beta \neq 0$ and $\mathbf{G_0}(S) \cong Z \oplus (Z/2Z)$ as claimed above.

By an accident of sorts, we are able to compute $\mathbf{K_0}(K[\Gamma])$ for fields of all characteristics. Indeed, as mentioned in Section 37, $\Gamma \cong Z_2 * Z_2$, the free product of two groups of order 2, and hence $K[\Gamma] \cong K[Z_2] \coprod_K K[Z_2]$. A result of [63] therefore yields

$$\bar{\mathbf{K}}_0(K[\Gamma]) \cong \bar{\mathbf{K}}_0(K[Z_2]) \oplus \bar{\mathbf{K}}_0(K[Z_2])$$

where, for any ring S, $\bar{\mathbf{K}}_0(S)$ is equal to $\mathbf{K}_0(S)$ modulo its cyclic subgroup generated by $[S]$. It then follows that

$$\mathbf{K}_0(K[\Gamma]) \cong \begin{cases} Z, & \text{if char } K = 2; \\ Z \oplus Z \oplus Z, & \text{if char } K \neq 2. \end{cases}$$

In fact, when char $K = 2$, [17] proves that all projective $K[\Gamma]$-modules are free.

Finally, if char $K \neq 2$, then the group ring $K[\Gamma]$ has finite global dimension, and hence $\mathbf{G}_0(K[\Gamma]) \cong \mathbf{K}_0(K[\Gamma])$ by Proposition 33.8. Thus we conclude that

$$\mathbf{G}_0(K[\Gamma]) \cong \begin{cases} Z \oplus (Z/2Z), & \text{if char } K = 2; \\ Z \oplus Z \oplus Z, & \text{if char } K \neq 2. \end{cases}$$

Notice, in particular, that $\mathbf{G}_0(K[\Gamma]) \not\cong \mathbf{K}_0(K[\Gamma])$ in the characteristic 2 case.

EXERCISES

1. Let R be a ring and let V be a right R-module. *Baer's criterion* asserts that V is injective if and only if every R-homomorphism $\sigma: I \to V$, with I a right ideal of R, extends to a map $\sigma^\star: R \to V$. Prove this.

2. Show that a finite direct sum of injective modules is injective. If R is Noetherian, use Baer's criterion to prove that an arbitrary direct sum of injectives is injective.

3. Let $R \subseteq S$ be rings and assume that $_RS$ is flat. If I is a right ideal of R, prove that the induced module satisfies $I \otimes_R S \cong IS \subseteq S$.

4. Let $\Gamma = \langle x, y \mid y^{-1}xy = x^{-1}, y^2 = 1 \rangle$ and let K be any field. If I is the augmentation ideal of $K[\Gamma]$, prove that

$$I = (1 - y)K[\Gamma] \oplus (1 - xy)K[\Gamma].$$

To show that the sum is direct, first observe that any element of $(1 - y)K[\Gamma]$ is uniquely writable as $\sum_i k_i(1 - y)x^i$ with $k_i \in K$. Then observe that any element of $(1 - xy)K[\Gamma]$ is annihilated on the left by $1 + xy$.

5. Let $R = K[\langle x \rangle] \subseteq K[\Gamma]$ and let $G_1 = \langle y \rangle$. Show that R^{G_1} is the polynomial ring $K[x + x^{-1}]$ and that $R = R^{G_1} \oplus xR^{G_1}$ is a free R^{G_1}-module on two generators.

6. Suppose that the finite group G acts on R. If R is right Noetherian and $\mathrm{tr}_G(R) = R^G$, prove that R^G is Noetherian. For this, note that if A is a right ideal of R^G, then $\mathrm{tr}_G(AR) = A\,\mathrm{tr}_G(R) = A$.

7. A ring R satisfies $\mathrm{Kdim}\,R \leq 1$ if and only if, for every nonzero right ideal I, R/I is an Artinian R-module. Now let the finite group G act on R and suppose that R^G is right Noetherian and $\mathrm{Kdim}\,R \leq 1$. If $R^G/\mathrm{tr}_G(R)$ is a right Artinian ring, prove that $\mathrm{Kdim}\,R^G \leq 1$. This is a result of [**77**].

40. Stably Free Modules

We close this book with two topics of interest. The first is concerned with $\mathbf{G_0}$ computations. Specifically, we continue the work of the preceding section and explicitly determine the Grothendieck group of the Zalesskii-Neroslavskii example. The second is related to $\mathbf{K_0}$ and concerns the existence of stably free modules which are not free.

To start with, suppose G acts as automorphisms on the group N. Then we say that the action is *fixed point free* if, for all $1 \neq g \in G$, we have $\mathbf{C}_N(g) = \langle 1 \rangle$. In the case of finite groups, this turns out to be an extremely important concept. Indeed if both G and N are nontrivial, then N is called a *Frobenius kernel*, G is a *Frobenius complement* and $N \rtimes G$ is a *Frobenius group*. Furthermore, it follows that N is necessarily nilpotent and that the structure of G is fairly tight. For example, the Sylow subgroups of G are either cyclic or quaternion and G is almost always solvable. A detailed discussion of such groups can be found in the book [**156**]. In the case of infinite groups, N need not be nilpotent. For example, the infinite dihedral group admits a fixed point free automorphism group of order 2 (see Exercise 1). We will not use any of these facts here; we just require an understanding of the notation.

The following result can be slightly generalized. However, its statement is already quite tedious.

Theorem 40.1. [115] *Let $K^t[\Gamma]$ be a twisted group algebra with Γ polycyclic-by-finite and with $\operatorname{char} K = p > 0$. Furthermore, let N be a torsion free normal subgroup of finite index in Γ, set $R = K^t[N]$ and let $G_1, G_2, \ldots, G_k, H_1, H_2, \ldots, H_m$ be representatives of the conjugacy classes of maximal finite subgroups of Γ. Assume that, for all $1 \le i \le k$ and $1 \le j \le m$, we have*

    i. *G_i is a nonidentity p-group which acts in a fixed point free manner on N, $K^t[G_i] \cong K[G_i]$ and R is a Noetherian right module over the fixed ring R^{G_i},*

    ii. *$K^t[H_j]$ is a division ring.*

Then

$$\bar{\mathbf{G}}_0(K^t[\Gamma]) \cong \oplus \sum_{i=1}^{k} Z/(|G_i| \cdot Z)$$

and

$$\mathbf{G}_0(K^t[\Gamma]) \cong Z \oplus \sum_{i=2}^{k} Z/(|G_i| \cdot Z)$$

where $|G_1| = \max \left\{ \, |G_i| \ \big| \ 1 \le i \le k \, \right\}$.

Proof. By Theorem 36.9, the induced module map determines an epimorphism

$$\oplus \sum_{i=1}^{k} \mathbf{G}_0(K^t[G_i]) \oplus \sum_{j=1}^{m} \mathbf{G}_0(K^t[H_j]) \to \mathbf{G}_0(K^t[\Gamma]).$$

Since $K^t[H_j]$ is a division ring, by assumption, $\mathbf{G}_0(K^t[H_j])$ is generated by $\left[K^t[H_j] \right]$ and hence its image in $\mathbf{G}_0(K^t[\Gamma])$ is generated by $\left[K^t[\Gamma] \right]$. In particular, if $k = 0$ then $\mathbf{G}_0(K^t[\Gamma])$ is the cyclic group generated by $\left[K^t[\Gamma] \right]$ and the result follows easily. Thus, for the remainder of this argument, we assume that $k \ge 1$. Since the image of $\mathbf{G}_0(K^t[G_1])$ also contains $\left[K^t[\Gamma] \right]$, it is now clear that the induced module map yields an epimorphism $\xi \colon \mathcal{A} \to \mathbf{G}_0(K^t[\Gamma])$, where $\mathcal{A} = \oplus \sum_{i=1}^{k} \mathbf{G}_0(K^t[G_i])$. Furthermore, since projective modules induce to projective modules, we also obtain an epimorphism $\bar{\xi} \colon \bar{\mathcal{A}} \to \bar{\mathbf{G}}_0(K^t[\Gamma])$ where $\bar{\mathcal{A}} = \oplus \sum_{i=1}^{k} \bar{\mathbf{G}}_0(K^t[G_i])$.

Note that, by (i) above, $K^t[G_i] \cong K[G_i]$ and G_i is a finite p-group. Thus, by Lemma 39.8, $\mathbf{G}_0(K^t[G_i])$ is the infinite cyclic group generated by $[W_i]$, where W_i is the principal module, and $\bar{\mathbf{G}}_0(K^t[G_i]) \cong Z/(|G_i| \cdot Z)$. In particular,

$$\bar{\mathcal{A}} \cong \oplus \sum_{i=1}^{k} Z/(|G_i| \cdot Z)$$

and the goal is to prove that $\bar{\xi}$ is an isomorphism. We of course already know that it is surjective.

To show that $\bar{\xi}$ is injective, we need a back map of sorts and for this we use Theorem 39.7(iii). To start with, since $K^t[G_i] \cong K[G_i]$, we see that $K^t[NG_i] = RG_i$ is a skew group ring of G_i over $R = K^t[N]$. In addition, $S = K^t[\Gamma]$ is a Noetherian K-algebra which is finitely generated as a right RG_i-module. It remains to consider the action of G_i on R and, by assumption, we know that R is a Noetherian module over the fixed ring R^{G_i}. Furthermore, the action of G_i on N is fixed point free. Thus, since $R = \oplus \sum_{y \in N} K\bar{y}$ and G_i permutes the summands with $y \neq 1$ in orbits of full size $|G_i|$, it follows easily that $R^{G_i} = K + \operatorname{tr}_{G_i}(R)$. Thus the hypothesis of Theorem 39.7 is satisfied and we conclude that there exists a map

$$\bar{\theta}_i \colon \bar{\mathbf{G}}_0(K^t[\Gamma]) \to \bar{\mathbf{G}}_0(K^t[G_i])$$

as described in that result. Hence, by combining these maps, we obtain a group homomorphism

$$\bar{\eta} \colon \bar{\mathbf{G}}_0(K^t[\Gamma]) \to \oplus \sum_{i=1}^{k} \bar{\mathbf{G}}_0(K^t[G_i]) = \bar{\mathcal{A}}.$$

It turns out that $\bar{\eta}$ is not quite the inverse of $\bar{\xi}$, but it is close enough.

To understand the combined map $\bar{\eta}\bar{\xi} \colon \bar{\mathcal{A}} \to \bar{\mathcal{A}}$, we must first observe that each G_i is self-normalizing. To start with, since N is torsion free, we have $N \cap G_i = \langle 1 \rangle$. Next since $N \triangleleft \Gamma$, the commutator group $[\mathbf{N}_N(G_i), G_i]$ satisfies

$$[\mathbf{N}_N(G_i), G_i] \subseteq N \cap G_i = \langle 1 \rangle.$$

Thus G_i centralizes $\mathbf{N}_N(G_i)$ and hence $\mathbf{N}_N(G_i) = \langle 1 \rangle$ since G_i acts in a fixed point free manner on N. Finally, $\mathbf{N}_\Gamma(G_i) \cap N = \mathbf{N}_N(G_i) = \langle 1 \rangle$

and $|\Gamma : N| < \infty$, so it follows that $\mathbf{N}_\Gamma(G_i)$ is finite. Thus, since $\mathbf{N}_\Gamma(G_i) \supseteq G_i$ and G_i is a maximal finite subgroup, we conclude that $\mathbf{N}_\Gamma(G_i) = G_i$ as required.

With this, we can use the Mackey decomposition to compute

$$W_{i,j} = (W_i^{|K^t[\Gamma]})_{|K^t[G_j]}.$$

To be precise, let $\mathcal{D}$ be a complete set of (G_i, G_j)-double coset representatives in Γ. Then, by Lemma 3.10, we have $W_{i,j} = \oplus \sum_{d \in \mathcal{D}} V(d)$ where

$$V(d) = \left((W_i \otimes \bar{d})_{|K^t[G_i^d \cap G_j]} \right)^{|K^t[G_j]}$$

and $W_i \otimes \bar{d}$ is a specific module for $K^t[G_i^d]$. In particular, if we let $p^{v(d)} = \dim_K V(d)$ then, since $\dim_K W_i = 1$, we see that

$$p^{v(d)} = \dim_K V(d) = |G_j : G_i^d \cap G_j| \le |G_j|.$$

Now $W_{i,j}$ is almost injective by Theorem 39.7(i) and

$$\mathrm{E}(W_{i,j}) = \oplus \sum_{d \in \mathcal{D}} \mathrm{E}\big(V(d)\big).$$

Thus it follows that almost all of the modules $V(d)$ are injective and therefore have dimension divisible by $|G_j|$ and hence by p. Since any $V(d)$ has $\dim_K V(d)$ composition factors all equal to W_j, it then follows from all of the above that

$$\theta_j(W_{i,j}) = a_{i,j}[W_j] + \mathbf{K}_0^c(K^t[G_j])$$

where $a_{i,j} \in Z$ is congruent modulo p to the number of $d \in \mathcal{D}$ with $v(d) = 0$.

Notice that $v(d) = 0$ if and only if $G_i^d \cap G_j = G_j$ and hence, since G_j is maximal, if and only if $G_i^d = G_j$. In particular, if $i \ne j$ then $a_{i,j} \equiv 0 \bmod p$ since G_i and G_j are not conjugate in Γ. On the other hand, if $i = j$ then we must have $d \in \mathbf{N}_\Gamma(G_i) = G_i$ and this corresponds to a unique (G_i, G_i)-double coset. In other words, $a_{i,i} \equiv 1 \bmod p$. Now, by Lemma 39.8, $\bar{\mathbf{G}}_0(K^t[G_i])$ is generated by

$w_i = [W_i] + \mathbf{K}_0^{\mathrm{c}}(K^t[G_i])$ and, viewing this as an element of $\bar{\mathcal{A}}$, we conclude that

$$\bar{\eta}\bar{\xi}(w_i) = \sum_{j=1}^{k} a_{i,j}w_j$$

$$\equiv w_i \mod p\bar{\mathcal{A}}.$$

In other words, $\bar{\eta}\bar{\xi}$ induces the identity map on $\bar{\mathcal{A}}/p\bar{\mathcal{A}}$ and, since $\bar{\mathcal{A}}$ is a finite additive p-group, this implies that $\bar{\eta}\bar{\xi}$ is an automorphism of $\bar{\mathcal{A}}$. Hence $\bar{\xi}$ is injective. We have therefore shown that $\bar{\xi}\colon \bar{\mathcal{A}} \to \bar{\mathbf{G}}_0(K^t[\Gamma])$ is an isomorphism so

$$\bar{\mathbf{G}}_0(K^t[\Gamma]) \cong \oplus \sum_{i=1}^{k} Z/(|G_i| \cdot Z)$$

as required.

Finally, since ξ is an epimorphism and $\mathbf{G}_0(K^t[G_i])$ is generated by $[W_i]$, we see that $\mathbf{G}_0(K^t[\Gamma])$ is generated by the elements $\alpha_i = [W_i^{|K^t[\Gamma]}]$. Furthermore, $|G_i| \cdot [W_i] = [K^t[G_i]]$ in $\mathbf{G}_0(K^t[G_i])$, so $|G_i|\alpha_i = [K^t[\Gamma]]$ in $\mathbf{G}_0(K^t[\Gamma])$. In particular, if $|G_1| \geq |G_i|$ for all i, then $\beta_i = \alpha_i - |G_1| \cdot |G_i|^{-1}\alpha_1 \in \mathbf{G}_0(K^t[\Gamma])$, $|G_i|\beta_i = 0$ and $\mathbf{G}_0(K^t[\Gamma])$ is generated by $\alpha_1, \beta_2, \beta_3, \ldots, \beta_k$. Now, if ρ_R denotes the reduced rank as an R-module, then $\rho_R\colon \mathbf{G}_0(K^t[\Gamma]) \to Z$ is a homomorphism and $\rho_R(K^t[\Gamma]) \neq 0$. Thus $\rho_R(\alpha_1) \neq 0$ and α_1 has infinite order. On the other hand, by mapping these generators to $\bar{\mathbf{G}}_0(K^t[\Gamma])$, we see that β_i has order precisely $|G_i|$ and that $\beta_2, \beta_3, \ldots, \beta_k$ are otherwise independent. Thus

$$\mathbf{G}_0(K^t[\Gamma]) \cong Z \oplus \sum_{i=2}^{k} Z/(|G_i| \cdot Z)$$

and the result follows. ∎

Recall that the Zalesskii-Neroslavskii example, as described in Theorem 38.8, depends on the field K of characteristic 2 and the element $\lambda \in K^{\bullet}$ of infinite multiplicative order. As a consequence of the above, we have

Corollary 40.2. [115] *If S is the Zalesskii-Neroslavskii example determined by K and λ, then*

$$\bar{G}_0(S) \cong \begin{cases} (Z/2Z)^3, & \text{if } \lambda \notin (K^\bullet)^2; \\ (Z/2Z)^4, & \text{if } \lambda \in (K^\bullet)^2. \end{cases}$$

Furthermore,

$$G_0(S) \cong \begin{cases} Z \oplus (Z/2Z)^2, & \text{if } \lambda \notin (K^\bullet)^2; \\ Z \oplus (Z/2Z)^3, & \text{if } \lambda \in (K^\bullet)^2. \end{cases}$$

Proof. We know that $S = K^t[\Gamma]$ where

$$\Gamma = \langle a, b, c \mid ab = ba, \ c^2 = 1, \ a^c = a^{-1}, \ b^c = b^{-1} \rangle$$
$$\cong A \rtimes Z_2$$

Here $A = \langle a, b \rangle$ is a free abelian group of rank 2 and $Z_2 = \langle c \rangle$. Note that, by Lemma 36.6(iv),

$$\mathrm{H}^1(\langle c \rangle, A) \cong \mathrm{ann}_A(1 + c)/A^{c-1} = A/A^2$$

and this group has order 4. Thus Γ has four conjugacy classes of maximal finite subgroups, all of order 2, and we can take as representatives of these classes the subgroups

$$G_1 = \langle c \rangle, \quad G_2 = \langle ac \rangle, \quad G_3 = \langle bc \rangle, \quad G_4 = \langle abc \rangle.$$

In $K^t[\Gamma]$ we have $\bar{a}\bar{b} = \lambda\bar{b}\bar{a}$, $\bar{c}^{-1}\bar{a}\bar{c} = \bar{a}^{-1}$, $\bar{c}^{-1}\bar{b}\bar{c} = \bar{b}^{-1}$ and $\bar{c}^2 = 1$. Thus $\bar{c}$, $\bar{a}\bar{c}$ and $\bar{b}\bar{c}$ all have square 1, so $K^t[G_i] = K[G_i]$ for $i = 1, 2, 3$. On the other hand,

$$(\bar{a}\bar{b}\bar{c})^2 = \bar{a}\bar{b}\bar{a}^{-1}\bar{b}^{-1} = \lambda$$

and hence

$$K^t[G_4] \cong \begin{cases} \text{a field}, & \text{if } \lambda \notin (K^\bullet)^2; \\ K[G_4], & \text{if } \lambda \in (K^\bullet)^2. \end{cases}$$

Finally observe that each G_i acts in a fixed point free manner on A and that $R = K^t[A]$ is a Noetherian module over the fixed ring R^{G_i}

by Proposition 38.11. Thus Theorem 40.1 applies and, in view of the dual nature of $K^t[G_4]$, it yields the appropriate formulas for $\bar{\mathbf{G}}_\mathbf{0}(S)$ and $\mathbf{G}_\mathbf{0}(S)$. ∎

We mention just one more consequence, this time for ordinary group algebras.

Corollary 40.3. [115] *Let $A \lhd \Gamma$ with A a finitely generated free abelian group and with Γ/A a finite p-group acting in a fixed point free manner on A. If $G_1, G_2, \ldots, G_k$ are representatives of the conjugacy classes of the maximal finite subgroups of Γ and if K is a field of characteristic p, then*

$$\bar{\mathbf{G}}_\mathbf{0}(K[\Gamma]) \cong \oplus \sum_{i=1}^{k} Z/(|G_i| \cdot Z)$$

and

$$\mathbf{G}_\mathbf{0}(K[\Gamma]) \cong Z \oplus \sum_{i=2}^{k} Z/(|G_i| \cdot Z)$$

where $|G_1| = \max \big\{ |G_i| \mid 1 \leq i \leq k \big\}$.

Proof. Certainly $G_i \cap A = \langle 1 \rangle$, so G_i is a finite p-group which acts in a fixed point free manner on A. Furthermore, since $R = K[A]$ is a finitely generated commutative K-algebra, it follows (see Exercise 3) that R is a Noetherian module over the fixed ring R^{G_i}. Thus Theorem 40.1 yields the result. Note that if some $G_i = \langle 1 \rangle$, then we must have $k = 1$ and $G_1 = \langle 1 \rangle$. In this case, just let $H_1 = \langle 1 \rangle$ in Theorem 40.1. ∎

Now we move on to our last topic. Let S be a right Noetherian ring and let P be a finitely generated projective S-module. We recall that P is stably free if $P \oplus S^n \cong S^m$ for some $m, n \geq 0$ and that, by Lemma 34.10, this occurs if and only if $[P]$ is in the cyclic subgroup of $\mathbf{K}_\mathbf{0}(S)$ generated by $[S]$. Of course, any free module is stably free. The question then is whether stably free, nonfree modules exist. The answer is "yes" for large classes of noncommutative rings and in particular for group rings. A slick approach to this problem is contained in [**197**] and we offer a small sample of the results of that interesting paper.

We start with two key special cases. The first is a generalized polynomial ring or *Ore extension* given by $S = R[x; \sigma, \delta]$. Specifically, this ring has the additive structure of the ordinary polynomial ring over R in one variable, but with multiplication given by $rx = xr^\sigma + \delta(r)$ for all $r \in R$. Here $\sigma \in \mathrm{Aut}(R)$ and δ is a σ-derivation. Second, we consider the skew group ring $S = R[x, x^{-1}; \sigma] = R\langle x\rangle$ of the infinite cyclic group $\langle x \rangle$ with $rx = xr^\sigma$. Notice that, in either case, if R is a domain then so is S and if R is right Noetherian then so also is S.

A polynomial $f(x) = \sum_{i=0}^{n} a_i x^i \in R[x; \sigma, \delta]$ is said to be *monic* if $a_n = 1$. Notice that this is independent of the side on which we write the coefficients since $1^\sigma = 1$. Similarly, $\deg f = n$ if $a_n \neq 0$ and this is also right-left symmetric.

Lemma 40.4. *Let R be a right Noetherian ring. Then the set of monic polynomials in $S = R[x; \sigma, \delta]$ is a right divisor set.*

Proof. The set of monic polynomials is certainly multiplicatively closed. Now let $f(x)$ be a monic polynomial in S. Then by long division, it follows that S/fS is generated by the images of all polynomials of degree less than that of f. In other words, S/fS is finitely generated and hence Noetherian as a right R-module. Thus if $g(x) \in S$, then the R-module

$$\left(fS + \sum_{i=0}^{\infty} gx^i R \right) / fS$$

is finitely generated, say by the images of $gx^0, gx^1, \ldots, gx^{n-1}$. Then $gx^n \in fS + \sum_{i=0}^{n-1} gx^i R$ and we conclude that $gh \in fS$ where $h(x)$ is a monic polynomial of degree n. ∎

Notice that if R is a division ring, then both $S = R[x; \sigma, \delta]$ and $S = R\langle x \rangle$ are principal right ideal rings and hence every right ideal is a free S-module. Thus if S is to have a nonfree ideal, then R must certainly contain a nonunit $r \neq 0$.

Theorem 40.5. [197] *Let R be a right Noetherian domain and let $S = R[x; \sigma, \delta]$ or $R[x, x^{-1}; \sigma]$. Suppose that there exists a nonunit $a \in R$ and some $b \in R$ with $S = aS + (x + b)S$. Furthermore, if S is the skew group ring, assume that $ba^{\sigma} \notin aR$. Then*

$$I = \{\, g \in S \mid ag \in (x + b)S \,\}$$

is a nonfree, stably free right ideal of S satisfying $I \oplus S \cong S \oplus S$.

Proof. The proofs of the two cases are quite parallel, but there are differences which have to be considered. Note that if S is the skew group ring $R[x, x^{-1}; \sigma]$, then the subring of S generated by R and x is $R[x; \sigma, \delta]$ with $\delta \equiv 0$. For obvious reasons, we call the elements of this subring the polynomials in S. Furthermore, since $\delta = 0$ in this case, the constant term of a polynomial is right-left symmetric.

Now let S be either ring. Then the hypothesis of the theorem guarantees that $a \neq 0$ and that S is a Noetherian domain. Furthermore, in the skew group ring case, $ba^{\sigma} \notin aR$ implies that $b \neq 0$. Consider the natural epimorphism

$$\varphi \colon aS \oplus (x + b)S \rightarrow aS + (x + b)S = S$$

and observe that, for each $g \in I$, there exists a unique $g^* \in S$ with $ag = (x + b)(-g^*)$. Then $ag \oplus (x + b)g^* \in \mathrm{Ker}(\varphi)$ and in this way we obtain a map from I to $\mathrm{Ker}(\varphi)$ which is easily seen to be an isomorphism. Since $aS_S \cong S_S \cong (x + b)S_S$, this yields a short exact sequence $0 \rightarrow I \rightarrow S \oplus S \rightarrow S \rightarrow 0$ which of course splits since S is projective. Thus $I \oplus S \cong S \oplus S$, so I is stably free, and it remains to show that I is not free. Since S is a Noetherian domain, this is equivalent to showing that I is not cyclic.

We first find some elements in I. To start with, we have

$$ax = xa^{\sigma} + \delta(a) = (x + b)a^{\sigma} + (\delta(a) - ba^{\sigma})$$

where again $\delta = 0$ in the skew group ring case. Since R is an Ore domain, there exist $r_1, r_2 \in R$ with $r_1 \neq 0$ and $(\delta(a) - ba^{\sigma})r_1 = ar_2$. This yields $a(xr_1 - r_2) = (x + b)a^{\sigma}r_1$ and hence $xr_1 - r_2 \in I$. Next, since R is right Noetherian, we can apply Lemma 40.4 to either

$R[x; \sigma, \delta]$ or to the polynomial subring of $R[x, x^{-1}; \sigma]$. In either case, since $x + b$ is monic, there exists a monic polynomial $f(x) \in S$ with $af \in (x + b)S$. Therefore $f \in I$ and I contains a monic polynomial.

Suppose, by way of contradiction, that $I = h(x)S$ is cyclic. We use the existence of the above elements of I to suitably narrow the possibilities for h. To start with, if $S = R\langle x \rangle$, then we can multiply h by an appropriate power of the unit x to make h a polynomial with nonzero constant term. Thus in either case, h is a polynomial. Next, since $xr_1 + r_2 \in I$ with $r_1 \neq 0$, it is clear that $\deg h(x) \leq 1$ at least if $S = R[x; \sigma, \delta]$. On the other hand, if $S = R[x, x^{-1}; \sigma]$ and $h(x)k(x) = xr_1 + r_2$ then, since h has a nonzero constant term, it follows that $k(x)$ is also a polynomial and again we have $\deg h(x) \leq 1$. Thus $h(x) = x\lambda + \mu$ for some $\lambda, \mu \in R$. Furthermore $h \in I$ implies that $ah(x) \in (x + b)S$ and hence $\lambda \neq 0$ since $b \neq 0$ when S is a skew group ring. Finally since $I = hS$ contains a monic polynomial, it follows easily that λ is a unit of R. Multiplying by λ^{-1}, we can therefore assume that $h(x) = x + \mu$.

Again $h(x) \in I$ implies that $ah(x) \in (x + b)S$ so

$$ah(x) = a(x + \mu) = xa^{\sigma} + \delta(a) + a\mu = (x + b)t \qquad (*)$$

for some $t \in S$. We can now complete the proof, handling the two cases separately. If S is the skew group ring then, since $b \neq 0$, $(*)$ implies first that t is a polynomial and then that $t \in R$. Thus, comparing coefficients yields $t = a^{\sigma}$ and then $ba^{\sigma} = bt = a\mu \in aR$, a contradiction. On the other hand, if S is the Ore extension, then we use $S = aS + (x+b)S$ to find elements $u, v \in S$ with $1 = au + (x+b)v$. In addition, by long division, $u = (x + \mu)w + c$ for some $w \in S$ and $c \in R$. Thus $(*)$ yields

$$1 = au + (x + b)v = a(x + \mu)w + (x + b)v + ac$$
$$= (x + b)(tw + v) + ac$$

and, by computing degrees, we have $tw + v = 0$ and then $ac = 1$. But R is a domain, so this implies that a is a unit of R, a contradiction, and the result follows. ∎

This yields a uniform and fairly simple approach to a number of diverse constructions. For example, we have

Corollary 40.6. [149] *Let D be a Noetherian domain containing elements α, β such that $[\alpha, \beta] = \alpha\beta - \beta\alpha = \gamma$ is a unit of D. If $S = D[x, y]$ is the polynomial ring over D in two variables, then S has a nonfree, stably free right ideal.*

Proof. Let $R = D[y]$ so that $S = R[x]$. Since $[\alpha, \beta] = \gamma$, we see that $[y + \alpha, x + \beta] = \gamma$ and γ is a unit of S. Thus since $y + \alpha$ is not a unit of R, the result follows from Theorem 40.5 with $a = y + \alpha$ and $b = \beta$. ∎

Note that the existence of α, β above is certainly guaranteed if D has a noncommutative division subring. Thus, if E is a noncommutative skew field, then the polynomial ring $S = E[x_1, x_2, \ldots, x_n]$ in $n \geq 2$ variables has a nonfree, stably free right ideal.

Corollary 40.7. [197] *Let D be a Noetherian domain and let $S = D[x, y \mid xy - yx = 1]$. Then S has a nonfree, stably free right ideal.*

Proof. If $R = D[y]$, then $S = R[x; \delta]$ is clearly an Ore extension. Thus since y is not a unit of R and $[x, y] = 1$, the result follows from Theorem 40.5 with $a = y$ and $b = 0$. ∎

In particular, let D be a Noetherian domain and define the nth *Weyl algebra* $A_n(D)$ inductively by $A_0(D) = D$ and

$$A_n(D) = A_{n-1}(D)[x, y \mid xy - yx = 1].$$

Then $A_n(D)$ is a Noetherian domain and, by the above, it has a nonfree, stably free right ideal if $n \geq 1$ (see [204]). Finally we have

Corollary 40.8. [8] *Let G be a nonabelian poly-Z group and let D be a Noetherian domain. Then the group ring $S = D[G]$ has a nonfree, stably free right ideal.*

Proof. We proceed by induction on the Hirsch number of G. Let

$$\langle 1 \rangle = G_0 \triangleleft G_1 \triangleleft \cdots \triangleleft G_n = G$$

be a subnormal series for G with each G_{i+1}/G_i infinite cyclic and set $H = G_{n-1}$. If $R = D[H]$ and $G = \langle H, x \rangle$, then certainly $S = R[x, x^{-1}; \sigma]$. There are two cases to consider.

Case 1. *H is nonabelian.*

Proof. By induction, $R = D[H]$ has a nonfree, stably free right ideal I. We show that the right ideal IS of S has similar properties. To start with, since S is a free left R-module, the embedding $I \hookrightarrow R$ yields $I \otimes_R S \hookrightarrow R \otimes_R S$. Thus since $R \otimes_R S \cong S$ via the map $r \otimes s \mapsto rs$, we conclude that $I \otimes_R S \cong IS$. In other words $IS \cong I^{|S}$ and, since I_R is stably free, it follows that IS_S is also stably free.

Now suppose, by way of contradiction, that IS is free and hence cyclic; say $IS = f(x)S$ for some $f(x) \in S$. Multiplying f by a suitable power of the unit $x \in S$, we can assume that f is a polynomial with nonzero constant term. Now $f(x)S \supseteq I \neq 0$ and say $f(x)g(x)$ is a nonzero element of $I \subseteq R$. Then by considering the lowest degree term of g, we conclude first that g is a polynomial and then that both f and g have degree 0. In other words, $f, g \in R$ and hence $fR = I$, a contradiction. ∎

Case 2. *H is abelian.*

Proof. Since H is a poly-Z group having Hirsch number $n - 1$, it is free abelian with generators $y_1, y_2, \ldots, y_{n-1}$. Now suppose $r \in R$ commutes with $r^{\sigma^{-1}}$ and observe that

$$(x + r^{\sigma^{-1}})(x - r)x^{-2} + rr^{\sigma^{-1}}x^{-2} = 1.$$

Thus the result will follow from Theorem 40.5 with $a = r$ and $b = r^{\sigma^{-1}}$ provided that $r^{\sigma^{-1}}r^{\sigma} \notin rR$. The goal then is to find such an element r.

Since G is nonabelian, x does not centralize H and hence say $y_1^x \neq y_1$. Suppose first that $y_1^x \neq y_1^{-1}$. Then $y_1^x, y_1^{x^{-1}} \notin \langle y_1 \rangle$ and we set $r = y_1 - 1$. Notice that r does indeed commute with its conjugate $r^{\sigma^{-1}} = y_1^{x^{-1}} - 1$ and that rR is the kernel of the homomorphism from $D[H]$ to the domain $D[H/\langle y_1 \rangle]$. Thus clearly $r^{\sigma^{-1}}r^{\sigma} \notin rR$. On the other hand, if $y_1^x = y_1^{-1}$ then we set $r = 1 + y_1 + y_1^3$. Again r commutes with its conjugate $r^{\sigma^{-1}} = 1 + y_1^{-1} + y_1^{-3}$ and it is easy to see that $r^{\sigma^{-1}} r^\sigma \notin rR$. Indeed if $r^{\sigma^{-1}} r^\sigma = rs$ for some $s \in R$ then, by projecting this equation into $D[\langle y_1 \rangle]$, it follows first that $s \in D[\langle y_1 \rangle]$ and then that s does not exist (see Exercise 4). Thus we have a suitable element r in all cases and the corollary is proved. ∎

Paper [**197**] then goes on to show that enveloping algebras of nonabelian, finite dimensional Lie algebras always have nonfree, stably free right ideals. Some aspects of this are considered in Exercises 5 and 6.

EXERCISES

1. Let $\Gamma = \langle x, y \mid y^{-1}xy = x^{-1}, y^2 = 1 \rangle$ be the infinite dihedral group. Show that $\Lambda = \langle x^2, y \rangle$ is a normal subgroup of Γ which is isomorphic to Γ and that conjugation by xy induces a fixed point free automorphism of order 2 on Λ.

2. Let $\Gamma = A \rtimes \langle x \rangle$ where $A = \langle a, b \rangle$ is free abelian, $x^4 = 1$ and $a^x = b$, $b^x = a^{-1}$. If K is a field of characteristic 2, compute $\bar{\mathbf{G}}_0(K[\Gamma])$ and $\mathbf{G}_0(K[\Gamma])$.

3. Let $R = K[a_1, a_2, \ldots, a_n]$ be a finitely generated commutative K-algebra and let G be a finite group of K-algebra automorphisms of R. Show that R is a Noetherian R^G-module. To this end, let S be the finitely generated K-subalgebra of R generated by the coefficients of the polynomials $f_i(x) = \prod_{g \in G}(x - a_i^g)$ for $i = 1, 2, \ldots, n$. Observe that S is a Noetherian subring of R^G and that R is a finitely generated S-module.

4. If $D[y]$ is any polynomial ring, show that $1 + y + y^3$ cannot divide $(1 + y^2 + y^3)^2$. Use this to deduce the appropriate facts about $r = 1 + y_1 + y_1^3$ in the proof of Corollary 40.8.

5. Let $S = R[x; \sigma, \delta]$ be an Ore extension with R a Noetherian domain and suppose that R contains a nonunit r with $\sum_{i=0}^{\infty} \delta^i(r)R = R$. Show that S has a nonfree, stably free right ideal. For this, observe that $rS + xS$ contains all $\delta^i(r)$.

6. Now let R be a commutative K-algebra freely generated by the finite dimensional vector space L and let δ be a nonzero K-derivation of R which stabilizes L. Use the previous exercise to prove that $S = R[x; \delta]$ has a nonfree, stably free right ideal. To start with, choose $a \in L$ with $\delta(a) \neq 0$ and define $a_i = \delta^i(a)$. Since L is finite dimensional and δ-stable, there exists n minimal with a_{n+1} in the K-linear span of $a_0, a_1, \ldots, a_n$. If $a_{n+1} = 0$, take $r = 1 + a_{n-1}a_n$ in the above. On the other hand, if $a_{n+1} \neq 0$ then we can write $a_{n+1} = \sum_{i=j}^{n} k_i a_i$ with $k_i \in K$ and $k_j \neq 0$. Now take $r = 1 + a_j$.

References

[1] S. A. Amitsur, *The identities of PI-rings*, Proc. Amer. Math. Soc. **4** (1953), 27–34.

[2] S. A. Amitsur, *Radicals of polynomial rings*, Canadian J. Math. **8** (1956), 355-361.

[3] S. A. Amitsur, *Algebras over infinite fields*, Proc. Amer. Math. Soc. **7** (1956), 35–48.

[4] S. A. Amitsur, *On rings of quotients*, Symposia Math. **8** (1972), 149-164.

[5] S. A. Amitsur and J. Levitzki, *Minimal identities for algebras*, Proc. Amer. Math. Soc. **1** (1950), 449–463.

[6] F. W. Anderson and K. R. Fuller, *"Rings and Categories of Modules,"* Springer, New York, 1974.

[7] V. A. Andrunakievitch and Ju. M. Rjabuhin, *Rings without nilpotent elements and completely prime ideals*, Doklady Akad. Nauk SSSR **180** (1968), 9–11.

[8] V. A. Artamonov, *Projective nonfree modules over group rings of solvable groups*, (in Russian), Mat. Sb. (N. S.) **116** (1981), 232–244.

[9] G. Azumaya, *New foundation of the theory of simple rings*, Proc. Japan Acad. **22** (1946), 325–332.

[10] H. Bass, *Torsion free and projective modules*, Trans. Amer. Math. Soc. **102** (1962), 319–327.

[11] S. S. Bedi and J. Ram, *Jacobson radical of skew polynomial rings and skew group rings*, Israel J. Math. **35** (1980), 327–338.

[12] K. I. Beidar, *Radicals of finitely generated algebras* (in Russian), Uspekhi Mat. Nauk **36** (1981), 203–204.

[13] A. Bell, *Localization and ideal theory in Noetherian strongly group-graded rings*, J. Algebra **105** (1987), 76–115.

[14] J. Bergen and M. Cohen, *Actions of commutative Hopf algebras*, Bull. London Math. Soc. **18** (1986), 159–164.

[15] J. Bergen and S. Montgomery, *Smash products and outer derivations*, Israel J. Math. **53** (1986), 321–345.

[16] G. M. Bergman, *The logarithmic limit-set of an algebraic variety*, Trans. Amer. Math. Soc. **157** (1971), 459–469.

[17] G. M. Bergman, *Modules over coproducts of rings*, Trans. Amer. Math. Soc. **200** (1974), 1–32.

[18] G. M. Bergman, *Homogeneous elements and prime ideals in Z-graded rings*, unpublished note, 1978.

[19] G. M. Bergman, see [**37**, Section 3].

[20] G. M. Bergman, *Example on the nontransitivity of Schelter-integral ring extensions*, unpublished note, 1980.

[21] G. M. Bergman, see [**163**, Corollary 3].

[22] G. M. Bergman and I. M. Isaacs, *Rings with fixed-point-free group actions*, Proc. London Math. Soc. (3) **27** (1973), 69–87.

[23] J. Bit-David and J. C. Robson, *Normalizing extensions I*, in "Proc. Ring Theory Conference (Antwerp, 1980)," Lecture Notes in Math. 825, Springer, Berlin, 1980.

[24] R. J. Blattner, M. Cohen and S. Montgomery, *Crossed products and inner actions of Hopf algebras*, Trans. Amer. Math. Soc. **298** (1986), 671–711.

[25] R. J. Blattner and S. Montgomery, *A duality theorem for Hopf module algebras*, J. Algebra **95** (1985), 153–172.

[26] K. A. Brown, *On zero divisors in group rings*, Bull. London Math. Soc. **8** (1976), 251–256.

[27] K. A. Brown, *The structure of modules over polycyclic groups*, Math. Proc. Cambridge Philos. Soc. **89** (1981), 257–283.

[28] H. Cartan, *Théorie de Galois pour les corps non commutatifs*, Ann. Sci. École Nor. Sup. (3) **64** (1947), 59–77.

[29] L. N. Childs and F. R. DeMeyer, *On automorphisms of separable algebras*, Pacific J. Math. **23** (1967), 25–34.

[30] W. Chin, *Prime ideals in differential operator rings and crossed products of infinite groups*, J. Algebra **106** (1987), 78–104.

[31] W. Chin and D. Quinn, *Rings graded by polycyclic-by-finite groups*, Proc. Amer. Math. Soc. **102** (1988), 235–241.

[32] G. H. Cliff, *Zero divisors and idempotents in group rings*, Canadian J. Math. **32** (1980), 596–602.

[33] G. H. Cliff and A. Weiss, *Moody's induction theorem*, Illinois J. Math. **32** (1988), 489–500.

[34] I. S. Cohen and A. Seidenberg, *Prime ideals and integral dependence*, Bull. Amer. Math. Soc. **52** (1946), 252–261.

[35] M. Cohen, *Semiprime Goldie centralizers*, Israel J. Math. **20** (1975), 37–45, (*Addendum*, **24** (1976), 89–93).

[36] M. Cohen and S. Montgomery, *Semisimple Artinian rings of fixed points*, Canadian Math. Bull. **18** (1975), 189–190.

[37] M. Cohen and S. Montgomery, *The normal closure of a semiprime ring*, in *"Ring Theory—Proceedings of the 1978 Antwerp Conference,"* pp. 43–59, Dekker, New York, 1979.

[38] M. Cohen and S. Montgomery, *Trace functions for finite automorphism groups of rings*, Arch. Math. **35** (1980), 516–527.

[39] M. Cohen and S. Montgomery, *Group-graded rings, smash products and group actions*, Trans. Amer. Math. Soc. **282** (1984), 237–258.

[40] M. Cohen and L. H. Rowen, *Group graded rings*, Comm. Algebra **11** (1983), 1253–1270.

[41] P. M. Cohn, *Rings with a weak algorithm*, Trans. Amer. Math. Soc. **109** (1963) 332–356.

[42] P. M. Cohn, *On the free product of associative rings III*, J. Algebra **8** (1968), 376–383.

[43] P. M. Cohn, *"Free Rings and Their Relations,"* Second Edition, Academic Press, London, 1985.

[44] I. G. Connell, *On the group ring*, Canadian J. Math. **15** (1963), 650–685.

[45] E. C. Dade, *Group-graded rings and modules*, Math. Z. **174** (1980), 241–262.

[46] F. DeMeyer and G. J. Janusz, *Finite groups with an irreducible representation of large degree*, Math. Z. **108** (1969), 145–153.

[47] J. Dieudonné, *La théorie de Galois des anneaux simples et semi-simples*, Comment. Math. Helv. **21** (1948), 154–184.

[48] W. Dicks, see [**43**, Theorem 6.10.4].

[49] W. Dicks and E. Formanek, *Poincaré series and a problem of S. Montgomery*, Linear Multilin. Algebra **12** (1982), 21–30.

[50] C. Faith, *Noetherian simple rings*, Bull. Amer. Math. Soc. **70** (1964), 730–731.

[51] C. Faith, *Galois subrings of Ore domains are Ore domains*, Bull. Amer. Math. Soc. **78** (1972), 1077–1080.

[52] D. R. Farkas, *Group rings: an annotated questionnaire*, Comm. Algebra **8** (1980), 585–602.

[53] D. R. Farkas, *Toward multiplicative invariant theory*, Contemporary Math. **43** (1985), 69–80.

[54] D. R. Farkas and P. Linnell, *Zero divisors in group rings: something old, something new*, Contemporary Math.

[55] D. R. Farkas and R. L. Snider, *K_0 and Noetherian group rings*, J. Algebra **42** (1976), 192–198.

[56] D. R. Farkas and R. L. Snider, *Noetherian fixed rings*, Pacific J. Math. **69** (1977), 347–353.

[57] D. R. Farkas and R. L. Snider, *Induced representations of polycyclic groups*, Proc. London Math. Soc. (3) **39** (1979), 193–207.

[58] J. W. Fisher and S. Montgomery, *Semiprime skew group rings*, J. Algebra **52** (1978), 241–247.

[59] J. W. Fisher and J. Osterburg, *Semiprime ideals in rings with finite group actions*, J. Algebra **50** (1978), 488–502.

[60] J. W. Fisher and J. Osterburg, *Finite group actions on noncommutative rings: a survey since 1970*, in "Ring Theory and Algebra III," Lecture Notes in Math. 55, Dekker, New York, 1980, pp. 357–393.

[61] E. Formanek, *The zero divisor question for supersolvable groups*, Bull. Australian Math. Soc. **9** (1973), 67–71.

[62] E. Formanek and A. I. Lichtman, *Ideals in group rings of free products*, Israel J. Math. **31** (1978), 101–104.

[63] S. Gersten, *On class groups of free products*, Ann. Math. **87** (1968), 392–398.

[64] A. W. Goldie, *The structure of Noetherian rings*, in *"Lectures on Rings and Modules,"* Lecture Notes in Math. 246, Springer, Berlin, 1972.

[65] A. W. Goldie and G. O. Michler, *Ore extensions and polycyclic group rings*, J. London Math. Soc. (2) **9** (1974/75), 337–345.

[66] O. Goldman, *Hilbert rings and the Hilbert Nullstellensatz*, Math. Z. **54** (1951), 136–140.

[67] A. Grothendieck, unpublished.

[68] P. Grzeszczuk, *On G-systems and G-graded rings*, Proc. Amer. Math. Soc. **95** (1985), 348–352.

[69] R. M. Guralnick, I. M. Isaacs and D. S. Passman, *Nonexistence of partial traces for group actions*, Rocky Mt. J. Math. **11** (1981), 395–405.

[70] D. Handelman and J. Lawrence, *Strongly prime rings*, Trans. Amer. Math. Soc. **211** (1975), 209–223.

[71] D. Handelman, J. Lawrence and W. Schelter, *Skew group rings*, Houston J. Math. **4** (1978) 175–198.

[72] A. G. Heinicke and J. C. Robson, *Normalizing extensions: prime ideals and incomparability*, J. Algebra **72** (1981), 237–268.

[73] I. N. Herstein, *"Noncommutative Rings,"* Carus Mathematical Monographs No. 15, Amer. Math. Soc., 1968.

[74] G. Higman, *Groups and rings having automorphisms without non-trivial fixed elements*, J. London Math. Soc. **32** (1957), 321–334.

[75] G. Hochschild, *Double vector spaces over division rings*, Am. J. Math. **71** (1949), 443–460.

[76] G. Hochschild, *Automorphisms of simple algebras*, Trans. Amer. Math. Soc. **69** (1950), 292–301.

[77] T. J. Hodges and J. Osterburg, *A rank two indecomposable projective module over a Noetherian domain of Krull dimension one*, Bull. London Math. Soc. **19** (1987), 139–144.

[78] R. B. Howlett and I. M. Isaacs, *On groups of central type*, Math. Z. **179** (1982), 555–569.

[79] I. M. Isaacs and D. S. Passman, *Groups with representations of bounded degree*, Canadian J. Math. **16** (1964), 299–309.

[80] N. Jacobson, *The fundamental theorem of the Galois theory for quasi-fields*, Ann. Math. **41** (1940), 1–7.

[81] N. Jacobson, *A note on divison rings*, Amer. J. Math. **69** (1947), 27–36.

[82] N. Jacobson, *"Structure of Rings,"* A. M. S. Colloq. Publ. 37, Amer. Math. Soc., Providence, 1956, (revised 1964).

[83] N. Jacobson, *"Theory of Fields and Galois Theory,"* Van Nostrand, Princeton, 1964.

[84] A. Joseph and L. W. Small, *An additivity principle for Goldie rank*, Israel J. Math. **31** (1978), 105–114.

[85] V. K. Kharchenko, *Galois extensions and quotient rings*, Algebra i Logika **13** (1974), 460–484, (English transl. (1975), 265–281).

[86] V. K. Kharchenko, *Galois subrings of simple rings*, Math. Notes **17** (1975), 533–536.

[87] V. K. Kharchenko, *Generalized identities with automorphisms*, Algebra i Logika **14** (1975), 215–237, (English transl. (1976), 132–148).

[88] V. K. Kharchenko, *Galois theory of semiprime rings*, Algebra i Logika **16** (1977), 313–363, (English transl. (1978), 208–258).

[89] V. K. Kharchenko, *Algebras of invariants of free algebras*, Algebra i Logika **17** (1978), 478–487, (English transl. (1979), 316–321).

[90] V. K. Kharchenko, *Constants of derivations of prime rings*, Izv. Akad. Nauk SSSR Ser. Mat. **45** (1981), 435–461, (English transl. Math. USSR Izv. **18** (1982), 381–400).

[91] V. K. Kharchenko, *Noncommutative invariants of finite groups and Noetherian varieties*, J. Pure Appl. Algebra **31** (1984), 83–90.

[92] V. K. Kharchenko, *"Automorphisms and Derivations of Associative Rings"* (in Russian).

[93] H. F. Kreimer, *On the Galois theory of separable algebras*, Pacific J. Math. **34** (1970), 727–740.

[94] P. H. Kropholler, P. A. Linnell and J. A. Moody, *Applications of a new K-theoretic theorem to soluble group rings*, Proc. Amer. Math. Soc.

[95] W. Krull, *Zum Dimensionsbegriff der Idealtheorie*, Math. Z. **42** (1937), 745–766.

[96] W. Krull, *Jacobsonsche Ringe, Hilbertscher Nullstellensatz, Dimensionstheorie*, Math. Z. **54** (1951), 354–387.

[97] D. R. Lane, *"Free Algebras of Rank Two and Their Automorphisms,"* Thesis, London Univ., 1976.

[98] J. Levitzki, *On automorphisms of certain rings*, Ann. Math. **36** (1935), 984–992.

[99] J. Lewin, *A note on zero divisors in group rings*, Proc. Amer. Math. Soc. **31** (1972), 357–359.

[100] J. Lewin, *The symmetric ring of quotients of a 2-fir*, Comm. Algebra **16** (1988), 1727–1732.

[101] A. I. Lichtman and W. S. Martindale, III, *The normal closure of the coproduct of domains over a division ring*, Comm. Algebra **13** (1985), 1643–1664.

[102] M. Lorenz, *Finite normalizing extensions of rings*, Math. Z. **176** (1981), 447–484.

[103] M. Lorenz, *The Goldie rank of prime supersolvable group algebras*, Mitt. Math. Sem. Giessen **149** (1981), 115–129.

[104] M. Lorenz, K_0 *of skew group rings and simple Noetherian rings without idempotents*, J. London Math. Soc. (2) **32** (1985), 41–50.

[105] M. Lorenz, *Goldie ranks of prime polycyclic crossed products*, in *"Perspectives in Ring Theory,"* Kluwer, Dordrecht.

[106] M. Lorenz, *Frobenius reciprocity and* G_0 *of skew group rings*, in *"Proc. of the Granada Conference–1986,"* Lecture Notes in Math. 1328, Springer, Berlin, pp. 165–173.

[107] M. Lorenz, S. Montgomery and L. W. Small, *Prime ideals in fixed rings II*, Comm. Algebra **10** (1982), 449–455.

[108] M. Lorenz and D. S. Passman, *Centers and prime ideals in group algebras of polycyclic-by-finite groups*, J. Algebra **57** (1979), 355–386.

[109] M. Lorenz and D. S. Passman, *Prime ideals in crossed products of finite groups*, Israel J. Math. **33** (1979), 89–132.

[110] M. Lorenz and D. S. Passman, *Integrality and normalizing extensions of rings*, J. Algebra **61** (1979), 289–297.

[111] M. Lorenz and D. S. Passman, *Addendum—Prime ideals in crossed products of finite groups*, Israel J. Math. **35** (1980), 311–322.

[112] M. Lorenz and D. S. Passman, *Two applications of Maschke's theorem*, Comm. Algebra **8** (1980), 1853–1866.

[113] M. Lorenz and D. S. Passman, *Observations on crossed products and fixed rings*, Comm. Algebra **8** (1980), 743–779.

[114] M. Lorenz and D. S. Passman, *Prime ideals in group algebras of polycyclic-by-finite groups*, Proc. London Math. Soc. (3) **43** (1981), 520–543.

[115] M. Lorenz and D. S. Passman, *The structure of G_0 for certain polycyclic group algebras and related algebras*, Contempory Math.

[116] W. S. Martindale, III, *Prime rings satisfying a generalized polynomial identity*, J. Algebra **12** (1969), 576–584.

[117] W. S. Martindale, III, *Fixed rings of automorphisms and the Jacobson radical*, J. London Math. Soc. (2) **17** (1978), 42–46.

[118] W. S. Martindale, III, *The normal closure of the coproduct of rings over a division ring*, Trans. Amer. Math. Soc. **293** (1986), 303–317.

[119] W. S. Martindale, III and S. Montgomery, *The normal closure of coproducts of domains*, J. Algebra **82** (1983), 1–17.

[120] H. Maschke, *Über den arithmetischen Charakter der Coefficienten der Substitutionen endlicher linearer Substitutionsgruppen*, Math. Ann. **50** (1898), 492–498.

[121] J. C. McConnell and J. C. Robson, *"Noncommutative Noetherian Rings,"* Wiley-Interscience, New York, 1987.

[122] J. E. McLaughlin, *A note on regular group rings*, Michigan Math. J. **5** (1958), 127–128.

[123] S. V. Mihovski, *On isomorphic crossed products of groups and rings*, C. R. Acad. Sci. Bulgaria, 1988.

[124] Y. Miyashita, *Finite outer Galois theory of non-commutative rings*, J. Fac. Sci. Hokkaido Univ. Ser. I **19** (1966), 115–134.

[125] Y. Miyashita, *On Galois extensions and crossed products*, J. Fac. Sci. Hokkaido Univ. Ser. I **21** (1970), 97–121.

[126] S. Montgomery, *The Jacobson radical and fixed rings of automorphisms*, Comm. Algebra **4** (1976), 459–465.

[127] S. Montgomery, *Outer automorphisms of semi-prime rings*, J. London Math. Soc. (2) **18** (1978), 209–220.

[128] S. Montgomery, *Automorphism groups of rings with no nilpotent elements*, J. Algebra **60** (1979), 238–248.

[129] S. Montgomery *"Fixed Rings of Finite Automorphism Groups of Associative Rings,"* Lecture Notes in Math. 818, Springer, Berlin, 1980.

[130] S. Montgomery, *X-inner automorphisms of filtered algebras*, Proc. Amer. Math. Soc. **83** (1981), 263–268.

[131] S. Montgomery, *Prime ideals in fixed rings*, Comm. Algebra **9** (1981), 423–449.

[132] S. Montgomery and D. S. Passman, *Crossed products over prime rings*, Israel J. Math. **31** (1978), 224–256.

[133] S. Montgomery and D. S. Passman, *X-inner automorphisms of group rings*, Houston J. Math. **7** (1981), 395–402.

[134] S. Montgomery and D. S. Passman, *Galois theory of prime rings*, J. Pure Appl. Algebra **31** (1984), 139–184.

[135] S. Montgomery and D. S. Passman, *Outer Galois theory of prime rings*, Rocky Mt. J. Math. **14** (1984), 305–318.

[136] S. Montgomery and D. S. Passman, *X-inner automorphisms of crossed products and semi-invariants of Hopf algebras*, Israel J. Math. **55** (1986), 33–57.

[137] S. Montgomery and D. S. Passman, *Prime ideals in fixed rings of free algebras*, Comm. Algebra **15** (1987), 2209–2234.

[138] S. Montgomery and L. W. Small, *Integrality and prime ideals in fixed rings of P. I. rings*, J. Pure Appl. Algebra **31** (1984), 185–190.

[139] S. Montgomery and M. Smith, *Algebras with a separable subalgebra whose centralizer satisfies a polynomial identity*, Comm. Algebra **3** (1975), 151–168.

[140] J. A. Moody, *Induction theorems for infinite groups*, Bull. Amer. Math. Soc. **17** (1987), 113–116.

[141] J. A. Moody, *Torsion-free solvable group rings are Ore domains*, unpublished note, 1987.

[142] J. A. Moody, *A Brauer induction theorem for G_0 of certain infinite groups*, J. Algebra.

[143] K. Morita, *Duality of modules and its applications to the theory of rings with minimum condition*, Sci. Rep. Tokyo Kyoiku Daigaku Sect. A **6** (1958), 85–142.

[144] T. Nakayama, *Galois theory for general rings with minimum condition*, J. Math. Soc. Japan **1** (1949), 203–216.

[145] T. Nakayama, *Galois theory of simple rings*, Trans. Amer. Math. Soc. **73** (1952), 276–292.

[146] T. Nakayama and G. Azumaya, *On irreducible rings*, Ann. Math. (2) **48** (1947), 949–965.

[147] E. Noether, *Nichtkommutative Algebra*, Math. Z. **37** (1933), 514–541.

[148] C. Năstăsescu, *Group rings of graded rings, Applications*, J. Pure Appl. Algebra **33** (1984), 315–335.

[149] M. Ojanguren and R. Sridharan, *Cancellation of Azumaya algebras*, J. Algebra **18** (1971), 501–505.

[150] J. Osterburg, *Fixed rings of simple rings*, Comm. Algebra **6** (1978), 1741–1750.

[151] J. Osterburg, *The influence of the algebra of the group*, Comm. Algebra **7** (1979), 1377–1396.

[152] J. Osterburg, *Smash products and G-Galois actions*, Proc. Amer. Math. Soc. **98** (1986), 217–221.

[153] A. Page, *Actions de groupes*, in *"Séminaire d'Algèbre P. Dubreil,"* Lecture Notes in Math. 740, Springer, Berlin, 1979.

[154] R. Paré and W. Schelter, *Finite extensions are integral*, J. Algebra **53** (1978), 477–479.

[155] D. S. Passman, *Nil ideals in group rings*, Michigan Math. J. **9** (1962), 375–384.

[156] D. S. Passman, *"Permutation Groups,"* Benjamin, New York, 1968.

[157] D. S. Passman, *Radicals of twisted group rings*, Proc. London Math. Soc. (3) **20** (1970), 409–437.

[158] D. S. Passman, *Linear identities in group rings*, Pacific J. Math. **36** (1971), 457–483.

[159] D. S. Passman, *Group rings satisfying a polynomial identity II*, Pacific J. Math. **39** (1971), 425–438.

[160] D. S. Passman, *Group rings satisfying a polynomial identity*, J. Algebra **20** (1972), 103–117.

[161] D. S. Passman, *"The Algebraic Structure of Group Rings,"* Wiley-Interscience, New York, 1977, (Krieger, Malabar, 1985).

[162] D. S. Passman, *Crossed products over semiprime rings*, Houston J. Math. **4** (1978), 583–592.

[163] D. S. Passman, *Fixed rings and integrality*, J. Algebra **68** (1981), 510–519.

[164] D. S. Passman, *Prime ideals in normalizing extensions*, J. Algebra **73** (1981), 556–572.

[165] D. S. Passman, *It's essentially Maschke's theorem*, Rocky Mt. J. Math. **13** (1983), 37–54.

[166] D. S. Passman, *Semiprime and prime crossed products*, J. Algebra **83** (1983), 158–178.

[167] D. S. Passman, *Infinite crossed products and group-graded rings*, Trans. Amer. Math. Soc. **284** (1984), 707–727.

[168] D. S. Passman, *Group rings of polycyclic groups*, in *"Group Theory: essays for Philip Hall,"* Academic Press, London, 1984, pp. 207–256.

[169] D. S. Passman, *Semiprime crossed products*, Houston J. Math. **11** (1985), 257–268.

[170] D. S. Passman, *On the Goldie rank of group algebras*, Comm. Algebra **13** (1985), 1305–1311.

[171] D. S. Passman, *Computing the symmetric ring of quotients*, J. Algebra **105** (1987), 207–235.

[172] D. S. Passman, *Prime ideals in polycyclic crossed products*, Trans. Amer. Math. Soc. **301** (1987), 737–759.

[173] D. S. Passman, *Prime ideals in enveloping rings*, Trans. Amer. Math. Soc. **302** (1987), 535–560.

[174] D. S. Passman, *Prime ideals in restricted enveloping rings*, Comm. Algebra **16** (1988), 1411–1436.

[175] K. R. Pearson and W. Stephenson, *A skew polynomial ring over a Jacobson ring need not be a Jacobson ring*, Comm. Algebra **5** (1977), 783–794.

[176] C. Procesi and L. W. Small, *Endomorphism rings of modules over PI-algebras*, Math. Z. **106** (1968), 178–180.

[177] S. D. Promislow, *A simple example of a torsion-free non unique product group*, Bull. London Math. Soc.

[178] E. Puczyłowski, *On fixed rings of automorphisms*, Proc. Amer. Math. Soc. **90** (1984), 517–518.

[179] D. Quillen, *Higher algebraic K-theory: I*, in *"Algebraic K-Theory I,"* Lecture Notes in Math. 341, Springer, Berlin, 1973, pp. 77–139.

[180] D. Quinn, *Group-graded rings and duality*, Trans. Amer. Math. Soc. **292** (1985), 155–167.

[181] D. Quinn, *Integrality over fixed rings*.

[182] A. Reid, *Semi-prime twisted group rings*, J. London Math. Soc. (2) **12** (1975/76), 413–418.

[183] E. Rips and Y. Segev, *Torsion-free groups without unique product property*, J. Algebra **108** (1987), 116-126.

[184] J. E. Roseblade, *Group rings of polycyclic groups*, J. Pure Appl. Algebra **3** (1973), 307–328.

[185] J. E. Roseblade, *Prime ideals in group rings of polycyclic groups*, Proc. London Math. Soc. (3) **36** (1978), 385–447.

[186] J. E. Roseblade *Five lectures on group rings*, in *"Proc. of Groups–St. Andrews 1985,"* London Math. Soc. Lecture Note Ser. 121, pp. 93–109, Cambridge Univ. Press, Cambridge, 1986.

[187] A. Rosenberg and D. Zelinsky, *Galois theory of continuous linear transformation rings*, Trans. Amer. Math. Soc. **79** (1955), 429–452.

[188] S. Rosset, *The Goldie rank of virtually polycyclic group rings*, in *"The Brauer Group,"* Lecture Notes in Math. 844, Springer, Berlin, 1981, pp. 35–45.

[189] S. Rosset, *Miscellaneous results on the Goldie rank conjecture*, unpublished note, 1981.

[190] J. J. Rotman, *"Notes on Homological Algebra,"* Van Nostrand Reinhold, New York, 1970.

[191] L. H. Rowen, *"Polynomial Identities in Ring Theory,"* Academic Press, New York, 1980.

[192] S. Schanuel, see [**10**, Proposition 2.1(b)].

[193] J. P. Serre, *Cohomologie des Groupes Discrets*, in *"Prospects in Mathematics,"* Annals of Math. Studies No. 70, Princeton, 1971, pp. 77–169.

[194] M. K. Smith, *Group algebras*, J. Algebra **18** (1971), 477–499.

[195] R. L. Snider, *The zero divisor conjecture for some solvable groups*, Pacific J. Math. **90** (1980), 191–196.

[196] J. T. Stafford, *A simple Noetherian ring not Morita equivalent to a domain*, Proc. Amer. Math. Soc. **68** (1978), 159–160.

[197] J. T. Stafford *Stably free, projective right ideals*, Compositio Math. **54** (1985), 63–78.

[198] H. Tominaga and T. Nagahara, *"Galois Theory of Simple Rings,"* Okayama Math. Lectures, Okayama Univ., 1970.

[199] K. H. Ulbrich, *Vollgraduierte Algebren*, Abhandlung Math. Sem. Univ. Hamburg **51** (1981), 136–148.

[200] M. Van den Bergh, *A duality theorem for Hopf algebras*, in *"Methods in Ring Theory,"* ASI Series C, No. 129, Reidel, Dordrecht, 1984, pp. 517–522.

[201] O. E. Villamayor, *On the semisimplicity of group algebras*, Proc. Amer. Math. Soc. **9** (1958), 621–627.

[202] O. E. Villamayor and D. Zelinsky, *Galois theory with infinitely many idempotents*, Nagoya Math. J. **35** (1969), 83–98.

[203] R. G. Walker, *Local rings and normalizing sets of elements*, Proc. London Math. Soc. (3) **24** (1972), 27–45.

[204] D. B. Webber, *Ideals and modules of simple Noetherian hereditary rings*, J. Algebra **16** (1970), 239–242.

[205] B. A. F. Wehrfritz, see [**57**, Theorem 2.1].

[206] C. Welsh, *Prime length in crossed products*.

[207] J. Wiegold, *Groups with boundedly finite classes of conjugate elements*, Proc. Royal Soc. London Ser. A **238** (1957), 389–401.

[208] A. E. Zalesskii and O. M. Neroslavskii, *There exist simple Noetherian rings with zero divisors but without idempotents*

(Russian with English summary), Comm. Algebra **5** (1977), 231–244.

[209] A. V. Zhuchin, *Cross products with identity*, Soviet Math. (Izv. VUZ) (no. 9) **29** (1985), 28–34.

Index

PURE AND APPLIED MATHEMATICS

*Presently out of print